Student Solutions Guide to Acc

Elementary Algebra

Second Edition

Larson/Hostetler

Carolyn F. Neptune
Johnson County Community College

D. C. Heath and Company
Lexington, Massachusetts   Toronto

Address editorial correspondence to:
D. C. Heath and Company
125 Spring Street
Lexington, MA 02173

International Standard Book Number: 0–669–41635–5

10   9   8   7   6   5   4   3

PREFACE

This *Student Solutions Guide* is a supplement to the textbook ELEMENTARY ALGEBRA, Second Edition by Roland E. Larson and Robert P. Hostetler. The *Student Solutions Guide* includes solutions for the odd-numbered exercises in the text, including the chapter reviews, chapter tests, and the cumulative tests.

I have written detailed workouts of the problems. The algebraic steps are clearly shown and explanatory comments are included where appropriate. There are usually several "correct" ways to arrive at a solution to a problem in mathematics. Therefore, you shouldn't be concerned if you have approached problems differently than I have.

I have made every effort to eliminate errors from this book, but I would sincerely appreciate and welcome your comments regarding corrections or suggested improvements for this supplement. You may contact me at the address shown below.

Preparing this solutions guide has been an interesting and enjoyable project for me, but many other individuals have made significant contributions to its completion. I want to give special thanks first to Ron Larson for his assistance while I was writing and to the people at D. C. Heath and Company for giving me this opportunity. In addition, I am grateful to my husband Harold for his unfailing patience, support, and understanding during these past months, to our daughter Jill and son Steve for their continual encouragement from afar, to my colleagues at Johnson County Community College for their interest and camaraderie, and even to faithful Bentley for his dogged companionship during late-night hours.

I hope you find the *Student Solutions Guide* to be a helpful supplement as you use the textbook, and I wish you well in your study of algebra.

Carolyn F. Neptune
Johnson County Community College
12345 College Boulevard
Overland Park, KS 66210

CONTENTS

CHAPTER P
Prerequisites: Arithmetic Review

P.1 | Real Numbers: Order and Absolute Value

7. Determine which of these numbers are (a) natural numbers, (b) integers, and (c) rational numbers $\{-3, 2, -\frac{3}{2}, \frac{9}{3}, 4.5\}$.

Solution

(a) natural numbers: $\{2, \frac{9}{3}\}$

(b) integers: $\{-3, 2, \frac{9}{3}\}$

(c) rational numbers: $\{-3, 2, -\frac{3}{2}, \frac{9}{3}, 4.5\}$

9. Place the correct inequality symbol ($<$ or $>$) between the two real numbers 2 and 5.

Solution

$2 < 5$ or $5 > 2$

11. Place the correct inequality symbol ($<$ or $>$) between the two real numbers $-\frac{9}{2}$ and -3.

Solution

$-\frac{9}{2} < -3$ or $-3 > -\frac{9}{2}$

13. Show the real numbers 4 and $-\frac{7}{2}$ as points on the real number line and place the correct inequality between the two numbers.

Solution

$4 > -\frac{7}{2}$ because 4 lies to the *right* of $-\frac{7}{2}$.

15. Show the real numbers -4.6 and 1.5 as points on the real number line and place the correct inequality between the two numbers.

Solution

$-4.6 < 1.5$ because -4.6 lies to the *left* of 1.5.

17. Determine the distance between a and zero when $a = -4$.

Solution

4: The distance between -4 and 0 is 4 units.

19. Find the opposite number of -3.

Solution

3: The opposite of -3 is 3; $-(-3) = 3$.

21. Evaluate $|-3.4|$.

Solution

$|-3.4| = 3.4$ because the distance between -3.4 and 0 is 3.4.

23. Evaluate $-|-23.6|$.

Solution

$-|-23.6| = -23.6$

Note: $|-23.6| = 23.6$

25. Place the correct symbol ($<$, $>$, or $=$) between the real numbers $|-4|$ and $|3|$.

Solution

$|-4| > |3|$ because $|-4| = 4$, $|3| = 3$ and 4 is greater than 3.

27. Place the correct symbol (<, >, or =) between the real numbers $-|-48.5|$ and $|-48.5|$.

Solution

$-|48.5| < |-48.5|$ because $-|-48.5| = -48.5$, $|-48.5| = 48.5$, and 48.5 is less than 48.5.

29. Plot the numbers $\frac{5}{2}$, π, -2, and $-|-3|$.

Solution

Note: $\pi \approx 3.14$

31. Find all real numbers whose distance from a is d when $a = 8$ and $d = 12.5$.

Solution

The number 12.5 units to the right of 8 is 20.5 because $8 + 12.5 = 20.5$.
The number 12.5 units to the left of 8 is -4.5 because $8 - 12.5 = -4.5$

33. Determine which of these numbers are (a) natural numbers, (b) integers, and (c) rational numbers, then plot them on the real number line.

$$\left\{-\frac{5}{2}, 6.5, -4.5, \frac{8}{4}, \frac{3}{4}\right\}$$

Solution

(a) natural numbers: $\left\{\frac{8}{4}\right\}$

(b) integers: $\left\{\frac{8}{4}\right\}$

(c) rational numbers: $\left\{-\frac{5}{2}, 6.5, -4.5, \frac{8}{4}, \frac{3}{4}\right\}$

35. Place the correct inequality symbol (< or >) between the two real numbers -4 and -1.

Solution

$-4 < -1$ or $-1 > -4$

37. Place the correct inequality symbol (< or >) between the two real numbers -2 and $\frac{3}{2}$.

Solution

$-2 < \frac{3}{2}$ or $\frac{3}{2} > -2$

39. Plot the real numbers $\frac{1}{3}$ and 4 as points on the real number line and place the correct inequality symbol between the two numbers.

Solution

$\frac{1}{3} < 4$ because $\frac{1}{3}$ lies to the *left* of 4.

41. Plot the real numbers -2π and -10 as points on the real number line and place the correct inequality symbol between the two numbers.

Solution

$-2\pi > -10$ because -2π lies to the *right* of -10.

Note: $-2\pi \approx -6.28$.

43. Plot the real numbers $\frac{7}{16}$ and $\frac{5}{8}$ as points on the real number line and place the correct inequality symbol between the two numbers.

Solution

$\frac{7}{16} < \frac{5}{8}$ because $\frac{7}{16}$ lies to the *left* of $\frac{5}{8}$.

Note: $\frac{5}{8} = \frac{10}{16}$

45. Plot the real numbers 0 and $-\frac{7}{16}$ as points on the real number line and place the correct inequality symbol between the two numbers.

Solution

$0 > -\frac{7}{16}$ because 0 lies to the *right* of $\frac{7}{16}$.

47. Determine the distance between a and 0 when $a = -3$.

Solution

3: The distance between -3 and 0 is 3 units.

49. Find the opposite of 5.

Solution

The opposite of 5 is -5.

51. Find the opposite of $-\frac{5}{2}$.

Solution

The opposite of $-\frac{5}{2}$ is $\frac{5}{2}$; $-\left(-\frac{5}{2}\right) = \frac{5}{2}$.

53. Evaluate $|7|$.

Solution

$|7| = 7$ because the distance between 7 and 0 is 7.

55. Evaluate $\left|-\frac{7}{2}\right|$.

Solution

$\left|-\frac{7}{2}\right| = \frac{7}{2}$ because the distance between $-\frac{7}{2}$ and 0 is $\frac{7}{2}$.

57. Evaluate $-|4.09|$.

Solution

$-|4.09| = -4.09$.

Note: $|4.09| = 4.09$

59. Evaluate $|-3.2|$.

Solution

$-|-3.2| = -3.2$.

Note: $|-3.2| = 3.2$.

61. Place the correct symbol ($<$, $>$, or $=$) between the two real numbers $|-15|$ and $|15|$.

Solution

$|-15| = |15|$ because $|-15| = 15$, $|15| = 15$, and $15 = 15$.

63. Place the correct symbol ($<$, $>$, or $=$) between the two real numbers $|32|$ and $|-50|$.

Solution

$|32| < |-50|$ because $|32| = 32$, $|-50| = 50$, and $32 < 50$.

65. Place the correct symbol ($<$, $>$, or $=$) between the two real numbers $\left|\frac{3}{16}\right|$ and $\left|\frac{3}{2}\right|$.

Solution

$\left|\frac{3}{16}\right| < \left|\frac{3}{2}\right|$ because $\left|\frac{3}{16}\right| = \frac{3}{16}$, $\left|\frac{3}{2}\right| = \left|\frac{3}{2}\right|$, and $\frac{3}{16} < \frac{3}{2}$.

Note: $\frac{3}{2} = \frac{24}{16}$, and $\frac{3}{16} < \frac{24}{16}$.

67. Place the correct symbol ($<$, $>$, or $=$) between the two real numbers $|-\pi|$ and $-|-2\pi|$.

Solution

$|-\pi| > -|-2\pi|$ because $|-\pi| = \pi$, $-|-2\pi| = -2\pi$, and $\pi > -2\pi$.

69. Show the numbers $\frac{3}{2}$, -2π, 3.2, and $|-4|$ on the real number line.

Solution

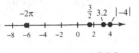

Note: $-2\pi \approx -6.28$.

71. Find all real numbers whose distance from a is given by d, when $a = -2$, and $d = 3.5$.

Solution

The number 3.5 units to the right of -2 is 1.5 because $-2 + 3.5 = 1.5$.

The number 3.5 units to the left of -2 is -5.5 because $-2 - 3.5 = -5.5$.

73. Decide whether the statement is true or false: The absolute value of any real number is positive. Explain your reasoning.

Solution

False; 0 is a real number and $|0| = 0$.

Note: 0 is *neither* positive nor negative.

75. Decide whether the statement is true or false: The absolute value of a rational number is a rational number. Explain your reasoning.

Solution

True; if a number can be expressed as the ratio of two integers, then its distance from 0 can also be expressed as a ratio of two integers.

77. Decide whether the statement is true or false: The opposite of a positive number is a negative number. Explain your reasoning.

Solution

True; positive numbers are to the right of 0 on thee number line; their opposites are to the left of 0 and are negative.

P.2 Integers and Prime Factorization

7. Find the product of $3 \cdot 5 \cdot 11$.

Solution

$3 \cdot 5 \cdot 11 = 165$

9. Find the product of $2 \cdot 2 \cdot 5 \cdot 5$.

Solution

$2 \cdot 2 \cdot 5 \cdot 5 = 100$

11. Is 2400 prime or composite?

Solution

2400 is composite; its prime factorization is $2400 = 2 \cdot 2 \cdot 2 \cdot 2 \cdot 2 \cdot 3 \cdot 5 \cdot 5$.

13. Is 3911 prime or composite?

Solution

3911 is prime; the divisibility tests yield no factors of 3911. By testing the remaining primes less than or equal to $\sqrt{3911} \approx 63$, you can conclude that 3911 is a prime number.

15. Write the prime factorization of 210.

Solution

$210 = 2 \cdot 3 \cdot 5 \cdot 7$

17. Write the prime factorization of 525.

Solution

$525 = 3 \cdot 5 \cdot 5 \cdot 7$

19. Find the greatest common factor of 20 and 45.

Solution

By prime factorization, $20 = 2 \cdot 2 \cdot 5$ and $45 = 3 \cdot 3 \cdot 5$. Thus, the greatest common factor is $2 \cdot 3$, or 6.

21. Find the greatest common factor of 18, 84, and 90.

Solution

By prime factorization, $18 = 2 \cdot 3 \cdot 3$, $84 = 2 \cdot 2 \cdot 3 \cdot 7$, and $90 = 2 \cdot 3 \cdot 3 \cdot 5$. Thus, the greatest common factor is 5.

23. List the first five integer multiples of 12.

Solution

The first five positive integer multiples of 12 are 12, 24, 36, 48, and 60.

25. Find the least common multiple of 10 and 18.

Solution

By prime factorization, $10 = 2 \cdot 5$ and $18 = 2 \cdot 3 \cdot 3$. Thus, the least common multiple is $2 \cdot 3 \cdot 3 \cdot 5$, or 90.

27. Find the least common multiple of 6, 9, and 14.

Solution

By prime factorization, $6 = 2 \cdot 3$, $9 = 3 \cdot 3$, and $14 = 2 \cdot 7$. Thus, the least common multiple is $2 \cdot 3 \cdot 3 \cdot 7$, or 126.

29. Verify that 63 and 1375 are relatively prime.

Solution

By prime factorization, $63 = 3 \cdot 3 \cdot 7$ and $1375 = 5 \cdot 5 \cdot 5 \cdot 11$. Because the numbers 63 and 1375 have no common prime factors, they are relatively prime.

31. *Writing* What is the only even prime number? Explain why there are no other even prime numbers.

Solution

The only even prime number is 2. There are no other even prime numbers because every other even number is divisible by itself, by 1, and by 2; all other even numbers are composites because they have more than two factors.

33. Find the product of $5 \cdot 17 \cdot 23$.

Solution

$5 \cdot 17 \cdot 23 = 1955$

35. Find the product of $5 \cdot 5 \cdot 13 \cdot 17$.

Solution

$5 \cdot 5 \cdot 13 \cdot 17 = 5525$

37. Find the product of $2 \cdot 2 \cdot 2 \cdot 7 \cdot 7 \cdot 7$.

Solution

$2 \cdot 2 \cdot 2 \cdot 7 \cdot 7 \cdot 7 = 2744$

39. Is 643 prime or composite?

Solution

643 is prime; the divisibility tests yield no factors of 643. By testing the remaining primes less than or equal to $\sqrt{643} \approx 25$, you can conclude that 643 is a prime number.

41. Is 8324 prime or composite?

Solution

8324 is composite; its prime factorization is $2 \cdot 2 \cdot 2081$.

43. Is 1321 prime or composite?

Solution

1321 is prime; the divisibility tests yield no factors of 1321. By testing the remaining primes less than or equal to $\sqrt{1321} \approx 36$, you can conclude that 1321 is a prime number.

45. Write the prime factorization of 120.

Solution

$120 = 2 \cdot 2 \cdot 2 \cdot 3 \cdot 5$

47. Write the prime factorization of 192.

Solution

$192 = 2 \cdot 2 \cdot 2 \cdot 2 \cdot 2 \cdot 2 \cdot 3$

49. Write the prime factorization of 2535.

Solution

$2535 = 3 \cdot 5 \cdot 13 \cdot 13$

51. Find the greatest common factor of 28 and 52.

Solution

By prime factorization, $28 = 2 \cdot 2 \cdot 7$ and $52 = 2 \cdot 2 \cdot 13$. Thus, the greatest common factor is $2 \cdot 2$, or 4.

53. Find the greatest common factor of 84, 98, and 192.

Solution

By prime factorization, $84 = 2 \cdot 2 \cdot 3 \cdot 7$, $98 = 2 \cdot 2 \cdot 7$, and $192 = 2 \cdot 2 \cdot 2 \cdot 2 \cdot 2 \cdot 2 \cdot 3$. Thus, the greatest common factor is 2.

55. Find the greatest common factor of 134, 225, 315, and 945.

Solution

By prime factorization, $134 = 2 \cdot 67$, $225 = 3 \cdot 3 \cdot 5 \cdot 5$, $315 = 3 \cdot 3 \cdot 5 \cdot 7$, and $945 = 3 \cdot 3 \cdot 3 \cdot 5 \cdot 7$. Because there are no common prime factors, the greatest common factor is 1.

57. List the first five positive integer multiples of 14.

Solution

The first five positive integer multiples of 14 are 14, 28, 42, 56, and 70.

59. Find the least common multiple of 15 and 10.

Solution

By prime factorization, $15 = 3 \cdot 5$ and $10 = 2 \cdot 5$. Thus, the least common multiple is $2 \cdot 3 \cdot 5$, or 30.

61. Find the least common multiple of 12, 20, and 25.

Solution

By prime factorization, $12 = 2 \cdot 2 \cdot 3$, $20 = 2 \cdot 2 \cdot 5$, and $25 = 5 \cdot 5$. Thus, the least common multiple is $2 \cdot 2 \cdot 3 \cdot 5 \cdot 5$, or 300.

63. Find the least common multiple of 4, 14, 28, and 49.

Solution

By prime factorization, $4 = 2 \cdot 2$, $14 = 2 \cdot 7$, $28 = 2 \cdot 2 \cdot 7$, and $49 = 7 \cdot 7$. Thus, the least common multiple is $2 \cdot 2 \cdot 7 \cdot 7$, or 196.

65. Decide whether the numbers 495 and 784 are relatively prime.

Solution

By prime factorization, $495 = 3 \cdot 3 \cdot 5 \cdot 11$ and $784 = 2 \cdot 2 \cdot 2 \cdot 2 \cdot 7 \cdot 7$. Thus, 495 and 784 are relatively prime because they have no common prime factors.

67. Decide whether the numbers 403 and 899 are relatively prime.

Solution

By prime factorization, $403 = 13 \cdot 31$ and $899 = 29 \cdot 31$. Thus, 403 and 899 are not relatively prime because they have a common factor of 31.

69. Decide whether the numbers 51, 85, and 199 are relatively prime.

Solution

By prime factorization, $51 = 3 \cdot 17$, $85 = 5 \cdot 17$, and $119 = 7 \cdot 17$. Thus, 51, 85, and 119 are not relatively prime because they have a common prime factor of 17.

71. *Investigation* The numbers 14, 15, and 16 are an example of three consecutive composite numbers. Is it possible to find ten consecutive composite numbers? If so, list an example. If not, explain why.

Solution

114, 115, 116, 117, 118, 119, 120, 121, 122, 123,

200, 201, 202, 203, 204, 205, 206, 207, 208, 209,

212, 213, 214, 215, 216, 217, 218, 219, 220, 221

(These are representative answers; other answers are possible.)

73. *The Sieve of Eratosthenes* Write the integers from 1 through 100 in ten lines of ten numbers each.

(a) Cross out the number 1. Cross out all multiples of 2 other than 2 itself. Do the same for 3, 5, and 7.

(b) Of what type are the remaining numbers? Explain why this is the only type of number left.

Solution

(a)
1	2	3	4	5	6	7	8	9	10
11	12	13	14	15	16	17	18	19	20
21	22	23	24	25	26	27	28	29	30
31	32	33	34	35	36	37	38	39	40
41	42	43	44	45	46	47	48	49	50
51	52	53	54	55	56	57	58	59	60
61	62	63	64	65	66	67	68	69	70
71	72	73	74	75	76	77	78	79	80
81	82	83	84	85	86	87	88	89	90
91	92	93	94	95	96	97	98	99	100

(b) The remaining numbers are prime.

Explanation: Every number on the list has a square root that is less than or equal to 10. The numbers 2, 3, 5, and 7 are the only prime numbers less than 10. Because all the multiples of these four prime numbers have been crossed out and the number 1 has been crossed out also, any number which remains on the list has no prime factors less than its square root. Thus, any number which remains on the list must be prime.

Mid-Chapter Quiz for Chapter P

1. Show the real numbers -2.5 and -4 as points on the real number line and place the correct inequality symbol ($<$ or $>$) between the two numbers.

Solution

$-2.5 > -4$

2. Show the real numbers $\frac{3}{16}$ and $\frac{3}{8}$ as points on the real number line and place the correct inequality symbol ($<$ or $>$) between the two numbers.

Solution

$\frac{3}{16} < \frac{3}{8}$

3. Show the real numbers -3.1 and 2.7 as points on the real number line and place the correct inequality symbol ($<$ or $>$) between the two numbers.

Solution

$-3.1 < 2.7$

4. Show the real numbers 2π and 6 as points on the real number line and place the correct inequality symbol ($<$ or $>$) between the two numbers.

Solution

$2\pi > 6$

Note: $2\pi \approx 6.28$

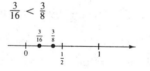

5. Evaluate $-|-0.75|$.

Solution

$-|-0.75| = -0.75$

Note: $|-0.75| = 0.75$

6. Evaluate $|25.2|$.

Solution

$|25.2| = 25.2$

7. Place the correct symbol ($<$, $>$, or $=$) between the real numbers $\left|\frac{7}{2}\right|$ and $|-3.5|$.

Solution

$\left|\frac{7}{2}\right| = |-3.5|$

Note: $\left|\frac{7}{2}\right| = \frac{7}{2}$ or 3.5, and $|-3.5| = 3.5$.

8. Place the correct symbol ($<$, $>$, or $=$) between the real numbers $\left|\frac{3}{4}\right|$ and $-|0.75|$.

Solution

$\left|\frac{3}{4}\right| > -|0.75|$

Note: $\left|\frac{3}{4}\right| = \frac{3}{4}$ or 0.75, and $-|0.75| = -0.75$, $0.75 > -0.75$.

9. Copy the number line, write the opposites of $-\frac{3}{2}$ and $\frac{5}{2}$, and plot the opposites on the number line.

Solution

The opposite of $-\frac{3}{2}$ is $\frac{3}{2}$; $-\left(-\frac{3}{2}\right) = \frac{3}{2}$.

The opposite of $\frac{5}{2}$ is $-\frac{5}{2}$.

10. Copy the number line, write the opposites of $-\frac{3}{4}$ and $\frac{9}{4}$, and plot the opposites on the number line.

Solution

The opposite of $-\frac{3}{4}$ is $\frac{3}{4}$; $-\left(-\frac{3}{4}\right) = \frac{3}{4}$.

The opposite of $\frac{9}{4}$ is $-\frac{9}{4}$.

11. Find the product of $7 \cdot 11 \cdot 11 \cdot 13$.

Solution

$7 \cdot 11 \cdot 11 \cdot 13 = 11,011$

12. Find the product of $5 \cdot 23 \cdot 29$.

Solution

$5 \cdot 23 \cdot 29 = 3335$

13. Determine whether the number 457 is prime or composite. Explain your reasoning.

Solution

457 is prime; the divisibility tests yield no factors of 457. By testing the remaining primes less than or equal to $\sqrt{457} \approx 21$, you can conclude that 457 is a prime number.

14. Determine whether the number 1341 is prime or composite. Explain your reasoning.

Solution

1341 is composite; its prime factorization is $3 \cdot 3 \cdot 149$.

15. Write the prime factorization of 354.

Solution

$354 = 2 \cdot 3 \cdot 59$

16. Write the prime factorization of 945.

Solution

$945 = 3 \cdot 3 \cdot 3 \cdot 5 \cdot 7$

17. Find the greatest common factor of 50 and 60.

Solution

By prime factorization, $50 = 2 \cdot 5 \cdot 5$ and $60 = 2 \cdot 2 \cdot 3 \cdot 5$. Thus, the greatest common factor is $2 \cdot 5$, or 10.

18. Find the greatest common factor of 276 and 1035.

Solution

By prime factorization, $276 = 2 \cdot 2 \cdot 3 \cdot 23$ and $1035 = 3 \cdot 3 \cdot 5 \cdot 23$. Thus, the greatest common factor is $3 \cdot 23$, or 69.

19. Find the least common multiple of 12 and 28.

Solution

By prime factorization, $12 = 2 \cdot 2 \cdot 3$ and $28 = 2 \cdot 2 \cdot 7$. Thus, the least common multiple is $2 \cdot 2 \cdot 3 \cdot 7$, or 84.

20. Find the least common multiple of 6, 10, 15, and 45.

Solution

By prime factorization, $6 = 2 \cdot 3$, $10 = 2 \cdot 5$, $15 = 3 \cdot 5$, and $45 = 3 \cdot 3 \cdot 5$. Thus, the least common multiple is $2 \cdot 3 \cdot 3 \cdot 5$, or 90.

P.3   Adding and Subtracting Integers

7. Evaluate the sum $2 + 7$. Sketch the addition on the real number line.

Solution

$2 + 7 = 9$

9. Evaluate the sum $-6 + 4$. Sketch the addition on the real number line.

Solution

$-6 + 4 = -2$

11. Find the sum of $-23 + 4$.

Solution

$-23 + 4 = -19$

13. Find the sum of $-10 + 6 + 34$.

Solution

$-10 + 6 + 34 = 30$

15. Find the sum of $32 + (-32) + (-16) = -16$.

Solution

$32 + (-32) + (-16) = -16$

17. Find the sum of $49 + (-|-17|)$.

Solution

$49 + (-|-17|) = 49 + (-17) = 32$

19. Write the subtraction problem $12 - 9$ as an addition problem and evaluate the result.

Solution

$12 - 9 = 12 + (-9) = 3$

21. Write the subtraction problem $-4 - (-4)$ as an addition problem and evaluate the result.

Solution

$-4 - (-4) = -4 + 4 = 0$

23. Evaluate the expression $55 - 20$.

Solution

$55 - 20 = 35$

25. Evaluate the expression $1000 - (-500)$.

Solution

$1000 - (-500) = 1500$

27. Evaluate the expression $-210 - 400$.

Solution

$-210 - 400 = -610$

29. Evaluate the expression $23 - (15 - 8)$.

Solution

$23 - (15 - 8) = 23 - 7 = 16$

31. Find the sum of 250 and -300.

Solution

$250 + (-300) = -50$

33. Subtract -120 from 380.

Solution

$380 - (-120) = 380 + 120 = 500$

35. *Temperature Change* The temperature at 6 A.M. was $-10°F$. By noon, the temperature had increased by $22°F$. What was the temperature at noon?

Solution

$-10 + 22 = 12$

The temperature at noon was $12°$ F.

37. *Flying Altitude* An airliner flying at an altitude of 31,000 feet is instructed to descend to an altitude of 24,000 feet. How many feet must the aircraft descend?

Solution

$31,000 - 24,000 = 7000$

The plane must descend 7000 feet.

39. Find the sum of $14 + (-14)$.

Solution

$14 + (-14) = 0$

41. Find the sum of $5 + |-3|$.

Solution

$5 + |-3| = 5 + 3 = 8$

43. Find the sum of $-18 + (-12)$.

Solution

$-18 + (-12) = -30$

45. Find the sum of $-32 + 16$.

Solution

$-32 + 16 = -16$

47. Find the sum of $-82 + (-36) + 82$.

Solution

$-82 + (-36) + 82 = -36$

49. Find the sum of $1200 + 1300 + (-275)$.

Solution

$1200 + 1300 + (-275) = 2225$

51. Find the sum of $1875 + (-3143) + 5826$.

Solution

$1875 + (-3143) + 5826 = 4558$

53. Find the sum of $|-890| + (-|-82|) + 90$.

Solution

$$|-890| + (-|-82|) + 90 = 890 + (-82) + 90$$
$$= 898$$

55. Perform the subtraction: $453 - 354$.

Solution

$453 - 354 = 99$

57. Perform the subtraction: $-714 - 320$.

Solution

$-714 - 320 = -1034$

59. Perform the subtraction: $-10 - (-4)$.

Solution

$-10 - (-4) = -10 + 4 = -6$

61. Perform the subtraction: $-942 - (-942)$.

Solution

$-942 - (-942) = -942 + 942 = 0$

63. Perform the subtraction: $|15| - |-7|$.

Solution

$|15| - |-7| = 15 - 7 = 8$

65. Perform the subtraction: $53 - (25 - 9)$.

Solution

$53 - (25 - 9) = 53 - 16 = 37$

67. Perform the subtraction: $-32 - [18 - (25 + 8)]$.

Solution

$$\begin{aligned} -32 - [18 - (25 + 8)] &= -32 - [18 - 33] \\ &= -32 - (-15) \\ &= -32 + 15 \\ &= -17 \end{aligned}$$

69. Perform the operation: $0 - (-12)$.

Solution

$$0 - (-12) = 0 + 12 = 12$$

71. Perform the operation: $-130 + 130$.

Solution

$$-130 + 130 = 0$$

73. Perform the operation: $72 - 85$.

Solution

$$72 - 85 = -13$$

75. Perform the operation: $-12 - 2 + |-3|$.

Solution

$$-12 - 2 + |-3| = -12 - 2 + 3 = -11$$

77. Perform the operation: $550 + (-1625) + (-4060) + 7132$.

Solution

$$550 + (-1625) + (-4060) + 7132 = 1997$$

79. Perform the operation: $34 - [54 - (-16 + 4) + 6$.

Solution

$$\begin{aligned} 34 - [54 - (-16 + 4) + 6] &= 34 - [54 - (-12) + 6] \\ &= 34 - [54 + 12 + 6] \\ &= 34 - 72 \\ &= -38 \end{aligned}$$

81. Find the sum of 72 and -37.

Solution

$$72 + (-37) = 35$$

83. Subtract 1500 from 2500.

Solution

$$2500 - 1500 = 1000$$

85. Subtract -750 from 800.

Solution

$$800 - (-750) = 800 + 750 = 1550$$

87. Find the absolute value of the sum of -45 and -80.

Solution

$$|-45 + (-80)| = |-125| = 125$$

89. What number must be added to 10 to obtain -5?

Solution

$$-5 - 10 = -15$$

Thus, -15 must be added to 10 to obtain -5.

91. *Banking*   You start a non-interest-earning checking account by depositing $500. During the first month, you write checks for $145 and $278. What is your balance at the end of the month?

Solution

$$500 - 145 - 278 = 77$$

Your balance at the end of the month is $77.

93. *Reading a Graph*   On Monday you purchased $800 worth of stock. The value of the stock during the remainder of the week is shown in the bar graph. Use the graph to complete the table showing the daily gains and losses during the week.

Solution

Tuesday: $821 - 800 = 21$
Wednesday: $837 - 821 = 16$
Thursday: $820 - 837 = -17$
Friday: $792 - 820 = -28$

Day	Daily Gain or Loss
Tuesday	Gained $21
Wednesday	Gained $16
Thursday	Lost $17
Friday	Lost $28

95. *Reading a Graph*   The bar graph gives the new AIDS cases reported in the U.S.

(a) Estimate the number of new cases in 1987.
(b) Estimate the total number of new cases reported in 1990, 1991, and 1992.
(c) Estimate the increase in the number of new cases from 1989 to 1990.

Solution

(a) 21,000
(b) $42,000 + 44,000 + 45,000 = 131,000$
(c) $42,000 - 34,000 = 8000$

P.4 | Multiplying and Dividing Integers

7. Write $3 \cdot 2$ as repeated addition and find the product.

Solution

$3 \cdot 2 = 2 + 2 + 2 = 6$

9. Find the product of 7×30.

Solution

$7 \times 30 = 210$

11. Find the product of $4(-8)$.

Solution

$4(-8) = -32$

13. Find the product of $5(-3)(-6)$.

Solution

$5(-3)(-6) = 90$

15. Find the product of $|-3|(-3)(4)$.

Solution

$|-3|(-3)(4) = 3(-3)(4) = -36$

17. Use the vertical multiplication algorithm to find the product of $75(-632)$.

Solution

$$\begin{array}{r} 632 \\ \times\ 75 \\ \hline 3160 \\ 4424 \\ \hline 47400 \end{array}$$

Thus, $75(-632) = -47,400$.

19. Perform the division in $27 \div 9$, if possible. If not possible, state the reason.

Solution

$27 \div 9 = 3$

21. Perform the division in $72 \div (-12)$, if possible. If not possible, state the reason.

Solution

$72 \div (-12) = -6$

23. Perform the division in $\frac{8}{0}$, if possible. If not possible, state the reason.

Solution

$\frac{8}{0}$ is undefined.

25. Perform the division in $\frac{-180}{-45}$, if possible. If not possible, state the reason.

Solution

$\frac{-180}{-45} = 4$

27. Use the long division algorithm to find $1440 \div -45$.

Solution

$$\begin{array}{r} 32 \\ 45\overline{)1440} \\ \underline{135} \\ 90 \\ \underline{90} \end{array}$$

Thus, $1440 \div -45 = -32$.

29. Use the long division algorithm to find $\frac{2209}{47}$.

Solution

$$\begin{array}{r} 47 \\ 47\overline{)2209} \\ \underline{188} \\ 329 \\ \underline{329} \end{array}$$

Thus, $\frac{2209}{47} = 47$.

31. Evaluate $\dfrac{5 - 3 + 16}{10 - 4}$.

Solution

$\dfrac{5 - 3 + 16}{10 - 4} = \dfrac{18}{6} = 3$

33. Evaluate $\dfrac{8(-45)}{18}$.

Solution

$\dfrac{8(-45)}{18} = \dfrac{-360}{18} = -20$

35. *Temperature Change*   The temperature measured by a weather balloon is decreasing approximately $3°$ for each 1000-foot increase in altitude. The balloon rises 8000 feet. Describe its temperature change.

Solution

$\left(\frac{8000}{1000}\right)(-3) = 8(-3) = -24°$

The temperature would decrease by approximately $24°$.

37. *Exam Scores*   A student has a total of 328 points after four 100-point exams. (a) What is the average number of points scored per exam? (b) The scores on the four exams are 87, 73, 77, and 91. Plot each of the scores and the average score on the real number line. (c) Find the difference between each score and the average score. Find the sum of these and give a possible explanation of the result.

Solution

(a) $\frac{328}{4} = 82$

The average number of points scored per exam is 82.

(b)
```
    73  77    82    87  91
  ‹─●┼──┼●┼──┼─●─┼──┼─●┼──┼─●┼─›
    72   76   80   84   88   92
```

(c) $(87 - 82) + (73 - 82) + (77 - 82) + (91 - 82) = 5 + (-9) + (-5) + (9) = 0$

The sum of these differences is 0.

39. Write 4×5 as repeated addition and find the product.

Solution

$4 \times 5 = 5 + 5 + 5 + 5 = 20$

41. Write $5 \times (-3)$ as repeated addition and find the product.

Solution

$5 \times (-3) = (-3) + (-3) + (-3) + (-3) + (-3)$
$\qquad\qquad = -15$

43. Find the product of $(-6)(-12)$.

Solution

$(-6)(-12) = 72$

45. Find the product of $(-6)(8)$.

Solution

$(-6)(8) = -48$

47. Find the product of $(310)(-32)$.

Solution

$(310)(-32) = -9920$

49. Find the product of $(-2)(-3)(-5)$.

Solution

$(-2)(-3)(-5) = -30$

51. Find the product of $|3(-5)(6)|$.

Solution

$|3(-5)(6)| = |-90| = 90$

53. Find the product of $(-3)(4)(0)(6)(8)$.

Solution

$(-3)(4)(0)(6)(8) = 0$

55. Use the vertical multiplication algorithm to find the product of 26×130.

Solution

```
   130
 × 26
 ────
   780
  260
 ────
  3380
```

57. Use the vertical multiplication algorithm to find the product of $(-72)(866)$.

Solution

```
   866
 × 72
 ────
  1732
 6062
 ─────
 62352
```

Thus, $(-72)(866) = -62,352$.

59. Perform the division in $54 \div 9$, if possible. If not possible, state the reason.

Solution

$54 \div 9 = 6$

61. Perform the division in $\frac{-81}{-3}$, if possible. If not possible, state the reason.

Solution

$\frac{-81}{-3} = 27$

63. Perform the division in $\frac{-58}{2}$, if possible. If not possible, state the reason.

Solution

$\frac{-58}{2} = -29$

65. Perform the division in $\frac{0}{81}$, if possible. If not possible, state the reason.

Solution

$\frac{0}{81} = 0$

67. Perform the division in $\frac{6}{-1}$, if possible. If not possible, state the reason.

Solution

$\frac{6}{-1} = -6$

69. Perform the division in $144 \div 0$, if possible. If not possible, state the reason.

Solution

$144 \div 0$ is undefined. Division by 0 is undefined.

71. Use the long division algorithm to find $-1248 \div 48$.

Solution

$$
\begin{array}{r}
26 \\
48 \overline{)1248} \\
96 \\
\hline
288 \\
288 \\
\hline
\end{array}
$$

Thus, $-1248 \div 48 = -26$.

73. Use the long division algorithm to find $\frac{2209}{47}$.

Solution

$$
\begin{array}{r}
47 \\
47 \overline{)2209} \\
188 \\
\hline
329 \\
329 \\
\hline
\end{array}
$$

Thus, $\frac{2209}{47} = 47$.

75. Use the long division algorithm to find $2750 \div 25$.

Solution

$$
\begin{array}{r}
110 \\
25 \overline{)2750} \\
25 \\
\hline
25 \\
25 \\
\hline
0 \\
0 \\
\hline
\end{array}
$$

Thus, $2750 \div 25 = 110$.

77. Evaluate the expression $\frac{25 - 10}{-2 - 3}$.

Solution

$$\frac{25 - 10}{-2 - 3} = \frac{15}{-5}$$
$$= -3$$

79. Evaluate the expression $\frac{|-200|}{10(-4)}$.

Solution

$\frac{|-200|}{10(-4)} = \frac{200}{-40} = -5$

81. Evaluate the expression $\frac{8 + 3 \cdot 4}{5}$.

Solution

$\frac{8 + 3 \cdot 4}{5} = \frac{8 + 12}{5} = \frac{20}{5} = 4$

83. Use a calculator to evaluate the expression $5(1650) - 3710$.

Solution

$5(1650) - 3710 = 8250 - 3710 = 4540$

85. Use a calculator to evaluate the expression $\frac{44,290}{515}$.

Solution

$\frac{44,290}{515} = 86$

87. Use a calculator to evaluate the expression $\dfrac{169{,}290}{162}$.

Solution

$$\dfrac{169{,}290}{162} = 1045$$

89. Find the product of $(-2)(532)(500)$ mentally. Explain your strategy.

Solution

$(-2)(532)(500) = -532{,}000$

You could multiply $-2(500)$ to obtain -1000; multiplying -1000 by 532 yields the result of $-532{,}000$.

91. *Savings Plan*   After you decide to save \$50 per month for 10 years, what is the total amount you will have deposited?

Solution

$(50)(12)(10) = 6000$

You will have deposited a total of \$6000.

93. *Loss Leaders*   To attract customers, a grocery store runs a sale on bananas. The bananas are *loss leaders*, which means the store loses money on the bananas but hopes to make it up on other items. The store sells 800 pounds at a loss of 26 cents per pound. What is the total loss?

Solution

$(800)(0.26) = 208$

The total loss on the bananas is \$208.

95. *Geometry*   Find the area of the floor of the building (see diagram in textbook).

Solution

$(32)(40) = 1280$

The area of the floor of the building is 1280 square feet.

97. *Geometry*   Find the volume of the prism. The volume of a rectangular prism is found by multiplying the length, width and height of the prism. (See figure in textbook.)

Solution

$(9)(6)(11) = 594$

The volume of the prism is 594 cubic inches.

99. *Temperature*   The temperature fell $28°$ in 4 hours. Find the average rate of change per hour.

Solution

$$\dfrac{-28}{4} = -7$$

The average rate of change in the temperature was $-7°$ per hour.

101. A nonzero product has 25 factors, 17 of which are negative. What is the sign of the product?

Solution

The product is negative because a product that has an odd number of negative factors is negative.

103. A number is divisible by 11 if the difference of the sum of the alternate digits and the sum of the remaining digits is divisible by 11. For example, 909,556,857 is divisible by 11 because

$$(9 + 9 + 5 + 8 + 7) - (0 + 5 + 6 + 5) = 22$$

is divisible by 11. Use your calculator to verify this result.

Solution

You could use a calculator to verify that $909{,}556{,}857 \div 11 = 82{,}686{,}987$.

Review Exercises for Chapter P

1. Plot the real numbers $-\frac{1}{10}$ and 4 on the real number line and place the correct symbol ($<$, $>$, or $=$) between the two numbers.

Solution

$-\frac{1}{10} < 4$ because $-\frac{1}{10}$ lies to the *left* of 4.

3. Plot the real numbers -3 and -7 on the real number line and place the correct symbol ($<$, $>$, or $=$) between the two numbers.

Solution

$-3 > -7$ because -3 lies to the *right* of -7.

5. Find the opposite of the number 152.

Solution

The opposite of 152 is -152.

7. Find the opposite of the number $-\frac{7}{3}$.

Solution

The opposite of $-\frac{7}{3}$ is $\frac{7}{3}$; $-\left(-\frac{7}{3}\right) = \frac{7}{3}$.

9. Evaluate the expression $|-8.5|$.

Solution

$|-8.5| = 8.5$

11. Evaluate the expression $-|-8.5|$.

Solution

$-|-8.5| = 8.5$

Note: $|-8.5| = -8.5$

13. Place the correct symbol ($<$, $>$, or $=$) between the two real numbers $|-84|$ and $|84|$.

Solution

$|-84| = |84|$ because $|-84| = 84$ and $|84| = 84$.

15. Place the correct symbol ($<$, $>$, or $=$) between the two real numbers $|\frac{3}{10}|$ and $-|\frac{4}{5}|$.

Solution

$|\frac{3}{10}| > -|\frac{4}{5}|$ because $|\frac{3}{10}| = \frac{3}{10}$, $-|\frac{4}{5}| = -\frac{4}{5}$, and $\frac{3}{10} > -\frac{4}{5}$.

17. Decide whether 839 is prime or composite.

Solution

839 is prime; the divisibility tests yield no factors of 839. By testing the remaining primes less than or equal to $\sqrt{839} \approx 29$, you can conclude the 839 is a prime number.

19. Decide whether 1764 is prime or composite.

Solution

1764 is composite; its prime factorization is $1764 = 2 \cdot 2 \cdot 3 \cdot 3 \cdot 7 \cdot 7$.

21. Write the prime factorization of the number 378.

Solution

$378 = 2 \cdot 3 \cdot 3 \cdot 3 \cdot 7$

23. Write the prime factorization of the number 1612.

Solution

$1612 = 2 \cdot 2 \cdot 13 \cdot 31$

25. Find the greatest common factor of 54 and 90.

Solution

By prime factorization, $54 = 2 \cdot 3 \cdot 3 \cdot 3$ and $90 = 2 \cdot 3 \cdot 3 \cdot 5$. Thus, the greatest common factor is $2 \cdot 3 \cdot 3$, or 18.

27. Find the greatest common factor of 63, 84, and 441.

Solution

By prime factorization, $63 = 3 \cdot 3 \cdot 7$, $84 = 2 \cdot 2 \cdot 3 \cdot 7$, and $441 = 3 \cdot 3 \cdot 7 \cdot 7$. Thus, the greatest common factor is $3 \cdot 7$, or 21.

29. Find the least common multiple of 16 and 18.

Solution

By prime factorization, $16 = 2 \cdot 2 \cdot 2 \cdot 2$ and $18 = 2 \cdot 3 \cdot 3$. Thus, the least common multiple is $2 \cdot 2 \cdot 2 \cdot 2 \cdot 3 \cdot 3$, or 144.

31. Find the least common multiple of 14, 20, and 24.

Solution

By prime factorization, $14 = 2 \cdot 7$, $20 = 2 \cdot 2 \cdot 5$, and $24 = 2 \cdot 2 \cdot 2 \cdot 3$. Thus, the least common multiple is $2 \cdot 2 \cdot 2 \cdot 3 \cdot 5 \cdot 7$, or 840.

33. Evaluate the expression $32 + 68$, if possible. If it is not possible, state the reason.

Solution

$32 + 68 = 100$

35. Evaluate the expression $16 + (-5)$, if possible. If it is not possible, state the reason.

Solution

$16 + (-5) = 11$

37. Evaluate the expression $350 - 125 + 15$, if possible. If it is not possible, state the reason.

Solution

$350 - 125 + 15 = 240$

39. Evaluate the expression $-114 + 76 - 230$, if possible. If it is not possible, state the reason.

Solution

$-114 + 76 - 230 = -268$

41. Evaluate the expression $|-86| - |124|$, if possible. If it is not possible, state the reason.

Solution

$|-86| - |124| = 86 - 124 = -38$

43. Evaluate the expression $122 - [45 - (32 + 8) - 23]$, if possible. If it is not possible, state the reason.

Solution

$$122 - [45 - (32 + 8) - 23] = 122 - [45 - 40 - 23]$$
$$= 122 - [-18]$$
$$= 122 + 18$$
$$= 140$$

45. Evaluate the expression 15×3, if possible. If it is not possible, state the reason.

Solution

$15 \times 3 = 45$

47. Evaluate the expression $-300(-5)$, if possible. If it is not possible, state the reason.

Solution

$-300(-5) = 1500$

49. Evaluate the expression $131(-6)(3)$, if possible. If it is not possible, state the reason.

Solution

$131(-6)(3) = -2358$

51. Evaluate the expression $\dfrac{-162}{9}$, if possible. If it is not possible, state the reason.

Solution

$\dfrac{-162}{9} = -18$

53. Evaluate the expression $815 \div 0$, if possible. If it is not possible, state the reason.

Solution

$815 \div 0$ is undefined. Division by 0 is undefined.

55. Evaluate the expression $\dfrac{78 - |-78|}{5}$, if possible. If it is not possible, state the reason.

Solution

$\dfrac{78 - |-78|}{5} = \dfrac{78 - 78}{5} = \dfrac{0}{5} = 0$

57. Evaluate the expression $\dfrac{54 - 4 \cdot 3}{6}$, if possible. If it is not possible, state the reason.

Solution

$$\dfrac{54 - 4 \cdot 3}{6} = \dfrac{54 - 12}{6}$$
$$= \dfrac{42}{6}$$
$$= 7$$

59. Evaluate the expression $\dfrac{6 \cdot 4 - 36}{4}$, if possible. If it is not possible, state the reason.

Solution

$$\dfrac{6 \cdot 4 - 36}{4} = \dfrac{24 - 36}{4}$$
$$= \dfrac{-12}{4}$$
$$= -3$$

61. Use the long division algorithm to find $33,768 \div (-72)$.

Solution

```
         4 6 9
72 ) 3 3 7 6 8
     2 8 8
     ‾‾‾‾‾
       4 9 6
       4 3 2
       ‾‾‾‾‾
         6 4 8
         6 4 8
         ‾‾‾‾‾
```

Thus, $33,768 \div (-72) = -469$.

63. Use a calculator to perform the operation $7(5207) - 52,318$.

Solution

$7(5207) - 52,318 = -15,869$

65. Use a calculator to perform the operation $\dfrac{345,582}{438}$.

Solution

$\dfrac{345,582}{438} = 789$

67. Subtract -549 from 613.

Solution

$613 - (-549) = 1162$

69. What must you add to 75 to obtain -27?

Solution

$-27 - 75 = -102$

You must add -102 to 75 to obtain -27.

71. *Reading a Table*   The ten NFL coaches with the most wins in regular season games through 1990 are listed in the table. Order the coaches in decreasing order according to the number of wins.

Solution

George Halas	319
Don Shula	269
Tom Landry	250
Earl Lambeau	225
Chuck Noll	177
Paul Brown	166
Bud Grant	158
Chuck Knox	155
Steve Owen	151
Hank Stram	131

73. *Unreadable Receipts*   A manufacturing company purchased components from a supplier at a cost of $9 each. The total on the receipt read $19,?02. (The digit represented by ? was unreadable.)

(a) Determine the missing digit and give a reason for your answer.

(b) How many components were purchased?

Solution

(a) The missing digit was a 6. The total on the receipt must be a multiple of 9 because the components were $9 each. According to the divisibility rule for 9, the sum of the digits must be divisible by 9. The readable digits in $19,?02 have a sum of 12. The smallest number that is divisible by 9 and larger than 12 is 18. Thus, the missing digit must by $18 - 12$, or 6.

(b) $\frac{19,602}{9} = 2178$

Thus, 2178 components were purchased.

75. *Reading a Table*   The costs of adult and student tickets for a concert are $25 and $10, respectively. The table gives the number of tickets sold for the first four days of sales. (a) Find the revenue from ticket sales for each day. (b) Find the revenue from ticket sales for each type of ticket. (c) Find the total revenue from ticket sales. Explain how you can obtain this result from parts (a) and (b). How could you use this to check your work?

Solution

(a) Day 1: $25(162) + 10(98) = \$5030$
     Day 2: $25(98) + 10(64) = \$3090$
     Day 3: $25(148) + 10(81) = \$4510$
     Day 4: $25(186) + 10(105) = \$5700$

(b) Adult Tickets: $25(162 + 98 + 148 + 186) = 25(594) = \$14,850$
     Student Tickets: $10(98 + 64 + 81 + 105) = 10(348) = \$3480$

(c) Total from (a): $5030 + 3090 + 4510 + 5700 = \$18,330$
     Total from (b): $14,850 + 3480 = \$18,330$

The total revenue from ticket sales can be determined by adding the daily ticket sales (from part a) or by adding the sales of the two types of tickets (from part b). These two totals should be the same.

77. *Think About It*   Your rotate the tires on your car, including the spare, so that all five tires are used equally. After 40,000 miles, how many miles has each tire been used?

Solution

$$\frac{40,000(4)}{5} = \frac{160,000}{5} = 32,000$$

Thus, each tire has been driven 32,000 miles.

79. *Think About It*   Which is smaller: $\frac{2}{3}$ or 0.6?

Solution

The smaller number is 0.6.

Note: $\frac{2}{3} = 0.666\ldots$ and $0.666\ldots > 0.6$

81. *True or False?*   The sum of two integers, one negative and one positive, is negative. Explain.

Solution

The statement is false because the sum of a positive and negative integer can be positive, negative or zero.

Note:

(a) If the absolute value of the negative integer is smaller than the positive integer, the sum is positive.

   Example: $12 + (-8) = 4$

(b) If the absolute value of the negative integer is larger than the positive integer, the sum is negative.

   Example: $12 + (-15) = -3$

(c) If the absolute value of the negative integer is equal to the positive integer, the sum is zero.

   Example: $12 + (-12) = 0$

83. *Reading a Graph*   The bar graph shows the popular votes (in millions) cast for president in the presidential elections from 1972 through 1992. (a) Estimate the total popular votes cast for the candidates in 1980. (b) Estimate the difference between the votes cast for the winning candidate and the independent candidate in 1972. (c) Estimate the total popular votes cast for each of the elections. Did the number of votes cast increase or decrease with time? (d) Describe how you can use the graph to determine whether the winning candidate received more than one-half of the popular vote. Did this always occur for the elections shown on the graph?

Solution

(a) $35,500,000 + 44,000,000 + 6,000,000 = 85,500,000$

   Approximately 85.5 million votes were cast.

(b) $47,000,000 - 1,000,000 = 46,000,000$

   The winning candidate had approximately 46 million more votes.

(c) 1972: $29,000,000 + 47,000,000 + 1,000,000 = 77,000,000$
        Approximately 77 million votes were cast.
     1976: $41,000,000 + 39,000,000 = 80,000,000$
        Approximately 80 million votes were cast.
     1980: $35,500,000 + 44,000,000 + 6,000,000 = 85,500,000$
        Approximately 85.5 million votes were cast.
     1984: $37,500,000 + 54,500,000 = 92,000,000$
        Approximately 92 million votes were cast.
     1988: $42,000,000 + 49,000,000 = 91,000,000$
        Approximately 91 million votes were cast.
     1992: $45,000,000 + 39,000,000 + 20,000,000 = 104,000,000$
        Approximately 104 million votes were cast.

   The number of votes cast generally increased with time, but the increase was inconsistent.

(d) The bar graph for the winning candidate must be taller than the combined, or stacked, bar graphs of the other candidates. No, the winning candidate did not receive more than one-half the popular vote in 1980 or in 1992.

Test for Chapter P

1. Which of the numbers $-10, 8, \frac{3}{4}, \frac{12}{4}, 6.5$ are (a) natural numbers? (b) integers? (c) rational numbers?

Solution

(a) Natural numbers: $\{8, \frac{12}{4}\}$

(b) Integers: $\{-10, 8, \frac{12}{4}\}$

(c) Rational numbers: $\{-10, 8, \frac{3}{4}, \frac{12}{4}, 6.5\}$

2. Place the correct inequality symbol ($<$ or $>$) between the real numbers $-\frac{3}{5}$ and $-|-2|$.

Solution

$-\frac{3}{5} > -|-2|$

3. Find all real numbers whose distance from 10 is 18.

Solution

The number 18 units to the right of 10 is 28.
The number 18 units to the left of 10 is -8.

4. Write the prime factorization of 234.

Solution

$234 = 2 \cdot 3 \cdot 3 \cdot 13$

5. Find the greatest common factor of 90 and 150.

Solution

The greatest common factor is $2 \cdot 3 \cdot 5$, or 30.

6. Find the least common multiple of 56 and 84.

Solution

The least common multiple is $2 \cdot 2 \cdot 2 \cdot 3 \cdot 7$, or 168.

7. Evaluate the expression $16 + (-20)$.

Solution

$16 + (-20) = -4$

8. Evaluate the expression $-50 - (-60)$.

Solution

$-50 - (-60) = -50 + 60 = 10$

9. Evaluate the expression $7 + |-3|$.

Solution

$7 + |-3| = 7 + 3 = 10$

10. Evaluate the expression $64 - (25 - 8)$.

Solution

$64 - (25 - 8) = 64 - 17 = 47$

11. Evaluate the expression $-5(32)$.

Solution

$-5(32) = -160$

12. Evaluate the expression $\frac{-72}{-9}$.

Solution

$\frac{-72}{-9} = 8$

13. Evaluate the expression $\dfrac{12 + 9}{7}$.

Solution

$\dfrac{12 + 9}{7} = \dfrac{21}{7} = 3$

14. Evaluate the expression $-\dfrac{(-2)(5)}{10}$.

Solution

$-\dfrac{(-2)(5)}{10} = -\dfrac{-10}{10} = -(-1) = 1$

15. The company's quarterly profits are shown in the bar graph at the right. What is the company's total profit for the year?

Solution

$513,200 + 136,500 + (-97,750) + (-101,500) = 450,450$

The total profit for the year $450,450.

16. A cord of wood is a pile 8 feet long, 4 feet wide, and 4 feet high. The volume of a rectangular solid is its length times its width times its height. Find the number of cubic feet in a cord of wood.

Solution

$(8)(4)(4) = 128$

There are 128 cubic feet in a cord of wood.

17. It is necessary to cut a 90-foot rope into six pieces of equal length. What is the length of each piece?

Solution

$90 \div 6 = 15$

Each piece of rope is 15 feet long.

18. In 1989 the expenditures for electricity and natural gas in commercial buildings in the United States were $55.9 million and $9.2 million, respectively. Calculate the difference in cost for these two energy sources.

Solution

$55.9 - 9.2 = 46.7$

The difference in cost is $ 46.7 million.

19. Consider the statement, "The sum of two negative integers is positive." Is the statement true or false? If it is false, suggest any change that would make it true.

Solution

The statement is false.

Possible changes:

　　The sum of two negative integers is *negative*.
　　The sum of two *positive* integers is positive.
　　The *product* of two negative integers is positive.
　　The *quotient* of two negative integers is positive.

20. In your own words, explain why the sum of two odd integers is an even integer.

Solution

Each odd integer can be expressed as an even integer plus 1. Thus, the sum of two odd integers can be expressed as the sum of two even integers plus 2. The sum of the two even integers is even; when 2 is added to the sum, the total remains even. For example, $13 = 12 + 1$ and $35 = 34 + 1$. Thus,

$$13 + 35 = (12 + 1) + (34 + 1)$$
$$= (12 + 34) + (1 + 1)$$
$$= 46 + 2$$
$$= 48.$$

CHAPTER ONE
The Real Number System

1.1   Adding and Subtracting Fractions

7. Write the fraction $\frac{2}{8}$ in reduced form.

Solution

$$\frac{2}{8} = \frac{(1)(2)}{(4)(2)} = \frac{1}{4}$$

9. Write the fraction $\frac{60}{192}$ in reduced form.

Solution

$$\frac{60}{192} = \frac{(5)(12)}{(16)(12)} = \frac{5}{16}$$

Note: This reducing could be done using several steps, such as the following.

$$\frac{60}{192} = \frac{(30)(2)}{(96)(2)} = \frac{(10)(3)}{(32)(3)} = \frac{(5)(2)}{(16)(2)} = \frac{5}{16}$$

11. *Geometry*   The figure in the textbook is divided into regions of equal area. Find the sum of the two fractions indicated by the shaded regions.

Solution

$$\frac{1}{5} + \frac{2}{5} = \frac{1+2}{5} = \frac{3}{5}$$

13. *Geometry*   The figure in the textbook is divided into regions of equal area. Find the sum of the two fractions indicated by the shaded regions.

Solution

$$\frac{2}{10} + \frac{4}{10} = \frac{2+4}{10} = \frac{6}{10} = \frac{(3)(2)}{(5)(2)} = \frac{3}{5}$$

15. Add $\frac{7}{15} + \frac{2}{15}$. Then simplify.

Solution

$$\frac{7}{15} + \frac{2}{15} = \frac{7+2}{15} = \frac{9}{15} = \frac{(3)(3)}{(5)(3)} = \frac{3}{5}$$

17. Add $\frac{3}{4} + \frac{5}{4}$. Then simplify.

Solution

$$\frac{3}{4} - \frac{5}{4} = \frac{3}{4} + \frac{-5}{4} = \frac{3+(-5)}{4} = \frac{-2}{4} = -\frac{(1)(2)}{(2)(2)} = -\frac{1}{2}$$

19. Write an equivalent fraction with the indicated denominator.

$$\frac{3}{8} = \frac{?}{16}$$

Solution

$$\frac{3}{8} = \frac{3(2)}{8(2)} = \frac{6}{16}$$

21. Add $\frac{1}{2} + \frac{1}{3}$. Then simplify.

Solution

$$\frac{1}{2} + \frac{1}{3} = \frac{1(3)}{2(3)}$$

$$= \frac{1(3)}{2(3)} + \frac{1(2)}{3(2)} = \frac{3}{6} + \frac{2}{6} = \frac{3+2}{6} = \frac{5}{6}$$

23. Subtract $\left|-\frac{1}{8}\right| - \frac{1}{6}$. Then simplify.

Solution

$$\left|-\frac{1}{8}\right| - \frac{1}{6} = \frac{1}{8} - \frac{1}{6} = \frac{1(3)}{8(3)} - \frac{1(4)}{6(4)} = \frac{3}{24} - \frac{4}{24} = \frac{3-4}{24} = \frac{-1}{24} = -\frac{1}{24}$$

25. Subtract $4 - \frac{8}{3}$. Then simplify.

Solution

$$4 - \frac{8}{3} = \frac{4}{1} + \frac{-8}{3}$$

$$= \frac{4(3)}{1(3)} + \frac{-8}{3}$$

$$= \frac{12}{3} + \frac{-8}{3}$$

$$= \frac{12 + (-8)}{3}$$

$$= \frac{4}{3}$$

27. Add and subtract $1 + \frac{2}{3} - \frac{5}{6}$. Then simplify.

Solution

$$1 + \frac{2}{3} - \frac{5}{6} = \frac{1}{1} + \frac{2}{3} + \frac{-5}{6}$$

$$= \frac{1(6)}{1(6)} + \frac{2(2)}{3(2)} + \frac{-5}{6}$$

$$= \frac{6}{6} + \frac{4}{6} + \frac{-5}{6}$$

$$= \frac{6 + 4 + (-5)}{6}$$

$$= \frac{5}{6}$$

29. Write the mixed number $4\frac{3}{5}$ as a fraction.

Solution

$$4\frac{3}{5} = \frac{4(5) + 3}{5} = \frac{23}{5}$$

31. Write the mixed number $3\frac{7}{10}$ as a fraction.

Solution

$$3\frac{7}{10} = \frac{3(10) + 7}{10} = \frac{37}{10}$$

33. Add $3\frac{1}{2} + 5\frac{2}{3}$. Then simplify.

Solution

$$3\frac{1}{2} + 5\frac{2}{3} = \frac{3(2) + 1}{2} + \frac{5(3) + 2}{3}$$

$$= \frac{7}{2} + \frac{17}{3}$$

$$= \frac{7(3)}{2(3)} + \frac{17(2)}{3(2)}$$

$$= \frac{21}{6} + \frac{34}{6}$$

$$= \frac{21 + 34}{6}$$

$$= \frac{55}{6}$$

35. Subtract $15\frac{5}{6} - 20\frac{1}{4}$. Then simplify.

Solution

$$15\frac{5}{6} - 20\frac{1}{4} = \frac{15(6) + 5}{6} - \frac{20(4) + 1}{4}$$

$$= \frac{95}{6} - \frac{81}{4}$$

$$= \frac{95(2)}{6(2)} - \frac{81(3)}{4(3)}$$

$$= \frac{190}{12} - \frac{243}{12}$$

$$= -\frac{53}{12}$$

37. Determine the unknown fractional part of the pie graph shown in the textbook.

Solution

$$1 - \frac{3}{10} - \frac{2}{5} = \frac{1}{1} - \frac{3}{10} - \frac{2}{5}$$

$$= \frac{1(10)}{1(10)} - \frac{3}{10} - \frac{2(2)}{5(2)} = \frac{10}{10} - \frac{3}{10} - \frac{4}{10} = \frac{10 - 3 - 4}{10} = \frac{3}{10}$$

Note: This problem could also be worked in two steps. First, add the two known fractions.

$$\frac{3}{10} + \frac{2}{5} = \frac{3}{10} + \frac{2(2)}{5(2)} = \frac{3}{10} + \frac{4}{10} = \frac{3 + 4}{10} = \frac{7}{10}$$

Then subtract the sum from 1.

$$1 - \frac{7}{10} = \frac{1}{1} - \frac{7}{10} = \frac{1(10)}{1(10)} - \frac{7}{10} = \frac{10}{10} - \frac{7}{10} = \frac{10 - 7}{10} = \frac{3}{10}$$

39. *Stock Price* On Monday, a stock closed at $52\frac{5}{8}$ per share. On Tuesday, it closed at $54\frac{1}{4}$ per share. Determine the increase in price.

Solution

$$54\frac{1}{4} - 52\frac{5}{8} = \frac{54(4)+1}{4} - \frac{52(8)+5}{8}$$

$$= \frac{217}{4} - \frac{421}{8}$$

$$= \frac{434}{8} - \frac{421}{8}$$

$$= \frac{13}{8} \text{ or } 1\frac{5}{8}$$

Thus, the increase in the stock price was $1\frac{5}{8}$ per share, or $1.625 per share.

43. What must you subtract from -15 to obtain 8?

Solution

$-15 - 8 = -23$

You must subtract -23 from -15 to obtain 8.

47. Write the fraction $\frac{12}{18}$ in reduced form.

Solution

$$\frac{12}{18} = \frac{2(6)}{3(6)} = \frac{2}{3}$$

51. Add $\frac{9}{11} + \frac{5}{11}$. Then simplify.

Solution

$$\frac{9}{11} + \frac{5}{11} = \frac{9+5}{11} = \frac{14}{11}$$

55. Add $-\frac{23}{11} + \frac{12}{11}$. Then simplify.

Solution

$$\frac{-23}{11} + \frac{12}{11} = \frac{-11}{11}$$

$$= -1$$

41. Find the greatest common factor of the numbers 28, 126, and 294.

Solution

By prime factorization, $28 = 2 \cdot 2 \cdot 7$, $126 = 2 \cdot 3 \cdot 3 \cdot 7$, and $294 = 2 \cdot 3 \cdot 7 \cdot 7$. Thus, the greatest common factor is $2 \cdot 7$ or 14.

45. *Profit and Loss* A company had a first-quarter loss of $312,500, a second-quarter profit of $275,500, a third-quarter profit of $297,750, and a fourth-quarter profit of $71,300. What was the profit for the year?

Solution

$-312,500 + 275,500 + 297,750 + 71,300 = 332,050$

Thus, the company's profit for the year was $332,050.

49. Write the fraction $\frac{28}{350}$ in reduced form.

Solution

$$\frac{28}{350} = \frac{2(14)}{25(14)} = \frac{2}{25}$$

53. Subtract $\frac{9}{16} - \frac{3}{16}$. Then simplify.

Solution

$$\frac{9}{16} - \frac{3}{16} = \frac{9}{16} + \frac{-3}{16}$$

$$= \frac{9+(-3)}{16} = \frac{6}{16} = \frac{(3)(2)}{(8)(2)} = \frac{3}{8}$$

Note: This problem can also be written as follows.

$$\frac{9}{16} - \frac{3}{16} = \frac{9-3}{16} = \frac{6}{16} = \frac{(3)(2)}{(8)(2)} = \frac{3}{8}$$

57. Add and subtract $\frac{13}{15} + \left|-\frac{11}{15}\right| - \frac{4}{15}$. Then simplify.

Solution

$$\frac{13}{15} + \left|-\frac{11}{15}\right| - \frac{4}{15} = \frac{13}{15} + \frac{11}{15} - \frac{4}{15}$$

$$= \frac{20}{15} = \frac{4(5)}{3(5)} = \frac{4}{3}$$

59. Subtract $\frac{1}{4} - \frac{1}{3}$. Then simplify.

Solution

$$\frac{1}{4} - \frac{1}{3} = \frac{1(3)}{4(3)} - \frac{1(4)}{3(4)}$$

$$= \frac{3}{12} - \frac{4}{12} = -\frac{1}{12}$$

61. Add $\frac{3}{16} + \frac{3}{8}$. Then simplify.

Solution

$$\frac{3}{16} + \frac{3}{8} = \frac{3}{16} + \frac{3(2)}{8(2)}$$

$$= \frac{3}{16} + \frac{6}{16} = \frac{3+6}{16} = \frac{9}{16}$$

63. Subtract $-\frac{5}{6} - \left(-\frac{3}{4}\right)$. Then simplify.

Solution

$$-\frac{5}{6} - \left(-\frac{3}{4}\right) = -\frac{5}{6} + \frac{3}{4}$$

$$= -\frac{5(2)}{6(2)} + \frac{3(3)}{4(3)}$$

$$= -\frac{10}{12} + \frac{9}{12} = -\frac{1}{12}$$

65. Subtract $-\frac{7}{8} - \frac{5}{6}$. Then simplify.

Solution

$$-\frac{7}{8} - \frac{5}{6} = -\frac{7(3)}{8(3)} - \frac{5(4)}{6(4)}$$

$$= -\frac{21}{24} - \frac{20}{24}$$

$$= -\frac{41}{24}$$

67. Add and subtract $\frac{5}{12} - \frac{3}{8} + \frac{5}{16}$. Then simplify.

Solution

$$\frac{5}{12} - \frac{3}{8} + \frac{5}{16} = \frac{5(4)}{12(4)} - \frac{3(6)}{8(6)} + \frac{5(3)}{16(3)}$$

$$= \frac{20}{48} - \frac{18}{48} + \frac{15}{48}$$

$$= \frac{17}{48}$$

69. Subtract $\frac{7}{4} - 3$. Then simplify.

Solution

$$\frac{7}{4} - 3 = \frac{7}{4} - \frac{3(4)}{4}$$

$$= \frac{7}{4} - \frac{12}{4}$$

$$= -\frac{5}{4}$$

71. Add and subtract $2 - \dfrac{25 - 15}{6} + \dfrac{3}{4}$. Then simplify.

Solution

$$2 - \frac{25-15}{6} + \frac{3}{4} = 2 - \frac{10}{6} + \frac{3}{4}$$

$$= \frac{2}{1} - \frac{5}{3} + \frac{3}{4}$$

$$= \frac{2(12)}{1(12)} - \frac{5(4)}{3(4)} + \frac{3(3)}{4(3)}$$

$$= \frac{24}{12} - \frac{20}{12} + \frac{9}{12}$$

$$= \frac{13}{12}$$

73. Write the mixed number $8\frac{2}{3}$ as a fraction.

Solution

$$8\frac{2}{3} = \frac{8(3) + 2}{3}$$

$$= \frac{26}{3}$$

75. Write the mixed number $-10\frac{5}{11}$ as a fraction.

Solution

$$-10\frac{5}{11} = -\frac{10(11) + 5}{11}$$

$$= -\frac{115}{11}$$

77. Subtract $1\frac{3}{16} - 2\frac{1}{4}$. Then simplify.

Solution

$$1\frac{3}{16} - 2\frac{1}{4} = \frac{19}{16} - \frac{9}{4} = \frac{19}{16} - \frac{9(4)}{4(4)}$$

$$= \frac{19}{16} - \frac{36}{16} = \frac{19-36}{16} = \frac{-17}{16} = -\frac{17}{16}$$

79. Subtract $-5\frac{2}{3} - 4\frac{5}{12}$. Then simplify.

Solution

$$-5\frac{2}{3} - 4\frac{5}{12} = -\frac{17}{3} - \frac{53}{12} = -\frac{17(4)}{3(4)} - \frac{53}{12} = -\frac{68}{12} - \frac{53}{12}$$

$$= \frac{-68 - 53}{12} = \frac{-121}{12} = -\frac{121}{12}$$

81. *Animal Feed* During the months of January, February, and March, an animal shelter bought $8\frac{3}{4}$ tons, $7\frac{1}{5}$ tons, and $9\frac{3}{8}$ tons of feed. Find the total amount of feed purchased during the first quarter of the year.

Solution

$$8\frac{3}{4} + 7\frac{1}{5} + 9\frac{3}{8} = \frac{35}{4} + \frac{36}{5} + \frac{75}{8}$$

$$= \frac{35(10)}{4(10)} + \frac{36(8)}{5(8)} + \frac{75(5)}{8(5)}$$

$$= \frac{350}{40} + \frac{288}{40} + \frac{375}{40}$$

$$= \frac{1013}{40}$$

$$= 25\frac{13}{40}$$

Thus, $25\frac{13}{40}$ tons, or 25.325 tons, of feed were purchased during the first quarter of the year.

83. *Volume* The fuel gauge on a gasoline tank indicates that the tank is $\frac{3}{8}$ full. What fraction of the tank is empty?

Solution

$1 - \frac{3}{8} = \frac{8}{8} - \frac{3}{8} = \frac{5}{8}$

Thus, $\frac{5}{8}$ of the gasoline tank is empty.

85. *Estimation* Mentally estimate the sum or difference of $\frac{3}{11} + \frac{7}{10}$ to the nearest integer.

Solution

The sum is approximately 1.

87. *Think About It* Determine the placement of the digits 3, 4, 5, and 6 in the following addition problem so that you obtain the specified sum.

$$\frac{?}{?} + \frac{?}{?} = \frac{13}{10}$$

Solution

$$\frac{4}{5} + \frac{3}{6} = \frac{13}{10}$$

Note: $\frac{4}{5} + \frac{3}{6} = \frac{4}{5} + \frac{1}{2} = \frac{8}{10} + \frac{5}{10} = \frac{13}{10}$

1.2   Multiplying and Dividing Fractions

7. Evaluate the expression $\frac{1}{2} \times \frac{3}{4}$. Write your result in reduced form.

Solution

$$\frac{1}{2} \times \frac{3}{4} = \frac{1 \cdot 3}{2 \cdot 4} = \frac{3}{8}$$

9. Evaluate the expression $\frac{2}{3} \times \left(-\frac{9}{16}\right)$. Write your result in reduced form.

Solution

$$\frac{2}{3} \times \left(-\frac{9}{16}\right) = -\frac{2 \cdot 9}{3 \cdot 16} = -\frac{\cancel{(2)}(3)\cancel{(3)}}{\cancel{(3)}(8)\cancel{(2)}} = -\frac{3}{8}$$

Note: The reducing could also be written this way.

$$\frac{2}{3} \times \left(-\frac{9}{16}\right) = -\frac{\overset{1}{\cancel{2}} \cdot \overset{3}{\cancel{9}}}{\underset{1}{\cancel{3}} \cdot \underset{8}{\cancel{16}}} = -\frac{3}{8}$$

11. Evaluate the expression $\left(-\frac{3}{2}\right)\left(-\frac{15}{16}\right)\left(\frac{12}{25}\right)$. Write your result in reduced form.

Solution

$$\left(-\frac{3}{2}\right)\left(-\frac{15}{16}\right)\left(\frac{12}{25}\right) = \frac{3}{2} \cdot \frac{15}{16} \cdot \frac{12}{25} = \frac{3 \cdot 15 \cdot 12}{2 \cdot 16 \cdot 25} = \frac{(3)\cancel{(5)}(3)\cancel{(4)}(3)}{(2)\cancel{(4)}(4)\cancel{(5)}(5)} = \frac{27}{40}$$

Note: The reducing could also be written this way.

$$\left(-\frac{3}{2}\right)\left(-\frac{5}{16}\right)\left(\frac{12}{25}\right) = \frac{3}{2} \cdot \frac{15}{16} \cdot \frac{12}{25} = \frac{3 \cdot \overset{3}{\cancel{15}} \cdot \overset{3}{\cancel{12}}}{2 \cdot \underset{4}{\cancel{16}} \cdot \underset{5}{\cancel{25}}} = \frac{27}{40}$$

13. Evaluate the expression $2\frac{3}{4} \times 3\frac{2}{3}$. Write your result in reduced form.

Solution

$$2\frac{3}{4} \times 3\frac{2}{3} = \left(\frac{11}{4}\right)\left(\frac{11}{3}\right) = \frac{121}{12}$$

15. Find the reciprocal of the number 7. Show that the product of the number and its reciprocal is 1.

Solution

The reciprocal of 7 is $\frac{1}{7}$.

$$7\left(\tfrac{1}{7}\right) = \tfrac{7}{1}\left(\tfrac{1}{7}\right) = \tfrac{7}{7} = 1$$

17. Find the reciprocal of the number $\frac{4}{7}$. Show that the product of the number and its reciprocal is 1.

Solution

The reciprocal of $\frac{4}{7}$ is $\frac{7}{4}$.

$$\tfrac{4}{7}\left(\tfrac{7}{4}\right) = \tfrac{28}{28} = 1$$

19. Find the reciprocal of the number $2\frac{1}{2}$. Show that the product of the number and its reciprocal is 1.

Solution

The reciprocal of $2\frac{1}{2}$, or $\frac{5}{2}$, is $\frac{2}{5}$.

$$\left(2\tfrac{1}{2}\right)\left(\tfrac{2}{5}\right) = \tfrac{5}{2}\left(\tfrac{2}{5}\right) = \tfrac{10}{10} = 1$$

21. Evaluate the expression $\frac{3}{8} \div \frac{3}{4}$. If it is not possible, explain why. If it is possible, write the result in reduced form.

Solution

$$\frac{3}{8} \div \frac{3}{4} = \frac{3}{8} \cdot \frac{4}{3} = \frac{3 \cdot 4}{8 \cdot 3} = \frac{(\cancel{3})(1)(\cancel{4})}{(2)(\cancel{4})(\cancel{3})} = \frac{1}{2}$$

23. Evaluate the expression $-\frac{5}{12} \div \frac{45}{32}$. If it is not possible, explain why. If it is possible, write the result in reduced. form.

Solution

$$-\frac{5}{12} \div \frac{45}{32} = -\frac{5}{12} \cdot \frac{32}{45}$$

$$= -\frac{5 \cdot 32}{12 \cdot 45} = -\frac{(\cancel{5})(8)(\cancel{4})}{(3)(\cancel{4})(9)(\cancel{5})} = -\frac{8}{27}$$

25. Evaluate the expression $\dfrac{-7/15}{-14/25}$. If it is not possible, explain why. If it is possible, write the result in reduced form.

Solution

$$\frac{-7/15}{-14/25} = -\frac{7}{15} \div \left(-\frac{14}{25}\right)$$

$$= -\frac{7}{15} \cdot \left(-\frac{25}{14}\right)$$

$$= \frac{7(25)}{15(14)}$$

$$= \frac{\cancel{7}(\cancel{5})(5)}{(\cancel{5})(3)(\cancel{7})(2)}$$

$$= \frac{5}{6}$$

27. Evaluate the expression $3\frac{3}{4} \div 2\frac{5}{8}$. If it is not possible, explain why. If it is possible, write the result in reduced form.

Solution

$$3\frac{3}{4} \div 2\frac{5}{8} = \frac{15}{4} \div \frac{21}{8}$$

$$= \frac{15}{4} \cdot \frac{8}{21}$$

$$= \frac{5(\cancel{3})(\cancel{4})(2)}{(\cancel{4})(\cancel{3})(7)}$$

$$= \frac{10}{7}$$

29. Write the fraction $\frac{3}{4}$ in decimal form. Use the bar notation for repeating digits.

Solution

$\frac{3}{4} = 0.75$

$$\begin{array}{r} .75 \\ 4\overline{)3.0} \\ \underline{2\ 8} \\ 20 \\ \underline{20} \\ 0 \end{array}$$

31. Write the fraction $\frac{5}{11}$ in decimal form. Use the bar notation for repeating digits.

Solution

$\frac{5}{11} = 0.\overline{45}$

$$\begin{array}{r} .4545\ldots = 0.\overline{45} \\ 11\overline{)5.0000} \\ \underline{4\ 4} \\ 60 \\ \underline{55} \\ 50 \\ \underline{44} \\ 60 \\ \underline{55} \\ 5 \end{array}$$

33. Evaluate the expression $1.21 + 4.06 - 3$. Round your answer to two decimal places.

Solution

$1.21 + 4.06 - 3 = 2.27$

35. Evaluate the expression $(-6.3)(9.05)$. Round your answer to two decimal places.

Solution

$(-6.3)(9.05) \approx -57.02$

$$
\begin{array}{r}
-6.3 \\
\times 9.05 \\
\hline
315 \\
567 \\
\hline
-57.015 \approx -57.02
\end{array}
$$
(rounded to two decimal places)

Note: To evaluate this expression on a standard scientific calculator or on a graphing calculator, use the following keystrokes.

Keystrokes	*Display*	
6.3 $\boxed{+/-}$ $\boxed{\times}$ 9.05 $\boxed{=}$	-57.105	Scientific
$\boxed{(-)}$ 6.3 $\boxed{\times}$ 9.05 $\boxed{\text{ENTER}}$	-57.105	Graphing

37. Evaluate the expression $4.69 \div 0.12$. Round your answer to two decimal places.

Solution

$4.69 \div 0.12 \approx 39.08$ (rounded to two decimal places)

$$
0.12\overline{)4.69} \;=\; 12\overline{)469}
$$

$$
\begin{array}{r}
39.0833\ldots \approx 39.08 \quad \text{(Rounded)} \\
12\overline{)469} \\
\underline{36} \\
109 \\
\underline{108} \\
100 \\
\underline{96} \\
40 \\
\underline{36} \\
40 \\
\underline{36} \\
4
\end{array}
$$

39. Evaluate the expression $\dfrac{(-15.1)(-6.02)}{-9.6}$. Round the answer to two decimal places.

Solution

$$
\frac{(-15.1)(-6.02)}{-9.6} = \frac{90.902}{-9.6} \approx -9.47 \quad \text{(rounded to two decimal places)}
$$

41. *Estimation* Each day for a week, you practiced the saxophone for $\frac{2}{3}$ hour.

(a) Explain how to use mental math to estimate the number of hours of practice in a week.

(b) Determine the actual number of hours you practiced during the week. Write the result in decimal form, rounding to one decimal place.

Solution

(a) Seven times $\frac{2}{3}$ hour is $\frac{14}{3}$ hours; this is approximately equal to 5, or $\frac{15}{3}$, hours. (*Note:* Estimated answers could vary.)

(b) $7\left(\frac{2}{3}\right) = \frac{7}{1}\left(\frac{2}{3}\right)$

$\qquad = \frac{14}{3}$

$\qquad = 4.666\ldots$

$\qquad \approx 4.7$ (rounded to one decimal place)

Thus, the actual number of hours you practiced during the week was approximately 4.7 hours.

43. *Walking Time* Your apartment is $\frac{3}{4}$ mile from the subway. If you walk at the rate of $3\frac{1}{4}$ miles per hour, how long will it take you to walk from your apartment to the subway?

Solution

$\dfrac{3}{4} \div 3\dfrac{1}{4} = \dfrac{3}{4} \div \dfrac{13}{4}$

$\qquad = \dfrac{3}{4} \cdot \dfrac{4}{13}$

$\qquad = \dfrac{3(\cancel{4})}{\cancel{4}(13)}$

$\qquad = \dfrac{3}{13}$

Thus, it will take $\frac{3}{13}$ hour, or approximately 14 minutes, to walk to the subway.

45. Evaluate the expression $-13 - 7$.

Solution

$-13 - 7 = -20$

47. Evaluate the expression $\frac{5}{16} - \frac{3}{10}$.

Solution

$\frac{5}{16} - \frac{3}{10} = \frac{25}{80} - \frac{24}{80} = \frac{1}{80}$

49. Evaluate the expression $-7\frac{3}{5} - 3\frac{1}{2}$.

Solution

$-7\frac{3}{5} - 3\frac{1}{2} = \frac{-38}{5} - \frac{7}{2}$

$\qquad = \frac{-76}{10} - \frac{35}{10}$

$\qquad = -\frac{111}{10}$

51. *Relay Race* In a $\frac{3}{4}$-mile relay race, the last change of runners occurs at the $\frac{2}{3}$-mile marker. How far does the last person run?

Solution

$\frac{3}{4} - \frac{2}{3} = \frac{9}{12} - \frac{8}{12}$

$\qquad = \frac{1}{12}$

The last runner runs $\frac{1}{12}$ of a mile.

53. Evaluate the expression $\frac{2}{3} \times \frac{1}{2}$.

Solution

$$\frac{2}{3} \times \frac{1}{2} = \frac{\cancel{2}(1)}{3\cancel{(2)}} = \frac{1}{3}$$

55. Evaluate the expression $\left(-\frac{7}{16}\right)\left(-\frac{12}{5}\right)$.

Solution

$$\left(-\frac{7}{16}\right)\left(-\frac{12}{5}\right) = \frac{7}{16} \cdot \frac{12}{5}$$
$$= \frac{7 \cdot 12}{16 \cdot 5}$$
$$= \frac{(7)(3)\cancel{(4)}}{\cancel{(4)}(4)(5)} = \frac{21}{20}$$

57. Evaluate the expression $\left(\frac{12}{11}\right)\left(-\frac{9}{44}\right)$.

Solution

$$\left(\frac{11}{12}\right)\left(-\frac{9}{44}\right) = -\frac{11(9)}{12(44)}$$
$$= -\frac{\cancel{11}\cancel{(3)}(3)}{(4)\cancel{(3)}(4)\cancel{(11)}}$$
$$= -\frac{3}{16}$$

59. Evaluate the expression $9\left(\frac{4}{15}\right)$.

Solution

$$9\left(\frac{4}{15}\right) = \frac{9}{1}\left(\frac{4}{15}\right)$$
$$= \frac{\cancel{(3)}(3)(4)}{\cancel{(3)}(5)}$$
$$= \frac{12}{5}$$

61. Evaluate the expression $\left(-\frac{3}{11}\right)\left(-\frac{11}{3}\right)$.

Solution

$$\left(-\frac{3}{11}\right)\left(-\frac{11}{3}\right) = \frac{(-3)(-11)}{(11)(3)}$$
$$= \frac{33}{33}$$
$$= 1$$

63. Evaluate the expression $2\frac{4}{5} \times 6\frac{2}{3}$.

Solution

$$2\frac{4}{5} \times 6\frac{2}{3} = \frac{14}{5} \cdot \frac{20}{3}$$
$$= \frac{14(20)}{5(3)}$$
$$= \frac{14\cancel{(5)}(4)}{\cancel{5}(3)}$$
$$= \frac{56}{3}$$

65. Evaluate the expression $\frac{3}{5} \div 0$.

Solution

$$\frac{3}{5} \div 0$$

Division by zero is undefined.

67. Evaluate the expression $\frac{15/8}{3/8}$.

Solution

$$\frac{15/8}{3/8} = \frac{15}{8} \div \frac{3}{8} = \frac{15}{8} \cdot \frac{8}{3} = \frac{15 \cdot 8}{8 \cdot 3}$$
$$= \frac{(5)\cancel{(3)}\cancel{(8)}}{\cancel{(8)}\cancel{(3)}(1)} = \frac{5}{1} = 5$$

69. Evaluate the expression $\frac{-5}{15/16}$.

Solution

$$\frac{-5}{15/16} = -\frac{5}{1} \div \frac{15}{16} = -\frac{5}{1} \cdot \frac{16}{15} = -\frac{5 \cdot 16}{1 \cdot 15} = -\frac{\cancel{(5)}(16)}{(1)(3)\cancel{(5)}} = -\frac{16}{3}$$

71. Evaluate the expression $-10 \div \frac{1}{9}$.

Solution

$$-10 \div \frac{1}{9} = -\frac{10}{1} \cdot \frac{9}{1}$$

$$= -\frac{90}{1}$$

$$= -90$$

73. Evaluate the expression $3\frac{3}{4} \div 1\frac{1}{2}$.

Solution

$$3\frac{3}{4} \div 1\frac{1}{2} = \frac{15}{4} \div \frac{3}{2}$$

$$= \frac{15}{4} \cdot \frac{2}{3}$$

$$= \frac{15(2)}{4(3)}$$

$$= \frac{5(3)(2)}{(2)(2)(3)}$$

$$= \frac{5}{2}$$

75. Write the fraction $\frac{7}{10}$ in decimal form. Use the bar notation for repeating digits.

Solution

$$\frac{7}{10} = 0.7$$

$$\begin{array}{r} .7 \\ 10\overline{)7.0} \\ \underline{7\ 0} \\ 0 \end{array}$$

77. Write the fraction $\frac{9}{16}$ in decimal form. Use the bar notation for repeating decimals.

Solution

$$\frac{9}{16} = 0.5625$$

$$\begin{array}{r} .5625 \\ 16\overline{)9.0000} \\ \underline{8\ 0} \\ 1\ 00 \\ \underline{96} \\ 40 \\ \underline{32} \\ 80 \\ \underline{80} \end{array}$$

79. Write the fraction $\frac{2}{3}$ in decimal form. Use the bar notation for repeating decimals.

Solution

$$\frac{2}{3} = 0.\overline{6}$$

$$\begin{array}{r} .666\ldots = 0.\overline{6} \\ 3\overline{)2.000} \\ \underline{1\ 8} \\ 20 \\ \underline{18} \\ 20 \\ \underline{18} \\ 2 \end{array}$$

81. Write the fraction $\frac{7}{12}$ in decimal form. Use the bar notation for repeating decimals.

Solution

$$\frac{7}{12} = 0.58\overline{3}$$

$$\begin{array}{r} .58333\ldots = 0.58\overline{3} \\ 12\overline{)7.00000} \\ \underline{6\ 0} \\ 1\ 00 \\ \underline{96} \\ 40 \\ \underline{36} \\ 40 \\ \underline{36} \\ 40 \\ \underline{36} \\ 4 \end{array}$$

83. Evaluate the expression $132.1 + (-25.45)$. Round the result to two decimal places.

Solution

$132.1 + (-25.45) = 106.65$

85. Evaluate the expression $(-0.09)(-0.45)$. Round the result to two decimal places.

Solution

$(-0.09)(-0.45) = 0.0405$

$$\approx 0.04$$

(rounded to two decimal places)

$$\begin{array}{r} 0.45 \\ \times 0.09 \\ \hline .0405 \end{array}$$

87. Evaluate the expression $\dfrac{-10.5}{0.75}$. Round the result to two decimal places.

Solution

$\dfrac{-10.5}{0.75} = -14$

$$0.75\overline{)10.50} \;=\; 75\overline{)1050}$$
$$\begin{array}{r} 14 \\ 75\overline{)1050} \\ \underline{75} \\ 300 \\ \underline{300} \end{array}$$

89. Evaluate the expression $\dfrac{(-0.2)(0.05)}{-1.5}$. Round the result to two decimal places.

Solution

$\dfrac{(-0.2)(0.05)}{-1.5} = \dfrac{-0.01}{-1.5}$

$$= 0.00666\ldots$$

$$= 0.00\overline{6}$$

$$\approx 0.01$$

91. *Profit or Loss?* Your company had a loss of $5519.80 in July, a profit of $2337.06 in August, and a profit of $3615.11 in September. Did your company have a profit or a loss for the third quarter? Explain.

Solution

$-5519.80 + 2337.06 + 3615.11 = 432.37$

The sum is positive; thus, the company had a profit of $432.37 for the quarter.

93. *Grocery Purchase* At a convenience store, you buy two gallons of milk at $2.23 per gallon and three loaves of bread at $1.23 per loaf. You give the clerk a twenty-dollar bill. How much change should you receive? (Assume there is no sales tax.)

Solution

The cost of the milk is $2(2.23) = \$4.46$, and the cost of the bread is $3(1.23) = \$3.69$. Thus, the total cost is $4.46 + 3.69 = \$8.15$. Your change is the difference: $20 - 8.15 = \$11.85$.

95. *Making Breadsticks*   You make 60 ounces of dough for breadsticks. If each breadstick requires $\frac{5}{4}$ ounces of dough, how many breadsticks can you make?

Solution

$$60 \div \frac{5}{4} = \frac{60}{1} \cdot \frac{4}{5}$$

$$= \frac{60(4)}{5}$$

$$= \frac{12(5)(4)}{5}$$

$$= 48$$

Thus, you can make 48 breadsticks.

97. *Annual Fuel cost*   The sticker on a new car gives the fuel efficiency as 22.3 miles per gallon. The average cost of fuel is $1.159 per gallon. Estimate the annual fuel cost for a car that will be driven approximately 12,000 miles per year.

Solution

The number of gallons needed to drive 12,000 miles in a car which gets 22.3 miles per gallon is

$$\frac{12,000}{22.3} \approx 538.117 \text{ gallons.}$$

At $1.159 per gallon, the annual fuel cost is

$$(538.117)(1.159) \approx \$623.68. \text{ (Rounded)}$$

(*Note:* More accurate answers are obtained if you round your answer *only* after all calculations are done.)

99. *Estimation*   Use mental math to decide whether $\left(5\frac{3}{4}\right) \times \left(4\frac{1}{8}\right)$ is less than 20. Explain your reasoning.

Solution

$\left(5\frac{3}{4}\right)\left(4\frac{1}{8}\right) > 20.$

The first number is greater than 5 and the second number is greater than 4. Therefore, their product must be greater than 5(4), or greater than 20.

101. *True or False?*   Determine whether the statement is true or false:

The reciprocal of every nonzero integer is an integer.

Solution

False. The reciprocal of the integer 5 is $\frac{1}{5}$, but $\frac{1}{5}$ is not an integer. In fact, 1 is the only integer that has an integer as its reciprocal.

103. *Think About It*   Think of different ways that you could correctly complete the following sentence. "If the product of two real numbers is 15, then . . . ."

Solution

"If the product of two real numbers is 15, then . . ."

   . . . the two numbers have the same sign.

   . . . the first number is the quotient of 15 divided by the second number.

   . . . the second number is the quotient of 15 divided by the first number.

Note: There are many correct answers.

Mid-Chapter Quiz for Chapter 1

1. Write the fraction $\frac{105}{126}$ in reduced form.

Solution

$$\frac{105}{126} = \frac{(3)(5)(7)}{(2)(3)(3)(7)}$$
$$= \frac{5}{6}$$

2. Write the fraction $\frac{3}{7}$ in decimal form rounded to three decimal places.

Solution

$$\frac{3}{7} \approx 0.429$$

$$\begin{array}{r} .4285 \approx 0.429 \\ 7\overline{)3.0000} \\ \underline{2\ 8} \\ 20 \\ \underline{14} \\ 60 \\ \underline{56} \\ 40 \\ \underline{35} \\ 5 \end{array}$$

3. Is it true that $\frac{1}{3} = 0.3$? Explain your answer.

Solution

No, $\frac{1}{3} \neq 0.3$. The number $\frac{1}{3}$ is not equal to 0.3. Instead, $\frac{1}{3} = 0.\overline{3}$.

4. Write the mixed number $-8\frac{4}{9}$ as a fraction.

Solution

$$-8\frac{4}{9} = -\frac{8(9) + 4}{9} = -\frac{76}{9}$$

5. Evaluate the expression $\frac{9}{11} - \frac{3}{11}$. Write fractions in reduced form.

Solution

$$\frac{9}{11} - \frac{3}{11} = \frac{9-3}{11} = \frac{6}{11}$$

6. Evaluate the expression $\frac{5}{8} + \frac{3}{4}$. Write fractions in reduced form.

Solution

$$\frac{5}{8} + \frac{3}{4} = \frac{5}{8} + \frac{6}{8} = \frac{11}{8}$$

7. Evaluate the expression $3 + \frac{5}{12}$. Write fractions in reduced form.

Solution

$$3 + \frac{5}{12} = \frac{3}{1} + \frac{5}{12}$$
$$= \frac{36}{12} + \frac{5}{12}$$
$$= \frac{41}{12}$$

8. Evaluate the expression $2 - \frac{3}{4} + 2\frac{4}{5}$. Write fractions in reduced form.

Solution

$$2 - \frac{3}{4} + 2\frac{4}{5} = \frac{2}{1} - \frac{3}{4} + \frac{14}{5}$$
$$= \frac{40}{20} - \frac{15}{20} + \frac{56}{20}$$
$$= \frac{81}{20}$$

9. Evaluate the expression $\frac{2}{5} \times \frac{5}{8}$. Write fractions in reduced form.

Solution

$$\frac{2}{5} \times \frac{5}{8} = \frac{2(5)}{(5)(8)} = \frac{2(5)}{(5)(4)(2)} = \frac{1}{4}$$

10. Evaluate the expression $\frac{9}{16} \times \frac{28}{54}$. Write fractions in reduced form.

Solution

$$\frac{9}{16} \times \frac{28}{54} = \frac{(9)(28)}{(16)(54)} = \frac{(9)(4)(7)}{(4)(4)(9)(6)} = \frac{7}{24}$$

11. Evaluate the expression $6\left(\frac{5}{21}\right)$. Write fractions in reduced form.

Solution

$$6\left(\frac{5}{21}\right) = \frac{6}{1}\left(\frac{5}{21}\right)$$

$$= \frac{(2)(\cancel{3})(5)}{(\cancel{3})(7)}$$

$$= \frac{10}{7}$$

12. Evaluate the expression $12\left(-\frac{9}{4}\right)\left(\frac{25}{81}\right)$. Write fractions in reduced form.

Solution

$$12\left(-\frac{9}{4}\right)\left(\frac{25}{81}\right) = -\frac{12}{1}\left(\frac{9}{4}\right)\left(\frac{25}{81}\right)$$

$$= -\frac{(\cancel{3})(\cancel{4})(\cancel{9})(25)}{1(\cancel{4})(\cancel{3})(3)(\cancel{9})}$$

$$= -\frac{25}{3}$$

13. Evaluate the expression $-\frac{5}{8} \div \frac{3}{16}$. Write fractions in reduced form.

Solution

$$-\frac{5}{8} \div \frac{3}{16} = -\frac{5}{8} \cdot \frac{16}{3}$$

$$= -\frac{(5)(16)}{(8)(3)}$$

$$= -\frac{(5)(\cancel{8})(2)}{(\cancel{8})(3)}$$

$$= -\frac{10}{3}$$

14. Evaluate the expression $\frac{114}{215} \div 2$. Write fractions in reduced form.

Solution

$$\frac{114}{215} \div 2 = \frac{114}{215} \cdot \frac{1}{2}$$

$$= \frac{\cancel{2}(57)}{(215)(\cancel{2})}$$

$$= \frac{57}{215}$$

15. Evaluate the expression $\dfrac{-9/25}{-21/10}$. Write fractions in reduced form.

Solution

$$\frac{-9/25}{-21/10} = -\frac{9}{25} \div -\frac{21}{10}$$

$$= -\frac{9}{25} \cdot \left(-\frac{10}{21}\right)$$

$$= \frac{(9)(10)}{(25)(21)}$$

$$= \frac{(3)(\cancel{3})(\cancel{5})(2)}{(5)(\cancel{5})(\cancel{3})(7)}$$

$$= \frac{6}{35}$$

16. Evaluate the expression $4\frac{3}{14} \div 3\frac{1}{2}$. Write fractions in reduced form.

Solution

$$4\frac{3}{14} \div 3\frac{1}{2} = \frac{59}{14} \div \frac{7}{2}$$

$$= \frac{59}{14} \cdot \frac{2}{7}$$

$$= \frac{59(\cancel{2})}{(\cancel{2})(7)(7)}$$

$$= \frac{59}{49}$$

17. Evaluate the expression $(3.8)(8.25)$. Round the answer to two decimal places.

Solution

$(3.8)(8.25) = 31.35$

$$\begin{array}{r} 8.25 \\ \times 3.8 \\ \hline 6600 \\ 2475 \\ \hline 31.35 \end{array}$$

18. Evaluate the expression $25.63 \div 2.7$. Round the answer to two decimal places.

Solution

$25.63 \div 2.7 \approx 9.49$

$$2.7\overline{)25.63} = 27\overline{)256.300}$$

$$\begin{array}{r} 9.492 \\ \hline 243 \\ \hline 13\ 3 \\ 10\ 8 \\ \hline 2\ 50 \\ 2\ 43 \\ \hline 70 \\ 54 \\ \hline 16 \end{array}$$

19. A $1\frac{1}{2}$-pound can of food costs \$3.29. What is the cost per ounce?

Solution

$$1\tfrac{1}{2}\text{pounds} = 1\tfrac{1}{2}(16)$$
$$= \tfrac{3}{2}(16)$$
$$= 24 \text{ ounces}$$

$\$3.29 \div 24 \approx 13.7\cent$ per ounce

The cost is approximately $14\cent$ per ounce.

20. One hundred shares of stock were bought for $\$35\frac{3}{8}$ per share and sold for $\$47\frac{1}{2}$ per share. Determine the profit from this investment.

Solution

$$100\left(47\tfrac{1}{2} - 35\tfrac{3}{8}\right) = 100(47.5 - 35.375)$$
$$= 100(12.125)$$
$$= \$1212.50$$

The profit from this investment is \$1212.50.

1.3   Exponents and Order of Operations

7. Rewrite $2 \cdot 2 \cdot 2 \cdot 2 \cdot 2$ in exponential form.

Solution

$$2 \cdot 2 \cdot 2 \cdot 2 \cdot 2 = 2^5$$

9. Rewrite $(-3)^6$ as a product.

Solution

$$(-3)^6 = (-3)(-3)(-3)(-3)(-3)(-3)$$

11. *True or False?*   Decide whether the statement "-2^4 is positive" is true or false. If it is false, state the reason.

Solution

False; -2^4 is not positive because
$-2^4 = -2 \cdot 2 \cdot 2 \cdot 2 = -16$.

13. Evaluate the expression 3^2. If it is not possible, state the reason.

Solution

$$3^2 = 3 \cdot 3 = 9$$

15. Evaluate the expression $\dfrac{1}{4^3}$. If it is not possible, state the reason.

Solution

$$\frac{1}{4^3} = \frac{1}{4 \cdot 4 \cdot 4} = \frac{1}{64}$$

17. Evaluate the expression $(-1.2)^3$. If it is not possible, state the reason.

Solution

$$(-1.2)^3 = (-1.2)(-1.2)(-1.2)$$
$$= -1.728$$

19. Evaluate the expression $\dfrac{1-3^2}{-2}$. If it is not possible, state the reason.

Solution

$$\frac{1-3^2}{-2} = \frac{1-9}{-2} = \frac{-8}{-2} = 4$$

21. Evaluate the expression 2.1×10^2. If it is not possible, state the reason.

Solution

$$2.1 \times 10^2 = 2.1 \times 100 = 210$$

23. Evaluate the expression $\dfrac{8.4}{10^3}$. If it is not possible, state the reason.

Solution

$$\frac{8.4}{10^3} = \frac{8.4}{1000} = 0.0084$$

25. Evaluate the expression $\dfrac{3^2+1}{0}$. If it is not possible, state the reason.

Solution

$$\frac{3^2+1}{0}$$

This expression is undefined because division by 0 is undefined.

27. Evaluate the expression $4 - [3(4-9) - |10-3|]$. Write fractional answers in reduced form.

Solution

$$4 - [3(4-9) - |10-3|] = 4 - [3(-5) - |7|]$$
$$= 4 - [-15 - 7]$$
$$= 4 - [-22]$$
$$= 4 + 22$$
$$= 26$$

29. Evaluate the expression $16 + 3 \cdot 4$. Write fractional answers in reduced form.

Solution

$$16 + 3 \cdot 4 = 16 + 12$$
$$= 28$$

31. Evaluate the expression $(16-5) \div (3-5)$. Write fractional answers in reduced form.

Solution

$$(16-5) \div (3-5) = 11 \div (-2)$$
$$= -\frac{11}{2}$$

33. Evaluate the expression $(-4)^2 - 3 \cdot 2^4$. Write fractional answers in reduced form.

Solution

$$(-4)^2 - 3 \cdot 2^4 = 16 - 3(16)$$
$$= 16 - 48$$
$$= -32$$

35. Evaluate the expression $4\left(-\frac{2}{3} + \frac{4}{3}\right)$. Write fractional answers in reduced form.

Solution

$$4\left(-\frac{2}{3} + \frac{4}{3}\right) = 4\left(\frac{-2+4}{3}\right)$$
$$= 4\left(\frac{2}{3}\right)$$
$$= \frac{4}{1} \cdot \frac{2}{3} = \frac{4 \cdot 2}{1 \cdot 3}$$
$$= \frac{8}{3}$$

37. Evaluate the expression $\dfrac{3 \cdot 6 - 4 \cdot 6}{5+1}$. Write fractional answers in reduced form.

Solution

$$\frac{3 \cdot 6 - 4 \cdot 6}{5+1} = \frac{18-24}{6}$$
$$= \frac{-6}{6}$$
$$= -1$$

39. Use a calculator to evaluate the expression $3.4^2 - 6(1.2)^3$. Round your result to two decimal places.

Solution

$$(3.4)^2 - 6(1.2)^3 = 11.56 - 6(1.728)$$
$$= 11.56 - 10.368$$
$$= 1.192$$
$$\approx 1.19 \quad \text{(Rounded to two decimal places)}$$

You could use the following calculator keystrokes.

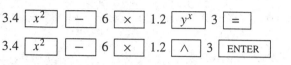

Keystrokes	*Display*	
3.4 $\boxed{x^2}$ $\boxed{-}$ 6 $\boxed{\times}$ 1.2 $\boxed{y^x}$ 3 $\boxed{=}$	1.192	Scientific
3.4 $\boxed{x^2}$ $\boxed{-}$ 6 $\boxed{\times}$ 1.2 $\boxed{\wedge}$ 3 $\boxed{\text{ENTER}}$	1.192	Graphing

41. Explain why the statement $4 \cdot 6^2 \neq 24^2$ is true. (The symbol $\neq$ means *is not equal to*.)

Solution

$4 \cdot 6^2 = 4 \cdot 6 \cdot 6 = 144$

$24^2 = 24 \cdot 24 = 576$

Therefore, $4 \cdot 6^2 \neq 24^2$.

43. *Geometry*   The land area of the earth is approximately 5.75×10^7 square miles. Evaluate this quantity.

Solution

$5.75 \times 10^7 = 5.75(10)(10)(10)(10)(10)(10)(10)$
$$= 57,500,000$$

45. *Geometry*   Find the area of the diagram.

Solution

The total area of the figure is the *sum* of the area of the upper rectangle and the area of the lower rectangle. The upper rectangle has length 3 and width 3. Its area is $3 \cdot 3 = 9$. The lower rectangle has length 9 and width 3. Its area is $9 \cdot 3 = 27$. Therefore, the total area is $9 + 27 = 36$ square units.

$$\text{Area} = 3 \cdot 3 + 9 \cdot 3$$
$$= 9 + 27 = 36 \text{ square units}$$

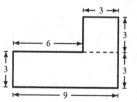

47. *Geometry*   The volume of a cube with each edge of length 7 inches is given by $V = 7 \cdot 7 \cdot 7$. Write the volume using exponential notation. What is the unit measure for the volume V?

Solution

$7 \cdot 7 \cdot 7 = 7^3$ cubic inches

The unit measure for the volume V is cubic inches.

49. Evaluate the expression $-\frac{0}{32}$. Write the result in reduced form.

Solution

$-\frac{0}{32} = 0$

51. Evaluate the expression $\left(-\frac{4}{3}\right)\left(-\frac{9}{16}\right)$. Write the result in reduced form.

Solution

$$\left(-\frac{4}{3}\right)\left(-\frac{9}{16}\right) = \frac{4(9)}{3(16)}$$

$$= \frac{\cancel{4}(\cancel{3})(3)}{\cancel{3}(\cancel{4})(4)}$$

$$= \frac{3}{4}$$

53. A pattern requires $2\frac{2}{3}$ yards of material to make a skirt and an additional $1\frac{1}{8}$ yards to make a vest. Find the total amount of material required.

Solution

$$2\frac{2}{3} + 1\frac{1}{8} = \frac{8}{3} + \frac{9}{8}$$

$$= \frac{64}{24} + \frac{27}{24}$$

$$= \frac{91}{24}$$

The total amount of material required is $\frac{91}{24}$, or $3\frac{19}{24}$ yards.

55. Rewrite $(-5)(-5)(-5)(-5)$ in exponential form.

Solution

$(-5)(-5)(-5)(-5) = (-5)^4$

57. Rewrite $\left(\frac{5}{8}\right)\left(\frac{5}{8}\right)\left(\frac{5}{8}\right)\left(\frac{5}{8}\right)\left(\frac{5}{8}\right)$ in exponential form.

Solution

$\left(\frac{5}{8}\right)\left(\frac{5}{8}\right)\left(\frac{5}{8}\right)\left(\frac{5}{8}\right)\left(\frac{5}{8}\right) = \left(\frac{5}{8}\right)^5$

59. Rewrite $(9.8)^3$ as a product.

Solution

$(9.8)^3 = (9.8)(9.8)(9.8)$

61. Rewrite $\left(-\frac{1}{2}\right)^5$ as a product.

Solution

$\left(-\frac{1}{2}\right)^5 = \left(-\frac{1}{2}\right)\left(-\frac{1}{2}\right)\left(-\frac{1}{2}\right)\left(-\frac{1}{2}\right)\left(-\frac{1}{2}\right)$

63. Is the value of -2^2 positive or negative?

Solution

$-2^2 = -2 \cdot 2 = -4$

The value is negative.

65. Is the value of -5^3 positive or negative?

Solution

$-5^3 = -5 \cdot 5 \cdot 5 = -125$

The value is negative.

67. Evaluate the expression 2^6. If it is not possible, state the reason.

Solution

$2^6 = (2)(2)(2)(2)(2)(2) = 64$

69. Evaluate the expression $(-5)^3$. If it is not possible, state the reason.

Solution

$(-5)^3 = (-5)(-5)(-5) = -125$

71. Evaluate the expression -5^3. If it is not possible, state the reason.

Solution

$-5^3 = -(5 \cdot 5 \cdot 5) = -125$

73. Evaluate the expression $\left(\frac{2}{3}\right)^4$. If it is not possible, state the reason.

Solution

$\left(\frac{2}{3}\right)^4 = \left(\frac{2}{3}\right)\left(\frac{2}{3}\right)\left(\frac{2}{3}\right)\left(\frac{2}{3}\right) = \frac{16}{81}$

75. Evaluate the expression $\dfrac{3^2 - 4^2}{0}$. If it is not possible, state the reason.

Solution

$\dfrac{3^2 - 4^2}{0}$ is undefined.

Division by zero is undefined.

77. Evaluate the expression $\dfrac{5^2 + 12^2}{13}$. If it is not possible, state the reason.

Solution

$$\dfrac{5^2 + 12^2}{13} = \dfrac{25 + 144}{13}$$

$$= \dfrac{169}{13}$$

$$= 13$$

79. Evaluate the expression 5.84×10^3. If it is not possible, state the reason.

Solution

$5.84 \times 10^3 = 5.84(10)(10)(10) = 5840$

81. Evaluate the expression $\dfrac{732}{10^2}$. If it is not possible, state the reason.

Solution

$\dfrac{732}{10^2} = \dfrac{732}{100} = 7.32$

83. Evaluate the expression $4 - 6 + 10$. Write fractional answers in reduced form.

Solution

$4 - 6 + 10 = -2 + 10 = 8$

85. Evaluate the expression $-|2 - (6 + 5)|$. Write fractional answers in reduced form.

Solution

$-|2 - (6 + 5)| = -|2 - 11| = -|-9| = -9$

87. Evaluate the expression $5 + (2^2 \cdot 3)$. Write fractional answers in reduced form.

Solution

$$5 + (2^2 \cdot 3) = 5 + (4 \cdot 3)$$

$$= 5 + 12$$

$$= 17$$

89. Evaluate the expression $(-6)^2 - (5^2 \cdot 4)$. Write fractional answers in reduced form.

Solution

$$(-6)^2 - (5^2 \cdot 4) = 36 - (25 \cdot 4)$$

$$= 36 - 100$$

$$= -64$$

91. Evaluate the expression $(45 \div 10) \cdot 2$. Write fractional answers in reduced form.

Solution

$(45 \div 10) \cdot 2 = (4.5)(2) = 9$

93. Evaluate the expression $\left(3 \cdot \frac{5}{9}\right) + 1 - \frac{1}{3}$. Write fractional answers in reduced form.

Solution

$$\left(3 \cdot \frac{5}{9}\right) + 1 - \frac{1}{3} = \left(\frac{3}{1} \cdot \frac{5}{9}\right) + \frac{1}{1} - \frac{1}{3}$$

$$= \frac{\cancel{(3)}(5)}{\cancel{(3)}(3)} + \frac{1}{1} - \frac{1}{3}$$

$$= \frac{5}{3} + \frac{3}{3} - \frac{1}{3}$$

$$= \frac{7}{3}$$

95. Evaluate the expression $\frac{3}{2}\left(\frac{2}{3}+\frac{1}{6}\right)$. Write fractional answers in reduced form.

Solution

$$\frac{3}{2}\left(\frac{2}{3}+\frac{1}{6}\right)=\frac{3}{2}\left(\frac{4}{6}+\frac{1}{6}\right)$$

$$=\frac{3}{2}\left(\frac{5}{6}\right)$$

$$=\frac{15}{12}$$

$$=\frac{\cancel{3}(5)}{4\cancel{(3)}}$$

$$=\frac{5}{4}$$

97. Evaluate the expression $\dfrac{\frac{7}{3}\left(\frac{2}{3}\right)}{\frac{28}{15}}$. Write fractional answers in reduced form.

Solution

$$\frac{\frac{7}{3}\left(\frac{2}{3}\right)}{\frac{28}{15}}=\frac{\frac{7}{3}\cdot\frac{2}{3}}{\frac{28}{15}}=\left(\frac{7}{3}\cdot\frac{2}{3}\right)\div\frac{28}{15}=\left(\frac{7\cdot2}{3\cdot3}\right)\div\frac{28}{15}$$

$$=\frac{14}{9}\div\frac{28}{15}=\frac{14}{9}\cdot\frac{15}{28}=\frac{14\cdot15}{9\cdot28}$$

$$=\frac{\cancel{(14)}\cancel{(3)}(5)}{(3)\cancel{(3)}\cancel{(14)}(2)}=\frac{5}{6}$$

99. Use a calculator to evaluate the expression $300\left(1+\dfrac{0.1}{12}\right)^{24}$. Round your result to two decimal places.

Solution

$$300\left(1+\frac{0.1}{12}\right)^{24}\approx366.12\quad\text{(Rounded to two decimal places)}$$

You could use the following calculator keystrokes.

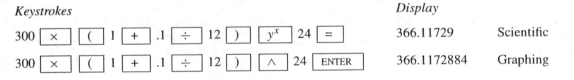

Keystrokes	*Display*	
300 × (1 + .1 ÷ 12) y^x 24 =	366.11729	Scientific
300 × (1 + .1 ÷ 12) ∧ 24 ENTER	366.1172884	Graphing

101. Is the statement $-3^2\neq(-3)(-3)$ true? Explain.

Solution

The statement is true.

$-3^2=-3\cdot3,$ or -9

$(-3)(-3)=9$

Thus, it is true that $-3^2\neq(-3)(-3)$.

103. *Interpreting a Pie Graph*   The portions of total expenses for a company are shown in the pie graph in the textbook. What portion of the total expenses is spent on utilities? If the total expenses are $450,000, how much is spent on utilities?

Solution

The portion of the expenses for all other categories (except utilities) is the sum

$$0.06 + 0.09 + 0.35 + 0.15 + 0.13 + 0.08 + 0.07 = 0.93.$$

Therefore, the portion of expenses that is spent on utilities is the difference

$$1 - 0.93 = 0.07.$$

If the total expenses are $450,000, the amount spent on utilities is the product

$$0.07(450,000) = \$31,500.$$

The amount spent on utilities is $31,500.

105. *Total Cost*   A car is purchased for $750 down and 48 monthly payments of $215 each. What is the total amount paid for the car?

Solution

$$750 + 48(215) = 750 + 10,320 = 11,070$$

Thus, the total amount paid for the car is $11,070.

1.4 | Algebra and Problem Solving

7. State the property of real numbers that justifies the statement $6(-3) = -3(6)$.

Solution

Commutative Property of Multiplication

9. State the property of real numbers that justifies the statement $0 + 15 = 15$.

Solution

Additive Identity Property

11. State the property of real numbers that justifies the statement $(2 \cdot 3)4 = 2(3 \cdot 4)$

Solution

Associative Property of Multiplication

13. State the property of real numbers that justifies the statement $7\left(\frac{1}{7}\right) = 1$

Solution

Multiplicative Inverse Property

15. State the property of real numbers that justifies the statement $(10 + 3) + 2 = 10 + (3 + 2)$

Solution

Associative Property of Addition

17. Use the Commutative Property of Addition or Multiplication to rewrite the expression $5(u + v) =$

Solution

$5(u + v) = 5(v + u)$ (using the Commutative Property of Addition)

   or

$5(u + v) = (u + v)5$ (using the Commutative Property of Multiplication)

19. Use the Distributive property to rewrite the expression $6(x + 2) = ?$

Solution

$6(x + 2) = 6x + 6 \cdot 2$ (or $6x + 12$)

21. Find (a) the additive inverse and (b) the multiplicative inverse of 50.

Solution

(a) Additive inverse: -50

(b) Multiplicative inverse: $\frac{1}{50}$

23. Find (a) the additive inverse and (b) the multiplicative inverse of -1.

Solution

(a) Additive inverse: 1

(b) Multiplicative inverse: -1

25. Rewrite the expression $10 + (8 + 2)$ using the Associative Property of Addition or Multiplication.

Solution

$10 + (8 + 2) = (10 + 8) + 2$

27. Rewrite the expression $(2 \cdot 3) \cdot 4$ using the Associative Property of Addition or Multiplication.

Solution

$(2 \cdot 3) \cdot 4 = 2 \cdot (3 \cdot 4)$

29. Simplify the expression $3(6 + 10)$.

Solution

$3(6 + 10) = 3(6) + 3(10)$ or $3(6 + 10) = 3(16)$

$\qquad\qquad\quad = 18 + 30 \qquad\qquad\qquad\qquad\qquad = 48$

$\qquad\qquad\quad = 48$

31. Simplify the expression $\frac{2}{3}(9z + 24)$.

Solution

$\frac{2}{3}(9z + 24) = \frac{2}{3} \cdot 9z + \frac{2}{3} \cdot 24 = 6z + 16$

33. Identify the property of real numbers used to justify each step in rewriting the following expression.

$$3 + 10(x + 1) = 3 + 10x + 10 = 3 + 10 + 10x = (3 + 10) + 10x = 13 + 10x$$

Solution

$3 + 10(x + 1) = 3 + 10x + 10$	Distributive Property
$= 3 + 10 + 10x$	Commutative Property of Addition
$= (3 + 10) + 10x$	Associative Property of Addition
$= 13 + 10x$	Addition of Real Numbers

35. Explain why the statement $5(x + 3) \neq 5x + 3$ is true.

Solution

$5(x + 3) = 5x + 5 \cdot 3 = 5x + 15$

Therefore, $5(x + 3) \neq 5x + 3$. The 5 should be multiplied by *both* terms in the parentheses.

37. Explain why the statement $\frac{8}{0} \neq 0$ is true.

Solution

$\frac{8}{0}$ is undefined because division by 0 is undefined. Therefore, $\frac{8}{0} \neq 0$.

39. *Sales Tax*   You purchase an item for x dollars. There is 6% sales tax, which implies that the total amount you must pay is $x + 0.06x$.

(a) Use the Distributive Property to rewrite the expression.

(b) How much must you pay if the item costs $25.95?

Solution

(a) $x + 0.06x = (1 + 0.06)x$, or $1.06x$

(b) $1.06(25.95) = 27.507$

You must pay $27.51 for the item.

41. Evaluate the expression $\dfrac{4 + 5 \cdot 2 - 14}{7}$.

Solution

$$\frac{4 + 5 \cdot 2 - 14}{7} = \frac{4 + 10 - 14}{7} = \frac{0}{7} = 0$$

43. Evaluate the expression $(-4)^2 - (30 \div 5)$

Solution

$$(-4)^2 - (30 \div 5) = 16 - 6 = 10$$

45. *Work Progress*   At the beginning of May, a construction project sign stated that $\frac{1}{6}$ of the project was complete. At the beginning of June, $\frac{2}{5}$ of the project was complete. What fraction of the work was completed during the month of May?

Solution

$$\frac{2}{5} - \frac{1}{6} = \frac{12}{30} - \frac{5}{30} = \frac{7}{30}$$

Therefore, $\frac{7}{30}$ of the work was completed during May.

47. State the property of real numbers that justifies the statement $x + 10 = 10 + x$.

Solution

Commutative Property of Addition

49. State the property of real numbers that justifies the statement $-16 + 16 = 0$.

Solution

Additive Inverse Property

51. State the property of real numbers that justifies the statement $4(3 \cdot 10) = (4 \cdot 3)10$.

Solution

Associative Property of Multiplication

53. State the property of real numbers that justifies the statement $10(6 - y) = 10 \cdot 6 - 10 \cdot y$.

Solution

Distributive Property

55. State the property of real numbers that justifies the statement $(4 + x)(2 - x) = 4(2 - x) + x(2 - x)$.

Solution

Distributive Property

57. State the property of real numbers that justifies the statement $x + (y + 3) = (x + y) + 3$.

Solution

Associative Property of Addition

59. Use the Commutative Property of Addition or Multiplication to rewrite the expression $(3 + x)7 = ?$.

Solution

$(3 + x)7 = 7(3 + x)$ (using the Commutative Property of Multiplication)

   or

$(3 + x)7 = (x + 3)7$ (using the Commutative Property of Addition)

61. Use the Associative Property of Addition or Multiplication to rewrite the expression $3x + (2y + 5) = ?$.

Solution

$3x + (2y + 5) = (3x + 2y) + 5$

63. Use the Distributive Property to rewrite the expression $(4 + y)25 = ?$

Solution

$(4 + y)25 = 4 \cdot 25 + 25y = 100 + 25y$

65. Find (a) the additive inverse and (b) the multiplicative inverse of the expression $2x$.

Solution

(a) Additive Inverse: $-2x$

(b) Multiplicative Inverse: $\dfrac{1}{2x}$

67. Find (a) the additive inverse and (b) the multiplicative inverse of the expression ab.

Solution

(a) Additive Inverse: $-ab$

(b) Multiplicative Inverse: $\dfrac{1}{ab}$

69. Rewrite the expression $(x + 3) + 2$ using the Associative Property of Addition or Multiplication.

Solution

$(x + 3) + 2 = x + (3 + 2)$, or $x + 5$

71. Rewrite the expression $2 \cdot (3y)$ using the Associative Property of Addition or Multiplication.

Solution

$2 \cdot (3y) = (2 \cdot 3)y$, or $6y$

73. Rewrite the expression $3(2x - 4)$ using the Distributive Property, and simplify the answer.

Solution

$3(2x - 4) = 3(2x) - 3(4) = 6x - 12$

75. Rewrite the expression $\frac{3}{5}(10y - 45)$ using the Distributive Property and simplify the answer.

Solution

$\frac{3}{5}(10y - 45) = \frac{3}{5}(10y) - \frac{3}{5}(45) = 6y - 27$

77. Identify the property of real numbers used to justify each step in rewriting the expression:

$$
\begin{aligned}
7x + 9 + 2x &= 7x + 2x + 9 &&\text{Commutative Property of Addition} \\
&= (7x + 2x) + 9 &&\text{Associative Property of Addition} \\
&= (7 + 2)x + 9 &&\text{Distributive Property} \\
&= 9x + 9 &&\text{Addition of Real Numbers} \\
&= 9(x + 1) &&\text{Distributive Property}
\end{aligned}
$$

79. *Geometry* Write an expression for the perimeter of the triangle shown in the figure. Use the properties of real numbers to simplify the expression.

Solution

$$(a - 2) + (b + 11) + (2c + 3) = a + b + 2c - 2 + 11 + 3$$
$$= a + b + 2c + 12$$

81. *Think About It* Determine whether the order in which the two activities are performed is "commutative." That is, do you obtain the same results regardless of which activity is performed first?

(a) "Drain the used oil from the engine."

(b) "Fill the crankcase with 5 quarts of new oil."

Solution

No, because the *order* in which these two actions are performed *does* affect the results.

Review Exercises for Chapter 1

1. Complete the statement $\dfrac{2}{3} = \dfrac{?}{15}$.

Solution

$$\frac{2}{3} = \frac{2(5)}{3(5)} = \frac{10}{15}$$

3. Complete the statement $\dfrac{6}{10} = \dfrac{?}{25}$.

Solution

$$\frac{6}{10} = \frac{3(2)}{5(2)} = \frac{3}{5} = \frac{3(5)}{5(5)} = \frac{15}{25}$$

5. Evaluate the expression $\frac{3}{25} + \frac{7}{25}$. Write the result in reduced form.

Solution

$$\frac{3}{25} + \frac{7}{25} = \frac{3+7}{25} = \frac{10}{25} = \frac{(2)(5)}{(5)(5)} = \frac{2}{5}$$

7. Evaluate the expression $\frac{27}{16} - \frac{15}{16}$. Write the result in reduced form.

Solution

$$\frac{27}{16} - \frac{15}{16} = \frac{27-15}{16} = \frac{12}{16} = \frac{(3)(4)}{(4)(4)} = \frac{3}{4}$$

9. Evaluate the expression $-\frac{5}{9} + \frac{2}{3}$. Write the result in reduced form.

Solution

$$-\frac{5}{9} + \frac{2}{3} = \frac{-5}{9} + \frac{2(3)}{3(3)} = -\frac{5}{9} + \frac{6}{9}$$
$$= \frac{-5+6}{9} = \frac{1}{9}$$

11. Evaluate the expression $\frac{25}{32} + \frac{7}{24}$. Write the result in reduced form.

Solution

$$\frac{25}{32} + \frac{7}{24} = \frac{25(3)}{32(3)} + \frac{7(4)}{24(4)}$$
$$= \frac{75}{96} + \frac{28}{96} = \frac{75+28}{96} = \frac{103}{96}$$

13. Evaluate the expression $5 - \frac{15}{4}$. Write the result in reduced form.

Solution

$$5 - \frac{15}{4} = \frac{5(4)}{1(4)} - \frac{15}{4} = \frac{20}{4} - \frac{15}{4} = \frac{20-15}{4} = \frac{5}{4}$$

15. Evaluate the expression $5\frac{3}{4} - 3\frac{5}{8}$. Write the result in reduced form.

Solution

$$5\frac{3}{4} - 3\frac{5}{8} = \frac{23}{4} - \frac{29}{8} = \frac{23(2)}{4(2)} - \frac{29}{8}$$
$$= \frac{46}{8} - \frac{29}{8} = \frac{46-29}{8} = \frac{17}{8}$$

17. Initially, a share of stock cost $\$35\frac{1}{4}$. The daily changes in closing values during the week are shown in the table. Determine the closing price of a share on Friday.

Solution

$$35\frac{1}{4} - \frac{3}{8} - \frac{1}{2} - \frac{1}{8} + 1\frac{1}{4} + \frac{1}{2} = \frac{144}{4} - \frac{3}{8} - \frac{1}{2} - \frac{1}{8} + \frac{5}{4} + \frac{1}{2}$$
$$= \frac{282}{8} - \frac{3}{8} - \frac{4}{8} - \frac{1}{8} + \frac{10}{8} + \frac{4}{8}$$
$$= \frac{288}{8}$$
$$= 36$$

Thus, the closing price on Friday was $36.
Note: These numbers could be written in decimal form.

$$35.25 - 0.375 - 0.5 - 1.125 + 1.25 + 0.5 = 36.$$

19. Find the reciprocal of the number 9.

Solution

The reciprocal of 9 is $\frac{1}{9}$.

21. Find the reciprocal of the number $-\frac{5}{3}$.

Solution

The reciprocal of $-\frac{5}{3}$ is $-\frac{3}{5}$.

23. Evaluate the expression $\frac{5}{8} \cdot \frac{-2}{15}$. If it is not possible, explain why.

Solution

$$\frac{5}{8} \cdot \frac{-2}{15} = \frac{5(-2)}{8 \cdot 15} = -\frac{(1)\cancel{(5)}\cancel{(2)}}{(4)\cancel{(2)}(3)\cancel{(5)}} = -\frac{1}{12}$$

25. Evaluate the expression $35\left(\frac{1}{35}\right)$. If it is not possible, explain why.

Solution

$$35\left(\frac{1}{35}\right) = \frac{35}{1} \cdot \frac{1}{35} = \frac{35 \cdot 1}{1 \cdot 35} = \frac{\cancel{(35)}(1)}{(1)\cancel{(35)}} = 1$$

27. Evaluate the expression $\frac{5}{14} \div \frac{15}{28}$. If it is not possible, explain why.

Solution

$$\frac{5}{14} \div \frac{15}{28} = \frac{5}{14} \cdot \frac{28}{15} = \frac{5 \cdot 28}{14 \cdot 15} = \frac{\cancel{(5)}\cancel{(14)}(2)}{\cancel{(14)}\cancel{(5)}(3)} = \frac{2}{3}$$

29. Evaluate expression $\dfrac{-3/4}{-7/8}$. If it is not possible, explain why.

Solution

$$\frac{-\frac{3}{4}}{-\frac{7}{8}} = \frac{\frac{3}{4}}{\frac{7}{8}} = \frac{3}{4} \div \frac{7}{8} = \frac{3}{4} \cdot \frac{8}{7}$$

$$= \frac{3 \cdot 8}{4 \cdot 7} = \frac{(3)(2)\cancel{(4)}}{\cancel{(4)}(7)} = \frac{6}{7}$$

31. Evaluate the expression $\dfrac{\frac{5}{9}}{0}$. If it is not possible, explain why.

Solution

$\dfrac{\frac{5}{9}}{0}$ is undefined. Division by zero is undefined.

33. Evaluate the expression $\dfrac{5.25}{0.25}$. If it is not possible, explain why.

Solution

$$\frac{5.25}{0.25} = 21$$

$$0.25\overline{)5.25} \;=\; 25\overline{)525}$$

with long division:
21; 50; 25; 25

35. Write the fraction $\frac{15}{8}$ in decimal form. (Use the bar notation for repeating digits.)

Solution

$$\frac{15}{8} = 1.875$$

37. Write the fraction $\frac{5}{12}$ in decimal form. (Use the bar notation for repeating digits.)

Solution

$$\frac{5}{12} = 0.41\overline{6}$$

39. A telephone call costs $0.64 for the first minute plus $0.72 for each additional minute. Find the cost of a five-minute call.

Solution

The cost of a five-minute call would be the sum of the $0.64 cost for the first minute plus the $0.72 cost of each of the *four* additional minutes.

$$0.64 + 4(0.72) = 0.64 + 2.88 = \$3.52$$

The cost of the call is $3.52.

41. A container of food weighing 22 ounces is purchased for $1.43. Find the cost per ounce.

Solution

The cost per ounce is the quotient of the cost divided by the number of ounces.

$$\frac{1.43}{22} = \$0.065 \text{ per ounce } \left(6\tfrac{1}{2}\cent \text{ per ounce}\right)$$

The food costs $0.065 per ounce.

43. Evaluate the expression 7^3.

Solution

$7^3 = 7 \cdot 7 \cdot 7 = 343$

47. Insert the correct symbol ($<$, $>$, or $=$) between the numbers 2^2 and 2^4.

Solution

$2^2 = 4$

$2^4 = 16$

$4 < 16$

Thus, $2^2 < 2^4$.

51. Use a calculator to evaluate the expression $(5.8)^4 - (3.2)^5$. Round your answer to two decimal places.

Solution

$$(5.8)^4 - (3.2)^5 = 11331.6496 - 335.54432$$

$$= 796.10528$$

$$\approx 796.11$$

55. After 3 years, the value of a \$16,000 car is given by $16,000\left(\frac{3}{4}\right)^3$.

(a) What is its value after 3 years?

(b) How much has the car depreciated during the 3 years?

Solution

(a) $16,000\left(\frac{3}{4}\right) = 6750$

(b) $16,000 - 6750 = 9250$

Thus, the car is worth \$6750 after three years. It has depreciated \$9250.

59. Evaluate the expression $3.2 - 1.5 + 11.4$. Write fractional results in reduced form.

Solution

$3.2 - 1.5 + 11.4 = 13.1$

45. Evaluate the expression $(-7)^3$.

Solution

$(-7)^3 = (-7)(-7)(-7) = -343$

49. Insert the correct symbol ($<$, $>$, or $=$) between the numbers $\frac{3}{4}$ and $\left(\frac{3}{4}\right)^2$.

Solution

$\frac{3}{4} = \frac{12}{16}$

$\left(\frac{3}{4}\right)^2 = \frac{9}{16}$

$\frac{12}{16} > \frac{9}{16}$

Thus, $\frac{3}{4} > \left(\frac{3}{4}\right)^2$.

53. Use a calculator to evaluate the expression $\dfrac{3000}{(1.05)^{10}}$. Round your answer to two decimal places.

Solution

$$\frac{3000}{(1.05)^{10}} \approx 1841.739761$$

$$\approx 1841.74$$

57. Enter any number between 0 and 1 in a calculator. Square the number, and then square the result. Continue this process. What number does the calculator display seem to be approaching?

Solution

The calculator display approaches 0. For example, here are the displays the calculator would show if you enter 0.25 and continue squaring.

0.25

0.0625

0.0039062

0.0000152

etc.

61. Evaluate the expression $\left(\frac{3}{5}\right)^4$. Write fractional results in reduced form.)

Solution

$$\left(\frac{3}{5}\right)^4 = \left(\frac{3}{5}\right)\left(\frac{3}{5}\right)\left(\frac{3}{5}\right)\left(\frac{3}{5}\right) = \frac{3 \cdot 3 \cdot 3 \cdot 3}{5 \cdot 5 \cdot 5 \cdot 5} = \frac{81}{625}$$

63. Evaluate the expression $240 - (4^2 \cdot 5)$. Write fractional results in reduced form.

Solution

$$240 - (4^2 \cdot 5) = 240 - (16 \cdot 5) = 240 - 80 = 160$$

65. Evaluate the expression $3^2(10 - 2^2)$. Write fractional results in reduced form.

Solution

$$3^2(10 - 2^2) = 9(10 - 4) = 9(6) = 54$$

67. Evaluate the expression $\left(\frac{3}{4}\right)\left(\frac{5}{6}\right) + 4$. Write fractional results in reduced form.

Solution

$$\left(\frac{3}{4}\right)\left(\frac{5}{6}\right) + 4 = \frac{3 \cdot 5}{4 \cdot 6} + 4$$

$$= \frac{\cancel{(3)}(5)}{(4)(2)\cancel{(3)}} + 4 = \frac{5}{8} + \frac{4}{1} = \frac{5}{8} + \frac{4(8)}{1(8)} = \frac{5}{8} + \frac{32}{8} = \frac{5 + 32}{8} = \frac{37}{8}$$

69. Evaluate the expression $\frac{3}{5}\left(\frac{3}{4} - \frac{2}{3}\right)$. Write fractional results in reduced form.

Solution

$$\frac{3}{5}\left(\frac{3}{4} - \frac{2}{3}\right) = \frac{3}{5}\left(\frac{9}{12} - \frac{8}{12}\right)$$

$$= \frac{3}{5}\left(\frac{1}{12}\right)$$

$$= \frac{3(1)}{5(12)}$$

$$= \frac{\cancel{3}(1)}{5\cancel{(3)}(4)}$$

$$= \frac{1}{20}$$

71. Evaluate the expression $\dfrac{\frac{3}{2} - \frac{1}{4}}{\frac{3}{8}}$. Write fractional results in reduced form.

Solution

$$\frac{\frac{3}{2} - \frac{1}{4}}{\frac{3}{8}} = \frac{\frac{3(2)}{2(2)} - \frac{1}{4}}{\frac{3}{8}} = \frac{\frac{6}{4} - \frac{1}{4}}{\frac{3}{8}} = \frac{\frac{6-1}{4}}{\frac{3}{8}} = \frac{\frac{5}{4}}{\frac{3}{8}}$$

$$= \frac{5}{4} \div \frac{3}{8} = \frac{5}{4} \cdot \frac{8}{3} = \frac{5 \cdot 8}{4 \cdot 3}$$

$$= \frac{(5)(2)\cancel{(4)}}{\cancel{(4)}(3)} = \frac{10}{3}$$

73. Evaluate the expression $\left(\frac{16}{25} - \frac{3}{5}\right) \div \frac{3}{10}$. Write fractional results in reduced form.

Solution

$$\left(\frac{16}{25} - \frac{3}{5}\right) \div \frac{3}{10} = \left(\frac{16}{25} - \frac{15}{25}\right) \div \frac{3}{10}$$

$$= \left(\frac{1}{25}\right) \div \frac{3}{10}$$

$$= \frac{1}{25} \cdot \frac{10}{3}$$

$$= \frac{1(10)}{25(3)}$$

$$= \frac{1\cancel{(5)}(2)}{5\cancel{(5)}(3)}$$

$$= \frac{2}{15}$$

75. Evaluate the expression $\dfrac{\frac{3}{8} + \frac{7}{8}}{\frac{5}{16}}$. Write fractional results in reduced form.

Solution

$$\frac{\frac{3}{8} + \frac{7}{8}}{\frac{5}{16}} = \frac{\frac{3+7}{8}}{\frac{5}{16}} = \frac{\frac{10}{8}}{\frac{5}{16}}$$

$$= \frac{10}{8} \cdot \frac{16}{5}$$

$$= \frac{10(16)}{8(5)}$$

$$= \frac{2\cancel{(5)}(2)\cancel{(8)}}{\cancel{8}\cancel{(5)}}$$

$$= 4$$

77. State the property of real numbers that justifies the statement $123 - 123 = 0$.

Solution

Additive Inverse Property

79. State the property of real numbers that justifies the statement $14(3) = 3(14)$.

Solution

Commutative Property of Multiplication

81. State the property of real numbers that justifies the statement $17 \cdot 1 = 17$

Solution

Multiplicative Identity Property

83. State the property of real numbers that justifies the statement $r + (2s + 3) = (r + 2s) + 3$

Solution

Associative Property of Addition

85. State the property of real numbers that justifies the statement $-2(7 + x) = -2 \cdot 7 + (-2)x$

Solution

Distributive Property

87. Sketch a diagram of a triangle. Label its base as $2x + 6$ and its height as 7. Write an expression for the area of the triangle. (The area of a triangle is one-half its base times its height.) Use the Distributive Property to rewrite the expression.

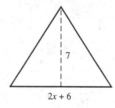

$$\tfrac{1}{2}(\text{base})(\text{height}) = \tfrac{1}{2}(2x + 6)(7)$$
$$= \left(\tfrac{1}{2} \cdot 2x + \tfrac{1}{2} \cdot 6\right)(7)$$
$$= (x + 3)(7)$$
$$= 7x + 21$$

Test for Chapter 1

1. Write the fraction $\frac{30}{72}$ in reduced form.

Solution

$$\frac{30}{72} = \frac{5\cancel{(6)}}{12\cancel{(6)}} = \frac{5}{12}$$

2. Write the fraction $\frac{5}{9}$ in decimal form rounded to three decimal places.

Solution

$$\frac{5}{9} = 0.\overline{5} \approx 0.556$$

3. Explain why the value of -3^4 is not equal to $(-3)^4$.

Solution

$$-3^4 = -3 \cdot 3 \cdot 3 \cdot 3 = -81$$
$$(-3)^4 = (-3)(-3)(-3)(-3) = 81$$

These expressions are not equal; the bases of the exponents are not the same.

4. State the order of operations for the expression $32 - 3 \cdot 2^3$.

Solution

$$\begin{aligned}
32 - 3 \cdot 2^3 &= 32 - 3 \cdot 8 && \text{exponentiation} \\
&= 32 - 24 && \text{multiplication} \\
&= 8 && \text{subtraction}
\end{aligned}$$

5. Evaluate the expression $\frac{5}{16} + \frac{3}{16}$. Write fractions in reduced form.

Solution

$$\frac{5}{16} + \frac{3}{16} = \frac{8}{16} = \frac{1}{2}$$

6. Evaluate the expression $\frac{5}{6} - \frac{1}{8}$. Write fractions in reduced form.

Solution

$$\frac{5}{6} - \frac{1}{8} = \frac{5(4)}{6(4)} - \frac{1(3)}{8(3)}$$

$$= \frac{20}{24} - \frac{3}{24}$$

$$= \frac{20 - 3}{24}$$

$$= \frac{17}{24}$$

7. Evaluate the expression $\left(-\frac{3}{4}\right)\left(-\frac{6}{15}\right)$. Write fractions in reduced form.

Solution

$$\left(-\frac{3}{4}\right)\left(-\frac{6}{15}\right) = \frac{3 \cdot 6}{4 \cdot 15}$$

$$= \frac{(\cancel{3})(3)(\cancel{2})}{(2)(\cancel{2})(5)(\cancel{3})}$$

$$= \frac{3}{10}$$

8. Evaluate the expression $-27\left(\frac{5}{6}\right)$. Write fractions in reduced form.

Solution

$$-27\left(\frac{5}{6}\right) = -\frac{27}{1}\left(\frac{5}{6}\right)$$

$$= -\frac{25(5)}{1(6)}$$

$$= -\frac{9(\cancel{3})(5)}{1(\cancel{3})(2)}$$

$$= -\frac{45}{2}$$

9. Evaluate the expression $\frac{7}{16} \div \frac{21}{28}$. Write fractions in reduced form.

Solution

$$\frac{7}{16} \div \frac{21}{28} = \frac{7}{16} \cdot \frac{28}{21}$$

$$= \frac{7 \cdot 28}{16 \cdot 21}$$

$$= \frac{(\cancel{7})(7)(\cancel{4})}{(4)(\cancel{4})(3)(\cancel{7})}$$

$$= \frac{7}{12}$$

10. Evaluate the expression $\frac{-8.1}{0.3}$. Write fractions in reduced form.

Solution

$$\frac{-8.1}{0.3} = -\frac{8.1}{0.3} = -27$$

$$0.3\overline{)8.1} = 3\overline{)81}$$
$$\phantom{0.3\overline{)8.1} = 3\overline{)}}\underline{6}$$
$$\phantom{0.3\overline{)8.1} = 3\overline{)}}21$$
$$\phantom{0.3\overline{)8.1} = 3\overline{)}}\underline{21}$$

11. Evaluate the expression $(-4)^3$. Write fractions in reduced form.

Solution

$$(-4)^3 = (-4)(-4)(-4) = -64$$

12. Evaluate the expression $-\left(\frac{2}{3}\right)^2$. Write the fractions in reduced form.

Solution

$$-\left(\frac{2}{3}\right)^2 = -\left(\frac{2}{3} \cdot \frac{2}{3}\right) = -\frac{4}{9}$$

13. Evaluate the expression $35 - (50 \div 5^2)$. Write the fractions in reduced form.

    Solution

    $$35 - (50 \div 5^2) = 35 - (50 \div 25)$$
    $$= 35 - 2$$
    $$= 33$$

14. Evaluate the expression $\frac{1}{4}\left(\frac{3}{5} - \frac{1}{10}\right)$. Write the fractions in reduced form.

    Solution

    $$\frac{1}{4}\left(\frac{3}{5} - \frac{1}{10}\right) = \frac{1}{4}\left(\frac{3(2)}{5(2)} - \frac{1}{10}\right) = \frac{1}{4}\left(\frac{6}{10} - \frac{1}{10}\right)$$
    $$= \frac{1}{4}\left(\frac{6-1}{10}\right) = \frac{1}{4}\left(\frac{5}{10}\right)$$
    $$= \frac{1 \cdot 5}{4 \cdot 10} = \frac{5}{40}$$
    $$= \frac{1(5)}{8(5)} = \frac{1}{8}$$

15. State the property of real numbers that justifies the statement $3(4 + 6) = 3 \cdot 4 + 3 \cdot 6$

    Solution

    Distributive Property

16. State the property of real numbers that justifies the statement $5 \cdot \frac{1}{5} = 1$

    Solution

    Multiplicative Inverse Property

17. State the property of real numbers that justifies the statement $3 + (4 + 8) = (3 + 4) + 8$.

    Solution

    Associative Property of Addition

18. State the property of real numbers that justifies the statement $3(x + 2) = (x + 2)3$.

    Solution

    Commutative Property of Multiplication

19. You purchase three boxes of cereal at $2.79 per box and two cans of pineapple at $0.59 per can. If you give the cashier $20, how much change should you receive? (Assume there is no sales tax.)

    Solution

    $3(2.79) + 2(0.59) = 8.37 + 1.18 = 9.55$

    $20 - 9.55 = 10.45$

    Thus, you should recieve $10.45 in change.

20. Copy the figure shown below. Then shade two-thirds of the figure. Write two different fractions that are represented by the shaded region. Which of these is in reduced form?

    Solution

    $\frac{2}{3}$ or $\frac{6}{9}$

    Other possible answers include $\frac{8}{12}, \frac{10}{15}, \frac{12}{18}$, etc. The answer in reduced form is $\frac{2}{3}$.

CHAPTER TWO
Fundamentals of Algebra

2.1 Algebraic Expressions and Exponents

7. Identify the variables and constants in the expression $x + 3$.

Solution

Variable: x
Constant: 3

9. Identify the variables and constants in the expression π.

Solution

Variable: none
Constant: π

11. Identify the terms of the expression $3x^2 + 5$.

Solution

$3x^2, 5$

13. Identify the coefficient of the term $-6x^2$.

Solution

-6

15. Rewrite the expression $2 \cdot u \cdot u \cdot u \cdot u$ in exponential form.

Solution

$2 \cdot u \cdot u \cdot u \cdot u = 2u^4$

(2 *is not* a factor of the base.)

17. Rewrite the expression $2u \cdot 2u \cdot 2u \cdot 2u$ in exponential form.

Solution

$2u \cdot 2u \cdot 2u \cdot 2u = (2u)^4$

(2 *is* a factor of the base.)

19. Rewrite the expression $3(a - b) \cdot 3(a - b) \cdot 3(a - b) \cdot 3(a - b) \cdot 3(a - b)$ in exponential form.

Solution

$3(a - b) \cdot 3(a - b) \cdot 3(a - b) \cdot 3(a - b) \cdot 3(a - b) = [3(a - b)]^5 \text{ or } 3^5(a - b)^5$

21. Expand the expression $2^2 x^4$ as a product of factors.

Solution

$2^2 x^4 = 2 \cdot 2 \cdot x \cdot x \cdot x \cdot x$

23. Expand the expression $(a^2)^3$ as a product of factors.

Solution

$(a^2)^3 = a^2 \cdot a^2 \cdot a^2$

$= a \cdot a \cdot a \cdot a \cdot a \cdot a$

25. Expand the expression $[3(r + s)^2][3(r + s)]^2$ as a product of factors.

Solution

$[3(r+s)^2][3(r+s)]^2 = 3(r+s)(r+s) \cdot 3(r+s) \cdot 3(r+s)$

27. Simplify the expression $u^2 \cdot u^4$.

Solution

$u^2 \cdot u^4 = u^{2+4} = u^6$

29. Simplify the expression $(-5z^3)z^2$.

Solution

$(-5z^3)z^2 = -5z^{3+2} = -5z^5$

31. Simplify the expression $(t^2)^4$.

Solution

$(t^2)^4 = t^{2(4)} = t^8$

33. Simplify the expression $5(uv)^5$.

Solution

$5(uv)^5 = 5u^5v^5$

35. Simplify the expression $(-2s)^3$.

Solution

$(-2s)^3 = (-2)^3s^3 = -8s^3$

37. Simplify the expression $(x - 2y)(x - 2y)^3$.

Solution

$(x - 2y)(x - 2y)^3 = (x - 2y)^{1+3} = (x - 2y)^4$

39. Simplify the expression $(-2x)^3(3x^2)^2 + 5x^2$.

Solution

$$(-2x)^3(3x^2)^2 + 5x^2 = (-2)^3 \cdot x^3 \cdot 3^2(x^2)^2 + 5x^2$$
$$= -8 \cdot x^3 \cdot 9 \cdot x^{2 \cdot 2} + 5x^2$$
$$= (-8)(9)(x^3 \cdot x^4) + 5x^2$$
$$= -72x^{3+4} + 5x^2$$
$$= -72x^7 + 5x^2$$

41. Simplify the expression $(u^2v^3)(uv^2)^4$.

Solution

$$(u^2v^3)(uv^2)^4 = u^2v^3 \cdot u^4(v^2)^4$$
$$= u^2v^3u^4v^8$$
$$= u^{2+4}v^{3+8}$$
$$= u^6v^{11}$$

43. Decide whether the expressions $x^5 \cdot x^3$ and x^{15} are equal.

Solution

$x^5 \cdot x^3 = x^{5+3} = x^8$

$x^5 \cdot x^3 \neq x^{15}$

No, they are *not* equal.

45. Decide whether the expressions $-3x^3$ and $-27x^3$ are equal.

Solution

$-3x^3 = -3 \cdot x \cdot x \cdot x$

$-27x^3 = -27 \cdot x \cdot x \cdot x$

$-3x^3 \neq -27x^3$

No, they are *not* equal.

47. Write the number 10,000 as a power of 10.

Solution

$10,000 = 10 \cdot 10 \cdot 10 \cdot 10 = 10^4$.

49. *Balance in an Account*   The balance in an account that has an annual interest rate of r, compounded quarterly for one year, is given by

$$P\left(1 + \frac{r}{4}\right)\left(1 + \frac{r}{4}\right)\left(1 + \frac{r}{4}\right)\left(1 + \frac{r}{4}\right).$$

Simplify this expression.

Solution

$$P\left(1 + \frac{r}{4}\right)\left(1 + \frac{r}{4}\right)\left(1 + \frac{r}{4}\right)\left(1 + \frac{r}{4}\right) = P\left(1 + \frac{r}{4}\right)^4$$

51. *Think About It* Find a positive integer n such that $n^3 + 1$ is a prime integer.

Solution

$n = 1$

If $n = 1$, $n^3 + 1 = 1^3 + 1$, or 2; 2 is a prime number.

Note: This is the only solution.

55. State the property of real numbers that justifies the statement $-5(3 + 6) = -5 \cdot 3 + (-5)6$.

Solution

Distributive Property

59. Identify the terms of the expression $\frac{5}{3} - 3y^3$.

Solution

$\frac{5}{3}, -3y^3$

63. Identify the terms of the expression $3(x + 5) + 10$.

Solution

$3(x + 5)$, 10

67. Identify the coefficient of the term $-\frac{1}{3}y$.

Solution

$-\frac{1}{3}$

71. Identify the coefficient of the term $-120x^2$.

Solution

-120

75. Write the expression $a \cdot a \cdot a \cdot b \cdot b$ in exponential form.

Solution

$a \cdot a \cdot a \cdot b \cdot b = a^3 b^2$

53. Evaluate the expression $36 \div 3^2 + 4^2$.

Solution

$$36 \div 3^2 + 4^2 = 36 \div 9 + 16$$
$$= 4 + 16$$
$$= 20$$

57. *Total Cost* You buy a pickup truck for $1800 down and 36 monthly payments of $625 each. (a) What is the total amount you will pay? (b) The cost of the pickup is $19,999. How much extra did you pay in finance charges and other fees?

Solution

(a) $1800 + 36(625) = 1800 + 22,500 = 24,300$
   Thus, you will pay a total of $24,300 for the pick-up truck.

(b) $24,300 - 19,999 = 4301$
   Thus, you paid $4301 in finance charges and other fees.

61. Identify the terms of the expression $2x - 3y + 1$.

Solution

$2x, -3y, 1$

65. Identify the terms of the expression $\dfrac{3}{x + 2} - 3x + 4$.

Solution

$\dfrac{3}{x + 2}, -3x, 4$

69. Identify the coefficient of the term $-\dfrac{3x}{2}$.

Solution

$-\frac{3}{2}$

73. Identify the coefficient of the term $-4.7u$.

Solution

-4.7

77. Write the expression $4 \cdot x \cdot x \cdot y \cdot x \cdot y$ in exponential form.

Solution

$$4 \cdot x \cdot x \cdot y \cdot x \cdot y = 4 \cdot x \cdot x \cdot x \cdot y \cdot y$$
$$= 4x^3 y^2$$

79. Write the expression $3 \cdot (x - y) \cdot (x - y) \cdot 3 \cdot 3$ in exponential form.

Solution

$$3 \cdot (x - y) \cdot (x - y) \cdot 3 \cdot 3 = 3 \cdot 3 \cdot 3 \cdot (x - y)(x - y)$$
$$= 3^3(x - y)^2$$

81. Write the expression $\left(\dfrac{x^2}{2}\right)\left(\dfrac{x^2}{2}\right)\left(\dfrac{x^2}{2}\right)$ in exponential form.

Solution

$$\left(\frac{x^2}{2}\right)\left(\frac{x^2}{2}\right)\left(\frac{x^2}{2}\right) = \left(\frac{x^2}{2}\right)^3$$

83. Expand the expression $4y^2z^3$ as a product of factors.

Solution

$$4y^2z^3 = 4 \cdot y \cdot y \cdot z \cdot z \cdot z$$

85. Expand the expression $(-2y)^3$ as a product of factors.

Solution

$$(-2y)^3 = (-2y)(-2y)(-2y) \text{ or}$$
$$(-2)(-2)(-2)y \cdot y \cdot y$$

87. Expand the expression $5x^3 \cdot x^4$ as a product of factors.

Solution

$$5x^3 \cdot x^4 = 5 \cdot x \cdot x \cdot x \cdot x \cdot x \cdot x \cdot x$$

89. Expand the expression $(ab)^3$ as a product of factors.

Solution

$$(ab)^3 = (ab)(ab)(ab) \text{ or}$$
$$a \cdot a \cdot a \cdot b \cdot b \cdot b$$

91. Expand the expression $(x + y)^2$ as a product of factors.

Solution

$$(x + y)^2 = (x + y)(x + y)$$

93. Expand the expression $\left(\dfrac{a}{3s}\right)^4$ as a product of factors.

Solution

$$\left(\frac{a}{3s}\right)^4 = \left(\frac{a}{3s}\right)\left(\frac{a}{3s}\right)\left(\frac{a}{3s}\right)\left(\frac{a}{3s}\right)$$

95. Simplify the expression $5x \cdot (x^6)$.

Solution

$$5x \cdot (x^6) = 5x^{1+6} = 5x^7$$

97. Simplify the expression $(-2x^2)(4x)$.

Solution

$$(-2x^2)(4x) = (-2)(4)(x^2 \cdot x)$$
$$= -8x^{2+1} = -8x^3$$

99. Simplify the expression $2b^4(-ab)(3b^2)$.

Solution

$$2b^4(-ab)(3b^2) = (2)(-1)(3)(a)(b^4 \cdot b \cdot b^2) = -6ab^{4+1+2} = -6ab^7$$

101. Simplify the expression $-2(3x)^2(3x) + 5x^2$.

Solution

$$-2(3x)^2(3x) + 5x^2 = -2(3x)^{2+1} + 5x^2$$
$$= -2(3x)^3 + 5x^2$$
$$= -2 \cdot 3^3 x^3 + 5x^2$$
$$= -2 \cdot 27x^3 + 5x^2$$
$$= -54x^3 + 5x^2$$

Note: This problem could also be worked in the following way.

$$-2(3x)^2(3x) + 5x^2 = -2 \cdot 3^2 x^2 \cdot 3x + 5x^2$$
$$= -2 \cdot 9 \cdot 3 \cdot x^2 \cdot x + 5x^2$$
$$= -54x^{2+1} + 5x^2$$
$$= -54x^3 + 5x^2$$

103. Simplify the expression $(a^2 b)^3 (ab^2)^4$.

Solution

$$(a^2 b)^3 (ab^2)^4 = (a^2)^3 \cdot b^3 \cdot a^4 (b^2)^4$$
$$= a^{2 \cdot 3} \cdot a^4 \cdot b^3 \cdot b^{2 \cdot 4}$$
$$= a^6 \cdot a^4 \cdot b^3 \cdot b^8$$
$$= a^{6+4} b^{3+8}$$
$$= a^{10} b^{11}$$

105. Simplify the expression $\dfrac{7x}{9y^2} \cdot \dfrac{7x^3}{9^2 y^3}$.

Solution

$$\frac{7x}{9y^2} \cdot \frac{7x^3}{9^2 y^3} = \frac{49x^{1+3}}{9^{1+2} y^{2+3}} = \frac{49x^4}{9^3 y^5} = \frac{49x^4}{729 y^5}$$

107. What power of 3 is 81?

Solution

4; $81 = 3^4$

109. Evaluate the expression $8 \cdot 10^3 + 3 \cdot 10^2 + 9 \cdot 10^1$.

Solution

$$8 \cdot 10^3 + 3 \cdot 10^2 + 9 \cdot 10^1 = 8 \cdot 10 \cdot 10 \cdot 10 + 3 \cdot 10 \cdot 10 + 9 \cdot 10$$
$$= 8(1000) + 3(100) + 9(10) = 8000 + 300 + 90 = 8390$$

111. *Reasonable Wages* You are offered a job for 25 days with the following pay schedule. On the first day the wages are 10¢, on the second day 20¢, the third day 40¢, and so on. Complete the table shown in the textbook which gives the wages for selected days in the 25-day assignment. Would you accept these wages?

Solution

t	1	5	10	20	25
$10(2^{t-1})$	\$ 0.10	\$ 1.60	\$ 51.20	\$ 52,428.80	\$ 1,677,721.60

Yes, most people would be happy to accept these wages.

113. *Moment of Inertia* The moment of inertia of a circular cylinder is given by

$$k\pi a^2 L\left(\frac{a^2}{2}\right).$$

Simplify this expression.

Solution

$$k\pi a^2 L\left(\frac{a^2}{2}\right) = \frac{k\pi a^2 L}{1}\left(\frac{a^2}{2}\right)$$

$$= \frac{k\pi a^{2+2} L}{2}$$

$$= \frac{k\pi a^4 L}{2} \text{ or } k\pi L\frac{a^4}{2}$$

115. Write the expression that corresponds to the following calculator steps. Then evaluate the expression.

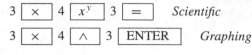

Solution

$3(4^3) = 192$

2.2 Basic Rules of Algebra

7. Identify the rule (or rules) of algebra illustrated by the equation $x + 2y = 2y + x$.

Solution

Commutative Property of Addition

9. Identify the rule (or rules) of algebra illustrated by the equation $(3x + 2y) + z = 3x + (2y + z)$.

Solution

Associative Property of Addition

11. Identify the rule (or rules) of algebra illustrated by the equation $16xy \cdot \dfrac{1}{16xy} = 1, xy \neq 0$.

Solution

Multiplicative Inverse Property

13. Identify the rule (or rules) of algebra illustrated by the equation $x(y + z) = xy + xz$.

Solution

Distributive Property

15. Complete the statement $(x + 1) - \blacksquare = 0$. State the rule of algebra that you used.

Solution

$(x + 1) - (x + 1) = 0$

Additive Inverse Property

17. Complete the statement $v(2) = \blacksquare$. State the rule of algebra that you used.

Solution

$v(2) = 2v$

Commutative Property of Multiplication

19. Complete the statement $(t + 5)(t - 2) = t(\blacksquare) + 5(\blacksquare)$. State the rule of algebra that you used.

Solution

$(t + 5)(t - 2) = t(t - 2) + 5(t - 2)$

Distributive Property

21. Use the Distributive Property to expand the expression $-5(2x - y)$.

Solution

$$-5(2x - y) = -5(2x) - (-5)(y)$$

$$= -10x + 5y$$

23. Use the Distributive Property to expand the expression $(x + 2)(3)$.

Solution

$$(x + 2)(3) = x(3) + 2(3)$$
$$= 3x + 6$$

25. Use the Distributive Property to expand the expression $x(x + xy + y^2)$.

Solution

$$x(x + xy + y^2) = x(x) + x(xy) + x(y^2)$$
$$= x^2 + x^2y + xy^2$$

27. Use the Distributive Property to expand the expression $z[z^2 - 2(z - 1)]$.

Solution

$$z[z^2 - 2(z - 1)] = z[z^2 - 2(z) - 2(-1)]$$
$$= z[z^2 - 2z + 2]$$
$$= z(z^2) - z(2z) + z(2)$$
$$= z^3 - 2z^2 + 2z$$

29. Identify the terms of the expression $6x^2 - 3xy + y^2$ and the coefficient of each term.

Solution

Term	Coefficient
$6x^2$	6
$-3xy$	-3
y^2	1

31. Identify the like terms of $16t^3 + 4 - 5 + 3t^3$.

Solution

In this expression, $16t^3$ and $3t^3$ are like terms, and the constants 4 and -5 are like terms.

33. Simplify the expression $3y - 5y$ by combining like terms.

Solution

$$3y - 5y = (3 - 5)y = -2y$$

35. Simplify the expression $x^2 - 2xy + 4 + xy$ by combining like terms.

Solution

$$x^2 - 2xy + 4 + xy = x^2 - 2xy + xy + 4$$
$$= x^2 + (-2 + 1)xy + 4$$
$$= x^2 + (-1)xy + 4$$
$$= x^2 - xy + 4$$

37. Simplify the expression $3\left(\dfrac{1}{x}\right) - \left(\dfrac{1}{x}\right) + 8$ by combining like terms.

Solution

$$3\left(\frac{1}{x}\right) - \frac{1}{x} + 8 = (3 - 1)\left(\frac{1}{x}\right) + 8$$
$$= 2\left(\frac{1}{x}\right) + 8 \quad \text{or} \quad \frac{2}{x} + 8$$

39. State why the two expressions $\frac{1}{2}x^2y$, $-\frac{5}{2}xy^2$ are not like terms.

Solution

$\frac{1}{2}x^2y$ and $-\frac{5}{2}xy^2$ are not like terms because their variable factors are not alike.

Note: $x^2y = x \cdot x \cdot y$ and $xy^2 = x \cdot y \cdot y$.

41. *Mental Math*   Use the Distributive Property to perform the required arithmetic *mentally* in $8(52) = 8(50 + 2)$.

Solution

$$8(52) = 8(50 + 2) = 8(50) + 8(2) = 400 + 16 = 416$$

43. *Geometry*   Write an expression for the perimeter of the triangle shown in the figure in the textbook. Use the Properties of Real Numbers to simplify the expression.

Solution

$$(x - 2) + (x + 11) + (2x + 3) = x - 2 + x + 11 + 2x + 3$$
$$= (1 + 1 + 2)x + (-2 + 11 + 3)$$
$$= 4x + 12$$

45. Simplify the expression $(-2x)^2x^4$.

Solution

$$(-2x)^2x^4 = (-2)^2x^2x^4$$
$$= 4x^{2+4} = 4x^6$$

47. Simplify the expression $5z^3(z^2)^2$.

Solution

$$5z^3(z^2)^2 = 5z^3z^{2(2)}$$
$$= 5z^3z^4 = 5z^{3+4} = 5z^7$$

49. *Balance in an Account*   The balance in an account that has an annual percentage rate of r, compounded annually for 4 years, is given by

$$P(1 + r)(1 + r)(1 + r)(1 + r).$$

Simplify the expression.

Solution

$$P(1 + r)(1 + r)(1 + r)(1 + r) = P(1 + r)^4$$

51. Identify the rule (or rules) of algebra illustrated by the equation $(9x)y = 9(xy)$.

Solution

Associative Property of Multiplication

53. Identify the rule (or rules) of algebra illustrated by the equation $(x^2 + y^2) \cdot 1 = x^2 + y^2$.

Solution

Multiplicative Identity Property

55. Identify the rule (or rules) of algebra illustrated by the equation $(x + 2)y = y(x + 2)$.

Solution

Commutative Property of Multiplication

57. Identify the rule (or rules) of algebra illustrated by the equation $x^2 + (y^2 - y^2) = x^2$.

Solution

Additive Inverse Property and Additive Identity Property

59. Complete the statement $5x \cdot \blacksquare = 1$. State the rule of algebra that you used.

Solution

$$5x \cdot \left(\frac{1}{5x}\right) = 1$$

Multiplicative Inverse Property

61. Complete the statement $(2z - 3) - \blacksquare = 0$. State the rule of algebra that you used.

Solution

$$(2z - 3) - (2z - 3) = 0$$

Additive Inverse Property

63. Complete the statement $12 + (8 - x) = \blacksquare -x$. State the rule of algebra that you used.

Solution

$$12 + (8 - x) = (12 + 8) - x$$

Associative Property of Addition

65. Use the Distributive Property to expand the expression $3(x^2y + x)$.

Solution

$$3(x^2y + x) = 3x^2y + 3x$$

67. Use the Distributive Property to expand the expression $(-1)(u - v)$.

Solution

$$(-1)(u - v) = (-1)u - (-1)v = -u + v$$

69. Use the Distributive Property to expand the expression $x(8 + x)$.

Solution

$$x(8 + x) = x(8) + x(x)$$
$$= 8x + x^2$$

71. Use the Distributive Property to expand the expression $2x(x + 9)$.

Solution

$$2x(x + 9) = 2x(x) + 2x(9)$$
$$= 2x^2 + 18x$$

73. Use the Distributive Property to expand the expression $-4y(3y - 4)$.

Solution

$$-4y(3y - 4) = -4y(3y) - (-4y)(4)$$
$$= -12y^2 + 16y$$

75. Use the Distributive Property to expand the expression $a(a^2 + ab + b^2)$.

Solution

$$a(a^2 + ab + b^2) = a(a^2) + a(ab) + a(b^2)$$
$$= a^3 + a^2b + ab^2$$

77. Identify the like terms of the equation $6x^2y + 2xy - 4x^2y$.

Solution

In this expression, $6x^2y$ and $-4x^2y$ are the like terms.

79. Simplify the expression $x + 5y - 3x - y$.

Solution

$$x + 5y - 3x - y = x - 3x + 5y - y$$
$$= (1 - 3)x + (5 - 1)y$$
$$= -2x + 4y$$

81. Simplify the expression $2x^2 + 9x^2 + 4$.

Solution

$$2x^2 + 9x^2 + 4 = (2 + 9)x^2 + 4 = 11x^2 + 4$$

83. Simplify the expression $3x^2 - x^2y + 4xy^2 + 3x^2y - xy^2 + y^2$.

Solution

$$3x^2 - x^2y + 4xy^2 + 3x^2y - xy^2 + y^2 = 3x^2 - x^2y + 3x^2y + 4xy^2 - xy^2 + y^2$$
$$= 3x^2 + (-1 + 3)x^2y + (4 - 1)xy^2 + y^2$$
$$= 3x^2 + 2x^2y + 3xy^2 + y^2$$

85. Simplify the expression $1.2\left(\dfrac{1}{x}\right) + 3.8\left(\dfrac{1}{x}\right) - 4x$.

Solution

$$1.2\left(\dfrac{1}{x}\right) + 3.8\left(\dfrac{1}{x}\right) - 4x = (1.2 + 3.8)\left(\dfrac{1}{x}\right) - 4x$$
$$= 5\left(\dfrac{1}{x}\right) - 4x \quad \text{or} \quad \dfrac{5}{x} - 4x$$

87. *True or False?* Decide whether $3(x - 4) \stackrel{?}{=} 3x - 4$ is true or false.

Solution

False

$$3(x - 4) = 3 \cdot x - 3 \cdot 4 = 3x - 12$$

Therefore, $3(x - 4) \neq 3x - 4$.

89. *True or False?* Decide whether $6x - 4x \stackrel{?}{=} 2x$ is true or false.

Solution

True

$$6x - 4x = (6 - 4)x = 2x$$

Therefore, $6x - 4x = 2x$.

91. *Mental Math* Use mental math to evaluate the expression $5(7.98) = 5(8 - 0.02)$.

Solution

$$5(7.98) = 5(8 - 0.02)$$
$$= 5(8) - 5(0.02)$$
$$= 40 - 0.10$$
$$= 39.9$$

93. *Geometric Model*   The figure shows two adjoining rectangles. Demonstrate the Distributive Property by filling in the blanks to express the combined area of the two rectangles in two ways.

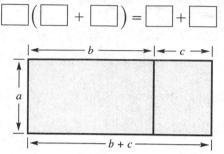

Solution

$a(b + c) = a \cdot b + a \cdot c$

The expression $a(b + c)$ represents the area of the entire figure, $a \cdot b$ represents the area of the rectangle on the left and $a \cdot c$ represents the area of the rectangle on the right. Thus, the area of the entire figure is the sum of the areas of the two smaller rectangles.

95. *Exploration*   You have defined a new mathematical operation using the symbol $\odot$. This operation is defined as $a \odot b = 2 \cdot a + b$. Is this operation commutative? Is it associative? Explain your reasoning.

Solution

$a \odot b = 2 \cdot a + b$ and $b \odot a = 2 \cdot b + a$

Thus, $a \odot b \neq b \odot a$, and so the operation is not commutative.

$a \odot (b \odot c) = a \odot (2b + c) = 2a + 2b + c$
$(a \odot b) \odot c = (2a + b) \odot c = 2(2a + b) + c = 4a + 2b + c$

Thus, $a \odot (b \odot c) \neq (a \odot b) \odot c$, and so the operation is not associative.

2.3   Rewriting and Evaluating Algebraic Expressions

7. Simplify the expression $-2(6x)$.

Solution

$-2(6x) = (-2 \cdot 6)x = -12x$

9. Simplify the expression $\dfrac{5x}{8} \cdot \dfrac{16}{5}$.

Solution

$$\frac{5}{8}x \cdot \frac{16}{5} = \left(\frac{5}{8} \cdot \frac{16}{5}\right)x = \left(\frac{5 \cdot 16}{8 \cdot 5}\right)x$$
$$= \left(\frac{\cancel{5} \cdot \cancel{8} \cdot 2}{\cancel{8} \cdot \cancel{5}}\right)x = 2x$$

11. Simplify the expression $x - (x + 4)$ by removing symbols of grouping and combining like terms.

Solution

$x - (x + 4) = x - x - 4$
$= -4$

13. Simplify the expression $3n + (n - 3)$ by removing symbols of grouping and combining like terms.

Solution

$3n + (n - 3) = 3n + n - 3$
$= 4n - 3$

15. Simplify the expression $2(x - 2y) + y$ by removing symbols of grouping and combining like terms.

Solution

$$2(x - 2y) + y = 2x - 4y + y$$
$$= 2x - 3y$$

17. Simplify the expression $3 - 2[6 + (4 - x)]$ by removing symbols of grouping and combining like terms.

Solution

$$3 - 2[6 + (4 - x)] = 3 - 2[6 + 4 - x]$$
$$= 3 - 2[10 - x]$$
$$= 3 - 20 + 2x$$
$$= 2x - 17 \text{ or } -17 + 2x$$

Note: This expression may also be simplified as follows:

$$3 - 2[6 + (4 - x)] = 3 - 2[6 + 4 - x]$$
$$= 3 - 12 - 8 + 2x$$
$$= 2x - 17$$

19. Simplify the expression $-3t(4 - t) + t(t + 1)$ by removing symbols of grouping and combining like terms.

Solution

$$-3t(4 - t) + t(t + 1) = -12t + 3t^2 + t^2 + t$$
$$= 4t^2 - 11t$$

21. Use the Distributive Property to simplify the expression $\frac{x}{2} + \frac{x}{3}$.

Solution

$$\frac{x}{2} + \frac{x}{3} = \frac{1}{2}x + \frac{1}{3}x = \left(\frac{1}{2} + \frac{1}{3}\right)x$$
$$= \left(\frac{1(3)}{2(3)} + \frac{1(2)}{3(2)}\right)x = \left(\frac{3}{6} + \frac{2}{6}\right)x$$
$$= \frac{5}{6}x$$

23. Use the Distributive Property to simplify the expression $\frac{2x}{3} - \frac{x}{3}$.

Solution

$$\frac{2x}{3} - \frac{x}{3} = \frac{2}{3}x - \frac{1}{3}x$$
$$= \left(\frac{2}{3} - \frac{1}{3}\right)x$$
$$= \frac{1}{3}x \quad \text{or} \quad \frac{x}{3}$$

25. Use the Distributive Property to simplify the expression $\frac{3x}{10} - \frac{x}{10} + \frac{4x}{5}$.

Solution

$$\frac{3x}{10} - \frac{x}{10} + \frac{4x}{5} = \frac{3}{10}x - \frac{1}{10}x + \frac{4}{5}x$$
$$= \left(\frac{3}{10} - \frac{1}{10} + \frac{4}{5}\right)x$$
$$= \left(\frac{3}{10} - \frac{1}{10} + \frac{4(2)}{5(2)}\right)x$$
$$= \left(\frac{3}{10} - \frac{1}{10} + \frac{8}{10}\right)x$$
$$= 1 \cdot x$$
$$= x$$

27. Evaluate the algebraic expression $2x - 1$ when (a) $x = \frac{1}{2}$ and (b) $x = 4$. If it is not possible, state the reason.

Solution

(a) When $x = \frac{1}{2}$, the value of $2x - 1$ is $2\left(\frac{1}{2}\right) - 1 = 1 - 1 = 0$.

(b) When $x = 4$, the value of $2x - 1$ is $2(4) - 1 = 8 - 1 = 7$.

29. Evaluate the algebraic expression $x - 3(x - y)$ when (a) $x = 3$, $y = 3$ and (b) $x = 4$, $y = -4$. If it is not possible, state the reason.

Solution

(a) When $x = 3$ and $y = 3$, the value of $x - 3(x - y)$ is $3 - 3(3 - 3) = 3 - 3(0) = 3 - 0 = 3$.

(b) When $x = 4$ and $y = -4$, the value of $x - 3(x - y)$ is $4 - 3(4 - (-4)) = 4 - 3(4 + 4) = 4 - 3(8) = 4 - 24 = -20$.

31. Evaluate the algebraic expression $\dfrac{5x}{y - 3}$ when (a) $x = 2$, $y = 4$ and (b) $x = 2$, $y = 3$. If it is not possible, state the reason.

Solution

(a) When $x = 2$ and $y = 4$, the value of $\dfrac{5x}{y - 3}$ is $\dfrac{5(2)}{4 - 3} = \dfrac{10}{1} = 10$.

(b) When $x = 2$ and $y = 3$, the value of $\dfrac{5x}{y - 3}$ is undefined because $\dfrac{5(2)}{3 - 3} = \dfrac{10}{0}$, and division by zero is undefined.

33. *Area of a Triangle*   Evaluate the algebraic expression $\frac{1}{2}bh$ when (a) $b = 3$, $h = 5$ and (b) $b = 2$, $h = 10$.

Solution

(a) When $b = 3$ and $h = 5$, the value of $\frac{1}{2}bh$ is $\frac{1}{2} \cdot 3 \cdot 5 = \frac{15}{2}$.

(b) When $b = 2$ and $h = 10$, the value of $\frac{1}{2}bh$ is $\frac{1}{2} \cdot 2 \cdot 10 = 10$.

35. *Finding a Pattern*   (a) Complete the table shown in the textbook by evaluating the expression $3x - 2$. (b) Use the table to find the increase in the value of the expression for each one-unit increase in x. (c) From the pattern in parts (a) and (b), predict the increase in the expression $\frac{2}{3}x + 4$ for each one-unit increase in x. Then verify your prediction.

Solution

(a)

x	-1	0	1	2	3	4
$3x - 2$	-5	-2	1	4	7	10

When $x = -1$, $\quad 3x - 2 = 3(-1) - 2 = -3 - 2 = -5$.
When $x = 0$, $\quad 3x - 2 = 3 \cdot 0 - 2 = 0 - 2 = -2$.
When $x = 1$, $\quad 3x - 2 = 3 \cdot 1 - 2 = 3 - 2 = 1$.
When $x = 2$, $\quad 3x - 2 = 3 \cdot 2 - 2 = 6 - 2 = 4$.
When $x = 3$, $\quad 3x - 2 = 3 \cdot 3 - 2 = 9 - 2 = 7$.
When $x = 4$, $\quad 3x - 2 = 3 \cdot 4 - 2 = 12 - 2 = 10$.

(b) For each one-unit increase in x, the value of the expression $3x - 2$ increases by 3, the coefficient of x.

(c) For each one-unit increase in x, the value of the expression $\frac{2}{3}x + 4$ increases by $\frac{2}{3}$, the coefficient of x.

37. *Balance in an Account*   The balance in an account with an initial deposit of P dollars, at an annual interest of rate of r for t years, is $P(1 + r)^t$. Find the balance for the given values of P, r, and t: $P = 10,000$, $r = 0.08$, $t = 10$.

Solution

If $P = 10,000$, $r = 0.08$, and $t = 10$, the value of $P(1 + r)^t$ is $10,000(1 + 0.08)^{10} \approx 21,589.25$.

39. *Geometry*   Find an expression for the area of the figure in the textbook. Then evaluate the expression for the given value of the variable $n = 80$.

Solution

$$A = \tfrac{1}{2}n(10 + n)$$
$$= 5n + \tfrac{1}{2}n^2$$

If n is 80, $5n + \tfrac{1}{2}n^2 = 5(80) + \tfrac{1}{2}(80)^2$
$$= 400 + \tfrac{1}{2}(6400)$$
$$= 400 + 3200$$
$$= 3600.$$

Thus, the area of the triangle is 3600 square units.

41. *Geometry*   Find an expression for the area of the figure in the textbook. Then evaluate the expression for the given values of the variables $x = 10$, $y = 3$.

Solution

$$A = (x + y)^2$$

If x is 10 and y is 3, $A = (10 + 3)^2 = (13)^2 = 169$. Thus, the area is 169 square units.

43. Use the Distributive Property to expand the expression $4(2x - 5)$.

Solution

$$4(2x - 5) = 8x - 20$$

45. Simplify the expression $\dfrac{5x}{3} - \dfrac{2x}{3} - 4$.

Solution

$$\frac{5x}{3} - \frac{2x}{3} - 4 = \frac{5}{3}x - \frac{2}{3}x - 4$$
$$= \left(\frac{5}{3} - \frac{2}{3}\right)x - 4$$
$$= \left(\frac{3}{3}\right)x - 4$$
$$= 1x - 4$$
$$= x - 4$$

47. *Profit*   A company had a loss of $1,530,000 during the first 6 months of the year. At the end of the year, the company ended up with an overall profit of $832,000. What was the profit during the last 2 quarters of the year?

Solution

$$832,000 - (-1,530,000) = 2,362,000$$

Thus, the company had a profit of $2,362,000 during the last two quarters of the year.

49. Simplify the expression $-7(5a)$.

Solution

$$-7(5a) = (-7 \cdot 5)a = -35a$$

51. Simplify the expression $(-2x)(-3x)$.

Solution

$$(-2x)(-3x) = (-2)(-3)(x \cdot x) = 6x^2$$

53. Simplify the expression $\dfrac{18a}{5} \cdot \dfrac{15}{6}$.

Solution

$$\dfrac{18a}{5} \cdot \dfrac{15}{6} = \dfrac{18a(15)}{5(6)}$$

$$= \dfrac{9(2)(a)(5)(3)}{5(3)(2)}$$

$$= 9a$$

55. Simplify the expression $(-1.5x^2)(4x^3)$.

Solution

$$(-1.5x^2)(4x^3) = (-1.5)(4)(x^2 \cdot x^3)$$

$$= -6x^{2+3}$$

$$= -6x^5$$

57. Simplify the expression $-3(x + 1) - 2$ by removing symbols of grouping and combining like terms.

Solution

$$-3(x + 1) - 2 = -3x - 3 - 2$$

$$= -3x - 5$$

59. Simplify the expression $\frac{2}{3}(12x + 15) + 16$ by removing symbols of grouping and combining like terms.

Solution

$$\tfrac{2}{3}(12x + 15) + 16 = 8x + 10 + 16 = 8x + 26$$

61. Simplify the expression $\frac{3}{8}(4 - y) - \frac{5}{2} + 10$ by removing symbols of grouping and combining like terms.

Solution

$$\dfrac{3}{8}(4 - y) - \dfrac{5}{2} + 10 = \dfrac{3}{8} \cdot \dfrac{4}{1} - \dfrac{3}{8}y - \dfrac{5}{2} + 10$$

$$= \dfrac{3 \cdot 4}{2 \cdot 4} - \dfrac{3}{8}y - \dfrac{5}{2} + 10$$

$$= \dfrac{3}{2} - \dfrac{5}{2} + 10 - \dfrac{3}{8}y$$

$$= -1 + 10 - \dfrac{3}{8}y$$

$$= 9 - \dfrac{3}{8}y$$

63. Simplify the expression $7x(2 - x) - 4x$ by removing symbols of grouping and combining like terms.

Solution

$$7x(2 - x) - 4x = 14x - 7x^2 - 4x$$

$$= -7x^2 + 10x$$

65. Simplify the expression $4x^2 + x(5 - x)$ by removing symbols of grouping and combining like terms.

Solution

$$4x^2 + x(5 - x) = 4x^2 + 5x - x^2$$

$$= 4x^2 - x^2 + 5x$$

$$= 3x^2 + 5x$$

67. Simplify the expression $3(r - 2s) - 5(3r - 5s)$ by removing symbols of grouping and combining like terms.

Solution

$$3(r - 2s) - 5(3r - 5s) = 3r - 6s - 15r + 25s$$

$$= 3r - 15r - 6s + 25s$$

$$= -12r + 19s$$

69. Use the Distributive Property to simplify the expression $\dfrac{4z}{5} + \dfrac{3z}{5}$.

Solution

$$\dfrac{4z}{5} + \dfrac{3z}{5} = \dfrac{4}{5}z + \dfrac{3}{5}z$$

$$= \left(\dfrac{4}{5} + \dfrac{3}{5}\right)z$$

$$= \dfrac{7}{5}z \quad \text{or} \quad \dfrac{7z}{5}$$

71. Use the Distributive Property to simplify the expression $\dfrac{x}{3} - \dfrac{5x}{4}$.

Solution

$$\dfrac{x}{3} - \dfrac{5x}{4} = \dfrac{1}{3}x - \dfrac{5}{4}x$$

$$= \left(\dfrac{1}{3} - \dfrac{5}{4}\right)x$$

$$= \left(\dfrac{1(4)}{3(4)} - \dfrac{5(3)}{4(3)}\right)x$$

$$= \left(\dfrac{4}{12} - \dfrac{15}{12}\right)x$$

$$= -\dfrac{11}{12}x \quad \text{or} \quad -\dfrac{11x}{12}$$

73. Evaluate the algebraic expression $2x^2 + 4x - 5$ when (a) $x = -2$ and (b) $x = 3$. If it is not possible, state the reason.

Solution

(a) When $x = -2$, the value of $2x^2 + 4x - 5$ is

$$2(-2)^2 + 4(-2) - 5 = 2 \cdot 4 + (-8) - 5$$

$$= 8 - 8 - 5$$

$$= -5.$$

(b) When $x = 3$, the value of $2x^2 + 4x - 5$ is

$$2(3)^2 + 4(3) - 5 = 2 \cdot 9 + 12 - 5$$

$$= 18 + 12 - 5$$

$$= 25.$$

75. Evaluate the algebraic expression $3x - 2y$ when (a) $x = 4$, $y = 3$ and (b) $x = \frac{2}{3}$, $y = 1$. If it is not possible, state the reason.

Solution

(a) When $x = 4$ and $y = 3$, the value of $3x - 2y$ is $3(4) - 2(3) = 12 - 6 = 6$.

(b) When $x = \frac{2}{3}$ and $y = 1$, the value of $3x - 2y$ is $3\left(\frac{2}{3}\right) - 2(1) = 2 - 2 = 0$.

77. Evaluate the algebraic expression $b^2 - 4ac$ when (a) $a = 2, b = -3, c = -1$ and (b) $a = -4, b = 6, c = -2$. If it is not possible, state the reason.

Solution

(a) When $a = 2$, $b = -3$, and $c = -1$, the value of $b^2 - 4ac$ is $(-3)^2 - 4(2)(-1) = 9 + 8 = 17$.

(b) When $a = -4$, $b = 6$, and $c = -2$, the value of $b^2 - 4ac$ is $6^2 - 4(-4)(-2) = 36 - 32 = 4$.

79. Evaluate the expression $\dfrac{x-2y}{x+2y}$ when (a) $x = 4$, $y = 2$ and (b) $x = 4$, $y = -2$. If it is not possible, state the reason.

Solution

(a) When $x = 4$ and $y = 2$, the value of $(x - 2y)/(x + 2y)$ is

$$\frac{4-2\cdot2}{4+2\cdot2} = \frac{4-4}{4+4}$$

$$= \frac{0}{8}$$

$$= 0.$$

(b) When $x = 4$ and $y = -2$, the value of $(x - 2y)/(x + 2y)$ is undefined because

$$\frac{4-2(-2)}{4+2(-2)} = \frac{4+4}{4-4} = \frac{8}{0}$$

and division by 0 is undefined.

81. *Finding a Pattern* (a) Complete the table in the textbook by evaluating $3 - 2x$. (b) Use the table to find the decrease in the value of the expression for each one-unit increase in x. (c) From the pattern in parts (a) and (b), predict the decrease in the expression $4 - \frac{3}{2}x$ for each one-unit increase in x. Then verify your prediction.

Solution

(a)

x	-1	0	1	2	3	4
$3 - 2x$	5	3	1	-1	-3	-5

(b) For each one-unit increase in x, the value of the expression $3 - 2x$ *decreases* by 2; Thus, the value changes by -2.

(c) You might notice that -2 is the coefficient of x in the expression $3 - 2x$. In the expression $4 - \frac{3}{2}x$, the coefficient of x is $-\frac{3}{2}$. You might predict that the value of this expression would *decrease* by $\frac{3}{2}$ for each one-unit increase in the value of x. Thus, the value changes by $-\frac{3}{2}$.

Verification:

x	-1	0	1	2	3	4
$4 - \frac{3}{2}x$	5.5	4	2.5	1	-0.5	-2

83. *Geometry*   The basic floor plan for a one-story house is shown in the figure.

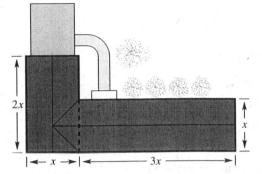

(a) Write a simplified algebraic expression for the amount of floor space in the house.

(b) Determine the amount of floor space if $x = 20$ feet. What units of measure are used for the amount of floor space?

Solution

(a) $2x \cdot x + 3x \cdot x = 2x^2 + 3x^2 = 5x^2$

The rectangle on the left has area of $2x \cdot x = 2x^2$.

The rectangle on the right has area of $3x \cdot x = 3x^2$.

The amount of floor space is the sum $2x^2 + 3x^2 = 5x^2$.

(b) If $x = 20$ feet, the value of $5x^2$ is $5(20)^2 = 5(400) = 2000$ square feet. The amount of floor space is measured in square feet.

85. *Geometry*   Use the formula for the area of a trapezoid, $\frac{1}{2}h(b_1 + b_2)$, to find the area of the trapezoidal house lot.

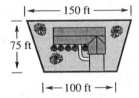

When $b_1 = 150$, $b_2 = 100$, and $h = 75$, the value of $\frac{1}{2}h(b_1 + b_2)$ is

$$\frac{1}{2}(75)(150 + 100) = \frac{1}{2}(75)(250) = 9375.$$

The area of the lot is 9375 square feet.

Mid-Chapter Quiz for Chapter 2

1. Identify the coefficients of the terms (a) $-5xy^2$ and (b) $\dfrac{5z}{16}$.

Solution

(a) The coefficient is -5.

(b) The coefficient is $\frac{5}{16}$.

2. Rewrite the expression in exponential form.

(a) $3y \cdot 3y \cdot 3y \cdot 3y$

(b) $2 \cdot (x - 3) \cdot (x - 3) \cdot 2 \cdot 2$

Solution

(a) $3y \cdot 3y \cdot 3y \cdot 3y = (3y)^4$

(b) $2 \cdot (x - 3)(x - 3)2 \cdot 2 = 2^3(x - 3)^2$

3. Simplify the expression $x^4 \cdot x^3$.

Solution

$x^4 \cdot x^3 = x^{4+3} = x^7$

4. Simplify the expression $(v^2)^5$.

Solution

$(v^2)^5 = v^{2(5)} = v^{10}$

5. Simplify the expression $(-3y)^2 y^3$.

Solution

$(-3y)^2 y^3 = (-3)^2 y^2 y^3 = 9y^{2+3} = 9y^5$

6. Simplify the expression $8(x - 4)^2(x - 4)^4$.

Solution

$8(x - 4)^2(x - 4)^4 = 8(x - 4)^{2+4} = 8(x - 4)^6$

7. Simplify the expression $\dfrac{2z^2}{3y} \cdot \dfrac{5z}{7y^3}$.

Solution

$\dfrac{2z^2}{3y} \cdot \dfrac{5z}{7y^3} = \dfrac{2z^2(5z)}{3y(7y^3)} = \dfrac{10z^3}{21y^4}$

8. Simplify the expression $\left(\dfrac{x}{y}\right)^2 \left(\dfrac{x}{y}\right)^5$.

Solution

$\left(\dfrac{x}{y}\right)^2 \left(\dfrac{x}{y}\right)^5 = \left(\dfrac{x}{y}\right)^{2+5} = \left(\dfrac{x}{y}\right)^7$

9. Identify the rule of algebra illustrated by the equation $-3(2y) = (-3 \cdot 2)y$.

Solution

Associative Property of Multiplication

10. Identify the rule of algebra illustrated by the equation $(x + 2)y = xy + 2y$.

Solution

Distributive Property

11. Identify the rule of algebra illustrated by the equation $3y \cdot \dfrac{1}{3y} = 1, \, y \neq 0$.

Solution

Multiplicative Inverse Property

12. Identify the rule of algebra illustrated by the equation $x - x^2 + 2 = -x^2 + x + 2$.

Solution

Commutative Property of Addition

13. Simplify the expression $y^2 - 3xy + y + 7xy$ by combining like terms.

Solution

$y^2 - 3xy + y + 7xy = y^2 + (-3 + 7)xy + y$
$$= y^2 + 4xy + y$$

14. Simplify the expression $10\left(\dfrac{1}{u}\right) - 7\left(\dfrac{1}{u}\right) + 3u$ by combining like terms.

Solution

$10\left(\dfrac{1}{u}\right) - 7\left(\dfrac{1}{u}\right) + 3u = (10 - 7)\left(\dfrac{1}{u}\right) + 3u$
$$= 3\left(\dfrac{1}{u}\right) + 3u$$

15. Simplify the expression $5(a - 2b) + 3(a + b)$ by removing symbols of grouping and combining like terms.

Solution

$$5(a - 2b) + 3(a + b) = 5a - 10b + 3a + 3b$$
$$= (5 + 3)a + (-10 + 3)b$$
$$= 8a - 7b$$

16. Simplify the expression $4x + 3[2 - 4(x + 6)]$ by removing symbols of grouping and combining like terms.

Solution

$$4x + 3[2 - 4(x + 6)] = 4x + 3[2 - 4x - 24]$$
$$= 4x + 3[-22 - 4x]$$
$$= 4x - 66 - 12x$$
$$= (4 - 12)x - 66$$
$$= -8x - 66$$

17. Evaluate the algebraic expression $x^2 - 3x$ when (a) $x = 3$, (b) $x = -2$, and (c) $x = 0$. If it is not possible, state the reason.

Solution

(a) If $x = 3$, the value of $x^2 - 3x$ is $3^2 - 3(3) = 9 - 9 = 0$.

(b) If $x = -2$, the value of $x^2 - 3x$ is $(-2)^2 - 3(-2) = 4 + 6 = 10$.

(c) If $x = 0$, the value of $x^2 - 3x$ is $0^2 - 3(0) = 0 - 0 = 0$.

18. Evaluate the algebraic expression $\dfrac{x}{y - 3}$ when (a) $x = 2$, $y = 4$, (b) $x = 0$, $y = -1$, and (c) $x = 5$, $y = 3$.

Solution

(a) If $x = 2$ and $y = 4$, the value of $\dfrac{x}{y - 3}$ is $\dfrac{2}{4 - 3} = \dfrac{2}{1} = 2$.

(b) If $x = 0$ and $y = -1$, the value of $\dfrac{x}{y - 3}$ is $\dfrac{0}{-1 - 3} = \dfrac{0}{-4} = 0$.

(c) If $x = 5$ and $y = 3$, the value of $\dfrac{x}{y - 3}$ is undefined because $\dfrac{5}{3 - 3} = \dfrac{5}{0}$ and division by zero is undefined.

19. Simplify the following expression for the moment of inertia of a cone of height h and radius r.

$$\left(\frac{1}{3}\pi r^2 h\right)\left(\frac{3}{10}r^2\right)$$

Solution

$$\left(\frac{1}{3}\pi r^2 h\right)\left(\frac{3}{10}r^2\right) = \frac{1(\cancel{3})}{\cancel{3}(10)}\pi r^{2+2} = \frac{1}{10}\pi r^4 h$$

20. Evaluate the expression $4 \cdot 10^4 + 5 \cdot 10^3 + 7 \cdot 10^2$.

Solution

$$4 \cdot 10^4 + 5 \cdot 10^3 + 7 \cdot 10^2 = 4 \cdot 10,000 + 5 \cdot 1000 + 7 \cdot 100$$
$$= 40,000 + 5000 + 700$$
$$= 45,700$$

2.4 Translating Expressions: Verbal to Algebraic

7. Translate the phrase "A number increased by 5" into an algebraic expression. (Let x represent the number.)

Solution

$x + 5$

9. Translate the phrase "Six less than a number" into an algebraic expression. (Let x represent the number.)

Solution

$x - 6$

11. Translate the phrase "twice a number" into an algebraic expression. (Let x represent the number.)

Solution

$2x$

13. Translate the phrase "The ratio of a number to 50" into an algebraic expression. (Let x represent the number.)

Solution

$\dfrac{x}{50}$

15. Translate the phrase "A number is tripled and the result is increased by 5" into an algebraic expression. (Let x represent the number.)

Solution

$3x + 5$

17. Write a verbal description of the algebraic expression $3x + \frac{1}{3}$. Use words only—do not use the variable. (There is more than one correct answer.)

Solution

A number is tripled and the result is increased by $\frac{1}{3}$.

The sum of three times a number and $\frac{1}{3}$

19. Write a verbal description of the algebraic expression $\dfrac{t + 1}{2}$. Use words only—do not use the variable. (There is more than one correct answer.)

Solution

A number is increased by 1 and the sum is divided by 2.
The sum of a number and 1 is divided by 2.

21. Write a verbal description of the algebraic expression $\frac{1}{2}(x + 1)$. Use words only—do not use the variable. (There is more than one correct answer.)

Solution

One-half the sum of a number and 1
The sum of a number and 1 is divided by 2.
The product of $\frac{1}{2}$ and the sum of a number and 1

23. Translate the following sentence into an algebraic expression. Simplify the expression.

"The sum of x and 3 is multiplied by x."

Solution

$(x + 3)x = x^2 + 3x$

25. Translate the following sentence into an algebraic expression. Simplify the expression.

"The difference of 9 and x is multiplied by 3."

Solution

$(9 - x)(3) = 27 - 3x$

27. *Total Amount of Money*   A cash register contains n dimes. Write an algebraic expression that represents the total amount of money (in dollars).

Solution

The amount of money is a product.

Verbal model: ⎡Value of dime⎤ · ⎡Number of dimes⎤

Labels: Value of dime = 0.10 (dollars)
                 Number of dimes = n

Algebraic expression: $0.10n$ (dollars)

29. *Weekly Paycheck*   You worked *n* hours in a week. Your hourly wage is $ 8.25 per hour. Write an algebraic expression that represents your gross pay for the week.

Solution

The gross pay is a product.

Verbal model: 　　　　 | Pay per hour | · | Number of hours |

Labels: 　　　　　　 Pay per hour = 8.25 (dollars)
　　　　　　　　　　 Number of hours = *n*

Algebraic expression: 　8.25*n* (dollars)

31. *Travel Time*   A truck travels 100 miles at an average speed of *r* miles per hour (see figure). Write an algebraic expression that represents the total travel time.

Solution

The travel time is a quotient.

Verbal model:
$$\frac{\boxed{\text{Distance traveled}}}{\boxed{\text{Average speed}}}$$

|←— 100 miles —→|

Labels: 　　　　　　 Distance traveled = 100 (miles)
　　　　　　　　　　 Average speed = *r* (miles per hour)

Algebraic expression: 　$\dfrac{100}{r}$ (hours)

33. *Computer Screen*   A computer screen has sides of length *s* inches (see figure). Write an algebraic expression that represents the area of the computer screen. What are the units of measure for the area of the screen?

Solution

The area of the screen is a product.

Verbal model: 　　　 | Length of screen | · | Width of screen |

Labels: 　　　　　　 Length of screen = *s* (inches)
　　　　　　　　　　 Width of screen = *s* (inches)

Algebraic expression: 　*s* · *s* or s^2 (square inches)

|← *s* →|

35. *Sum*   Write an expression that represents the sum of three consecutive integers, the first of which is n.

Solution

This expression is a sum.

Verbal model: | First consecutive integer | + | Second consecutive integer | + | Third consecutive integer |

Labels: First consecutive integer $= n$
Second consecutive integer $= n + 1$
Third consecutive integer $= n + 2$

Algebraic expression: $n + (n + 1) + (n + 2) = 3n + 3$

37. *Finding a Pattern*   (a) Complete the table shown in the textbook. The third row in the table consists of the difference between consecutive entries of the second row. (b) Describe the pattern of the third row. What would the pattern be if the algebraic expression were $an + b$ rather than $2n - 1$?

Solution

(a)

n	0	1	2	3	4	5
$2n - 1$	-1	1	3	5	7	9
Differences		2	2	2	2	2

(b) All entries in the third row are 2's. The value of the expression $2n - 1$ increases by 2 for each increase of 1 in the value of n. Note that 2 is the coefficient of n in the expression $2n - 1$. Thus, if the algebraic expression were $an + b$, the differences would all be equal to a.

39. Simplify the algebraic expression $-y^2(y^2 + 4) + 6y^2$.

Solution

$$-y^2(y^2 + 4) + 6y^2 = -y^4 - 4y^2 + 6y^2$$
$$= -y^4 + (-4y^2 + 6y^2)$$
$$= -y^4 + 2y^2$$

41. Evaluate the expression $x^2 - y^2$ when (a) $x = 4$, $y = 3$ and (b) $x = -5$, $y = 3$.

Solution

(a) If $x = 4$ and $y = 3$, the value of $x^2 - y^2 = 4^2 - 3^2 = 16 - 9 = 7$.

(b) If $x = -5$ and $y = 3$, the value of $x^2 - y^2 = (-5)^2 - 3^2 = 25 - 9 = 16$.

43. *Parcel Service*   You are shipping 133 pounds of a product. The parcel service you are using will not accept packages that weigh more than 30 pounds. What is the fewest number of packages you can use to ship the product?

Solution

$\frac{133}{30} = 4\frac{13}{30}$

Thus, four packages will not be sufficient. The fewest number of packages required to ship the product is 5.

45. Match the verbal phrase "Twelve decreased by 3 times a number" with the corresponding algebraic expression.

Solution

(d) $12 - 3x$

47. Match the verbal phrase "Eleven times a number plus $\frac{1}{3}$" with the corresponding algebraic expression.

Solution

(e) $11x + \frac{1}{3}$

49. Match the verbal phrase "The difference between 3 times a number and 12" with the corresponding algebraic expression.

Solution

(b) $3x - 12$

51. Translate the phrase "25 more than a number" into an algebraic expression. (Let x represent the number.)

Solution

$x + 25$

53. Translate the phrase "One-fourth of a number" into an algebraic expression. (Let x represent the number.)

Solution

$\frac{1}{4}x$ or $\frac{x}{4}$

55. Translate the phrase "A number divided by 3" into an algebraic expression. (Let x represent the number.)

Solution

$\frac{x}{3}$

57. Translate the phrase "Five times a number, plus 8" into an algebraic expression. (Let x represent the number.)

Solution

$5x + 8$

59. Translate the phrase "Ten times the sum of a number and 4" into an algebraic expression. (Let x represent the number.)

Solution

$10(x + 4)$

61. Translate the phrase "The absolute value of the sum of a number and 4" into an algebraic expression. (Let x represent the number.)

Solution

$|x + 4|$

63. Translate the phrase "The square of a number increased by 1" into an algebraic expression. (Let x represent the number.)

Solution

$x^2 + 1$

65. Write a verbal description of the algebraic expression $x - 10$. Use words only; do not use the variable. (There is more than one correct answer.)

Solution

10 less than a number
A number decreased by 10
The difference of a number and 10

67. Write a verbal description of the algebraic expression $7x + 4$. Use words only; do not use the variable. (There is more than one correct answer.)

Solution

The product of seven and a number is increased by 4.
Four more than the product of 7 and a number
Seven times a number is increased by 4.

69. Write a verbal description of the algebraic expression $3(2 - x)$. Use words only; do not use the variable. (There is more than one correct answer.)

Solution

Three times the difference of 2 and a number

71. Write a verbal description of the algebraic expression $x^2 + 5$. Use words only; do not use the variable. (There is more than one correct answer.)

Solution

Five more than the square of a number
The square of a number, increased by 5

73. Translate the following sentence into an algebraic expression. Simplify the expression.

"The sum of 6 and n is multiplied by 5."

Solution

$(6 + n)5 = 30 + 5n$

75. Translate the following sentence into an algebraic expression. Simplify the expression.

"The product of eight times the sum of x and 24 is divided by 2."

Solution

$$\frac{8(x + 24)}{2}$$

77. *Sales Tax*   The sales tax on a purchase of L dollars is 6%. Write an algebraic expression that represents the total amount of sales tax. (To find 6% of a quantity, multiply the quantity by 0.06.)

Solution

The amount of sales tax is a product.

Verbal model: $\boxed{\begin{array}{c}\text{Percent of}\\\text{sales tax}\end{array}} \cdot \boxed{\begin{array}{c}\text{Amount of}\\\text{purchase}\end{array}}$

Labels: Percent of sales tax $= 0.06$ (in decimal form)
Amount of purchase $= L$ (dollars)

Algebraic expression: $0.06L$ (dollars)

79. *Camping Fee*   A campground charges $15 for adults and $2 for children. Write an algebraic expression that represents the total camping fee for m adults and n children.

Solution

Verbal model: $\boxed{\begin{array}{c}\text{Fee per}\\\text{parent}\end{array}} \cdot \boxed{\begin{array}{c}\text{Number of}\\\text{parents}\end{array}} + \boxed{\begin{array}{c}\text{Fee per}\\\text{child}\end{array}} \cdot \boxed{\begin{array}{c}\text{Number of}\\\text{children}\end{array}}$

Labels: Fee per parent $= 15$ (dollars)
Number of parents $= m$
Fee per child $= 2$ (dollars)
Number of children $= n$

Algebraic expression: $15m + 2n$

81. *Geometry*   Write an algebraic expression that represents the perimeter of the picture frame shown in the figure in the textbook.

Solution

The perimeter is a sum of products.

Verbal model: $\boxed{2} \cdot \boxed{\text{Length of frame}} + \boxed{2} \cdot \boxed{\text{Width of frame}}$

Labels: Length of frame $= 1.5w$
Width of frame $= w$

Algebraic expression: $2(1.5w) + 2 \cdot w = 3w + 2w = 5w$

83. *Geometry*   Write an algebraic expression that represents the area of the region shown in the textbook. Then use the rules of algebra to simplify the expression.

Solution

The area of a rectangle is the product of the length and width of the rectangle.

$3x(6x - 1) = 18x^2 - 3x$

85. *Geometry*   Write an algebraic expression that represents the area of the region shown in the textbook. Then use the rules of algebra to simplify the expression.

Solution

The area of a triangle is one-half the product of the base and height of the triangle.

$$\frac{1}{2}(12)(5x^2 + 2) = 6(5x^2 + 2)$$
$$= 30x^2 + 12$$

87. *Sum*   Write an algebraic expression that represents the sum of two consecutive odd integers, the first of which is $2n + 1$.

Solution

This expression is a sum.

Verbal model: | First consecutive odd integer | $+$ | Second consecutive odd integer |

Labels: First consecutive odd integer $= 2n + 1$
Second consecutive odd integer $= 2n + 3$

Algebraic expression: $(2n + 1) + (2n + 3)$ or $4n + 4$

(*Note:* Each successive odd integer is 2 *more* than the previous odd integer; $(2n + 1) + 2 = 2n + 3$.)

89. *Finding a Pattern*   (a) Complete the table shown in the textbook. The third row in the table consists of the difference between consecutive entries of the second row. (b) Describe the pattern of the third row. What would the pattern be if the algebraic expression were $an + b$, rather than $2n + 5$?

Solution

(a)

n	0	1	2	3	4	5
$2n + 5$	5	7	9	11	13	15
Differences	2	2	2	2	2	

(b) All entries in the third row are 2's. The value of the expression $2n + 5$ increases by 2 for each increase of 1 in the value of n. Note that 2 is the coefficient of n in the expression $2n + 5$. Thus, if the algebraic expression were $an + b$, the differences would all be equal to a.

91. *Exploration*   Find a and b so that the expression $an + b$ would yield the following table.

n	0	1	2	3	4	5
$an + b$	1	5	9	13	17	21

Solution

The differences in the last row are all 4's. Thus, the coefficient of n must be 4. However, $4(0) = 4, 4(1) = 4, 4(2) = 8, 4(3) = 12$, etc. In each instance, the value of $an + b$ is one more than four times n. Therefore the expression $an + b$ must be $4n + 1$. This indicates that $a = 4$ and $b = 1$.

2.5 Introduction to Equations

7. Decide whether the value of x is a solution to the equation $2x - 6 = 0$ when (a) $x = 3$ and (b) $x = 1$.

Solution

(a) $x = 3$

$$2(3) - 6 \stackrel{?}{=} 0$$

$$6 - 6 \stackrel{?}{=} 0$$

$$0 = 0$$

3 *is* a solution.

(b) $x = 1$

$$2(1) - 6 \stackrel{?}{=} 0$$

$$2 - 6 \stackrel{?}{=} 0$$

$$-4 \neq 0$$

1 *is not* a solution.

9. Decide whether the value of x is a solution to the equation $x + 5 = 2x$ when (a) $x = -1$ and (b) $x = 5$.

Solution

(a) $x = -1$

$$-1 + 5 \stackrel{?}{=} 2(-1)$$

$$4 \neq -2$$

-1 *is not* a solution.

(b) $x = 5$

$$5 + 5 \stackrel{?}{=} 2(5)$$

$$10 = 10$$

5 *is* a solution.

11. Decide whether the value of x is a solution to the equation $x^2 - 4 = x + 2$ when (a) $x = 3$ and (b) $x = -2$.

Solution

(a) $x = 3$

$$3^2 - 4 \stackrel{?}{=} 3 + 2$$

$$9 - 4 \stackrel{?}{=} 3 + 2$$

$$5 = 5$$

3 *is* a solution.

(b) $x = -2$

$$(-2)^2 - 4 \stackrel{?}{=} -2 + 2$$

$$4 - 4 \stackrel{?}{=} 0$$

$$0 = 0$$

-2 *is* a solution.

13. Decide whether the value of x is a solution to the equation $\dfrac{2}{x} - \dfrac{1}{x} = 1$ when (a) $x = 3$ and (b) $x = \dfrac{1}{3}$.

Solution

(a) $x = 3$

$$\frac{2}{3} - \frac{1}{3} \stackrel{?}{=} 1$$

$$\frac{2-1}{3} \stackrel{?}{=} 1$$

$$\frac{1}{3} \neq 1$$

3 *is not* a solution.

(b) $x = \dfrac{1}{3}$

$$\frac{2}{1/3} - \frac{1}{1/3} \stackrel{?}{=} 1$$

$$\left(2 \div \frac{1}{3}\right) - \left(1 \div \frac{1}{3}\right) \stackrel{?}{=} 1$$

$$\left(\frac{2}{1} \cdot \frac{3}{1}\right) - \left(\frac{1}{1} \cdot \frac{3}{1}\right) \stackrel{?}{=} 1$$

$$6 - 3 \stackrel{?}{=} 1$$

$$3 \neq 1$$

$\frac{1}{3}$ *is not* a solution.

15. Use a calculator to decide whether the value of x is a solution of the equation $2x^2 - x = 10$ when (a) $x = 2.5$ and (b) $x = -1.09$.

Solution

(a) $x = 2.5$

$$2(2.5)^2 - 2.5 \stackrel{?}{=} 10$$

$$10 = 10$$

2.5 *is* a solution.

(b) $x = -1.09$

$$2(-1.09)^2 - (-1.09) \stackrel{?}{=} 10$$

$$3.4662 \neq 10$$

-1.09 *is not* a solution.

17. Use a calculator to decide whether the value of x is a solution of the equation $x^3 - 1.728 = 0$ when (a) $x = \frac{6}{5}$ and (b) $x = -\frac{6}{5}$.

Solution

(a) $x = \frac{6}{5}$

$$\left(\frac{6}{5}\right)^3 - 1.728 \stackrel{?}{=} 0$$

$$0 = 0$$

$\frac{6}{5}$ *is* a solution.

(b) $x = -\frac{6}{5}$

$$\left(-\frac{6}{5}\right)^3 - 1.728 \stackrel{?}{=} 0$$

$$-3.456 \neq 0$$

$-\frac{6}{5}$ *is not* a solution.

19. Justify each step of the following solution of the equation $4x = -28$.

$$4x = -28$$

$$\frac{4x}{4} = \frac{-28}{4}$$

$$x - 7$$

Solution

$4x = -28$	Given equation
$\dfrac{4x}{4} = \dfrac{-28}{4}$	Divide both sides by 4.
$x = -7$	Simplify. (solution)

21. Justify each step of the following solution of the equation $2x - 2 = x + 3$.

$$2x - 2 = x + 3$$

$$-x + 2x - 2 = -x + x + 3$$

$$x - 2 = 3$$

$$x - 2 + 2 = 3 + 2$$

$$x = 5$$

Solution

$2x - 2 = x + 3$	Given equation
$-x + 2x - 2 = -x + x + 3$	Subtract x from each side.
$x - 2 = 3$	Combine like terms.
$x - 2 + 2 = 3 + 2$	Add 2 to both sides.
$x = 5$	Combine like terms. (solution)

23. *Mental Math*   Solve the equation $x + 4 = 6$ mentally and check your solution.

Solution

$x = 2$

Check: $2 + 4 = 6$

25. *Mental Math*   Solve the equation $3x = 30$ mentally and check your solution.

Solution

$x = 10$

Check: $3(10) = 30$

27. *Test Score*   After your instructor added 6 points to each student's test score, your score is 94. What was your original score?

Solution

Verbal model: $\boxed{\text{Original score}} + \boxed{\text{Additional points}} = \boxed{\text{Final score}}$

Labels: Original score $= x$ (points)
Additional points $= 6$ (points)
Final Score $= 94$ (points)

Equation: $x + 6 = 94$

Note: There are other equivalent ways to write this equation. Here are some other possibilities:

$94 - x = 6$

$94 = x + 6$

$x = 94 - 6$

There are also equivalent ways of writing equations for the *other* exercises in this section.

29. Four times the sum of a number and 6 is 100. What is the number?

Solution

Verbal model: $\boxed{4} \cdot \boxed{\text{Sum of number and 6}} = \boxed{100}$
Label: Number $= x$
Equation: $4(x + 6) = 100$

Note: The parentheses are necessary to indicate that 4 is multiplied by the *sum* of x and 6. (*If* the equation were written as $4x + 6 = 100$, the 4 would be multiplied by the x only.)

31. *Fund Raising*   A student group is selling boxes of greeting cards at a profit of \$ 1.75 each. The group needs \$2000 to have enough money for a trip to Washington, D.C. How many boxes does the group need to sell to earn \$2000?

Solution

Verbal model: $\boxed{\text{Profit per box}} \cdot \boxed{\text{Number of boxes}} = \boxed{\text{Amount earned}}$

Labels: Profit per box $= 1.75$ (dollars)
Number of boxes $= x$
Amount earned $= 2000$ (dollars)

Equation: $1.75x = 2000$

33. *Average Speed*   After traveling for 3 hours, your family is still 25 miles from completing a 160-mile trip (see figure). What was your average speed during the first 3 hours?

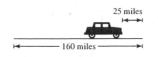
25 miles

160 miles

Solution

Verbal model: $\boxed{\begin{array}{c}\text{Distance}\\\text{traveled}\end{array}} + \boxed{\begin{array}{c}\text{Remaining}\\\text{distance}\end{array}} = \boxed{\begin{array}{c}\text{Total}\\\text{distance}\end{array}}$

$\boxed{\begin{array}{c}\text{Travel}\\\text{rate}\end{array}} \cdot \boxed{\begin{array}{c}\text{Travel}\\\text{time}\end{array}} + \boxed{\begin{array}{c}\text{Remaining}\\\text{distance}\end{array}} = \boxed{\begin{array}{c}\text{Total}\\\text{distance}\end{array}}$

Labels: Travel rate $= x$ (miles per hour)
Travel time $= 3$ (hours)
Remaining distance $= 25$ (miles)
Total distance $= 160$ (miles)

Equation: $x \cdot 3 + 25 = 160$ or $3x + 25 = 160$

Note: Remember the formula $d = rt$; distance equals rate times time.

35. Write a verbal description of the equation $x + 8 = 25$. Use words only; do not use the variable. (There is more than one correct answer.)

Solution

The sum of a number and 8 is 25.

37. Write a verbal description of the equation $2(x - 5) = 12$. Use words only; do not use the variable. (There is more than one correct answer.)

Solution

Twice the difference of a number and 5 is 12.

39. Simplify the algebraic expression $3x - 2(x - 5)$.

Solution

$$3x - 2(x - 5) = 3x - 2x + 10$$
$$= x + 10$$

41. Translate the phrase "Four more than twice a number" into an algebraic expression. (Let x represent the number.)

Solution

$4 + 2x$ or $2x + 4$

43. *Volume*   At the beginning of the day, a gasoline tank was full. The tank holds 20 gallons. At the end of the day the fuel gauge indicates that the tank is $\frac{5}{8}$ full. How many gallons of gasoline were used?

Solution

$$20 - 20\left(\frac{5}{8}\right) = \frac{20}{1} - \frac{20(5)}{8} = \frac{20(8)}{1(8)} - \frac{20(5)}{8}$$
$$= \frac{160}{8} - \frac{100}{8} = \frac{160 - 100}{8}$$
$$= \frac{60}{8} = \frac{15(4)}{2(4)}$$
$$= \frac{15}{2} \text{ or } 7\frac{1}{2}$$

Thus, 7.5 gallons of gasoline were used.

45. *Geometry*   Write expressions for the perimeter and area of the figure in the text book.

Solution

Perimeter: The perimeter of a square is 4 times the length of a side.

$$4\left(\frac{3x}{2}\right) = \frac{4}{1}\left(\frac{3x}{2}\right)$$
$$= \frac{12x}{2} = \frac{6x(2)}{1(2)} = 6x$$

Area: The area of a square is the product of the length of a side times itself; the area is the square of the length of a side.

$$\left(\frac{3x}{2}\right)^2 = \frac{3^2 x^2}{2^2} = \frac{9x^2}{4}$$

47. Decide whether the value of x is a solution of the equation $2x + 4 = 2$ when (a) $x = 0$ and (b) $x = -1$.

Solution

(a) $x = 0$

$$2(0) + 4 \stackrel{?}{=} 2$$

$$0 + 4 \stackrel{?}{=} 2$$

$$4 \neq 2$$

0 *is not* a solution.

(b) $x = -1$

$$2(-1) + 4 \stackrel{?}{=} 2$$

$$-2 + 4 \stackrel{?}{=} 2$$

$$2 = 2$$

-1 *is* a solution.

49. Decide whether the value of x is a solution to the equation $x + 3 = 2(x - 4)$ when (a) $x = 11$ and (b) $x = -5$.

Solution

(a) $x = 11$

$$11 + 3 \stackrel{?}{=} 2(11 - 4)$$

$$14 \stackrel{?}{=} 2(7)$$

$$14 = 14$$

11 *is* a solution.

(b) $x = -5$

$$-5 + 3 \stackrel{?}{=} 2(-5 - 4)$$

$$-2 \stackrel{?}{=} 2(-9)$$

$$-2 \neq -18$$

-5 *is not* a solution.

51. Decide whether the value of x is a solution to the equation $2x + 10 = 7(x + 1)$ when (a) $x = \frac{3}{5}$ and (b) $x = \frac{2}{3}$.

Solution

(a) $x = \frac{3}{5}$

$$2\left(\tfrac{3}{5}\right) + 10 \stackrel{?}{=} 7\left(\tfrac{3}{5} + 1\right)$$

$$\tfrac{6}{5} + \tfrac{10}{1} \stackrel{?}{=} 7\left(\tfrac{3}{5} + \tfrac{5}{5}\right)$$

$$\tfrac{6}{5} + \tfrac{50}{5} \stackrel{?}{=} \tfrac{7}{1}\left(\tfrac{8}{5}\right)$$

$$\tfrac{56}{5} = \tfrac{56}{5}$$

$\frac{3}{5}$ *is* a solution.

(b) $x = \frac{2}{3}$

$$2\left(\tfrac{2}{3}\right) + 10 \stackrel{?}{=} 7\left(\tfrac{2}{3} + 1\right)$$

$$\tfrac{4}{3} + \tfrac{10}{1} \stackrel{?}{=} 7\left(\tfrac{2}{3} + \tfrac{3}{3}\right)$$

$$\tfrac{4}{3} + \tfrac{30}{3} \stackrel{?}{=} \tfrac{7}{1}\left(\tfrac{5}{3}\right)$$

$$\tfrac{34}{3} \neq \tfrac{35}{3}$$

$\frac{2}{3}$ *is not* a solution.

53. Decide whether the value of x is a solution of the equation $\dfrac{4}{x} + \dfrac{2}{x} = 1$ when (a) $x = 0$ and (b) $x = 6$.

Solution

(a) $x = 0$

$$\tfrac{4}{0} + \tfrac{2}{0} \stackrel{?}{=} 1$$

Division by 0 is undefined. 0 *is not* a solution.

(b) $x = -1$

$$\tfrac{4}{6} + \tfrac{2}{6} \stackrel{?}{=} 1$$

$$\tfrac{6}{6} = 1$$

6 *is* a solution.

55. Use a calculator to determine whether the value of x is a solution of the equation $x + 3 = 3.5x$ when (a) $x = 1.2$ and (b) $x = 4.8$.

Solution

(a) $x = 1.2$

$$1.2 + 3 \overset{?}{=} 3.5(1.2)$$

$$4.2 = 4.2$$

1.2 *is* a solution.

(b) $x = 4.8$

$$4.8 + 3 \overset{?}{=} 3.5(4.8)$$

$$7.8 \neq 16.8$$

4.8 *is not* a solution.

57. Use a calculator to determine whether the value of x is a solution of the equation $x(22 - 5x) = 17$ when (a) $x = 1$ and (b) $x = 3.4$.

Solution

(a) $x = 1$

$$1[22 - 5(1)] \overset{?}{=} 17$$

$$1[22 - 5] \overset{?}{=} 17$$

$$1(17) = 17$$

1 *is* a solution.

(b) $x = 3.4$

$$3.4[22 - 5(3.4)] \overset{?}{=} 17$$

$$3.4[22 - 17] \overset{?}{=} 17$$

$$3.4(5) \overset{?}{=} 17$$

$$17 = 17$$

3.5 *is* a solution.

59. Justify each step of the following solution.

$$\frac{2}{3}x = 12$$

$$\frac{3}{2}x\left(\frac{2}{3}x\right) = \frac{3}{2}(12)$$

$$x = 18$$

Solution

$$\frac{2}{3}x = 12 \qquad \text{Given equation}$$

$$\frac{3}{2}x\left(\frac{2}{3}x\right) = \frac{3}{2}(12) \qquad \text{Multiply both sides by 3/2.}$$

$$x = 18 \qquad \text{Simplify. (solution)}$$

61. Justify each step of the following solution.

$$x = -2(x + 3)$$

$$x = -2x - 6$$

$$2x + 2 = 2x - 2x - 6$$

$$3x = 0 - 6$$

$$3x = -6$$

$$\frac{3x}{3} = \frac{-6}{3}$$

$$x = -2$$

Solution

$$x = -2(x + 3) \qquad \text{Given equation}$$

$$x = -2x - 6 \qquad \text{Distributive Property}$$

$$2x + x = 2x - 2x - 6 \qquad \text{Add } 2x \text{ to each side.}$$

$$3x = 0 - 6 \qquad \text{Combine like terms.}$$

$$3x = -6 \qquad \text{Additive identity property}$$

$$\frac{3x}{3} = \frac{-6}{3} \qquad \text{Divide each side by 3.}$$

$$x = -2 \qquad \text{Simplify. (solution)}$$

63. *Computer Purchase* Construct an equation for the following word problem. Do not solve the equation.

You have $3650 saved for the purchase of a new computer that will cost $4532. How much more must you save?

Solution

Verbal model: | Amount saved | + | Additional savings needed | = | Computer cost |

Labels: Amount saved = $3650
               Additional savings needed = x (dollars)
               Computer cost = $4532

Equation: $3650 + x = 4532$

Note: Remember that there are other equivalent ways of writing this equation.

65. *Travel Costs* Construct an equation for the following word problem. Do not solve the equation.

A company pays its sales representatives 25 cents per mile for the use of their personal cars. A sales representative submitted a bill to be reimbursed for $105.75 for driving. How many miles did the sales representative drive?

Solution

Verbal model: | Cost per mile | · | Number of miles | = | Cost for driving |

Labels: Cost per mile = 0.25 (dollars per mile)
               Number of miles = x
               Cost for driving = 105.75 (dollars)

Equation: $0.25x = 105.75$

67. Construct an equation for the following word problem. Do not solve the equation.

The sum of a number and 12 is 45. What is the number?

Solution

Verbal model: | Unknown number | + | 12 | = | 45 |

Label: Unknown number = x

Equation: $x + 12 = 45$

69. Construct an equation for the following word problem. Do not solve the equation.

Find a number such that two times the number decreased by 14 equals the number divided by 3.

Solution

Verbal model: $2 \cdot$ | Number | $- 14 = \dfrac{\boxed{\text{Number}}}{3}$

Labels: Number = x

Equation: $2x - 14 = \dfrac{x}{3}$

71. Construct an equation for the following word problem. Do not solve the equation.

The sum of three consecutive even integers is 18. Find the first number.

Solution

Verbal model: ┌─────────────┐ ┌─────────────┐ ┌─────────────┐
│First consecutive│ + │Second consecutive│ + │Third consecutive│ = │18│
│even integer │ │even integer │ │even integer │
└─────────────┘ └─────────────┘ └─────────────┘

Labels: First consecutive even integer $= 2n$
Second consecutive even integer $= 2n + 2$
Third consecutive even integer $= 2n + 4$

Equation: $2n + (2n + 2) + (2n + 4) = 18$

73. *Dimensions of a Mirror*   Construct an equation for the following word problem. Do not solve the equation.

The width of a rectangular mirror is one-third its length, as shown in the figure. The perimeter of the mirror is 96 inches. What are the dimensions of the mirror?

Solution

Verbal model: │2│ · │Length of mirror│ + │2│ · │Width of mirror│ = │Perimeter of mirror│

Labels: Length of mirror $= l$ (inches)

Width of mirror $= \dfrac{l}{3}$ (inches)

Perimeter of mirror $= 96$ (inches)

Equation: $2l + 2\left(\dfrac{l}{3}\right) = 96$

Note: The width could also be written as $\frac{1}{3}l$.

75. *Price of a Product*   Construct an equation for the following word problem. Do not solve the equation.

The price of a product has increased by $45 over the past year. It is now selling for $375. What was the price one year ago?

Solution

Verbal model: │Original price│ + │Increase in price│ = │Current price│

Labels: Original price $= x$ (dollars)
Increase in price $= 45$ (dollars)
Current price $= 375$ (dollars)

Equation: $x + 45 = 375$

77. *Annual Depreciation* Construct an equation for the following word problem. Do not solve the equation.

A corporation buys equipment with an initial purchase price of $750,000. It is estimated that its useful life will be 3 years and at that time its value will be $75,000. The total depreciation is divided equally among the three years. Determine the amount of depreciation declared each year.

Solution

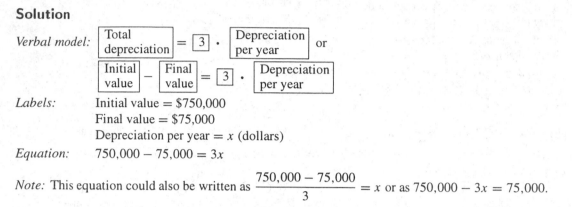

Verbal model:

Labels: Initial value = $750,000
Final value = $75,000
Depreciation per year = x (dollars)

Equation: $750,000 - 75,000 = 3x$

Note: This equation could also be written as $\dfrac{750,000 - 75,000}{3} = x$ or as $750,000 - 3x = 75,000$.

79. *Average Speed* After traveling for 4 hours, you are still 24 miles from completing a 200-mile trip. It will take one-half hour to travel the last 24 miles. Find the average speed during the first 4 hours of the trip.

Solution

Verbal model:

| Distance traveled | + | Remaining distance | = | Total distance | or

| Rate of travel | · | Time of travel | + | Remaining distance | = | Total distance |

Labels: Rate of travel = r (miles per hour)
Time of travel = 4 (hours)
Remaining distance = 24 (miles)
Total distance = 200 (miles)

Equation: $r(4) + 24 = 200$
$4r + 24 = 200$

Note: The one-half hour required to travel the last 24 miles is unnecessary information.

81. Write a verbal description of the equation $9 - x^2 = 0$. Use words only; do not use the variable. (There is more than one correct answer.)

Solution

Nine decreased by the square of a number is equal to 0.

83. Write a verbal description of the equation $10(x - 3) = 8x$. Use words only; do not use the variable. (There is more than one correct answer.)

Solution

The product of 10 and the difference of a number and 3 is 8 times the number.

If a number is decreased by 3 and the difference is multiplied by 10, the result is 8 times the number.

85. Simplify the following expression. State the units of the simplified value.

$$\frac{3 \text{ dollars}}{\text{unit}} \cdot (5 \text{ units})$$

Solution

$$\frac{3 \text{ dollars}}{\text{unit}} \cdot (5 \text{ units}) = 15 \text{ dollars}$$

87. Simplify the following expression. State the units of the simplified value.

$$\frac{3 \text{ dollars}}{\text{pound}} \cdot (5 \text{ pounds})$$

Solution

$$\frac{3 \text{ dollars}}{\text{pound}} \cdot (5 \text{ pounds}) = 15 \text{ dollars}$$

89. Simplify the following expression. State the units of the simplified value.

$$\frac{5 \text{ feet}}{\text{second}} \cdot \frac{60 \text{ seconds}}{\text{minute}} \cdot (20 \text{ minutes})$$

Solution

$$\frac{5 \text{ feet}}{\cancel{\text{second}}} \cdot \frac{60 \cancel{\text{ seconds}}}{\cancel{\text{minute}}} \cdot (20 \cancel{\text{ minutes}}) = 6000 \text{ feet}$$

Review Exercises for Chapter 2

1. Identify the terms and coefficients of $4 - \frac{1}{2}x^3$.

Solution

Terms: $\quad 4, -\frac{1}{2}x^3$

Coefficients: $\quad 4, -\frac{1}{2}$

3. Identify the terms and coefficients of $y^2 - 10yz + \frac{2}{3}z^2$.

Solution

Terms: $\quad y^2, -10yz, \frac{2}{3}z^2$

Coefficients: $\quad 1, -10, \frac{2}{3}$

5. Rewrite the product $5z \cdot 5z \cdot 5z$ in exponential form.

Solution

$5z \cdot 5z \cdot 5z = (5z)^3$ or $5^3 z^3$

7. Rewrite the product $a(b - c) \cdot a(b - c)$ in exponential form.

Solution

$a(b - c) \cdot a(b - c) = [a(b - c)]^2$ or $a^2(b - c)^2$

9. Simplify the expression $(x^3)^2$.

Solution

$(x^3)^2 = x^{3 \cdot 2} = x^6$

11. Simplify the expression $t^4(-2t^2)$.

Solution

$t^4(-2t^2) = -2t^{4+2} = -2t^6$

13. Simplify the expression $(xy)(-5x^2y^3)$.

Solution

$$(xy)(-5x^2y^3) = -5x^{1+2}y^{1+3}$$
$$= -5x^3y^4$$

15. Simplify the expression $(-2y^2)^3(8y)$.

Solution

$$(-2y^2)^3(8y) = (-2)^3(y^2)^3(8y)$$
$$= -8y^{2 \cdot 3}(8y)$$
$$= -8 \cdot 8 \cdot y^6 \cdot y$$
$$= -64y^{6+1}$$
$$= -64y^7$$

17. *Depreciation*   You pay P dollars for new equipment. Its value after 5 years is given by

$$P\left(\tfrac{9}{10}\right)\left(\tfrac{9}{10}\right)\left(\tfrac{9}{10}\right)\left(\tfrac{9}{10}\right)\left(\tfrac{9}{10}\right).$$

Simplify this expression.

Solution

$$P\left(\tfrac{9}{10}\right)\left(\tfrac{9}{10}\right)\left(\tfrac{9}{10}\right)\left(\tfrac{9}{10}\right)\left(\tfrac{9}{10}\right) = P\left(\tfrac{9}{10}\right)^5$$

19. Identify the rule of algebra illustrated by the equation

$$xy \cdot \frac{1}{xy} = 1.$$

Solution

Multiplicative Inverse Property

21. Identify the rule of algebra illustrated by the equation $(x - y)(2) = 2(x - y)$.

Solution

Commutative Property of Multiplication

23. Identify the rule of algebra illustrated by the equation $2x + (3y - z) = (2x + 3y) - z$.

Solution

Associative Property of Addition

25. Use the Distributive Property to expand the expression $4(x + 3y)$.

Solution

$4(x + 3y) = 4x + 4 \cdot 3y = 4x + 12y$

27. Use the Distributive Property to expand the expression $-5(2u - 3v)$.

Solution

$-5(2u - 3v) = (-5)(2u) - (-5)(3v) = -10u + 15v$

29. Use the Distributive Property to expand the expression $x(8x + 5y)$.

Solution

$x(8x + 5y) = x(8x) + x(5y) = 8x^2 + 5xy$

31. Simplify the expression $-(-a + 3b)$.

Solution

$-(-a + 3b) = a - 3b$

Note: The sign of *each* term is changed.

The expression $-(-a + 3b)$ can be written as

$(-1)(-a + 3b) = (-1)(-a) + (-1)(3b) = a - 3b$.

33. Simplify the expression $3a - 5a$.

Solution

$3a - 5a = (3 - 5)a = -2a$

35. Simplify the expression $3p - 4q + q + 8p$.

Solution

$$3p - 4q + q + 8p = 3p + 8p - 4q + q$$
$$= (3 + 8)p + (-4 + 1)q$$
$$= 11p + (-3)q$$
$$= 11p - 3q$$

37. Simplify the expression $\frac{1}{4}s - 6t + \frac{7}{2}s + t$.

Solution

$$\frac{1}{4}s - 6t + \frac{7}{2}s + t = \frac{1}{4}s + \frac{7}{2}s - 6t + t$$
$$= \left(\frac{1}{4} + \frac{14}{4}\right)s + (-6 + 1)t$$
$$= \frac{15}{4}s - 5t$$

39. Simplify the expression $x^2 + 3xy - xy + 4$.

Solution

$$x^2 + 3xy - xy + 4 = x^2 + (3 - 1)xy + 4$$
$$= x^2 + 2xy + 4$$

41. Simplify the expression $5x - 5y + 3xy - 2x + 2y$.

Solution

$$5x - 5y + 3xy - 2x + 2y = 5x - 2x + 3xy - 5y + 2y$$
$$= (5 - 2)x + 3xy + (-5 + 2)y$$
$$= 3x + 3xy - 3y$$

43. Simplify the expression $5\left(1 + \dfrac{r}{n}\right)^2 - 2\left(1 + \dfrac{r}{n}\right)^2$.

Solution

$$5\left(1 + \frac{r}{n}\right)^2 - 2\left(1 + \frac{r}{n}\right)^2 = (5 - 2)\left(1 + \frac{r}{n}\right)^2 = 3\left(1 + \frac{r}{n}\right)^2$$

45. Simplify the expression $5(u - 4) + 10$.

Solution

$$5(u - 4) + 10 = 5u - 5 \cdot 4 + 10$$
$$= 5u - 20 + 10$$
$$= 5u - 10$$

47. Simplify the expression $3s - (r - 2s)$.

Solution

$$3s - (r - 2s) = 3s - r + 2s$$
$$= 3s + 2s - r$$
$$= (3 + 2)s - r$$
$$= 5s - r$$

49. Simplify the expression $10 - [8(5 - x) + 2]$.

Solution

$$10 - [8(5 - x) + 2] = 10 - [40 - 8x + 2]$$
$$= 10 - 40 + 8x - 2$$
$$= 8x + 10 - 40 - 2$$
$$= 8x - 32$$

51. Simplify the expression $2[x + 2(y - x)]$.

Solution

$$2[x + 2(y - x)] = 2[x + 2y - 2x]$$
$$= 2x + 4y - 4x$$
$$= (2 - 4)x + 4y$$
$$= -2x + 4y$$

53. Simplify the expression $-3(1 - 10z) + 2(1 - 10z)$.

Solution

$$-3(1 - 10z) + 2(1 - 10z) = (-3 + 2)(1 - 10z)$$
$$= -1(1 - 10z)$$
$$= -1 + 10z$$

Note: The parentheses could be removed first.

$$-3(1 - 10z) + 2(1 - 10z) = -3 \cdot 1 - (-3)(10z) + 2 \cdot 1 - 2(10z)$$
$$= -3 + 30z + 2 - 20z = -3 + 2 + 30z - 20z$$
$$= -1 + (30 - 20)z = -1 + 10z$$

55. Simplify the expression $\frac{1}{3}(42 - 18z) - 2(8 - 4z)$.

Solution

$$\frac{1}{3}(42 - 18z) - 2(8 - 4z) = 14 - 6z - 16 + 8z$$
$$= -6z + 8z + 14 - 16$$
$$= (-6 + 8)z + 14 - 16$$
$$= 2z - 2$$

57. Evaluate the expression $x^2 - 2x + 5$ when (a) $x = 0$ and (b) $x = 2$.

Solution

(a) When $x = 0$, the value of $x^2 - 2x + 5$ is $0^2 - 2 \cdot 0 + 5 = 0 - 0 + 5 = 5$.

(b) When $x = 2$, the value of $x^2 - 2x + 5$ is $2^2 - 2 \cdot 2 + 5 = 4 - 4 + 5 = 5$.

59. Evaluate the expression $y^2 - y(y + 1)$ when (a) $y = -1$ and (b) $y = 2$.

Solution

(a) When $y = -1$, the value of the expression $y^2 - y(y + 1)$ is $(-1)^2 - (-1)(-1 + 1) = 1 + 1(0) = 1 + 0 = 1$.

(b) When $y = 2$, the value of the expression $y^2 - y(y + 1)$ is $2^2 - 2(2 + 1) = 4 - 2(3) = 4 - 6 = -2$.

61. Translate the phrase "Two thirds of a number, plus 5" into an algebraic expression. (Let x represent the number.)

Solution

$\frac{2}{3}x + 5$

63. Translate the phrase "Ten less than twice a number" into an algebraic expression. (Let x represent the number.)

Solution

$2x - 10$

65. Translate the phrase "Fifty, increased by the product of 7 and a number" into an algebraic expression. (Let x represent the number.)

Solution

$50 + 7x$

67. Translate the phrase "The sum of a number and 10 divided by 8" into an algebraic expression. (Let x represent the number.)

Solution

$\dfrac{x + 10}{8}$

69. Translate the phrase "The sum of the square of a number and 64" into an algebraic expression. (Let x represent the number.)

Solution

$x^2 + 64$

71. Write a verbal description of the algebraic expression $x + 3$ without using the variable. (There is more than one correct answer.)

Solution

The sum of a number and three *or* a number increased by three

73. Write a verbal description of the algebraic expression $(y - 2)/3$ without using the variable. (There is more than one correct answer.)

Solution

A number is decreased by two and the result is divided by three *or* two is subtracted from a number and the result is divided by three.

75. *Income Tax*   The income tax rate on a taxable income of I dollars is 28%. Write an algebraic expression that represents the total amount of income tax. (To find 28% of a quantity, multiply the quantity by 0.28.)

Solution

The amount of tax is a product.

Verbal model: | Percent of income tax | · | Taxable income |

Labels: Percent of income tax $= 0.28$ (in decimal form)
Taxable income $= I$ (dollars)

Algebraic expression: $0.28I$ (dollars)

77. *Area*   The front of a built-in refrigerator has a width of w feet and a height that is 3 feet greater than the width (see figure). Write an algebraic expression that represents the area of the front of the refrigerator.

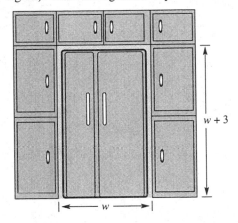

$w + 3$

w

Solution

The area of the front of the refrigerator is a product.

Verbal model: | Width of refrigerator | $\cdot$ | Height of refrigerator |

Labels: Width of refrigerator $= w$ (feet)
Height of refrigerator $= w + 3$ (feet)

Algebraic expression: $w(w + 3)$ or $w^2 + 3w$ (square feet)

79. *Sum*   Write an algebraic expression that represents the sum of three consecutive odd integers, the first of which is $2n - 1$.

Solution

The expression is a sum.

Verbal description: | First consecutive odd integer | $+$ | Second consecutive odd integer | $+$ | Third consecutive odd integer |

Labels: First consecutive odd integer $= 2n - 1$
Second consecutive odd integer $= 2n + 1$
Third consecutive odd integer $= 2n + 3$

Algebraic expression: $(2n - 1) + (2n + 1) + (2n + 3)$ or $6n + 3$

(*Note:* Each successive odd integer is 2 *more* than the previous odd integer; $(2n - 1) + 2 = 2n + 1$ and $(2n + 1) + 2 = 2n + 3$.)

81. *Geometry*   The face of a tape deck has the dimensions shown in the figure in the textbook. Find an algebraic expression that represents the area of the face of the tape deck. (*Hint:* The area is given by the difference of the areas of two rectangles.)

Solution

The area is a difference of products.

Verbal description: | Length of larger rectangle | $\cdot$ | Width of larger rectangle | $-$ | Length of smaller rectangle | $\cdot$ | Width of smaller rectangle |

Labels: Length of larger rectangle $= 6x$
Width of larger rectangle $= 2x$
Length of smaller rectangle $= 3x$
Width of smaller rectangle $= x$

Algebraic expression: $6x \cdot 2x - 3x \cdot x$ or $12x^2 - 3x^2$ or $9x^2$ (square units)

83. *Finding a Pattern* (a) Complete the table, shown in the textbook, in which the third and fourth rows are the differences between consecutive entries of the preceding rows. (b) Describe the patterns for the third and fourth rows of the table.

Solution

(a)

n	0	1	2	3	4	5
$n^2 + 3n + 2$	2	6	12	20	30	42
Differences:		4	6	8	10	12
Differences:			2	2	2	2

When $n = 0$, the value of $n^2 + 3n + 2$ is $0^2 + 3 \cdot 0 + 2 = 0 + 0 + 2 = 2$.
When $n = 1$, the value of $n^2 + 3n + 2$ is $1^2 + 3 \cdot 1 + 2 = 1 + 3 + 2 = 6$.
When $n = 2$, the value of $n^2 + 3n + 2$ is $2^2 + 3 \cdot 2 + 2 = 4 + 6 + 2 = 12$.
When $n = 3$, the value of $n^2 + 3n + 2$ is $3^2 + 3 \cdot 3 + 2 = 9 + 9 + 2 = 20$.
When $n = 4$, the value of $n^2 + 3n + 2$ is $4^2 + 3 \cdot 4 + 2 = 16 + 12 + 2 = 30$.
When $n = 5$, the value of $n^2 + 3n + 2$ is $5^2 + 3 \cdot 5 + 2 = 25 + 15 + 2 = 42$.

(b) In the third row, the differences are consecutive even integers. In the fourth row, each difference is two.

85. Decide whether the value of x is a solution of the equation $5x + 6 = 36$ when (a) $x = 3$ and (b) $x = 6$.

Solution

(a) $x = 3$

$5(3) + 6 \overset{?}{=} 36$

$15 + 6 \overset{?}{=} 36$

$21 \neq 36$

3 *is not* a solution.

(b) $x = 6$

$5(6) + 6 \overset{?}{=} 36$

$30 + 6 \overset{?}{=} 36$

$36 = 36$

6 *is* a solution.

87. Decide whether the value of x is a solution of the equation $3x - 12 = x$ when (a) $x = -1$ and (b) $x = 6$.

Solution

(a) $x = -1$

$3(-1) - 12 \overset{?}{=} -1$

$-3 - 12 \overset{?}{=} -1$

$-15 \neq -1$

-1 *is not* a solution.

(b) $x = 6$

$3(6) - 12 \overset{?}{=} 6$

$18 - 12 \overset{?}{=} 6$

$6 = 6$

6 *is* a solution.

89. Determine whether the value of x is a solution of the equation $4(2 - x) = 3(2 + x)$ when (a) $x = \frac{2}{7}$ and (b) $x = -\frac{2}{3}$.

Solution

(a) $4\left(2 - \frac{2}{7}\right) \overset{?}{=} 3\left(2 + \frac{2}{7}\right)$

$\quad 4\left(\frac{14}{7} - \frac{2}{7}\right) \overset{?}{=} 3\left(\frac{14}{7} + \frac{2}{7}\right)$

$\quad\quad 4\left(\frac{12}{7}\right) \overset{?}{=} 3\left(\frac{16}{7}\right)$

$\quad\quad\quad \frac{48}{7} = \frac{48}{7}$

$\frac{2}{7}$ *is* a solution.

(b) $4\left(2 - \left(-\frac{2}{3}\right)\right) \overset{?}{=} 3\left(2 + \left(-\frac{2}{3}\right)\right)$

$\quad 4\left(\frac{6}{3} + \frac{2}{3}\right) \overset{?}{=} 3\left(\frac{6}{3} - \frac{2}{3}\right)$

$\quad\quad 4\left(\frac{8}{3}\right) \overset{?}{=} 3\left(\frac{4}{3}\right)$

$\quad\quad\quad \frac{32}{3} \neq 4$

$-\frac{2}{3}$ *is not* a solution.

91. Decide whether the value of x is a solution of the equation $\dfrac{4}{x} - \dfrac{2}{x} = 5$ when (a) $x = -1$ and (b) $x = \dfrac{2}{5}$.

Solution

(a) $\dfrac{4}{-1} - \dfrac{2}{-1} \overset{?}{=} 5$

$\quad -4 - (-2) \overset{?}{=} 5$

$\quad\quad -4 + 2 \overset{?}{=} 5$

$\quad\quad\quad -2 \neq 5$

-1 *is not* a solution.

(b) $\dfrac{4}{2/5} - \dfrac{2}{2/5} \overset{?}{=} 5$

$\quad 4\left(\frac{5}{2}\right) - 2\left(\frac{5}{2}\right) \overset{?}{=} 5$

$\quad\quad 10 - 5 \overset{?}{=} 5$

$\quad\quad\quad 5 = 5$

$\frac{2}{5}$ *is* a solution.

93. Decide whether the value of x is a solution of the equation $x(x - 7) = -12$ when (a) $x = 3$ and (b) $x = 4$.

Solution

(a) $3(3 - 7) \overset{?}{=} -12$

$\quad 3(-4) \overset{?}{=} -12$

$\quad\quad -12 = -12$

$\quad 3$ *is* a solution.

(b) $4(4 - 7) \overset{?}{=} -12$

$\quad 4(-3) \overset{?}{=} -12$

$\quad\quad -12 = -12$

$\quad 4$ *is* a solution.

95. *Sum*   Write an expression that represents the statement "The sum of a number and its reciprocal is $\frac{37}{6}$. (Identify the letters you choose as labels.)

Solution

Verbal model: $\boxed{\text{Unknown number}}$ + $\boxed{\text{Reciprocal of number}}$ = $\boxed{\dfrac{37}{6}}$

Labels: Unknown number = x

Reciprocal = $\dfrac{1}{x}$

Equation: $x + \dfrac{1}{x} = \dfrac{37}{6}$

97. *Geometry*   Write an equation that represents the following statement. (Identify the letters you choose as labels.)

The area of the shaded region in the figure is 24 square inches.

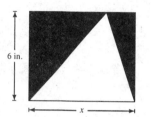

Solution

Verbal model: $\boxed{\text{Area of rectangle}}$ − $\boxed{\text{Area of triangle}}$ = $\boxed{\text{Area of shaded region}}$

Labels: Area of rectangle = $x(6)$ or $6x$ (square inches)
Area of triangle $\frac{1}{2}(x)(6)$ or $3x$ (square inches)
Shaded area = 24 (square inches)

Equation: $6x - \frac{1}{2}(x)(6) = 24$ or $6x - 3x = 24$
This equation could also be simplified to $3x = 24$.

Note: The area of a rectangle is the product of its length and width ($A = lw$). The area of a triangle is one-half the product of its base and height $\left(A = \frac{1}{2}bh\right)$.

Test for Chapter 2

1. Identify the terms and coefficients of the expression $2x^2 - 7xy + 3y^3$.

Solution

Terms: $2x^2, -7xy, 3y^3$
Coefficients: $2, -7, 3$

2. Rewrite the product $x \cdot (x + y) \cdot x \cdot (x + y) \cdot x$ in exponential form.

Solution

$x \cdot (x + y) \cdot x \cdot (x + y) \cdot x = x^3(x + y)^2$

3. Identify the rule of algebra demonstrated in the expression $(5x)y = 5(xy)$.

Solution

Associative Property of Multiplication

4. Identify the rule of algebra demonstrated in the expression $2 + (x - y) = (x - y) + 2$.

Solution

Commutative Property of Addition

5. Identify the rule of algebra demonstrated in the expression $7xy - 7xy = 0$.

Solution

Additive Inverse Property

6. Identify the rule of algebra demonstrated in the expression $1 \cdot (x + 5) = (x + 5)$.

Solution

Multiplicative Identity Property

7. Use the Distributive Property to expand the expression $3(x + 8)$.

Solution

$3(x + 8) = 3x + 24$

8. Use the Distributive Property to expand the expression $-y(3 - 2y)$.

Solution

$-y(3 - 2y) = -3y + 2y^2$

9. Simplify the expression $(c^2)^4$.

Solution

$$(c^2)^4 = c^{2 \cdot 4}$$
$$= c^8$$

10. Simplify the expression $-5uv(2u^3)$.

Solution

$$-5uv(2u^3) = -5 \cdot 2 \cdot u \cdot u^3 \cdot v$$
$$= -10u^{1+3}v$$
$$= -10u^4v$$

11. Simplify the expression $3b - 2a + a - 10b$.

Solution

$$3b - 2a + a - 10b = (-2 + 1)a + (3 - 10)b$$
$$= -a - 7b$$

12. Simplify the expression $15(u - v) - 7(u - v)$.

Solution

$$15(u - v) - 7(u - v) = (15 - 7)(u - v)$$
$$= 8(u - v)$$
$$= 8u - 8v$$

Note: This problem can also be worked as follows.

$$15(u - v) - 7(u - v) = 15u - 15v - 7u + 7v$$
$$= (15 - 7)u + (-15 + 7)v$$
$$= 8u - 8v$$

13. Simplify the expression $3z - (4 - z)$.

Solution

$$3z - (4 - z) = 3z - 4 + z$$
$$= 3z + z - 4$$
$$= 4z - 4$$

14. Simplify the expression $2[10 - (t + 1)]$.

Solution

$$2[10 - (t + 1)] = 2[10 - t - 1]$$
$$= 2[10 - 1 - t]$$
$$= 2[9 - t]$$
$$= 18 - 2t$$

15. Evaluate the expression when $x = 3$ and $y = -12$.

(a) $x^3 - 2$ (b) $x^2 + 4(y + 2)$

Solution

(a) When $x = 3$, the value of $x^3 - 2$ is $3^3 - 2 = 27 - 2 = 25$.

(b) When $x = 3$ and $y = -12$, the value of $x^2 + 4(y + 2)$ is $3^2 + 4(-12 + 2) = 9 + 4(-10) = 9 - 40 = -31$.

16. Explain why it is not possible to evaluate $(a + 2b)/(3a - b)$ when $a = 2$ and $b = 6$.

Solution

When $a = 2$ and $b = 6$, the value of $(a + 2b)/(3a - b)$ is undefined because

$$\frac{2 + 2(6)}{3(2) - 6} = \frac{2 + 12}{6 - 6} = \frac{14}{0}$$

and division by zero is undefined.

17. Translate the phrase, "one-fifth of a number, increased by two," into an algebraic expression. Let n represent the number.

Solution

$\frac{1}{5}n + 2$ or $\frac{n}{5} + 2$

18. (a) Write expressions for the perimeter and area of the rectangle shown in the textbook.

(b) Simplify the expressions.

(c) Identify the unit of measure for each expression.

(d) Evaluate each expression when $w = 12$ feet.

Solution

(a) The perimeter of a rectangle is twice the length plus twice the width.

Perimeter: $2(2w - 4) + 2(w)$

The area of a rectangle is the product of the length and the width.

Area: $(2w - 4)w$

(b) Perimeter: $2(2w - 4) + 2w = 4w - 8 + 2w = 6w - 8$
Area: $(2w - 4)w = 2w^2 - 4w$

(c) The perimeter is measured in units of length, such as feet, meters, inches, etc. The area is measured in square units, such as square centimeters, square inches, square yards, etc.

(d) When $w = 12$ feet, the perimeter $6w - 8 = 6(12) - 8$

$$= 72 - 8$$
$$= 64 \text{ feet.}$$

When $w = 12$ feet, the area $2w^2 - 4w = 2(12)^2 - 4(12)$

$$= 2(144) - 48$$
$$= 288 - 48$$
$$= 240 \text{ square feet.}$$

19. Write an algebraic expression for the income from a concert if the prices of the tickets for adults and children were $3 and $2, respectively. Let n represent the number of adults in attendance and m represent the number of children.

Solution

The concert income is a sum of products.

Verbal description: | Price of adult's ticket | · | Number of adults | $+$ | Price of child's ticket | · | Number of children |

Labels: Price of adult's ticket = 3 (dollars)
Number of adults = n
Price of child's ticket = 2 (dollars)
Number of children = m

Algebraic expression: $3n + 2m$ (dollars)

20. Determine whether the values of x are solutions of $6(3 - x) - 5(2x - 1) = 7$ when (a) $x = -2$ and (b) $x = 1$.

Solution

(a) $6(3 - (-2)) - 5(2(-2) - 1) \stackrel{?}{=} 7$

$6(3 + 2) - 5(-4 - 1) \stackrel{?}{=} 7$

$6(5) - 5(-5) \stackrel{?}{=} 7$

$30 + 25 \stackrel{?}{=} 7$

$55 \neq 7$

-2 *is not* a solution.

(b) $6(3 - 1) - 5(2 \cdot 1 - 1) \stackrel{?}{=} 7$

$6(2) - 5(2 - 1) \stackrel{?}{=} 7$

$12 - 5(1) \stackrel{?}{=} 7$

$12 - 5 \stackrel{?}{=} 7$

$7 = 7$

1 *is* a solution.

CHAPTER THREE
Linear Equations and Problem Solving

 3.1 Solving Linear Equations

7. Solve the equation $x + 6 = 14$ mentally.

Solution

$x = 8$

9. Solve the equation $4s = 12$ mentally.

Solution

$s = 3$

11. Justify each step of the following solution.

$$5x + 15 = 0$$
$$5x + 15 - 15 = 0 - 15$$
$$5x = -15$$
$$\frac{5x}{5} = \frac{-15}{5}$$
$$x = -3$$

Solution

$5x + 15 = 0$	Given equation
$5x + 15 - 15 = 0 - 15$	Subtract 15 from both sides.
$5x = -15$	Combine like terms.
$\dfrac{5x}{5} = \dfrac{-15}{5}$	Divide both sides by 5.
$x = -3$	Simplify (solution)

13. Solve the equation $8x - 2 = 20$ and check your solution.

Solution

$$8x - 2 = 20$$
$$8x - 2 + 2 = 20 + 2$$
$$8x = 22$$
$$\frac{8x}{8} = \frac{22}{8}$$
$$x = \frac{22}{8}$$
$$x = \frac{11}{4}$$

15. Solve the equation $10 - 4x = -6$ and check your solution.

Solution

$$10 - 4x = -6$$
$$10 - 10 - 4x = -6 - 10$$
$$-4x = -16$$
$$\frac{-4x}{-4} = \frac{-16}{-4}$$
$$x = 4$$

17. Solve the equation $-5x = 30$ and check your solution.

Solution

$$-5x = 30$$
$$\frac{-5x}{-5} = \frac{30}{-5}$$
$$x = -6$$

19. Solve the equation $6x - 4 = 0$ and check your solution.

Solution

$$6x - 4 = 0$$

$$6x - 4 + 4 = 0 + 4$$

$$6x = 4$$

$$\frac{6x}{6} = \frac{4}{6}$$

$$x = \frac{4}{6}$$

$$x = \frac{2}{3}$$

21. Solve the equation $4 - 7x = 5x$ and check your solution.

Solution

$$4 - 7x = 5x$$

$$4 - 7x + 7x = 5x + 7x$$

$$4 = 12x$$

$$\frac{4}{12} = \frac{12x}{12}$$

$$\frac{1}{3} = x$$

23. Solve the equation $15x - 3 = 15 - 3x$ and check your solution.

Solution

$$15x - 3 = 15 - 3x$$

$$15x + 3x - 3 = 15 + 3x - 3x$$

$$18x - 3 = 15$$

$$18x - 3 + 3 = 15 + 3$$

$$18x = 18$$

$$\frac{18x}{18} = \frac{18}{18}$$

$$x = 1$$

25. Solve the equation $-6t = 0$ and check your solution.

Solution

$$-6t = 0$$

$$\frac{-6t}{-6} = \frac{0}{-6}$$

$$t = 0$$

27. Solve the equation $t - \frac{1}{3} = \frac{1}{2}$ and check your solution.

Solution

$$t - \tfrac{1}{3} = \tfrac{1}{2}$$

$$t - \tfrac{1}{3} + \tfrac{1}{3} = \tfrac{1}{2} + \tfrac{1}{3}$$

$$t = \tfrac{3}{6} + \tfrac{2}{6}$$

$$t = \tfrac{5}{6}$$

29. Solve the equation $2s + \frac{3}{2} = 2s + 2$ and check your solution.

Solution

$$2s + \tfrac{3}{2} = 2s + 2$$

$$2s - 2s + \tfrac{3}{2} = 2s - 2s + 2$$

$$\tfrac{3}{2} = 2 \text{ (False)}$$

This equation has no solution.

31. Solve the equation $2y - 18 = -5y - 4$ and check your solution.

Solution

$$2y - 18 = -5y - 4$$
$$2y + 5y - 18 = -5y + 5y - 4$$
$$7y - 18 = -4$$
$$7y - 18 + 18 = -4 + 18$$
$$7y = 14$$
$$\frac{7y}{7} = \frac{14}{7}$$
$$y = 2$$

33. Solve the equation $0.234x + 1 = 2.805$. Round the solution to two decimal places.

Solution

$$0.234x + 1 = 2.805$$
$$0.234x + 1 - 1 = 2.805 - 1$$
$$0.234x = 1.805$$
$$\frac{0.234x}{0.234} = \frac{1.805}{0.234}$$
$$x \approx 7.71 \text{ (Rounded)}$$

35. Solve the following equation. Round the solution to two decimal places.

$$\frac{x}{3.155} = 2.850$$

Solution

$$\frac{x}{3.155} = 2.850$$
$$3.155\left(\frac{x}{3.155}\right) = 3.155(2.850)$$
$$x \approx 8.99 \text{ (Rounded)}$$

37. *Geometry* · The length of a rectangular tennis court is 6 feet more than twice the width (see figure). The length is 78 feet. What is the width?

Solution

Verbal model: $\boxed{\text{Length of court}} = \boxed{2} \cdot \boxed{\text{Width of court}} + \boxed{6}$

Labels: Length = 78 (feet)
Width = w (feet)

Equation: $$78 = 2w + 6$$
$$78 - 6 = 2w + 6 - 6$$
$$72 = 2w$$
$$\frac{72}{2} = \frac{2w}{2}$$
$$36 = w$$

The width of the court is 36 feet.

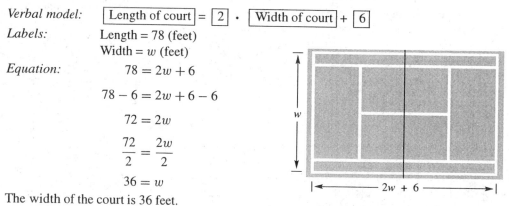

39. Find a number such that the sum of that number and 45 is 75.

Solution

$$x + 45 = 75$$
$$x + 45 - 45 = 75 - 45$$
$$x = 30$$

41. Evaluate the expression $5 + 4 \cdot 3$.

Solution

$$5 + 4 \cdot 3 = 5 + 12 = 17$$

43. Simplify the expression $-3(3x - 2y) + 5y$.

Solution

$$-3(3x - 2y) + 5y = -9x + 6y + 5y$$
$$= -9x + 11y$$

45. Translate the following sentence into an algebraic expression. (Let x represent the number.)

A number is decreased by 10 and the difference is doubled.

Solution

$(x - 10)2$ or $2(x - 10)$

47. Solve the equation $x - 9 = 4$ mentally.

Solution

$x = 13$

49. Solve the equation $7y = 28$ mentally.

Solution

$y = 4$

51. Justify each step of the following solution.

$$-2x = 8$$
$$\frac{-2x}{-2} = \frac{8}{-2}$$
$$x = -4$$

Solution

$$-2x = 8 \qquad \text{Given equation}$$
$$\frac{-2x}{-2} = \frac{8}{-2} \qquad \text{Divide both sides by } -2.$$
$$x = -4 \qquad \text{Simplify (solution)}$$

53. Solve the equation $-14x = 42$ and check your solution.

Solution

$$-14x = 42$$
$$\frac{-14x}{-14} = \frac{42}{-14}$$
$$x = -3$$

55. Solve the equation $25x - 4 = 46$ and check your solution.

Solution

$$25x - 4 = 46$$
$$25x - 4 + 4 = 46 + 4$$
$$25x = 50$$
$$\frac{25x}{25} = \frac{50}{25}$$
$$x = 2$$

57. Solve the equation $3y - 2 = 2y$ and check your solution.

Solution

$$3y - 2 = 2y$$
$$3y - 3y - 2 = 2y - 3y$$
$$-2 = -y$$
$$(-1)(-2) = (-1)(-y)$$
$$2 = y$$

59. Solve the equation $4 - 5t = 16 + t$ and check your solution.

Solution

$$4 - 5t = 16 + t$$
$$4 - 5t - t = 16 + t - t$$
$$4 - 6t = 16$$
$$4 - 4 - 6t = 16 - 4$$
$$-6t = 12$$
$$\frac{-6t}{-6} = \frac{12}{-6}$$
$$t = -2$$

61. Solve the equation $-3t = 0$ and check your solution.

Solution

$$-3t = 0$$
$$\frac{-3t}{-3} = \frac{0}{-3}$$
$$t = 0$$

63. Solve the equation $-3t + 5 = -3t$ and check your solution.

Solution

$$-3t + 5 = -3t$$
$$-3t + 3t + 5 = -3t + 3t$$
$$5 = 0 \text{ (False)}$$

The original equation has no solution.

65. Solve the equation $2x + 4 = -3x + 6$ and check your solution.

Solution

$$2x + 4 = -3x + 6$$
$$2x + 3x + 4 = -3x + 3x + 6$$
$$5x + 4 = 6$$
$$5x + 4 - 4 = 6 - 4$$
$$5x = 2$$
$$\frac{5x}{5} = \frac{2}{5}$$
$$x = \frac{2}{5}$$

67. Solve the equation $2x = -3x$ and check your solution.

Solution

$$2x = -3x$$
$$2x + 3x = -3x + 3x$$
$$5x = 0$$
$$\frac{5x}{5} = \frac{0}{5}$$
$$x = 0$$

69. Solve the equation $2x - 5 + 10x = 3$ and check your solution.

Solution

$$2x - 5 + 10x = 3$$
$$12x - 5 = 3$$
$$12x - 5 + 5 = 3 + 5$$
$$12x = 8$$
$$\frac{12x}{12} = \frac{8}{12}$$
$$x = \frac{2}{3}$$

71. Solve the equation $x/3 = 10$ and check your solution.

Solution

$$\frac{x}{3} = 10$$

$$3\left(\frac{x}{3}\right) = 3 \cdot 10$$

$$x = 30$$

Note: $3\left(\dfrac{x}{3}\right) = \dfrac{3}{1} \cdot \dfrac{x}{3} = \dfrac{3x}{3} = x$

73. Solve the equation $x - \frac{1}{3} = \frac{4}{3}$ and check your solution.

Solution

$$x - \frac{1}{3} = \frac{4}{3}$$

$$x - \frac{1}{3} + \frac{1}{3} = \frac{4}{3} + \frac{1}{3}$$

$$x = \frac{5}{3}$$

75. Solve the equation $3x + \frac{1}{4} = \frac{3}{4}$ and check your solution.

Solution

$$3x + \frac{1}{4} = \frac{3}{4}$$

$$3x + \frac{1}{4} - \frac{1}{4} = \frac{3}{4} - \frac{1}{4}$$

$$3x = \frac{2}{4}$$

$$3x = \frac{1}{2}$$

$$\frac{1}{3}(3x) = \frac{1}{3}\left(\frac{1}{2}\right)$$

$$x = \frac{1}{6}$$

77. Solve the equation $0.02x - 0.96 = 1.50$. Round the solution to two decimal places.

Solution

$$0.02x - 0.96 = 1.50$$

$$0.02x - 0.96 + 0.96 = 1.50 + 0.96$$

$$0.02x = 2.46$$

$$\frac{0.02x}{0.02} = \frac{2.46}{0.02}$$

$$x = 123.00$$

79. Solve the following equation. Round the solution to two decimal places.

$$\frac{x}{3.25} + 1 = 2.08$$

Solution

$$\frac{x}{3.25} + 1 = 2.08$$

$$\frac{x}{3.25} + 1 - 1 = 2.08 - 1$$

$$\frac{x}{3.25} = 1.08$$

$$\frac{x}{3.25}(3.25) = 1.08(3.25)$$

$$x = 3.51$$

81. The sum of two consecutive odd integers is 72. Find the two integers.

Solution

Verbal model: | First consecutive odd integer | + | Second consecutive odd integer | = | 72 |

Labels: First consecutive odd integer $= 2n + 1$
              Second consecutive odd integer $= 2n + 3$

Equation: $(2n + 1) + (2n + 3) = 72$

$$2n + 1 + 2n + 3 = 72$$

$$4n + 4 = 72$$

$$4n + 4 - 4 = 72 - 4$$

$$4n = 68$$

$$\frac{4n}{4} = \frac{68}{4}$$

$$n = 17$$

$$2n + 1 = 2(17) + 1 = 34 + 1 = 35$$

$$2n + 3 = 2(17) + 3 = 34 + 3 = 37$$

The integers are 35 and 37.

83. *Geometry* The perimeter of a rectangle is 240 inches. Find the dimensions of the rectangle if the length is twice the width.

Solution

Verbal model: 2 | Length | + 2 | Width | = | Perimeter |

Labels: Width $= w$ (inches)
              Length $= 2w$ (inches)
              Perimeter $= 240$ (inches)

Equation: $2(2w) + 2w = 240$

$$4w + 2w = 240$$

$$6w = 240$$

$$w = 40 \text{ and } 2w = 80$$

Thus, the width of the rectangle is 40 inches and the length is 80 inches.

85. *Construction*   You are asked to cut a 12-foot board into three pieces. Two pieces are to have the same length and the third is to be twice as long as the others. How long are the pieces?

Solution

Verbal model: $\boxed{\text{First board length}} + \boxed{\text{Second board length}} + \boxed{\text{Third board length}} = \boxed{\text{Original board length}}$

Labels: First board length $= x$ (feet)
Second board length $= x$ (feet)
Third board length $= 2x$ (feet)
Original board length $= 12$ (feet)

Equation: $x + x + 2x = 12$

$$4x = 12$$

$$x = 3 \text{ and } 2x = 6$$

Thus, the first two pieces are 3 feet long and the third piece is 6 feet long.

87. *Car Repair*   The bill for the repair of your car was $415. The cost for parts was $265. The cost for labor was $25 per hour. How many hours did the repair work take?

Solution

Verbal model: $\boxed{\text{Cost for parts}} + \boxed{\text{Labor cost per hour}} \cdot \boxed{\text{Number of hours of labor}} = \boxed{\text{Total bill}}$

Labels: Cost for parts $= 265$ (dollars)
Labor cost per hour $= 25$ (dollars per hour)
Number of hours of labor $= x$ (hours)
Total bill $= 415$ (dollars)

Equation: $265 + 25x = 415$

$$265 - 265 + 25x = 415 - 265$$

$$25x = 150$$

$$\frac{25x}{25} = \frac{150}{25}$$

$$x = 6$$

Thus, the repair work took 6 hours.

89. *Finding a Pattern* The length of a rectangle is t times its width, as shown in the figure in the textbook. The rectangle has a perimeter of 1200 meters, which implies that $2w + 2(tw) = 1200$ where w is the width of the rectangle.

(a) Complete the table shown in the textbook.

(b) Use the completed table to draw a conclusion concerning the area of a rectangle of given perimeter as its length increases relative to its width.

Solution

(a)

t	1	1.5	2	3	4	5
Width	300	240	200	150	120	100
Length	300	360	400	450	480	500
Area	90,000	86,400	80,000	67,500	57,600	50,000

$t = 1$:
$$P = 2w + 2(tw)$$
$$1200 = 2w + 2(1 \cdot w)$$
$$1200 = 2w + 2w$$
$$1200 = 4w$$
$$\frac{1200}{4} = \frac{4w}{4}$$
$$300 = w$$
$$\text{Length} = tw = 1(300) = 300$$
$$\text{Area} = (300)(300) = 90{,}000$$

$t = 1.5$:
$$1200 = 2w + 2(1.5w)$$
$$1200 = 2w + 3w$$
$$1200 = 5w$$
$$\frac{1200}{5} = \frac{5w}{5}$$
$$240 = w$$
$$\text{Length} = tw = 1.5(240) = 360$$
$$\text{Area} = (240)(360) = 86{,}400$$

$t = 2$:
$$1200 = 2w + 2(2w)$$
$$1200 = 2w + 4w$$
$$1200 = 6w$$
$$\frac{1200}{6} = \frac{6w}{6}$$
$$200 = w$$
$$\text{Length} = tw = 2(200) = 400$$
$$\text{Area} = (400)(200) = 80{,}000$$

$t = 3$:
$$1200 = 2w + 2(3w)$$
$$1200 = 2w + 6w$$
$$1200 = 8w$$
$$\frac{1200}{8} = \frac{8w}{8}$$
$$150 = w$$
$$\text{Length} = tw = 3(150) = 450$$
$$\text{Area} = (450)(150) = 67{,}500$$

$t = 4$:
$$1200 = 2w + 2(4w)$$
$$1200 = 2w + 8w$$
$$1200 = 10w$$
$$\frac{1200}{10} = \frac{10w}{10}$$
$$120 = w$$
$$\text{Length} = tw = 4(120) = 480$$
$$\text{Area} = 480(120) = 57{,}600$$

$t = 5$:
$$1200 = 2w + 2(5w)$$
$$1200 = 2w + 10w$$
$$1200 = 12w$$
$$\frac{1200}{12} = \frac{12w}{12}$$
$$100 = w$$
$$\text{Length} = tw = 5(100) = 500$$
$$\text{Area} = 500(100) = 50{,}000$$

(b) The area of a rectangle of a given perimeter *decreases* as its length increases relative to its width.

3.2 Percents and the Percent Equation

7. Complete the table (shown in the textbook) showing the equivalent forms of a percent.

Solution

Percent	Parts out of 100	Decimal	Fraction
40%	40	0.4	$\frac{2}{5}$

(a) 40% means 40 parts out of 100.

(b) *Verbal model:* $\boxed{\text{Decimal}} \cdot \boxed{100\%} = \boxed{\text{Percent}}$

   Label: Decimal $= x$

   Equation: $x(100\%) = 40\%$

$$x = \frac{40\%}{100\%}$$

$$x = 0.4$$

(c) *Verbal model:* $\boxed{\text{Fraction}} \cdot \boxed{100\%} = \boxed{\text{Percent}}$

   Label: Fraction $= x$

   Equation: $x(100\%) = 40\%$

$$x = \frac{40\%}{100\%}$$

$$x = \frac{2}{5}$$

9. Complete the table (shown in the textbook) showing the equivalent forms of a percent.

Solution

Percent	Parts out of 100	Decimal	Fraction
15.5%	15.5	0.155	$\frac{31}{200}$

(a) *Verbal model:* $\boxed{\text{Decimal}} \cdot \boxed{100\%} = \boxed{\text{Percent}}$

   Label: Percent $= x$

   Equation: $(0.155)(100\%) = x$

$$15.5\% = x$$

(b) 15.5% means 15.5 parts out of 100.

(c) *Verbal model:* $\boxed{\text{Fraction}} \cdot \boxed{100\%} = \boxed{\text{Percent}}$

   Label: Fraction $= x$

   Equation: $x(100\%) = 15.5\%$

$$x = \frac{15.5\%}{100\%}$$

$$x = \frac{155\%}{1000\%} = \frac{31}{200}$$

11. Change 12.5% to a decimal.

Solution

Verbal model: $\boxed{\text{Decimal}} \cdot \boxed{100\%} = \boxed{\text{Percent}}$

Label: Decimal $= x$

Equation: $x(100\%) = 12.5\%$

$$x = \frac{12.5\%}{100\%}$$

$$x = 0.125$$

13. Change 250% to a decimal.

Solution

Verbal model: $\boxed{\text{Decimal}} \cdot \boxed{100\%} = \boxed{\text{Percent}}$

Label: Decimal $= x$

Equation: $x(100\%) = 250\%$

$$x = \frac{250\%}{100\%}$$

$$x = 2.5$$

15. Change 0.075 to a percent.

Solution

Verbal model: $\boxed{\text{Decimal}} \cdot \boxed{100\%} = \boxed{\text{Percent}}$

Label: Percent $= x$

Equation: $0.075(100\%) = x$

$$7.5\% = x$$

17. Change 0.62 to a percent.

Solution

Verbal model: $\boxed{\text{Decimal}} \cdot \boxed{100\%} = \boxed{\text{Percent}}$

Label: Percent $= x$

Equation: $0.62(100\%) = x$

$$62\% = x$$

19. Change $\frac{4}{5}$ to a percent.

Solution

Verbal model: $\boxed{\text{Fraction}} \cdot \boxed{100\%} = \boxed{\text{Percent}}$

Label: Percent $= x$

Equation: $\frac{4}{5}(100\%) = x$

$$80\% = x$$

21. Change $\frac{7}{20}$ to a percent.

Solution

Verbal model: $\boxed{\text{Fraction}} \cdot \boxed{100\%} = \boxed{\text{Percent}}$

Label: Percent $= x$

Equation: $\frac{7}{20}(100\%) = x$

$35\% = x$

23. What percent of the figure is shaded?

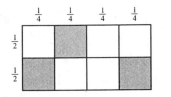

Solution

$\frac{3}{8}$ of the figure is shaded.

$\frac{3}{8}(100\%) = 37\frac{1}{2}\%$ or 37.5%

25. What is 30% of 150?

Solution

Verbal model: $\boxed{\text{What number}} = \boxed{30\% \text{ of } 150}$ $(a = pb)$

Label: $a =$ unknown number

Percent equation: $a = 0.30(150)$

$a = 45$

Therefore, 45 is 30% of 150.

27. What is $\frac{3}{4}\%$ of 56?

Solution

Verbal model: $\boxed{\text{What number}} = \boxed{\frac{3}{4}\% \text{ of } 56}$ $(a = pb)$

Label: $a =$ unknown number

Percent equation: $a = 0.0075(56)$ *Note:* $\frac{3}{4}\% = 0.75\% = 0.0075$

$a = 0.42$

Therefore, 0.42 is $\frac{3}{4}\%$ of 56.

29. $12\frac{1}{2}\%$ of what number is 275?

Solution

Verbal model: $\boxed{275} = \boxed{12\frac{1}{2}\% \text{ of what number}}\ (a = pb)$

Label: $b =$ unknown number

Percent equation: $275 = 0.125b$

$$\frac{275}{0.125} = b$$

$$2200 = b$$

Therefore, 275 is $12\frac{1}{2}\%$ of 2200.

31. 1000 is what percent of 200?

Solution

Verbal model: $\boxed{1000} = \boxed{\text{What percent of 200}}\ (a = pb)$

Label: $p =$ unknown percent (in decimal form)

Percent equation: $1000 = p(200)$

$$\frac{1000}{200} = p$$

$$5 = p$$

Therefore, 1000 is 500% of 200.

33. *Rent Payment* You spend 17% of your monthly income of $2500 for rent. What is your monthly rent payment?

Solution

Verbal model: $\boxed{\text{Rent}} = \boxed{\text{Percent of income}}\ (a = pb)$

Labels: $a =$ rent (dollars)
$p = 0.17$ (percent in decimal form)
$b =$ income $= \$2500$

Percent equation: $a = 0.17(2500)$

$$a = 425$$

Therefore, the monthly rent payment is $425.

35. *Snowfall*   During the winter, there were 120 inches of snow. Of that, 86 inches fell in December. What percent of the snow fell in December?

Solution

Verbal model: $\boxed{\text{December snowfall}} = \boxed{\text{Percent of winter snowfall}}\ (a = pb)$

Labels: $a =$ December snowfall $= 86$ (inches)
$p =$ unknown percent (in decimal form)
$b =$ winter snowfall $= 120$ (inches)

Percent equation: $86 = p(120)$

$$\frac{86}{120} = p$$

$$0.71\overline{6} = p$$

Therefore, $71.\overline{6}\%$ or $71\frac{2}{3}\%$ of the snow fell in December.

37. *Defective Parts*   A quality control engineer tested several parts and found two to be defective. The engineer reported that 2.5% were defective. How many were tested?

Solution

Verbal model: $\boxed{\text{Defective parts}} = \boxed{\text{Percent of sample parts}}\ (a = pb)$

Labels: $a = $ defective parts $= 2$
$p = 0.025$ (percent in decimal form)
$b = $ sample parts

Percent equation: $2 = 0.025b$

$$\frac{2}{0.025} = b$$

$$80 = b$$

There were 80 parts in the sample.

39. *Membership Drive*   Because of a membership drive for a public television station, the current membership is 125% of what it was a year ago. The current number is 7815. How many members did the station have last year?

Solution

Verbal model: $\boxed{\text{Current membership}} = \boxed{\text{Percent of membership one year ago}}\ (a = pb)$

Labels: $a = $ current membership $= 7815$
$p = 1.25$ (percent in decimal form)
$b = $ membership one year ago

Percent equation: $7815 = 1.25b$

$$\frac{7815}{1.25} = b$$

$$6252 = b$$

The membership one year ago was 6252.

41. *Graphical Estimation*   Every year, approximately 8000 Americans suffer spinal cord injuries. The graph in the textbook classifies the major causes of these injuries. Estimate the number of Americans who enter each of these classifications annually. (Source: *U.S. News & World Report,* January 24, 1994)

Solution

Vehicular accidents: 45% of 8000 $= 0.45(8000) = 3600$
Falls: 22% of 8000 $= 0.22(8000) = 1760$
Acts of violence: 16% of 8000 $= 0.16(8000) = 1280$
Sports injuries: 13% of 8000 $= 0.13(8000) = 1040$
Other: 4% of 8000 $= 0.04(8000) = \ \ 320$

43. Solve the linear equation $2x - 5 = x + 9$. Write a justification for each step of your solution. Then check your solution.

Solution

$2x - 5 = x + 9$	Original equation
$2x - x - 5 = x - x + 9$	Subtract x from each side.
$x - 5 = 9$	Combine like terms.
$x - 5 + 5 = 9 + 5$	Add 5 to each side.
$x = 14$	Combine like terms (solution).

45. Solve the linear equation $2x + \frac{3}{2} = \frac{3}{2}$. Write a justification for each step of your solution. Then check your solution.

Solution

$2x + \dfrac{3}{2} = \dfrac{3}{2}$	Given equation
$2x + \dfrac{3}{2} - \dfrac{3}{2} = \dfrac{3}{2} - \dfrac{3}{2}$	Subtract 3/2 from each side.
$2x = 0$	Combine like terms.
$\dfrac{2x}{2} = \dfrac{0}{2}$	Divide both sides by 2.
$x = 0$	Simplify (solution).

47. *Telephone Charge* A telephone company charges $1.37 for the first minute and $0.95 for each additional minute. Find the cost of a 15-minute call.

Solution

Verbal model: $\boxed{\begin{array}{c}\text{Cost of}\\ \text{first minute}\end{array}} + \boxed{\begin{array}{c}\text{Cost per}\\ \text{additional minute}\end{array}} \cdot \boxed{\begin{array}{c}\text{Number of}\\ \text{additional minutes}\end{array}} = \boxed{\text{Total cost}}$

Labels: Cost of first minute $= 1.37$ (dollars)
Cost per additional minute $= 0.95$ (dollars per minute)
Number of additional minutes $= 15 - 1 = 14$ (minutes)
Total cost $= x$

Equation: $1.37 + 0.95(14) = x$

$1.37 + 13.30 = x$

$14.67 = x$

Thus, the cost of a 15-minute call is $14.67.

49. Complete the table in the textbook showing the equivalent forms of a percent.

Solution

Percent	Parts out of 100	Decimal	Fraction
63%	63	0.63	$\frac{63}{100}$

(a) 63 parts out of 100 means 63%.

(b) *Verbal model:* $\boxed{\text{Decimal}} \cdot \boxed{100\%} = \boxed{\text{Percent}}$

    Label: Decimal $= x$

    Equation: $x(100\%) = 63\%$

$$x = \frac{63\%}{100\%}$$

$$x = 0.63$$

(c) *Verbal model:* $\boxed{\text{Fraction}} \cdot \boxed{100\%} = \boxed{\text{Percent}}$

    Label: Fraction $= x$

    Equation: $x(100\%) = 63\%$

$$x = \frac{63\%}{100\%}$$

$$x = \frac{63}{100}$$

51. Complete the table in the textbook showing the equivalent forms of a percent.

Solution

Percent	Parts out of 100	Decimal	Fraction
60%	60	0.6	$\frac{3}{5}$

(a) *Verbal model:* $\boxed{\text{Fraction}} \cdot \boxed{100\%} = \boxed{\text{Percent}}$

    Label: Percent $= x$

    Equation: $\frac{3}{5}(100\%) = x$

$$60\% = x$$

(b) 60% means 60 parts out of 100.

(c) *Verbal model:* $\boxed{\text{Decimal}} \cdot \boxed{100\%} = \boxed{\text{Percent}}$

    Label: Decimal $= x$

    Equation: $x(100\%) = 60\%$

$$x = \frac{60\%}{100\%}$$

$$x = 0.6$$

53. Change $\frac{3}{4}\%$ to a decimal. (Round your result to two decimal places.)

Solution

Verbal model: | Decimal | $\cdot$ | 100% | $=$ | Percent |

Label: Decimal $= x$

Equation: $x(100\%) = \frac{3}{4}\%$

$$x = \frac{\frac{3}{4}\%}{100\%}$$

$$x = \frac{0.75\%}{100\%}$$

$$x = 0.0075$$

55. Change 125% to a decimal. (Round your result to two decimal places.)

Solution

Verbal model: | Decimal | $\cdot$ | 100% | $=$ | Percent |

Label: Decimal $= x$

Equation: $x(100\%) = 125\%$

$$x = \frac{125\%}{100\%}$$

$$x = 1.25$$

57. Change 0.20 to a percent.

Solution

Verbal model: | Decimal | $\cdot$ | 100% | $=$ | Percent |

Label: Decimal $= x$

Equation: $0.20(100\%) = x$

$$20\% = x$$

59. Change 2.5 to a percent.

Solution

Verbal model: | Decimal | $\cdot$ | 100% | $=$ | Percent |

Label: Decimal $= x$

Equation: $2.5(100\%) = x$

$$250\% = 0$$

61. Change $\frac{1}{4}$ to a percent.

Solution

Verbal model: | Fraction | $\cdot$ | 100% | $=$ | Percent |

Label: Decimal $= x$

Equation: $\frac{1}{4}(100\%) = x$

$$25\% = x$$

63. Change $\frac{5}{6}$ to a percent.

Solution

Verbal model: $\boxed{\text{Fraction}} \cdot \boxed{100\%} = \boxed{\text{Percent}}$

Label: Decimal $= x$

Equation: $\frac{5}{6}(100\%) = x$

$83\frac{1}{3}\% = x$

65. What is 9.5% of 816?

Solution

Verbal model: $\boxed{\text{What number}} = \boxed{9.5\% \text{ of } 816}$ $(a = pb)$

Label: $a =$ unknown number

Percent equation: $a = 0.095(816)$

$a = 77.52$

Therefore, 77.52 is 9.5% of 816.

67. What is 200% of 88?

Solution

Verbal model: $\boxed{\text{What number}} = \boxed{200\% \text{ of } 88}$ $(a = pb)$

Label: $a =$ unknown number

Percent equation: $a = (2.00)(88)$

$a = 176$

Therefore, 176 is 200% of 88.

69. 43% of what number is 903?

Solution

Verbal model: $\boxed{903} = \boxed{43\% \text{ of what number}}$ $(a = pb)$

Label: $b =$ unknown number

Percent equation: $903 = 0.43b$

$\dfrac{903}{0.43} = b$

$2,100 = b$

Therefore, 903 is 43% of 2100.

71. 450% of what number is 594?

Solution

Verbal model: $\boxed{594} = \boxed{450\% \text{ of what number}} (a = pb)$

Label: b = unknown number

Percent equation: $594 = 4.50b$

$$\frac{594}{4.50} = b$$

$$132 = b$$

Therefore, 594 is 450% of 132.

73. 0.6% of what number is 2.16?

Solution

Verbal model: $\boxed{2.16} = \boxed{0.6\% \text{ of what number}} (a = pb)$

Label: b = unknown number

Percent equation: $2.16 = 0.006b$

$$\frac{2.16}{0.006} = b$$

$$360 = b$$

Therefore, 2.16 is 0.6% of 360.

75. 576 is what percent of 800?

Solution

Verbal model: $\boxed{576} = \boxed{\text{What percent of 800}} (a = pb)$

Label: p = unknown percent (in decimal form)

Percent equation: $576 = p(800)$

$$\frac{576}{800} = p$$

$$0.72 = p$$

Therefore, 576 is 72% of 800.

77. 45 is what percent of 360?

Solution

Verbal model: $\boxed{45} = \boxed{\text{What percent of 360}} (a = pb)$

Label: p = unknown percent (in decimal form)

Percent equation: $45 = p(360)$

$$\frac{45}{360} = p$$

$$0.125 = p$$

Therefore, 45 is 12.5% of 360.

79. What percent of the figure is shaded?

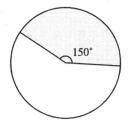

Solution

$\frac{150}{360}$ of the figure is shaded.

$$\frac{150}{360} = \frac{5}{12}$$

$$\frac{5}{12}(100\%) = 41\frac{2}{3}\% \approx 41.67\%$$

81. *Cost of Housing* You budget 30% of your annual after-tax income for housing. Your after-tax income is $32,500. What amount can you spend on housing?

Solution

Verbal model: | Budget for housing | = | Percent of income |

Labels: Budget for housing = a (dollars)
Percent = 0.30 (in decimal form)
Income = 32,500 (dollars)

Equation: $a = 0.30(32,500)$

$a = 9750$

Thus, you can spend $9750 on housing.

83. *Original Price* A coat sells for $250 during a 20% storewide clearance sale. What was the original price of the coat?

Solution

Verbal model: | Original price | − | Percent of original price | = | Sale price |

Labels: Original price = x (dollars)
Percent = 0.20 (in decimal form)
Sale price = 250 (dollars)

Equation: $x - 0.20x = 250$

$0.80x = 250$

$$\frac{0.80x}{0.80} = \frac{250}{0.80}$$

$x = 312.50$

Thus, the original price of the coat was $312.50.

85. *Eligible Voters* The news media reported that 6432 votes were cast in the last election and this represented 63% of the eligible voters of a district. How many eligible voters are in the district?

Solution

Verbal model: $\boxed{\text{Votes cast}} = \boxed{\text{Percent of eligible voters}}$ $(a = pb)$

Labels: a = votes cast = 6432
$p = 0.63$ (percent in decimal form)
b = eligible voters

Percent equation: $6432 = 0.63b$

$$\frac{6432}{0.63} = b$$

$$10{,}210 \approx b$$

There are approximately 10,210 voters in the district.

87. *Interpreting a Table* The table in the textbook shows the number of women scientists in the United States as a percentage of the total number of scientists in each field in 1983 and 1992. (Source: U.S. Bureau of Labor and Statistics)

(a) Find the total number of mathematical and computer scientists (men and women) in 1992.

(b) Find the total number of chemists (men and women) in 1983.

(c) Explain how the number of women in biology can increase while the percent of women in biology decreases.

Solution

(a) *Verbal model:* $\boxed{\begin{array}{c}\text{Number of female}\\\text{math/computer scientists}\end{array}} = \boxed{\begin{array}{c}\text{Percent of all}\\\text{math/computer scientists}\end{array}}$ $(a = pb)$

Labels: a = female math/computer scientists = 313,200
$p = 0.335$ (percent in decimal form)
b = unknown number of math/computer scientists

Equation: $313{,}200 = 0.335(b)$

$$\frac{313{,}200}{0.335} = b$$

$$934{,}925 \approx b$$

Thus, the total number of mathematical and computer scientists in the U.S. in 1992 was approximately 934,925.

(b) *Verbal model:* $\boxed{\begin{array}{c}\text{Number of female}\\\text{chemists in 1983}\end{array}} = \boxed{\begin{array}{c}\text{Percent of all}\\\text{chemists in 1983}\end{array}}$ $(a = pb)$

Labels: a = female chemists = 22,800
$p = 0.233$ (percent in decimal form)
b = unknown number of chemists

Equation: $22{,}800 = 0.233(b)$

$$\frac{22{,}800}{0.233} = b$$

$$97{,}854 \approx b$$

Thus, the total number of chemists in the U.S. in 1983 was approximately 97,854.

(c) The number of women in biology increased from 1983 to 1992, but the percentage of women in biology decreased during the same period. This is because the number of men in biology increased *faster* than the number of women during this period.

89. *Estimation* Figure (a) was put into a photocopier and reduced to produce figure (b). Estimate the percent of reduction. (See figures in textbook.)

Solution

Verbal model: $\boxed{\text{Second height}} = \boxed{\text{What percent of first height}}$ $(a = pb)$

Labels: $a = $ second height $\approx 1\frac{1}{2}$ (inches)
$p = $ percent in decimal form
$b = $ first height ≈ 2 (inches)

Percent equation: $1\frac{1}{2} = p(2)$

$1.5 = 2p$

$\dfrac{1.5}{2} = p$

$0.75 = p$

The second figure is approximately 75% of the first figure. (The original figure has been reduced by 25%.)

3.3   More About Solving Linear Equations

7. Solve the following equation mentally.

$$\frac{x}{10} = \frac{1}{5}$$

Solution

$x = 2$

9. Solve the following equation mentally.

$$\frac{z+2}{3} = 4$$

Solution

$z = 10$

11. Solve the equation $2(x - 3) = 4$ and check your result.

Solution

$2(x - 3) = 4$

$2x - 6 = 4$

$2x - 6 + 6 = 4 + 6$

$2x = 10$

$\dfrac{2x}{2} = \dfrac{10}{2}$

$x = 5$

13. Solve the equation $3 - (2x - 4) = 3$ and check your result.

Solution

$3 - (2x - 4) = 3$

$3 - 2x + 4 = 3$

$-2x + 7 = 3$

$-2x + 7 - 7 = 3 - 7$

$-2x = -4$

$\dfrac{-2x}{-2} = \dfrac{-4}{-2}$

$x = 2$

15. Solve the equation $8(t - 3) = 0$ and check your result.

Solution

$$8(t - 3) = 0$$
$$8t - 24 = 0$$
$$8t - 24 + 24 = 0 + 24$$
$$8t = 24$$
$$\frac{8t}{8} = \frac{24}{8}$$
$$t = 3$$

17. Solve the equation $7 = 3(x + 2) - 3(x - 5)$ and check your result.

Solution

$$7 = 3(x + 2) - 3(x - 5)$$
$$7 = 3x + 6 - 3x + 15$$
$$7 = 21 \quad \text{(False)}$$

Thus, this equation has no solution.

19. Solve the equation $0.6(x + 4) = 2(x + 4)$ and check your result.

Solution

$$0.6(x + 4) = 2(x + 4)$$
$$0.6x + 2.4 = 2x + 8$$
$$10(0.6x + 2.4) = 10(2x + 8)$$
$$6x + 24 = 20x + 80$$
$$6x - 20x + 24 = 20x - 20x + 80$$
$$-14x + 24 = 80$$
$$-14x + 24 - 24 = 80 - 24$$
$$-14x = 56$$
$$\frac{-14x}{-14} = \frac{56}{-14}$$
$$x = -4$$

21. Solve the equation $2(3x + 5) - 7 = 3(5x - 2)$ and check your result.

Solution

$$2(3x + 5) - 7 = 3(5x - 2)$$
$$6x + 10 - 7 = 15x - 6$$
$$6x + 3 = 15x - 6$$
$$6x - 15x + 3 = 15x - 15x - 6$$
$$-9x + 3 = -6$$
$$-9x + 3 - 3 = -6 - 3$$
$$-9x = -9$$
$$\frac{-9x}{-9} = \frac{-9}{-9}$$
$$x = 1$$

23. Solve the following equation and check your result.

$$\frac{x}{2} = \frac{3}{2}$$

Solution

$$\frac{x}{2} = \frac{3}{2}$$
$$2\left(\frac{x}{2}\right) = 2\left(\frac{3}{2}\right)$$
$$x = 3$$

25. Solve the following equation and check your result.

$$\frac{6x}{25} = \frac{3}{5}$$

Solution

$$\frac{6x}{25} = \frac{3}{5}$$
$$30x = 75 \qquad \text{(Cross-multiply)}$$
$$\frac{30x}{30} = \frac{75}{30}$$
$$x = \frac{75}{30} = \frac{5}{2}$$

27. Solve the following equation and check your result.

$$\frac{5x}{4} + \frac{1}{2} = 0$$

Solution

$$\frac{5x}{4} + \frac{1}{2} = 0$$

$$4\left(\frac{5x}{4} + \frac{1}{2}\right) = 4(0)$$

$$4\left(\frac{5x}{4}\right) + 4\left(\frac{1}{2}\right) = 0$$

$$5x + 2 = 0$$

$$5x + 2 - 2 = 0 - 2$$

$$5x = -2$$

$$\frac{5x}{5} = \frac{-2}{5}$$

$$x = -\frac{2}{5}$$

29. Solve the following equation and check your result.

$$\frac{100 - 4u}{3} = \frac{5u + 6}{4} + 6$$

Solution

$$\frac{100 - 4u}{3} = \frac{5u + 6}{4} + 6$$

$$12\left(\frac{100 - 4u}{3}\right) = 12\left(\frac{5u + 6}{4} + 6\right)$$

$$\frac{12}{1}\left(\frac{100 - 4u}{3}\right) = \frac{12}{1}\left(\frac{5u + 6}{4}\right) + 12(6)$$

$$4(100 - 4u) = 3(5u + 6) + 72$$

$$400 - 16u = 15u + 18 + 72$$

$$400 - 16u = 15u + 90$$

$$400 - 16u - 15u = 15u - 15u + 90$$

$$400 - 31u = 90$$

$$400 - 400 - 31u = 90 - 400$$

$$-31u = -310$$

$$\frac{-31u}{-31} = \frac{-310}{-31}$$

$$u = 10$$

31. Solve the following equation by first cross-multiplying.

$$\frac{x - 2}{5} = \frac{2}{3}$$

Solution

$$\frac{x - 2}{5} = \frac{2}{3}$$

$$3(x - 2) = 5(2)$$

$$3x - 6 = 10$$

$$3x - 6 + 6 = 10 + 6$$

$$3x = 16$$

$$\frac{3x}{3} = \frac{16}{3}$$

$$x = \frac{16}{3}$$

33. Solve the following equation by first cross-multiplying.

$$\frac{x}{4} = \frac{1 - 2x}{3}$$

Solution

$$\frac{x}{4} = \frac{1 - 2x}{3}$$

$$3(x) = 4(1 - 2x)$$

$$3x = 4 - 8x$$

$$3x + 8x = 4 - 8x + 8x$$

$$11x = 4$$

$$\frac{11x}{11} = \frac{4}{11}$$

$$x = \frac{4}{11}$$

35. *Time to Complete a Task* Two people can complete 80% of a task in t hours, where t must satisfy the equation $\dfrac{t}{10} + \dfrac{t}{15} = 0.8$.
Solve this equation for t.

Solution

$$\frac{t}{10} + \frac{t}{15} = 0.8$$

$$30\left(\frac{t}{10} + \frac{t}{15}\right) = 30(0.8)$$

$$30\left(\frac{t}{10}\right) + 30\left(\frac{t}{15}\right) = 24$$

$$3t + 2t = 24$$

$$5t = 24$$

$$\frac{5t}{5} = \frac{24}{5}$$

$$t = \frac{24}{5} \text{ or } 4.8 \text{ hours}$$

37. *Balancing a Seesaw* Find the position of the fulcrum so the seesaw shown in the figure will balance. ($W_1 = 90$, $W_2 = 60$, and $a = 10$.)

Solution

Use the equation $W_1 x = W_2(a - x)$.

$$W_1 x = W_2(a - x)$$

$$90x = 60(10 - x)$$

$$90x = 600 - 60x$$

$$90x + 60x = 600 - 60x + 60x$$

$$150x = 600$$

$$\frac{150x}{150} = \frac{600}{150}$$

$$x = 4 \text{ feet}$$

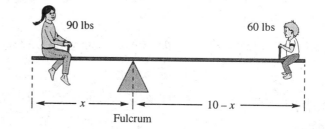

90 lbs   60 lbs

x $10 - x$

Fulcrum

39. Solve the equation $3x - 42 = 0$.

Solution

$$3x - 42 = 0$$

$$3x - 42 + 42 = 0 + 42$$

$$3x = 42$$

$$\frac{3x}{3} = \frac{42}{3}$$

$$x = 14$$

41. Solve the equation $2 - 3x = 14 + x$.

Solution

$$2 - 3x = 14 + x$$

$$2 - 3x - x = 14 + x - x$$

$$2 - 4x = 14$$

$$2 - 2 - 4x = 14 - 2$$

$$-4x = 12$$

$$\frac{-4x}{-4} = \frac{12}{-4}$$

$$x = -3$$

43. What is $\frac{1}{2}$% of 6000?

Solution

Verbal model: $\boxed{\text{What number}} = \boxed{\frac{1}{2}\text{% of } 6000}$ $(a = pb)$

Labels: $a =$ unknown number

Equation: $a = 0.005(6000)$

  $a = 30$

Therefore, 30 is $\frac{1}{2}$%, or 0.5%, of 6000.

45. Solve the equation $\frac{3}{2}x = 9$ mentally.

Solution

$x = 6$

47. Solve the equation $2(y - 4) = 12$ mentally.

Solution

$y = 10$

49. Solve the equation $7(x + 5) = 49$ and check your result.

Solution

$$7(x + 5) = 49$$
$$7x + 35 = 49$$
$$7x + 35 - 35 = 49 - 35$$
$$7x = 14$$
$$\frac{7x}{7} = \frac{14}{7}$$
$$x = 2$$

51. Solve the equation $4 - (z + 6) = 8$ and check your result.

Solution

$$4 - (z + 6) = 8$$
$$4 - z - 6 = 8$$
$$-z - 2 = 8$$
$$-z - 2 + 2 = 8 + 2$$
$$-z = 10$$
$$(-1)(-z) = (-1)(10)$$
$$z = -10$$

53. Solve the equation $-3(t + 5) = 0$ and check your result.

Solution

$$-3(t + 5) = 0$$
$$-3t - 15 = 0$$
$$-3t - 15 + 15 = 0 + 15$$
$$-3t = 15$$
$$\frac{-3t}{-3} = \frac{15}{-3}$$
$$t = -5$$

55. Solve the equation $-3(t + 5) = 6$ and check your result.

Solution

$$-3(t + 5) = 6$$
$$-3t - 15 = 6$$
$$-3t - 15 + 15 = 6 + 15$$
$$-3t = 21$$
$$\frac{-3t}{-3} = \frac{21}{-3}$$
$$t = -7$$

57. Solve the equation $7x - 2(x - 2) = 12$ and check your result.

Solution

$$7x - 2(x - 2) = 12$$
$$7x - 2x + 4 = 12$$
$$5x + 4 = 12$$
$$5x + 4 - 4 = 12 - 4$$
$$5x = 8$$
$$\frac{5x}{5} = \frac{8}{5}$$
$$x = \frac{8}{5}$$

59. Solve the equation $6 = 3(y + 1) - 4(1 - y)$ and check your result.

Solution

$$6 = 3(y + 1) - 4(1 - y)$$
$$6 = 3y + 3 - 4 + 4y$$
$$6 = 7y - 1$$
$$6 + 1 = 7y - 1 + 1$$
$$7 = 7y$$
$$\frac{7}{7} = \frac{7y}{7}$$
$$1 = y$$

61. Solve the equation $7(2x - 1) = 4(1 - 5x) + 6$ and check your result.

Solution

$$7(2x - 1) = 4(1 - 5x) + 6$$
$$14x - 7 = 4 - 20x + 6$$
$$14x - 7 = 10 - 20x$$
$$14x + 20x - 7 = 10 - 20x + 20x$$
$$34x - 7 = 10$$
$$34x - 7 + 7 = 10 + 7$$
$$34x = 17$$
$$\frac{34x}{34} = \frac{17}{34}$$
$$x = \frac{1}{2}$$

63. Solve the following equation and check your result.

$$\frac{y}{5} = \frac{3}{5}$$

Solution

$$\frac{y}{5} = \frac{3}{5}$$
$$5\left(\frac{y}{5}\right) = 5\left(\frac{3}{5}\right)$$
$$y = 3$$

65. Solve the following equation and check your result.

$$\frac{y}{5} = -\frac{3}{10}$$

Solution

$$\frac{y}{5} = -\frac{3}{10}$$
$$5\left(\frac{y}{5}\right) = 5\left(-\frac{3}{10}\right)$$
$$y = -\frac{3}{2}$$

Note: This could also be solved using cross-multiplication.

67. Solve the following equation and check your result.

$$\frac{t + 4}{6} = \frac{2}{3}$$

Solution

$$\frac{t + 4}{6} = \frac{2}{3}$$
$$3(t + 4) = 12 \qquad \text{(Cross-multiply)}$$
$$3t + 12 = 12$$
$$3t + 12 - 12 = 12 - 12$$
$$3t = 0$$
$$\frac{3t}{3} = \frac{0}{3}$$
$$t = 0$$

69. Solve the equation $0.2x - 0.5x = 1$ and check your result.

Solution

$$0.2x - 0.5x = 1$$

$$-0.3x = 1$$

$$10(-0.3x) = 10(1)$$

$$-3x = 10$$

$$\frac{-3x}{-3} = \frac{10}{-3}$$

$$x = -\frac{10}{3}$$

71. Solve the equation $0.24(z + 5) - 0.03(z + 24) = 0$ and check your result.

Solution

$$0.24(z + 5) - 0.03(z + 24) = 0$$

$$0.24z + 1.20 - 0.03z - 0.72 = 0$$

$$0.21z + 0.48 = 0$$

$$100(0.21z + 0.48) = 100(0)$$

$$21z + 48 = 0$$

$$21z + 48 - 48 = 0 - 48$$

$$21z = -48$$

$$\frac{21z}{21} = \frac{-48}{21}$$

$$z = -\frac{16}{7}$$

73. Solve the following equation by first cross-multiplying.

$$\frac{5x - 4}{4} = \frac{2}{3}$$

Solution

$$\frac{5x - 4}{4} = \frac{2}{3}$$

$$3(5x - 4) = 4(2)$$

$$15x - 12 = 8$$

$$15x - 12 + 12 = 8 + 12$$

$$15x = 20$$

$$\frac{15x}{15} = \frac{20}{15}$$

$$x = \frac{4}{3}$$

75. Solve the following equation by first cross-multiplying.

$$\frac{10 - x}{2} = \frac{x + 4}{5}$$

Solution

$$\frac{10 - x}{2} = \frac{x + 4}{5}$$

$$5(10 - x) = 2(x + 4)$$

$$50 - 5x = 2x + 8$$

$$50 - 5x - 2x = 2x - 2x + 8$$

$$50 - 7x = 8$$

$$50 - 50 - 7x = 8 - 50$$

$$-7x = -42$$

$$\frac{-7x}{-7} = \frac{-42}{-7}$$

$$x = 6$$

77. *Fireplace Construction*   A fireplace is 93 inches wide. Each brick in the fireplace has a length of 8 inches and there is $\frac{1}{2}$ inch of mortar between adjoining bricks. Let n be the number of bricks per row. (See figure in textbook.)

(a) Explain why the number of bricks per row is the solution of the equation $8n + \frac{1}{2}(n-1) = 93$.

(b) Find the number of bricks per row in the fireplace.

Solution

(a) Each brick is 8 inches long, so the n bricks in each row have a combined length of $8n$ inches. There is $\frac{1}{2}$ inch of mortar between adjoining bricks and there are $n-1$ mortar joints between the n bricks; therefore, the mortar joints in each row have a combined length of $\frac{1}{2}(n-1)$ inches. Thus, the $8n$ inches of bricks plus the $\frac{1}{2}(n-1)$ inches of mortar joints equal the 93 inches of the width of the fireplace.

(b)
$$8n + \tfrac{1}{2}(n-1) = 93$$
$$8n + \tfrac{1}{2}n - \tfrac{1}{2} = 93$$
$$2\left(8n + \tfrac{1}{2}n - \tfrac{1}{2}\right) = 2(93)$$
$$16n + n - 1 = 186$$
$$17n - 1 = 186$$
$$17n - 1 + 1 = 186 + 1$$
$$17n = 187$$
$$n = 11$$

Thus, there are 11 bricks per row in the fireplace.

In Exercises 79 and 81, solve the equation $p_1 x + p_2(a - x) = p_3 a$ for x.

79. *Mixture Problem*   Determine the number of quarts of a 10% solution that must be mixed with a 30% solution to obtain 100 quarts of a 25% solution. ($p_1 = 0.1$, $p_2 = 0.3$, $p_3 = 0.25$, and $a = 100$.)

Solution
$$p_1 x + p_2(a - x) = p_3 a$$
$$0.1x + 0.3(100 - x) = 0.25(100)$$
$$0.1x + 30 - 0.3x = 25$$
$$-0.2x + 30 = 25$$
$$-0.2x + 30 - 30 = 25 - 30$$
$$-0.2x = -5$$
$$\frac{-0.2x}{-0.2} = \frac{-5}{-0.2}$$
$$x = 25 \text{ quarts}$$

81. *Mixture Problem*   An 8-quart automobile cooling system is filled with coolant that is 40% antifreeze. Determine the amount that must be withdrawn and replaced with pure antifreeze so that the 8 quarts of coolant will be 50% antifreeze. ($p_1 = 1$, $p_2 = 0.4$, $p_3 = 0.5$, and $a = 8$.)

Solution

$$p_1 x + p_2(a - x) = p_3 a$$

$$1(x) + 0.4(8 - x) = 0.5(8)$$

$$x + 3.2 - 0.4x = 4$$

$$0.6x + 3.2 = 4$$

$$0.6x + 3.2 - 3.2 = 4 - 3.2$$

$$0.6x = 0.8$$

$$\frac{0.6x}{0.6} = \frac{0.8}{0.6}$$

$$x = \frac{4}{3} \text{ or } 1.\overline{3} \text{ quarts}$$

83. *Exploration*   Review Exercises 79 and 81, and describe what p_1, p_2, p_3, and a represent.

Solution

In Exercises 79–81, p_1 and p_2 represent the concentrations of two solutions which are being mixed together; in Exercise 82, they represent the two prices. In Exercises 79–81, p_3 represents the concentration of the resulting mixture; in Exercise 82, it represents the price of the mixture. In all four exercises, a represents the quantity of the resulting mixture.

Mid-Chapter Quiz for Chapter 3

1. Solve the equation $120 - 3y = 0$.

Solution

$$120 - 3y = 0$$

$$120 - 120 - 3y = 0 - 120$$

$$-3y = -120$$

$$\frac{-3y}{-3} = \frac{-120}{-3}$$

$$y = 40$$

2. Solve the equation $10(y - 8) = 0$.

Solution

$$10(y - 8) = 0$$

$$10y - 80 = 0$$

$$10y - 80 + 80 = 0 + 80$$

$$10y = 80$$

$$\frac{10y}{10} = \frac{80}{10}$$

$$y = 8$$

3. Solve the equation $3x + 1 = x + 20$.

Solution

$$3x + 1 = x + 20$$

$$3x - x + 1 = x - x + 20$$

$$2x + 1 = 20$$

$$2x + 1 - 1 = 20 - 1$$

$$2x = 19$$

$$\frac{2x}{2} = \frac{19}{2}$$

$$x = \frac{19}{2}$$

4. Solve the equation $6x + 8 = 8 - 2x$.

Solution

$$6x + 8 = 8 - 2x$$

$$6x + 2x + 8 = 8 - 2x + 2x$$

$$8x + 8 = 8$$

$$8x + 8 - 8 = 8 - 8$$

$$8x = 0$$

$$\frac{8x}{8} = \frac{0}{8}$$

$$x = 0$$

5. Solve the equation $-10x + \frac{2}{3} = \frac{7}{3} - 5x$.

Solution

$$-10x + \frac{2}{3} = \frac{7}{3} - 5x$$

$$3\left(-10x + \frac{2}{3}\right) = 3\left(\frac{7}{3} - 5x\right)$$

$$-30x + 2 = 7 - 15x$$

$$-30x + 15x + 2 = 7 - 15x + 15x$$

$$-15x + 2 = 7$$

$$-15x + 2 - 2 = 7 - 2$$

$$-15x = 5$$

$$\frac{-15x}{15} = \frac{5}{-15}$$

$$x = -\frac{1}{3}$$

6. Solve the following equation.

$$\frac{x}{5} + \frac{x}{8} = 1$$

Solution

$$\frac{x}{5} + \frac{x}{8} = 1$$

$$40\left(\frac{x}{5} + \frac{x}{8}\right) = 40(1)$$

$$40\left(\frac{x}{5}\right) + 40\left(\frac{x}{8}\right) = 40$$

$$8x + 5x = 40$$

$$13x = 40$$

$$\frac{13x}{13} = \frac{40}{13}$$

$$x = \frac{40}{13}$$

7. Solve the following equation.

$$\frac{9 + x}{3} = 15$$

Solution

$$\frac{9 + x}{3} = 15$$

$$3\left(\frac{9 + x}{3}\right) = 3(15)$$

$$9 + x = 45$$

$$9 - 9 + x = 45 - 9$$

$$x = 36$$

8. Solve the equation $4 - 0.3(1 - x) = 7$.

Solution

$$4 - 0.3(1 - x) = 7$$

$$4 - 0.3 + 0.3x = 7$$

$$3.7 + 0.3x = 7$$

$$10(3.7 + 0.3x) = 10(7)$$

$$37 + 3x = 70$$

$$37 - 37 + 3x = 70 - 37$$

$$3x = 33$$

$$\frac{3x}{3} = \frac{33}{3}$$

$$x = 11$$

9. Solve the equation $32.86 - 10.5x = 11.25$. Round the solution to two decimal places. In your own words, explain how to check the solution.

Solution

$$32.86 - 10.5x = 11.25$$

$$32.86 - 32.86 - 10.5x = 11.25 - 32.86$$

$$-10.5x = -21.61$$

$$\frac{-10.5x}{-10.5} = \frac{-21.61}{-10.5}$$

$$x \approx 2.06$$

Comment: Replace the variable x by the number 2.06. Simplify the left-hand side of the equation, and check whether it is *approximately* equal to 11.25, the number on the right-hand side. (The two numbers will not be *equal* because of the rounding.)

10. Solve the equation $(x/5.45) + 3.2 = 12.6$. Round the solution to two decimal places. In your own words, explain how to check the solution.

Solution

$$\frac{x}{5.45} + 3.2 = 12.6$$

$$\frac{x}{5.45} + 3.2 - 3.2 = 12.6 - 3.2$$

$$\frac{x}{5.45} = 9.4$$

$$5.45\left(\frac{x}{5.45}\right) = 5.45(9.4)$$

$$x = 51.23$$

Comment: Replace the variable x by the number 51.23. Simplify the left-hand side of the equation, and check whether it is equal to 12.6, the number on the right-hand side.

11. What is 62% of 25?

Solution

Verbal model: $\boxed{\text{What number}} = \boxed{62\% \text{ of } 25}$ $(a = pb)$

Labels: Unknown number $= a$

Equation: $a = 0.62(25)$

$$a = 15.5$$

Therefore, 62% of 25 is 15.5.

12. What is $\frac{1}{2}$% of 8400?

Solution

Verbal model: $\boxed{\text{What number}} = \boxed{\frac{1}{2}\% \text{ of } 8400}$ $(a = pb)$

Labels: $a =$ unknown number

Equation: $a = 0.005(8400)$

$$a = 42$$

Therefore, 42 is $\frac{1}{2}$%, or 0.5%, of 8400.

13. 300 is what percent of 150?

Solution

Verbal model: $\boxed{300} = \boxed{\text{What percent of 150}} \;(a = pb)$

Label: $p = \text{unknown percent}$

Equation:
$$\frac{300}{150} = p$$
$$2 = p$$
$$200\% = p$$

Therefore, 300 is 200% of 150.

14. 145.6 is 32% of what number?

Solution

Verbal model: $\boxed{145.6} = \boxed{32\% \text{ of what number}} \;(a = pb)$

Label: $b = \text{unknown number}$

Equation:
$$145.6 = 0.32(b)$$
$$\frac{145.6}{0.32} = b$$
$$455 = b$$

Therefore, 145.6 is 32% of 455.

15. The perimeter of a rectangle is 60 meters. Find the dimensions of the rectangle if the length is one and one-half times the width.

Solution

Verbal model: $2 \boxed{\text{Length}} + 2 \boxed{\text{Width}} = \boxed{\text{Perimeter}}$

Labels:
Width $= w$ (meters)
Length $= \frac{3}{2}w$ (meters)
Perimeter $= 60$ (meters)

Equation:
$$2\left(\tfrac{3}{2}w\right) + 2w = 60$$
$$3w + 2w = 60$$
$$5w = 60$$
$$w = 12 \text{ and } \tfrac{3}{2}w = 18$$

Therefore, the length of the rectangle is 18 meters and the width is 12 meters.

16. You have two jobs. In the first job, you work 40 hours a week and earn $7.50 per hour. In the second job you earn $6.00 per hour and can work as many hours as you want. If you want to earn $360 a week, how many hours must you work at the second job?

Solution

Verbal model: $\boxed{\begin{array}{c}\text{Hourly pay} \\ \text{at job 1}\end{array}} \cdot \boxed{\begin{array}{c}\text{Hours} \\ \text{at job 1}\end{array}} + \boxed{\begin{array}{c}\text{Hourly pay} \\ \text{at job 2}\end{array}} \cdot \boxed{\begin{array}{c}\text{Hours} \\ \text{at job 2}\end{array}} = \boxed{\begin{array}{c}\text{Weekly} \\ \text{pay}\end{array}}$

Labels: Hourly pay at job 1 = 7.50 (dollars per hour)
Hours at job 1 = 40 (hours)
Hourly pay at job 2 = 6 (dollars per hour)
Hours at job 2 = x (hours)
Weekly pay = 360 (dollars)

Equation: $7.50(40) + 6(x) = 360$

$$300 + 6x = 360$$

$$6x = 60$$

$$x = 10$$

Therefore, you must work 10 hours per week at the second job.

17. A region has an area of 42 square meters. It must be divided into three subregions so that the second has twice the area of the first, and the third has twice the area of the second. Determine the area of each subregion.

Solution

Verbal model: $\boxed{\text{First area}} + \boxed{\text{Second area}} + \boxed{\text{Third area}} = \boxed{\text{Total area}}$

Labels: First area = x (square meters)
Second area = $2x$ (square meters)
Third area = $2(2x) = 4x$ (square meters)
Total area = 42 (square meters)

Equation: $x + 2x + 4x = 42$

$$7x = 42$$

$$x = 6 \text{ and } 2x = 12, \ 4x = 24$$

Therefore, the area of the first subregion is 6 square meters, the area of the second subregion is 12 square meters, and the area of the third subregion is 24 square meters.

18. To get an A in a course you must have an average of at least 90 points for three tests of 100 points each. For the first two tests, your scores are 84 and 93. What must you score on the third test to earn a 90% average for the course?

Solution

Verbal model: $\frac{1}{3}\left(\boxed{\text{First score}} + \boxed{\text{Second score}} + \boxed{\text{Third score}} \right) = 0.90(100)$

Labels: First score = 84 (points)
Second score = 93 (points)
Third score = x (points)

Equation:

$$\frac{1}{3}(84 + 93 + x) = 0.90(100)$$

$$\frac{1}{3}(84) + \frac{1}{3}(93) + \frac{1}{3}x = 90$$

$$28 + 31 + \frac{1}{3}x = 90$$

$$59 + \frac{1}{3}x = 90$$

$$\frac{1}{3}x = 31$$

$$3\left(\frac{1}{3}x\right) = 3(31)$$

$$x = 93$$

Therefore, you must score 93 on the third test to earn a 90% average.

19. The price of a television set is approximately 108% of what it was two years ago. The current price is $535. What was the approximate price two years ago?

Solution

Verbal model: $\boxed{\text{Current price}} = \boxed{108\% \text{ of price two years ago}}$ $(a = pb)$

Labels: Current price = 535 (dollars)
Price two years ago = b (dollars)

Equation:

$$535 = 1.08(b)$$

$$\frac{535}{1.08} = b$$

$$495.37 \approx b$$

Therefore, two years ago the price was approximately $495.37.

20. The figure in the textbook shows the economic costs per year for Alzheimer's. What percent of the total cost is the value of time of unpaid caregivers? (Source: *American Journal of Public Health*)

Solution

Verbal model: $\boxed{\text{Value of time of unpaid caregivers}} = \boxed{\text{What percent}} \cdot \boxed{\text{Total cost}}$

Labels: Value of time of unpaid caregivers = 37 (billions of dollars)
Unknown percent = p (percent in decimal form)
Total cost = 22 + 24 + 37 (billions of dollars)

Equation: $37 = p(22 + 24 + 37)$

$37 = p(83)$

$\frac{37}{83} = p$

$0.446 \approx p$

Therefore, the value of time of unpaid caregivers is approximately 44.6% of the total economic costs.

3.4 Ratio and Proportion

5. Write the ratio 36 to 9 as a fraction in reduced form.

Solution

$36 \text{ to } 9 = \frac{36}{9} = \frac{4}{1}$

7. Write the ratio 14 : 21 as a fraction in reduced form.

Solution

$14 : 21 = \frac{14}{21} = \frac{2}{3}$

9. Express the ratio 36 inches to 24 inches as a fraction in reduced form. (Use the same units for both quantities.)

Solution

$\frac{36 \text{ inches}}{24 \text{ inches}} = \frac{36}{24} = \frac{3}{2}$

11. Express the ratio 1 quart to 1 gallon as a fraction in reduced form. (Use the same units for both quantities.)

Solution

$\frac{1 \text{ quart}}{1 \text{ gallon}} = \frac{1 \text{ quart}}{4 \text{ quarts}} = \frac{1}{4}$

13. Express the ratio 75 centimeters to 2 meters as a fraction in reduced form. (Use the same units for both quantities.)

Solution

$\frac{75 \text{ centimeters}}{2 \text{ meters}} = \frac{75 \text{ centimeters}}{200 \text{ centimeters}}$

$= \frac{75}{200}$

$= \frac{3}{8}$

15. Express the ratio 90 minutes to 2 hours as a fraction in reduced form. (Use the same units for both quantities.)

Solution

$\frac{90 \text{ minutes}}{2 \text{ hours}} = \frac{90 \text{ minutes}}{120 \text{ minutes}}$

$= \frac{90}{120}$

$= \frac{3}{4}$

17. *Study Hours* Express the following statement as a ratio in reduced form. You study 6 hours per day and are in class 3 hours per day. Find the ratio of the number of study hours to class hours. (Use the same units for both quantities.)

Solution

$\frac{6 \text{ hours}}{3 \text{ hours}} = \frac{6}{3} = \frac{2}{1}$

19. *Price-Earnings Ratio* Express the following statement as a ratio in reduced form. The ratio of the price of a stock to its earnings is called the *price-earnings ratio*. A certain stock sells for $78 per share and earns $6.50 per share. What is the price-earnings ratio?

Solution

$\frac{78 \text{ dollars}}{6.50 \text{ dollars}} = \frac{78}{6.50} = \frac{12}{1}$

21. Find the unit price (in $/oz) for a 20-ounce can of pineapple for 79¢.

Solution

Verbal model: $\boxed{\dfrac{\text{Unit}}{\text{price}}} = \boxed{\dfrac{\text{Total price}}{\text{Total units}}}$

Unit price: $\dfrac{79 \text{ cents}}{20 \text{ oz}} = \dfrac{\$0.79}{20} = \$0.0395$ per ounce

23. Find the unit price (in $/oz) for a 1-pound, 4-ounce loaf of bread for $1.29.

Solution

Verbal model: $\boxed{\dfrac{\text{Unit}}{\text{price}}} = \boxed{\dfrac{\text{Total price}}{\text{Total units}}}$

Total units: 1 pound + 4 ounces = $1(16 \text{ oz}) + 4 \text{ oz} = 16 \text{ oz} + 4 \text{ oz} = 20 \text{ oz}$

Unit price: $\dfrac{\$1.29}{20 \text{ oz}} = \$0.0645$ per ounce

25. Which product has the smaller unit price: a $27\frac{3}{4}$-ounce can of spaghetti sauce for $1.19, or a 32-ounce jar for $1.45?

Solution

The unit price for the smaller jar is

$$\text{Unit price} = \dfrac{\text{Total price}}{\text{Total units}} = \dfrac{\$1.19}{27\frac{3}{4} \text{ oz}} = \dfrac{\$1.19}{27.75 \text{ oz}} \approx \$0.0429 \text{ per ounce.}$$

The unit price for the larger jar is

$$\text{Unit price} = \dfrac{\text{Total price}}{\text{Total units}} = \dfrac{\$1.45}{32 \text{ oz}} \approx \$0.0453 \text{ per ounce.}$$

The $27\frac{3}{4}$-jar has the smaller unit price.

27. Solve the following proportion.

$$\dfrac{3}{5} = \dfrac{y}{20}$$

Solution

$$\dfrac{3}{5} = \dfrac{y}{20}$$

$$20\left(\dfrac{3}{5}\right) = 20\left(\dfrac{y}{20}\right) \qquad \text{(Multiply both sides by 20.)}$$

$$12 = y$$

29. Solve the following proportion.

$$\dfrac{8}{3} = \dfrac{t}{6}$$

Solution

$$\dfrac{8}{3} = \dfrac{t}{6}$$

$$6\left(\dfrac{8}{3}\right) = 6\left(\dfrac{t}{6}\right)$$

$$16 = t$$

31. Solve the following proportion.

$$\frac{x+6}{3} = \frac{x-5}{2}$$

Solution

$$\frac{x+6}{3} = \frac{x-5}{2}$$

$$2(x+6) = 3(x-5) \quad \text{(Cross-multiply.)}$$

$$2x + 12 = 3x - 15$$

$$2x - 3x + 12 = 3x - 3x - 15$$

$$-x + 12 = -15$$

$$-x + 12 - 12 = -15 - 12$$

$$-x = -27$$

$$(-1)(-x) = (-1)(-27)$$

$$x = 27$$

33. *Gasoline Cost* A car uses 20 gallons of gasoline for a trip of 360 miles. How many gallons would be used on a trip of 400 miles?

Solution

Verbal model:
$$\boxed{\frac{\text{Gallons for shorter trip}}{\text{Miles for shorter trip}}} = \boxed{\frac{\text{Gallons for longer trip}}{\text{Miles for longer trip}}}$$

Labels:
Gallons for shorter trip $= 20$
Miles for shorter trip $= 360$
Gallons for longer trip $= x$
Miles for longer trip $= 400$

Proportion:
$$\frac{20}{360} = \frac{x}{400}$$

$$400\left(\frac{20}{360}\right) = 400\left(\frac{x}{400}\right)$$

$$22.\overline{2} = x$$

On the 400-mile trip, $22.\overline{2}$ gallons (or $22\frac{2}{9}$ gallons) would be used.

35. *Amount of Gasoline* The gasoline-to-oil ratio for a two-cycle engine is 40 to 1. How much gasoline is required to produce a mixture that contains one-half pint of oil?

Solution

Verbal model: $\boxed{\text{Gas to oil ratio}} = \boxed{\dfrac{\text{Pints of gas in mixture}}{\text{Pints of oil in mixture}}}$

Labels: Gas to oil ratio $= \dfrac{40}{1}$
Pints of gas in mixture $= x$
Pints of oil in mixture $= \dfrac{1}{2}$

Proportion: $\dfrac{40}{1} = \dfrac{x}{1/2}$

$40 = x \left(\dfrac{2}{1} \right)$ *Note:* $x \div \dfrac{1}{2} = x \left(\dfrac{2}{1} \right).$

$40 = 2x$

$\dfrac{40}{2} = \dfrac{2x}{2}$

$20 = x$

Thus, 20 pints of gasoline or ($2\frac{1}{2}$ gallons) are required for the mixture.

37. *Estimation* Use the map to estimate the distance between Philadelphia and Pittsburgh.

Solution

Verbal model: $\boxed{\dfrac{\text{Inches on map scale}}{\text{Miles represented on scale}}} = \boxed{\dfrac{\text{Inches between cities on mappt}}{\text{Miles between cities}}}$

Labels: Inches on map scale $\approx \dfrac{11}{16}$
Miles represented on scale $= 100$
Inches between cities on map $\approx 1\frac{11}{16}$
Miles between cities $= x$

Proportion: $\dfrac{\frac{11}{16}}{100} = \dfrac{1\frac{11}{16}}{x}$

$\dfrac{0.6875}{100} = \dfrac{1.6875}{x}$

$0.6875x = 1.6875$ (Cross-multiply.)

$x = \dfrac{1.6875}{0.6875}$

$x \approx 245$

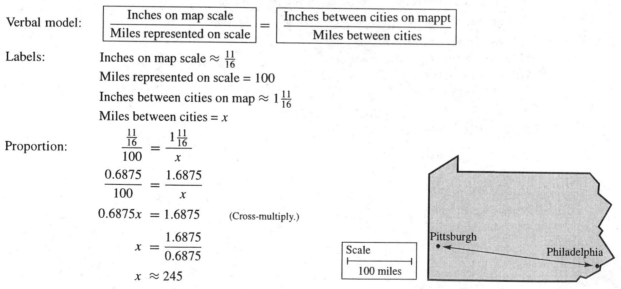

The approximate distance between the cities is 245 miles.

39. *Similar Triangles* The two triangles shown in the textbook are similar. Find the length of x. Use the fact that if two triangles are similar, their corresponding sides are proportional.

Solution

$$\frac{1}{2} = \frac{x}{5}$$

$$5\left(\frac{1}{2}\right) = 5\left(\frac{x}{5}\right) \qquad \text{(Multiply both sides by 5.)}$$

$$\frac{5}{2} = x$$

Note: There are several ways to set up a proportion to solve this question. For example, we could use this proportion instead.

$$\frac{2}{1} = \frac{5}{x}$$

$$2x = 5 \qquad \text{(Cross-multiply.)}$$

$$\frac{2x}{2} = \frac{5}{2}$$

$$x = \frac{5}{2}$$

Some people find proportions easier to solve when the x is in the numerator of one of the ratios, as it was in our original solution. The proportion could also be written as

$$\frac{2}{5} = \frac{1}{x} \text{ or } \frac{5}{2} = \frac{x}{1}.$$

41. Solve the linear equation $50 - z = 15$ and check your solution.

Solution

$$50 - z = 15$$

$$50 - 50 - z = 15 - 50$$

$$-z = -35$$

$$\frac{-z}{-1} = \frac{-35}{-1}$$

$$z = 35$$

43. Solve the following linear equation and check your solution.

$$\frac{x}{6} + \frac{x}{3} = 1$$

Solution

$$\frac{x}{6} + \frac{x}{3} = 1$$

$$18\left(\frac{x}{6} + \frac{x}{3}\right) = 18(1)$$

$$18\left(\frac{x}{6}\right) + 18\left(\frac{x}{3}\right) = 18$$

$$3x + 6x = 18$$

$$9x = 18$$

$$\frac{9x}{9} = \frac{18}{9}$$

$$x = 2$$

45. *Geometry* The perimeter of an isosceles triangle is 120 inches. The third side is $\frac{1}{2}$ the length of the two sides of equal length. Find the lengths of the sides of the triangle.

Solution

Verbal model: | Length of first side | + | Length of second side | + | Length of third side | = | Perimeter of triangle |

Labels: Length of first side $= x$ (inches)
               Length of second side $= x$ (inches)
               Length of third side $= \frac{1}{2}x$ (inches)
               Perimeter $= 120$ (inches)

Equation:

$$x + x + \tfrac{1}{2}x = 120$$

$$2\left(x + x + \tfrac{1}{2}x\right) = 2(120)$$

$$2x + 2x + x = 240$$

$$5x = 240$$

$$x = 48 \text{ and } \tfrac{1}{2}x = 24$$

Therefore, the two equal sides are 48 inches long, and the third side is 24 inches long.

47. Write the ratio 27 to 54 as a fraction in reduced form.

Solution

27 to 54 $= \frac{27}{54} = \frac{1}{2}$

49. Write the ratio 144 : 16 as a fraction in reduced form.

Solution

$144 : 16 = \frac{144}{16} = \frac{9}{1}$

51. Express the ratio 15 feet to 12 feet as a fraction in reduced form. (Use the same units for both quantities.)

Solution

$$\frac{15 \text{ feet}}{12 \text{ feet}} = \frac{15}{12} = \frac{5}{4}$$

53. Express the ratio 7 nickels to 3 quarters as a fraction in reduced form. (Use the same units for both quantities.)

Solution

$$\frac{7 \text{ nickels}}{3 \text{ quarters}} = \frac{7 \text{ nickels}}{15 \text{ nickels}} = \frac{7}{15}$$

Note: This problem could also be done using cents as the common ratio.

$$\frac{7 \text{ nickels}}{3 \text{ quarters}} = \frac{35 \text{ cents}}{75 \text{ cents}} = \frac{35}{75} = \frac{7}{15}$$

55. Express the ratio 3 hours to 90 minutes as a fraction in reduced form. (Use the same units for both quantities.)

Solution

$$\frac{3 \text{ hours}}{90 \text{ minutes}} = \frac{180 \text{ minutes}}{90 \text{ minutes}} = \frac{180}{90} = \frac{2}{1}$$

57. Express the ratio 2 meters to 75 centimeters as a fraction in reduced form. (Use the same units for both quantities.)

Solution

$$\frac{2 \text{ meters}}{75 \text{ centimeters}} = \frac{200 \text{ centimeters}}{75 \text{ centimeters}} = \frac{200}{75} = \frac{8}{3}$$

59. Express the ratio $5\frac{1}{2}$ pints to 2 quarts as a fraction in reduced form. (Use the same units for both quantities.)

Solution

$$\frac{5\frac{1}{2} \text{ pints}}{2 \text{ quarts}} = \frac{11 \text{ cups}}{8 \text{ cups}} = \frac{11}{8}$$

61. *Student-Teacher Ratio* Express the following statement as a ratio in reduced form. There are 2921 students and 127 faculty members at your school. Find the ratio of the number of students to the number of faculty. (Use the same units for both quantities.)

Solution

$$\frac{2921 \text{ students}}{127 \text{ faculty members}} = \frac{23 \text{ people}}{1 \text{ people}} = \frac{23}{1}$$

63. *Gear Ratio* Express the following statement as a ratio in reduced form. The *gear ratio* of two gears is the ratio of the number of teeth in one gear to the number of teeth in the other gear. Find the gear ratio of the larger gear to the smaller gear for the gears shown in the figure in the textbook. (Use the same units for both quantities.)

Solution

$$\frac{45 \text{ teeth}}{30 \text{ teeth}} = \frac{45}{30} = \frac{3}{2}$$

65. *Geometry* Find the ratio of the area of the larger pizza to the area of the smaller pizza in the figure in the textbook. (*Note:* The area of a circle is $A = \pi r^2$.)

Solution

Verbal model: $\boxed{\text{Unknown ratio}} = \boxed{\dfrac{\text{Area of larger pizza}}{\text{Area of smaller pizza}}}$

Labels: Unknown ratio $= x$
Area of larger pizza $= \pi(10)^2$
Area of smaller pizza $= \pi(7)^2$

Equation: $x = \dfrac{\pi(10)^2}{\pi(7)^2} = \dfrac{100\pi}{49\pi} = \dfrac{100}{49}$

Note: The ratio could also be determined by using the decimal approximations of π.

$$x = \frac{\pi(10)^2}{\pi(7)^2} = \frac{(3.14)(10)^2}{(3.14)(7)^2} = \frac{100}{49}$$

67. Which product has the smaller unit price: a 10-ounce package of frozen green beans for 59¢, or a 16-ounce package for 89¢?

Solution

The unit price for the smaller package is

$$\text{Unit price} = \frac{\text{Total price}}{\text{Total units}} = \frac{\$0.59}{10 \text{ oz}} = \$0.059 \text{ per ounce.}$$

The unit price for the larger package is

$$\text{Unit price} = \frac{\text{Total price}}{\text{Total units}} = \frac{\$0.89}{16 \text{ oz}} \approx \$0.056 \text{ per ounce.}$$

The 16-ounce package has the smaller unit price.

69. Which product has the smaller unit price: a 2-liter bottle (67.6 ounces) of soft drink for $1.09, or six 12-ounce cans for $1.69?

Solution

The unit price for the 2-liter bottle is

$$\text{Unit price} = \frac{\text{Total price}}{\text{Total units}} = \frac{\$1.09}{67.6 \text{ oz}} \approx \$0.016 \text{ per ounce.}$$

The unit price for the cans is

$$\text{Unit price} = \frac{\text{Total price}}{\text{Total units}} = \frac{\$1.69}{6(12) \text{ oz}} = \frac{\$1.69}{72 \text{ oz}} \approx \$0.023 \text{ per ounce.}$$

The 2-liter bottle has the smaller unit price.

71. Solve the following proportion.

$$\frac{t}{4} = \frac{25}{2}$$

Solution

$$\frac{t}{4} = \frac{25}{2}$$

$$4\left(\frac{t}{4}\right) = 4\left(\frac{25}{2}\right) \qquad \text{(Multiply both sides by 4.)}$$

$$t = 50$$

73. Solve the following proportion.

$$\frac{x}{5} = \frac{2}{3}$$

Solution

$$\frac{x}{5} = \frac{2}{3}$$

$$5\left(\frac{x}{5}\right) = 5\left(\frac{2}{3}\right) \qquad \text{(Multiply both sides by 5.)}$$

$$x = \frac{10}{3}$$

75. Solve the following proportion.

$$\frac{x+1}{5} = \frac{3}{10}$$

Solution

$$\frac{x+1}{5} = \frac{3}{10}$$

$$10(x+1) = 15 \qquad \text{(Cross-multiply.)}$$

$$10x + 10 = 15$$

$$10x + 10 - 10 = 15 - 10$$

$$10x = 5$$

$$\frac{10x}{10} = \frac{5}{10}$$

$$x = \frac{1}{2}$$

77. Solve the following proportion.

$$\frac{x+2}{8} = \frac{x-1}{3}$$

Solution

$$\frac{x+2}{8} = \frac{x-1}{3}$$

$$3(x+2) = 8(x-1)$$

$$3x + 6 = 8x - 8$$

$$3x - 8x + 6 = 8x - 8x - 8$$

$$-5x + 6 = -8$$

$$-5x + 6 - 6 = -8 - 6$$

$$-5x = -14$$

$$\frac{-5x}{-5} = \frac{-14}{-5}$$

$$x = \frac{14}{5}$$

79. *Building Material* One hundred cement blocks are needed to build a 16-foot wall. How many blocks are needed to build a 40-foot wall?

Solution

Verbal model: $\boxed{\dfrac{\text{Blocks for smaller wall}}{\text{Length of smaller wall}}} = \boxed{\dfrac{\text{Blocks for larger wall}}{\text{Length of smaller wall}}}$

Labels: Blocks for smaller wall $= 100$
Length of smaller wall $= 16$ (feet)
Blocks for larger wall $= x$
Length of larger wall $= 40$ (feet)

Proportion: $\dfrac{100}{16} = \dfrac{x}{40}$

$$40\left(\frac{100}{16}\right) = 40\left(\frac{x}{40}\right)$$

$$250 = x$$

Thus, 250 blocks are needed to build a 40-foot wall.

81. *Real Estate Taxes* The tax on a property with an assessed value of $65,000 is $825. Find the tax on a property with an assessed value of $90,000.

Solution

Verbal model: $\boxed{\dfrac{\text{Tax on first property}}{\text{Value of first property}}} = \boxed{\dfrac{\text{Tax on second property}}{\text{Value of second property}}}$

Labels: Tax on first property = $825
Value of first property = $65,000
Tax on second property = x (dollars)
Value of second property = $90,000

Proportion: $$\frac{825}{65,000} = \frac{x}{90,000}$$

$$90,000\left(\frac{825}{65,000}\right) = 90,000\left(\frac{x}{90,000}\right)$$

$$\$1142 \approx x$$

The taxes on the property with an assessed value of $90,000 are approximately $1142.

83. *Pounds of Sand* The ratio of cement to sand in an 80-pound bag of dry mix is 1 to 4. Find the number of pounds of sand in the bag. (*Note:* Dry mix is composed of only cement and sand.)

Solution

Verbal model: $\boxed{\begin{array}{c}\text{Ratio of}\\ \text{cement to sand}\end{array}} = \boxed{\dfrac{\text{Pounds of cement}}{\text{Pounds of sand}}}$

Labels: Ratio of cement to sand = $\frac{1}{4}$
Pounds of cement = $80 - x$
Pounds of sand = x

Proportion: $$\frac{1}{4} = \frac{80 - x}{x}$$

$$x = 4(80 - x) \qquad \text{(Cross-multiply.)}$$

$$x = 320 - 4x$$

$$x + 4x = 320 - 4x + 4x$$

$$5x = 320$$

$$\frac{5x}{5} = \frac{320}{5}$$

$$x = 64$$

There are 64 pounds of sand in the bag.

85. *Shadow Length* In the figure below, how long is the man's shadow? (*Hint:* Use similar triangles to create a proportion.) Use the fact that if two triangles are similar, their corresponding sides are proportional.

Solution

Verbal model: $\boxed{\dfrac{\text{Height of smaller triangle}}{\text{Base of smaller triangle}}} = \boxed{\dfrac{\text{Height of larger triangle}}{\text{Base of larger triangle}}}$

Labels: Height of smaller triangle (person's height) = 6 (feet)
Base of smaller triangle (length of shadow) = x (feet)
Height of larger triangle (height of streetlight) = 15 (feet)
Base of larger triangle = $10 + x$ (feet)

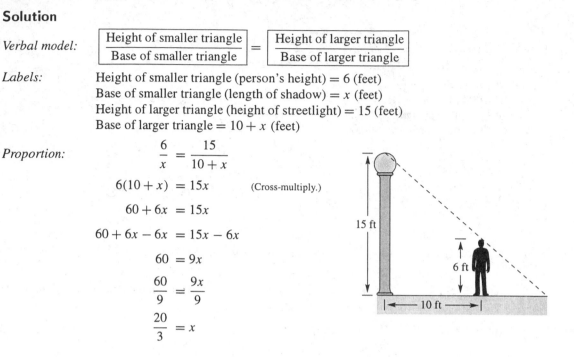

Proportion:

$$\frac{6}{x} = \frac{15}{10 + x}$$

$$6(10 + x) = 15x \qquad \text{(Cross-multiply.)}$$

$$60 + 6x = 15x$$

$$60 + 6x - 6x = 15x - 6x$$

$$60 = 9x$$

$$\frac{60}{9} = \frac{9x}{9}$$

$$\frac{20}{3} = x$$

The length of the shadow is $6\frac{2}{3}$ feet (or 6 feet, 8 inches).

87. *Estimation* Use the Consumer Price Index to estimate the price in 1992 of a watch that cost $58 in 1973.

Solution

Verbal model: $\boxed{\dfrac{\text{Price in 1992}}{\text{Price in 1973}}} = \boxed{\dfrac{\text{Index in 1992}}{\text{Index in 1973}}}$

Labels: Price in 1992 = x (dollars)
Price in 1973 = 58 (dollars)
Index in 1992 = 140.3
Index in 1973 = 44.4

Proportion:

$$\frac{x}{58} = \frac{140.3}{44.4}$$

$$58\left(\frac{x}{58}\right) = 58\left(\frac{140.3}{44.4}\right)$$

$$x = \frac{58(140.3)}{44.4}$$

$$x \approx 183.27$$

Therefore, the 1992 price of the watch was approximately $183.27.

89. *Estimation*   Use the Consumer Price Index to estimate the price in 1950 of a coat that cost $225 in 1990.

Solution

Verbal model: $\boxed{\dfrac{\text{Price in 1950}}{\text{Price in 1990}}} = \boxed{\dfrac{\text{Index in 1950}}{\text{Index in 1990}}}$

Labels: Price in 1950 $= x$ (dollars)
Price in 1990 $= 225$ (dollars)
Index in 1950 $= 24.1$
Index in 1990 $= 130.7$

Proportion: $\dfrac{x}{225} = \dfrac{24.1}{130.7}$

$$225\left(\frac{x}{225}\right) = 225\left(\frac{24.1}{130.7}\right)$$

$$x = \frac{225(24.1)}{130.7}$$

$$x \approx 41.49$$

Therefore, the 1950 price of the coat was approximately $41.49.

3.5   Linear Inequalities and Applications

7. Describe the inequality $x \geq 3$ verbally and sketch its graph.

Solution

x is greater than or equal to 3.

9. Write the inequality (shown on the graph in the textbook) symbolically.

Solution

$-2 \leq x < 1$

11. Write the inequality "x greater than 0 and less than or equal to 6" symbolically.

Solution

$0 < x \leq 6$

13. Determine whether the value of x is a solution of the inequality $5x - 12 > 0$ when (a) $x = 3$, (b) $x = -3$, (c) $x = \frac{5}{2}$, and (d) $x = \frac{3}{2}$.

Solution

(a) $x = 3$

$5(3) - 12 \overset{?}{>} 0$

$15 - 12 \overset{?}{>} 0$

$3 > 0$

3 *is* a solution.

(b) $x = -3$

$5(-3) - 12 \overset{?}{>} 0$

$-15 - 12 \overset{?}{>} 0$

$-27 \not> 0$

-3 *is not* a solution.

(c) $x = \frac{5}{2}$

$5\left(\frac{5}{2}\right) - 12 \overset{?}{>} 0$

$\frac{25}{2} - \frac{24}{2} \overset{?}{>} 0$

$\frac{1}{2} > 0$

$\frac{5}{2}$ *is* a solution.

(d) $x = \frac{3}{2}$

$5\left(\frac{3}{2}\right) - 12 \overset{?}{>} 0$

$\frac{15}{2} - \frac{24}{2} \overset{?}{>} 0$

$-\frac{9}{2} \not> 0$

$\frac{3}{2}$ *is not* a solution.

15. Determine whether the value of x is a solution of the inequality $0 < \frac{1}{4}(x - 2) < 2$ when (a) $x = 4$, (b) $x = 10$, (c) $x = 0$, and (d) $x = \frac{7}{2}$.

Solution

(a) $x = 4$

$$0 \overset{?}{<} \frac{4 - 2}{4} \overset{?}{<} 2$$

$$0 \overset{?}{<} \frac{2}{4} \overset{?}{<} 2$$

$$0 < \frac{1}{2} < 2$$

4 *is* a solution.

(b) $x = 10$

$$0 \overset{?}{<} \frac{10 - 2}{4} \overset{?}{<} 2$$

$$0 \overset{?}{<} \frac{8}{4} \overset{?}{<} 2$$

$$0 < 2 \not< 2$$

10 *is not* a solution.

(c) $x = 0$

$$0 \overset{?}{<} \frac{0 - 2}{4} \overset{?}{<} 2$$

$$0 \overset{?}{<} \frac{-2}{4} \overset{?}{<} 2$$

$$0 \not< -\frac{1}{2} < 2$$

0 *is not* a solution.

(d) $x = \frac{7}{2}$

$$0 \overset{?}{<} \frac{\frac{7}{2} - 2}{4} \overset{?}{<} 2$$

$$0 \overset{?}{<} \frac{\frac{7}{2} - \frac{4}{2}}{4} \overset{?}{<} 2$$

$$0 \overset{?}{<} \frac{\frac{3}{2}}{4} \overset{?}{<} 2$$

$$0 \overset{?}{<} \left(\frac{3}{2} \div 4\right) \overset{?}{<} 2$$

$$0 \overset{?}{<} \frac{3}{2} \cdot \frac{1}{4} \overset{?}{<} 2$$

$$0 < \frac{3}{8} < 2$$

$\frac{7}{2}$ *is* a solution.

17. Match the inequality $x < 4$ with its graph. [The graphs are labeled (a), (b), (c), (d), (e), and (f), and are shown in the textbook.]

Solution

(b)

19. Match the inequality $-3 \leq x < 2$ with its graph. [The graphs are labeled (a), (b), (c), (d), (e), and (f), and are shown in the textbook.]

Solution

(c)

21. Match the inequality $8 > x > \frac{3}{2}$ with its graph. [The graphs are labeled (a), (b), (c), (d), (e), and (f), and are shown in the textbook.]

Solution

(d)

23. Use inequality notation to denote the statement "x is nonnegative."

Solution

$x \geq 0$

25. Use inequality notation to denote the statement "z is at least 3."

Solution

$z \geq 3$

27. Solve and graph the inequality $t - 3 \geq 2$.

Solution

$$t - 3 \geq 2$$

$$t - 3 + 3 \geq 2 + 3$$

$$t \geq 5$$

29. Solve and graph the inequality $4x < 12$.

Solution

$$4x < 12$$

$$\frac{4x}{4} < \frac{12}{4}$$

$$x < 3$$

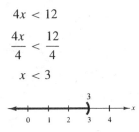

31. Solve and graph the inequality $-\frac{3}{4}x > -3$.

Solution

$$-\frac{3}{4}x > -3$$

$$4\left(-\frac{3}{4}x\right) > 4(-3)$$

$$-3x > -12$$

$$\frac{-3x}{-3} < \frac{-12}{-3} \qquad \text{(Reverse inequality)}$$

$$x < 4$$

33. Solve and graph the inequality $4 - 2x < 3$.

Solution

$$4 - 2x < 3$$

$$4 - 4 - 2x < 3 - 4$$

$$-2x < -1$$

$$\frac{-2x}{-2} > \frac{-1}{-2} \qquad \text{(Reverse inequality)}$$

$$x > \frac{1}{2}$$

35. Solve and graph the inequality $6 < 3(y + 1) - 4(1 - y)$.

Solution

$$6 < 3(y + 1) - 4(1 - y)$$

$$6 < 3y + 3 - 4 + 4y$$

$$6 < 7y - 1$$

$$6 + 1 < 7y - 1 + 1$$

$$7 < 7y$$

$$\frac{7}{7} < \frac{7y}{7}$$

$$1 < y \text{ or } y > 1$$

37. Solve and graph the following inequality.

$$\frac{x}{5} - \frac{x}{2} \le 1$$

Solution

$$\frac{x}{5} - \frac{x}{2} \le 1$$

$$10\left(\frac{x}{5} - \frac{x}{2}\right) \le 10(1)$$

$$10\left(\frac{x}{5}\right) - 10\left(\frac{x}{2}\right) \le 10$$

$$2x - 5x \le 10$$

$$-3x \le 10$$

$$\frac{-3x}{-3} \ge \frac{10}{-3} \quad \begin{array}{l}\text{(Reverse} \\ \text{inequality)}\end{array}$$

$$x \ge \frac{-10}{3}$$

39. Solve and graph the following inequality.

$$-4 < \frac{2x - 3}{3} < 4$$

Solution

$$-4 < \frac{2x - 3}{3} < 4$$

$$3(-4) < 3\left(\frac{2x - 3}{3}\right) < 3(4)$$

$$-12 < 2x - 3 < 12$$

$$-12 + 3 < 2x - 3 + 3 < 12 + 3$$

$$-9 < 2x < 15$$

$$-\frac{9}{2} < \frac{2x}{2} < \frac{15}{2}$$

$$-\frac{9}{2} < x < \frac{15}{2}$$

41. *Annual Operating Budget* A utility company has a fleet of vans. The annual operating cost C (in dollars) per van is $C = 0.32m + 2300$ where m is the number of miles traveled by a van in a year. What number of miles will yield an annual operating cost that is less than $10,000?

Solution

$$0.32m + 2300 = C \text{ and } C < \$10,000$$

$$0.32m + 2300 < 10,000$$

$$0.32m + 2300 - 2300 < 10,000 - 2300$$

$$0.32m < 7700$$

$$\frac{0.32m}{0.32} < \frac{7700}{0.32}$$

$$m < 24,062.5$$

The annual operating cost will be less than $10,000 if the number of miles is less than 24,062.5.

43. *Planet Distances* Mars is farther from the sun than Venus, and Venus is farther from the sun than Mercury, as shown in the figure. What can be said about the relationship between the distances from the sun to Mars and Mercury?

Sun   Mercury   Venus   Earth   Mars

Solution

Mars is farther from the sun than Mercury. (This illustrates the transitive property of inequalities.)

45. Place the correct inequality symbol between the two real numbers.

(a) $-\frac{1}{2}$ $\quad -7$
(b) $-\frac{1}{3}$ $\quad -\frac{1}{6}$

Solution

(a) $-\frac{1}{2} > -7$
(b) $-\frac{1}{3} < -\frac{1}{6}$

47. Solve the following equation and check your solution.

$$\frac{9+x}{3} = 15$$

Solution

$$\frac{9+x}{3} = 15$$

$$3\left(\frac{9+x}{3}\right) = 3(15)$$

$$9 + x = 45$$

$$9 - 9 + x = 45 - 9$$

$$x = 36$$

49. Solve the equation $4 - 3(1 - x) = 7$ and check your solution.

Solution

$$4 - 3(1 - x) = 7$$

$$4 - 3 + 3x = 7$$

$$1 + 3x = 7$$

$$1 - 1 + 3x = 7 - 1$$

$$3x = 6$$

$$\frac{3x}{3} = \frac{6}{3}$$

$$x = 2$$

51. *Geometry* Write expressions for the perimeter and area of the triangle shown in the textbook. Then simplify the expressions.

Solution

$$\text{Perimeter} = (x^2 + 2x) + (5x - 3) + x^2$$

$$= x^2 + 2x + 5x - 3 + x^2$$

$$= 2x^2 + 7x - 3$$

$$\text{Area} = \frac{1}{2}(5x - 3)(x^2)$$

$$= \frac{(5x - 3)x^2}{2}$$

$$= \frac{5x^3 - 3x^2}{2}$$

53. Determine whether the value of x is a solution of the inequality $3 - \frac{1}{2}x \geq 0$ when (a) $x = 10$, (b) $x = 6$, (c) $x = -\frac{3}{4}$, and (d) $x = 0$.

Solution

(a) $x = 10$

$3 - \frac{1}{2}(10) \overset{?}{\geq} 0$

$3 - 5 \overset{?}{\geq} 0$

$-2 \not\geq 0$

10 *is not* a solution.

(b) $x = 6$

$3 - \frac{1}{2}(6) \overset{?}{\geq} 0$

$3 - 3 \overset{?}{\geq} 0$

$0 \geq 0$

6 *is* a solution.

(c) $x = -\frac{3}{4}$

$3 - \frac{1}{2}\left(-\frac{3}{4}\right) \overset{?}{\geq} 0$

$3 + \frac{3}{8} \overset{?}{\geq} 0$

$3\frac{3}{8} \geq 0$

$-\frac{3}{4}$ *is* a solution.

(d) $x = 0$

$3 - \frac{1}{2}(0) \overset{?}{\geq} 0$

$3 - 0 \overset{?}{\geq} 0$

$3 \geq 0$

0 *is* a solution.

55. Determine whether the value of x is a solution of the inequality $-2 < \frac{1}{3}(x-1) \le 5$ when (a) $x = 0$, (b) $x = 25$, (c) $x = 16$, and (d) $x = -5$.

Solution

(a) $x = 0$

$$-2 \overset{?}{<} \tfrac{1}{3}(0-1) \overset{?}{\le} 5$$

$$-2 \overset{?}{<} \tfrac{1}{3}(-1) \overset{?}{\le} 5$$

$$-2 < -\tfrac{1}{3} \le 5$$

0 *is* a solution.

(b) $x = 25$

$$-2 \overset{?}{<} \tfrac{1}{3}(25-1) \overset{?}{\le} 5$$

$$-2 \overset{?}{<} \tfrac{1}{3}(24) \overset{?}{\le} 5$$

$$-2 < 8 \not\le 5$$

25 *is not* a solution.

(c) $x = 16$

$$-2 \overset{?}{<} \tfrac{1}{3}(16-1) \overset{?}{\le} 5$$

$$-2 \overset{?}{<} \tfrac{1}{3}(15) \overset{?}{\le} 5$$

$$-2 < 5 \le 5$$

16 *is* a solution.

(d) $x = -5$

$$-2 \overset{?}{<} \tfrac{1}{3}(-5-1) \overset{?}{\le} 5$$

$$-2 \overset{?}{<} \tfrac{1}{3}(-6) \overset{?}{\le} 5$$

$$-2 \not< -2 \le 5$$

-5 *is not* a solution.

57. Solve and graph the inequality $x + 4 \le 6$.

Solution

$$x + 4 \le 6$$

$$x + 4 - 4 \le 6 - 4$$

$$x \le 2$$

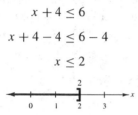

59. Solve and graph the inequality $-10x < 40$.

Solution

$$-10x < 40$$

$$\frac{-10x}{-10} > \frac{40}{-10} \qquad \text{(Reverse inequality)}$$

$$x > -4$$

61. Solve and graph the inequality $-3n > -9$.

Solution

$$-3n > -9$$

$$\frac{-3n}{-3} < \frac{-9}{-3} \qquad \text{(Reverse inequality)}$$

$$n < 3$$

63. Solve and graph the inequality $\frac{2}{3}x \le 12$.

Solution

$$\frac{2}{3}x \le 12$$

$$3\left(\frac{2}{3}\right)x \le 3(12)$$

$$2x \le 36$$

$$\frac{2x}{2} \le \frac{36}{2}$$

$$x \le 18$$

65. Solve and graph the inequality $2x - 5 > 7$.

Solution

$$2x - 5 > 7$$
$$2x - 5 + 5 > 7 + 5$$
$$2x > 12$$
$$\frac{2x}{2} > \frac{12}{2}$$
$$x > 6$$

67. Solve and graph the inequality $5 - x \leq 1$.

Solution

$$5 - x \leq 1$$
$$5 - 5 - x \leq 1 - 5$$
$$-x \leq -4$$
$$(-1)(-x) \geq (-1)(-4) \quad \text{(Reverse inequality)}$$
$$x \geq 4$$

69. Solve and graph the inequality $2x - 5 > 6 - x$.

Solution

$$2x - 5 > -x + 6$$
$$2x + x - 5 > -x + x + 6$$
$$3x - 5 > 6$$
$$3x - 5 + 5 > 6 + 5$$
$$3x > 11$$
$$\frac{3x}{3} > \frac{11}{3}$$
$$x > \frac{11}{3}$$

71. Solve and graph the inequality $10(1 - y) < 3 - 2y$.

Solution

$$10(1 - y) < 3 - 2y$$
$$10 - 10y < 3 - 2y$$
$$10 - 10y + 2y < 3 - 2y + 2y$$
$$10 - 8y < 3$$
$$10 - 10 - 8y < 3 - 10$$
$$-8y < -7$$
$$\frac{-8y}{-8} > \frac{-7}{-8} \quad \text{(Reverse inequality)}$$
$$y > \frac{7}{8}$$

73. Solve and graph the inequality $6(3 - z) \geq 5(3 + z)$.

Solution

$$6(3 - z) \geq 5(3 + z)$$

$$18 - 6z \geq 15 + 5z$$

$$18 - 6z - 5z \geq 15 + 5z - 5z$$

$$18 - 11z \geq 15$$

$$18 - 18 - 11z \geq 15 - 18$$

$$-11z \geq -3$$

$$\frac{-11z}{-11} \leq \frac{-3}{-11} \qquad \text{(Reverse inequality)}$$

$$z \leq \frac{3}{11}$$

75. Solve and graph the following inequality.

$$\frac{5x}{4} + \frac{1}{2} > 0$$

Solution

$$\frac{5x}{4} + \frac{1}{2} > 0$$

$$4\left(\frac{5x}{4} + \frac{1}{2}\right) > 4(0)$$

$$4\left(\frac{5x}{4}\right) + 4\left(\frac{1}{2}\right) > 0$$

$$5x + 2 > 0$$

$$5x + 2 - 2 > 0 - 2$$

$$5x > -2$$

$$\frac{5x}{5} > \frac{-2}{5}$$

$$x > -\frac{2}{5}$$

77. Solve and graph the inequality $1 < 2x + 3 < 9$.

Solution

$$1 < 2x + 3 < 9$$

$$1 - 3 < 2x + 3 - 3 < 9 - 3$$

$$-2 < 2x < 6$$

$$\frac{-2}{2} < \frac{2x}{2} < \frac{6}{2}$$

$$-1 < x < 3$$

79. Solve and graph the following inequality.

$$6 > \frac{x - 2}{-3} > -2$$

Solution

$$6 > \frac{x - 2}{-3} > -2$$

$$-3(6) < -3\left(\frac{x - 2}{-3}\right) < -3(-2) \qquad \text{(Reverse inequality)}$$

$$-18 < x - 2 < 6$$

$$-18 + 2 < x - 2 + 2 < 6 + 2$$

$$-16 < x < 8$$

81. Solve and graph the inequality $\frac{3}{4} > x + 1 > \frac{1}{4}$.

Solution

$$\frac{3}{4} > x + 1 > \frac{1}{4}$$

$$4\left(\frac{3}{4}\right) > 4(x + 1) > 4\left(\frac{1}{4}\right)$$

$$3 > 4x + 4 > 1$$

$$3 - 4 > 4x + 4 - 4 > 1 - 4$$

$$-1 > 4x > -3$$

$$\frac{-1}{4} > \frac{4x}{4} > \frac{-3}{4}$$

$$-\frac{1}{4} > x > -\frac{3}{4} \text{ or } -\frac{3}{4} < x < -\frac{1}{4}$$

Alternate Solution:

$$\frac{3}{4} > x + 1 > \frac{1}{4}$$

$$\frac{3}{4} - 1 > x + 1 - 1 > \frac{1}{4} - 1$$

$$\frac{3}{4} - \frac{4}{4} > x > \frac{1}{4} - \frac{1}{4}$$

$$-\frac{1}{4} > x > -\frac{3}{4} \text{ or } -\frac{3}{4} < x < -\frac{1}{4}$$

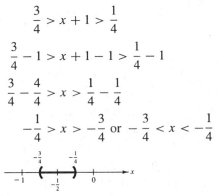

83. Use inequality notation to denote the statement "x is at least 4."

Solution

$x \geq 4$

85. Use inequality notation to denote the statement "y is no more than 25."

Solution

$y \leq 25$

87. *Budgets* Department A's budget is less than Department B's budget, and Department B's budget is less than Department C's budget. What can you say about the relationship between the budgets of Department A and Department C? Identify the property of inequalities that is demonstrated.

Solution

Department A's budget < Department C's budget
Transitive Property

89. *Telephone Cost* The cost for a long-distance telephone call is $0.46 for the first minute and $0.31 for each additional minute. The total cost of the call cannot exceed $4. Find the interval of time that is available for the call.

Solution

Verbal model: | Cost for first minute | + | Cost for additional minutes | $\leq$ | $4 |

Labels:
Cost for first minute = $0.46
Number of *additional* minutes = x
Cost per additional minute = $0.31
Cost for additional minutes = $0.31x$ (dollars)

Inequality:
$$0.46 + 0.31x \leq 4$$
$$0.46 - 0.46 + 0.31x \leq 4 - 0.46$$
$$0.31x \leq 3.54$$
$$\frac{0.31x}{0.31} \leq \frac{3.54}{0.31}$$
$$x \leq 11.4 \quad \text{(Rounded)}$$
$$x + 1 \leq 11.4 + 1$$
$$x + 1 \leq 12.4$$

Note: The variable x represents the number of *additional* minutes *after* the first minute. Thus, $x + 1$ represents the *total* length of the call.

The call will cost no more than $4 if the call is no more than 12.4 minutes long. (If any *portion* of a minute is billed at the *full* minute rate, the call must be no more than 12 minutes long to keep the cost from exceeding $4.)

91. *Distance* The minimum and maximum speeds on a highway are 45 miles per hour and 65 miles per hour. You travel nonstop for 4 hours on this highway. Assuming that you stay within the speed limits, give an interval for the distance you travel.

Solution

Distance = (Rate)(Time)
$$(45)4 \leq \text{distance} \leq (65)4$$
$$180 \leq \text{distance} \leq 260$$

Therefore, the distance you travel is from 180 to 260 miles.

Review Exercises for Chapter 3

1. Solve the equation $y - 25 = 10$ mentally.

Solution

$y = 35$

3. Solve the following equation mentally.

$$\frac{x}{4} = 7$$

Solution

$x = 28$

5. Justify each step of the following solution.

$$10x - 12 = 18$$

$$10x - 12 + 12 = 18 + 12$$

$$10x = 30$$

$$\frac{10x}{10} = \frac{30}{10}$$

$$x = 3$$

Solution

$10x - 12 = 18$	Given equation
$10x - 12 + 12 = 18 + 12$	Add 12 to each side.
$10x = 30$	Combine like terms.
$\dfrac{10x}{10} = \dfrac{30}{10}$	Divide by 10 on each side.
$x = 3$	Simplify (solution).

7. Solve the equation $10x = 50$ and check your solution.

Solution

$$10x = 50$$

$$\frac{10x}{10} = \frac{50}{10}$$

$$x = 5$$

9. Solve the equation $8x + 7 = 39$ and check your solution.

Solution

$$8x + 7 = 39$$

$$8x + 7 - 7 = 39 - 7$$

$$8x = 32$$

$$\frac{8x}{8} = \frac{32}{8}$$

$$x = 4$$

11. Solve the equation $24 - 7x = 3$ and check your solution.

Solution

$$24 - 7x = 3$$

$$24 - 24 - 7x = 3 - 24$$

$$-7x = -21$$

$$\frac{-7x}{-7} = \frac{-21}{-7}$$

$$x = 3$$

13. Solve the equation $15x - 4 = 16$ and check your solution.

Solution

$$15x - 4 = 16$$

$$15x - 4 + 4 = 16 + 4$$

$$15x = 20$$

$$\frac{15x}{15} = \frac{20}{15}$$

$$x = \frac{20}{15}$$

$$x = \frac{4}{3}$$

15. Solve the equation $3x - 2(x + 5) = 10$ and check your solution.

Solution

$$3x - 2(x + 5) = 10$$

$$3x - 2x - 10 = 10$$

$$x - 10 = 10$$

$$x - 10 + 10 = 10 + 10$$

$$x = 20$$

17. Solve the equation $2x + 3 = 2x - 2$ and check your solution.

Solution

$$2x + 3 = 2x - 2$$

$$2x - 2x + 3 = 2x - 2x - 2$$

$$3 = 2 \quad \text{(False)}$$

The equation has *no* solution.

19. Solve the following equation and check your solution.

$$\frac{x}{5} = 4$$

Solution

$$\frac{x}{5} = 4$$

$$5\left(\frac{x}{5}\right) = 5(4)$$

$$x = 20$$

21. Solve the equation $\frac{2}{3}x - \frac{1}{6} = \frac{9}{2}$ and check your solution.

Solution

$$\frac{2}{3}x - \frac{1}{6} = \frac{9}{2}$$

$$6\left(\frac{2}{3}x - \frac{1}{6}\right) = 6\left(\frac{9}{2}\right)$$

$$6\left(\frac{2}{3}x\right) - 6\left(\frac{1}{6}\right) = 27$$

$$4x - 1 = 27$$

$$4x - 1 + 1 = 27 + 1$$

$$4x = 28$$

$$\frac{4x}{4} = \frac{28}{4}$$

$$x = 7$$

23. Solve the following equation and check your solution.

$$\frac{x}{3} = \frac{1}{9}$$

Solution

$$\frac{x}{3} = \frac{1}{9}$$

$$3\left(\frac{x}{3}\right) = 3\left(\frac{1}{9}\right)$$

$$x = \frac{1}{3}$$

25. Solve the following equation and check your solution.

$$\frac{u}{10} + \frac{u}{5} = 6$$

Solution

$$\frac{u}{10} + \frac{u}{5} = 6$$

$$10\left(\frac{u}{10} + \frac{u}{5}\right) = 10(6)$$

$$10\left(\frac{u}{10}\right) + 10\left(\frac{u}{5}\right) = 60$$

$$u + 2u = 60$$

$$3u = 60$$

$$\frac{3u}{3} = \frac{60}{3}$$

$$u = 20$$

27. Solve the equation $516x - 875 = 3250$. Round your result to two decimal places.

Solution

$$516x - 875 = 3250$$

$$516x - 875 + 875 = 3250 + 875$$

$$516x = 4125$$

$$\frac{516x}{516} = \frac{4125}{516}$$

$$x \approx 7.99 \text{ (Rounded)}$$

29. Solve the following equation. Round your result to two decimal places.

$$\frac{x}{4.625} = 48.5$$

Solution

$$\frac{x}{4.625} = 48.5$$

$$4.625\left(\frac{x}{4.625}\right) = 4.625(48.5)$$

$$x \approx 224.31 \text{ (Rounded)}$$

31. *Driving Distances*   On a 1200-mile trip, you drive about $1\frac{1}{2}$ times as much as your friend. Approximate the number of miles each of you drives.

Solution

Verbal model: | Distance that you drive | $+$ | Distance that your friend drives | $=$ | Total distance |

Labels: Distance that you drive $= x$ (miles)
Distance that your friend drives $= \frac{3}{2}x$ (miles)
Total distance $= 1200$ (miles)

Equation:

$$x + \frac{3}{2}x = 1200$$

$$2\left(x + \frac{3}{2}x\right) = 2(1200)$$

$$2x + 3x = 2400$$

$$5x = 2400$$

$$x = 480 \text{ and } \frac{3}{2}x = 720$$

Therefore, you drive 480 miles and your friend drives 720 miles.

33. Complete the table shown in the textbook.

Solution

Percent	Parts out of 100	Decimal	Fraction
35%	35	0.35	$\frac{7}{20}$

Parts out of 100: 35% means 35 parts out of 100.

Decimal: *Verbal model:* $\boxed{\text{Decimal}} \cdot \boxed{100\%} = \boxed{\text{Percent}}$

           Label: Decimal $= x$

           Equation: $x(100\%) = 35\%$

$$x = \frac{35\%}{100\%}$$

$$x = 0.35$$

Fraction: *Verbal model:* $\boxed{\text{Fraction}} \cdot \boxed{100\%} = \boxed{\text{Percent}}$

           Label: Fraction $= x$

           Equation: $x(100\%) = 35\%$

$$x = \frac{35\%}{100\%}$$

$$x = \frac{35}{100}$$

$$x = \frac{7}{20}$$

35. What is 125% of 16?

Solution

Verbal model: $\boxed{\text{What number}} = \boxed{125\% \text{ of } 16}$ $(a = pb)$

Label: $a =$ unknown number

Percent equation: $a = (1.25)(16)$

                     $a = 20$

Therefore, 20 is 125% of 16.

37. $37\frac{1}{2}\%$ of what number is 150?

Solution

Verbal model: $\boxed{150} = \boxed{37\frac{1}{2}\% \text{ of what number}}$ $(a = pb)$

Label: $b =$ unknown number

Percent equation: $150 = 0.375b$

$$\frac{150}{0.375} = b$$

$$400 = b$$

Therefore, 150 is $37\frac{1}{2}\%$ of 400.

39. 150 is what percent of 250?

Solution

Verbal model: $\boxed{150} = \boxed{\text{What percent of 250}}$ $(a = pb)$

Label: $p =$ unknown percent (in decimal form)

Percent equation: $150 = p(250)$

$$\frac{150}{250} = p$$

$$0.6 = p$$

Therefore, 150 is 60% of 250.

41. *Revenue* The revenues for a corporation (in millions of dollars) in the years 1993 and 1994 were $4521.4 and $4679.0, respectively. Determine the percentage increase in revenue from 1993 to 1994.

Solution

Verbal model: $\boxed{\text{Revenue increase}} = \boxed{\text{What percent of 1993 revenue}}$ $(a = pb)$

Labels: $a =$ revenue increase $= 4679.00 - 4521.40 = \$157.60$
$p =$ unknown percent (in decimal form)
$b = 1993$ revenue $= \$4521.40$

Percent equation: $157.60 = p(4521.40)$

$$\frac{157.60}{4521.40} = p$$

$$0.035 \approx p$$

Therefore, the percentage increase in revenue was approximately 3.5%.

43. Express the ratio 18 inches to 4 yards as a fraction in reduced form. (Use the same units of measure for both quantities.)

Solution

$$\frac{18 \text{ inches}}{4 \text{ yards}} = \frac{18 \text{ inches}}{4(36) \text{ inches}} = \frac{18}{144} = \frac{1}{8}$$

45. Express the ratio 2 hours to 90 minutes as a fraction in reduced form. (Use the same units of measure for both quantities.)

Solution

$$\frac{2 \text{ hours}}{90 \text{ minutes}} = \frac{2(60) \text{ minutes}}{90 \text{ minutes}} = \frac{120}{90} = \frac{4}{3}$$

47. Solve the following proportion.

$$\frac{7}{16} = \frac{z}{8}$$

Solution

$$\frac{7}{16} = \frac{z}{8}$$

$$(7)(8) = 16z$$

$$56 = 16z$$

$$\frac{56}{16} = z$$

$$\frac{7}{2} = z$$

49. Solve the following proportion.

$$\frac{x+2}{4} = \frac{x-1}{3}$$

Solution

$$\frac{x+2}{4} = \frac{x-1}{3}$$

$$3(x+2) = 4(x-1)$$

$$3x + 6 = 4x - 4$$

$$-x + 6 = -4$$

$$-x = -10$$

$$x = 10$$

51. *Real Estate Taxes* The tax on property with an assessed value of $75,000 is $1150. Find the tax on property with an assessed value of $110,000. Use a proportion to solve the problem.

Solution

Verbal model:
$$\boxed{\dfrac{\text{Tax on first property}}{\text{Value of first property}}} = \boxed{\dfrac{\text{Tax on second property}}{\text{Value of second property}}}$$

Labels:
Tax on first property = $1150
Value of first property = $75,000
Tax on second property = x (dollars)
Value of second property = $110,000

Equation:
$$\frac{1150}{75,000} = \frac{x}{110,000}$$

$$110,000\left(\frac{1150}{75,000}\right) = \left(\frac{x}{110,000}\right)110,000$$

$$1687 \approx x$$

The taxes would be approximately $1687.

53. *Map Distance* The scale represents 100 miles on the map. Use the map to approximate the distance between St. Petersburg and Tallahassee.

Solution

Verbal model:
$$\boxed{\dfrac{\text{Inches on map scale}}{\text{Miles represented on scale}}} = \boxed{\dfrac{\text{Inches between cities on map}}{\text{Miles between cities}}}$$

Labels:
Inches on map scale $\approx \frac{7}{16}$
Miles represented on scale = 100
Inches between cities on map $\approx \frac{15}{16}$
Miles between cities = x

Proportion:
$$\frac{7/16}{100} = \frac{15/16}{x}$$

$$\frac{7}{16}x = 100\left(\frac{15}{16}\right) \quad \text{(Cross-multiply.)}$$

$$0.4375x = 93.75$$

$$x = \frac{93.75}{0.4375}$$

$$x \approx 214$$

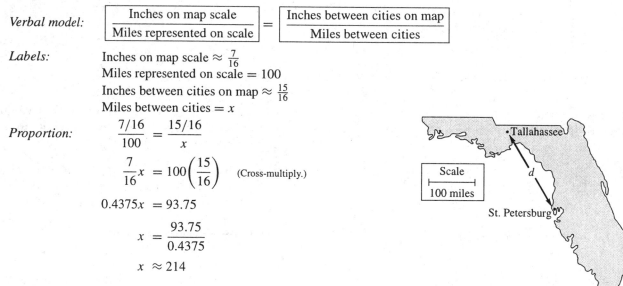

The cities are approximately 214 miles apart.

55. Solve and graph the inequality $x + 5 \geq 7$.

Solution

$$x + 5 \geq 7$$

$$x + 5 - 5 \geq 7 - 5$$

$$x \geq 2$$

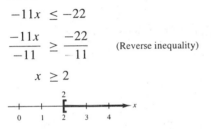

57. Solve and graph the inequality $3x - 8 < 1$.

Solution

$$3x - 8 < 1$$

$$3x - 8 + 8 < 1 + 8$$

$$3x < 9$$

$$\frac{3x}{3} < \frac{9}{3}$$

$$x < 3$$

59. Solve and graph the inequality $-11x \leq -22$.

Solution

$$-11x \leq -22$$

$$\frac{-11x}{-11} \geq \frac{-22}{-11} \qquad \text{(Reverse inequality)}$$

$$x \geq 2$$

61. Solve and graph the inequality $\frac{4}{5}x > 8$.

Solution

$$\frac{4}{5}x > 8$$

$$5\left(\frac{4}{5}x\right) > 5(8)$$

$$4x > 40$$

$$\frac{4x}{4} > \frac{40}{4}$$

$$x > 10$$

Note: This could be done in one step.

$$\frac{4}{5}x > 8$$

$$\frac{5}{4}\left(\frac{4}{5}x\right) > \frac{5}{4}(8)$$

$$x > 10$$

63. Solve and graph the inequality $14 - \frac{1}{2}t < 12$.

Solution

$$14 - \tfrac{1}{2}t \ < 12$$
$$2\left(14 - \tfrac{1}{2}t\right) \ < 2(12)$$
$$28 - t \ < 24$$
$$28 - 28 - t \ < 24 - 28$$
$$-t \ < -4$$
$$(-1)(-t) \ > (-1)(-4) \quad \text{(Reverse inequality)}$$
$$t \ > 4$$

65. Solve and graph the inequality $3(1 - y) \geq 2(4 + y)$.

Solution

$$3(1 - y) \ \geq 2(4 + y)$$
$$3 - 3y \ \geq 8 + 2y$$
$$3 - 3y - 2y \ \geq 8 + 2y - 2y$$
$$3 - 5y \ \geq 8$$
$$3 - 3 - 5y \ \geq 8 - 3$$
$$-5y \ \geq 5$$
$$\frac{-5y}{-5} \ \leq \frac{5}{-5} \quad \text{Reverse inequality)}$$
$$y \ \leq -1$$

67. Solve and graph the following inequality.

$$-2 < \frac{x}{3} \leq 2$$

Solution

$$-2 < \frac{x}{3} \leq 2$$
$$3(-2) < 3\left(\frac{x}{3}\right) \leq 3(2)$$
$$-6 < x \leq 6$$

69. Solve and graph the following inequality.

$$3 > \frac{x + 1}{-2} > 0$$

Solution

$$3 \ > \frac{x + 1}{-2} > 0$$
$$-2(3) \ < -2\left(\frac{x + 1}{-2}\right) < -2(0) \quad \text{(Reverse inequality)}$$
$$-6 \ < x + 1 < 0$$
$$-6 - 1 \ < x + 1 - 1 < 0 - 1$$
$$-7 \ < x < -1$$

71. Write an inequality that represents the statement "z is at least 10."

Solution

$z \geq 10$

73. Write an inequality that represents the statement "y is more than 8 but less than 12."

Solution

$8 < y < 12$

75. Write an inequality that represents the statement "The volume V is less than 12 cubic feet."

Solution

$V < 12$

Test for Chapter 3

1. Solve the equation $4x - 3 = 18$ and check your solution.

Solution

$$4x - 3 = 18$$

$$4x - 3 + 3 = 18 + 3$$

$$4x = 21$$

$$\frac{4x}{4} = \frac{21}{4}$$

$$x = \frac{21}{4}$$

2. Solve the equation $10 - (2 - x) = 2x + 1$ and check your solution.

Solution

$$10 - (2 - x) = 2x + 1$$

$$10 - 2 + x = 2x + 1$$

$$8 + x = 2x + 1$$

$$8 + x - 2x = 2x - 2x + 1$$

$$8 - x = 1$$

$$8 - 8 - x = 1 - 8$$

$$-x = -7$$

$$(-1)(-x) = (-1)(-7)$$

$$x = 7$$

3. Solve the following equation and check your solution.

$$\frac{5x}{4} = \frac{5}{2} + x$$

Solution

$$\frac{5x}{4} = \frac{5}{2} + x$$

$$4\left(\frac{5x}{4}\right) = 4\left(\frac{5}{2} + x\right)$$

$$5x = 4\left(\frac{5}{2}\right) + 4x$$

$$5x = 10 + 4x$$

$$5x - 4x = 10 + 4x - 4x$$

$$x = 10$$

4. Solve the following equation and check your solution.

$$\frac{t + 2}{3} = \frac{2t}{5}$$

Solution

$$\frac{t + 2}{3} = \frac{2t}{5}$$

$$5(t + 2) = 3(2t)$$

$$5t + 10 = 6t$$

$$5t - 5t + 10 = 6t - 5t$$

$$10 = t$$

5. Solve $4.08(x + 10) = 9.50(x - 2)$. Round the result to two decimal places.

Solution

$$4.08(x + 10) = 9.50(x - 2)$$

$$4.08x + 40.8 = 9.50x - 19$$

$$4.08x - 9.50x + 40.8 = 9.50x - 9.50x - 19$$

$$-5.42x + 40.8 = -19$$

$$-5.42x + 40.8 - 40.8 = -19 - 40.8$$

$$-5.42x = -59.8$$

$$\frac{-5.42x}{-5.42} = \frac{-59.8}{-5.42}$$

$$x \approx 11.03 \qquad \text{(Rounded)}$$

6. When the sum of a number and 6 is divided by 8, the result is 7. Find the number.

Solution

$$\frac{x + 6}{8} = 7$$

$$8\left(\frac{x + 6}{8}\right) = 8(7)$$

$$x + 6 = 56$$

$$x + 6 - 6 = 56 - 6$$

$$x = 50$$

7. The bill (including parts and labor) for the repair of a home appliance is $134. The cost for parts is $62. How many hours were spent repairing the appliance if the cost of labor was $18 per hour?

Solution

Verbal model: $\boxed{\text{Cost for parts}} + \boxed{\text{Labor cost per hour}} \cdot \boxed{\text{Hours of labor}} = \boxed{\text{Total bill}}$

Labels: Cost for parts = 62 (dollars)
Labor cost per hour = 18 (dollars/hour)
Hours of labor = x (hours)
Total bill = 134 (dollars)

Equation:

$$62 + 18x = 134$$

$$62 - 62 + 18x = 134 - 62$$

$$18x = 72$$

$$\frac{18x}{18} = \frac{72}{18}$$

$$x = 4$$

Therefore, 4 hours were spent repairing the appliance.

8. Express the fraction $\frac{3}{8}$ as a percent and a decimal.

Solution

(a) *Verbal model:* $\boxed{\text{Fraction}} \cdot \boxed{100\%} = \boxed{\text{Percent}}$

Label: Percent = x

Equation:

$$\frac{3}{8}(100\%) = x$$

$$37.5\% = x$$

(b) *Verbal model:* $\boxed{\text{Decimal}} \cdot \boxed{100\%} = \boxed{\text{Percent}}$

Label: Decimal = x

Equation:

$$x(100\%) = 37.5\%$$

$$x = \frac{37.5\%}{100\%} = 0.375$$

9. 324 is 27% of what number?

Solution

Verbal model: $\boxed{324} = \boxed{27\% \text{ of what number}}$ $(a = pb)$

Labels: b = unknown number

Percent equation:

$$324 = 0.27b$$

$$\frac{324}{0.27} = b$$

$$1200 = b$$

Therefore, 324 is 27% of 1200.

10. 90 is what percent of 250?

Solution

Verbal model: $\boxed{90} = \boxed{\text{What percent of 250}}$ $(a = pb)$

Labels: $p = $ unknown percent (in decimal form)

Percent equation: $90 = p(250)$

$$\frac{90}{250} = p$$

$$0.36 = p$$

Therefore, 90 is 36% of 250.

11. The price of a product increased by 20% during the past year. It is now selling for $240. What was the price 1 year ago?

Solution

Verbal model: $\boxed{\begin{array}{c}\text{Original}\\\text{price}\end{array}} + \boxed{\begin{array}{c}\text{Increase of 20\% of}\\\text{the original price}\end{array}} = \boxed{\begin{array}{c}\text{Current}\\\text{price}\end{array}}$

Labels: Original price $= x$ (dollars)
Increase $= 0.20x$ (dollars)
Current price $= 240$ (dollars)

Equation: $x + 0.20x = 240$

$$1.20x = 240$$

$$100(1.20x) = 100(240)$$

$$120x = 24,000$$

$$x = 200$$

Therefore, the original price of the product was $200.

12. Express the ratio of 40 inches to 2 yards as a fraction in reduced form. Use the same units of measure for both quantities, and explain how you made this conversion.

Solution

$$\frac{40 \text{ inches}}{2 \text{ yards}} = \frac{40 \text{ inches}}{2(36) \text{ inches}} = \frac{40}{72} = \frac{5}{9}$$

13. Solve the following proportion.

$$\frac{5}{8} = \frac{x}{12}$$

Solution

$$\frac{5}{8} = \frac{x}{12}$$

$$12\left(\frac{5}{8}\right) = 12\left(\frac{x}{12}\right)$$

$$\frac{60}{8} = x$$

$$\frac{15}{2} = x$$

(The answer can also be written as 7.5.)

14. Solve the following proportion.

$$\frac{2x}{3} = \frac{x + 14}{5}$$

Solution

$$\frac{2x}{3} = \frac{x + 14}{5}$$

$$2x(5) = 3(x + 14) \qquad \text{(Cross-multiply.)}$$

$$10x = 3x + 42$$

$$7x = 42$$

$$x = \frac{42}{7}$$

$$x = 6$$

15. On the map shown in the textbook, 1 centimeter represents 55 miles. Approximate the distance between the two indicated cities.

Solution

The distance between Columbus and Akron measures 2 centimeters.

$$\frac{1 \text{ centimeter}}{55 \text{ miles}} = \frac{2 \text{ centimeters}}{x \text{ miles}}$$

$$\frac{1}{55} = \frac{2}{x}$$

$$1 \cdot x = 55 \cdot 2$$

$$x = 110$$

The distance between Columbus and Akron is approximately 110 miles.

---

16. Solve and graph the inequality $x + 3 \leq 7$.

Solution

$$x + 3 \leq 7$$

$$x + 3 - 3 \leq 7 - 3$$

$$x \leq 4$$

17. Solve and graph the inequality $-\frac{2x}{3} > 4$.

Solution

$$-\frac{2x}{3} > 4$$

$$3\left(-\frac{2x}{3}\right) > 3(4)$$

$$-2x > 12$$

$$\frac{-2x}{-2} < \frac{12}{-2} \qquad \text{(Reverse inequality)}$$

$$x < -6$$

---

18. Solve and graph the inequality $-3 < 2x - 1 \leq 3$.

Solution

$$-3 < 2x - 1 \leq 3$$

$$-3 + 1 < 2x - 1 + 1 \leq 3 + 1$$

$$-2 < 2x \leq 4$$

$$\frac{-2}{2} < \frac{2x}{2} \leq \frac{4}{2}$$

$$-1 < x \leq 2$$

19. Solve and graph the inequality $2 \geq \frac{3 - x}{2} > -1$.

Solution

$$2 \geq \frac{3 - x}{2} > -1$$

$$2(2) \geq 2\left(\frac{3 - x}{2}\right) > 2(-1)$$

$$4 \geq 3 - x > -2$$

$$4 - 3 \geq 3 - 3 - x > -2 - 3$$

$$1 \geq -x > -5$$

$$(-1)(1) \leq (-1)(-x) < (-1)(-5)$$

$$-1 \leq x < 5$$

20. Use inequality notation to express each phrase mathematically.

(a) y is no more than 10.   (b) t is greater than or equal to 10.

Solution

(a) $y \leq 10$ (b) $t \geq 10$

Cumulative Test for Chapters P–3

1. Place the correct symbol ($<$ or $>$) between the numbers $-\frac{3}{4}$ and $\left|-\frac{7}{8}\right|$.

Solution

$-\frac{3}{4} < \left|-\frac{7}{8}\right|$

Note: $\left|-\frac{7}{8}\right| = \frac{7}{8}$ and $-\frac{3}{4} < \frac{7}{8}$

2. Evaluate the expression $(-200)(2)(-3)$.

Solution

$(-200)(2)(-3) = 1200$

3. Evaluate the expression $\frac{3}{8} - \frac{5}{6}$.

Solution

$$\frac{3}{8} - \frac{5}{6} = \frac{3 \cdot 3}{8 \cdot 3} - \frac{5 \cdot 4}{6 \cdot 4}$$
$$= \frac{9}{24} - \frac{20}{24}$$
$$= \frac{9 - 20}{24}$$
$$= -\frac{11}{24}$$

4. Evaluate the expression $-\frac{2}{9} \div \frac{8}{75}$.

Solution

$$-\frac{2}{9} \div \frac{8}{75} = -\frac{2}{9} \cdot \frac{75}{8}$$
$$= -\frac{2 \cdot 75}{9 \cdot 8}$$
$$= -\frac{\cancel{(2)}\cancel{(3)}(25)}{(3)\cancel{(3)}\cancel{(2)}(4)}$$
$$= -\frac{25}{12}$$

5. Evaluate the expression $-(-2)^3$.

Solution

$-(-2)^3 = -(-8) = 8$

6. Use exponential form to write the product
$3 \cdot (x + y) \cdot (x + y) \cdot 3 \cdot 3$.

Solution

$3 \cdot (x + y) \cdot (x + y) \cdot 3 \cdot 3 = 3^3(x + y)^2$

7. Use the Distributive Property to expand $-2x(x - 3)$.

Solution

$-2x(x - 3) = -2x^2 + 6x$

8. Identify the rule of algebra illustrated by
$2 + (3 + x) = (2 + 3) + x$.

Solution

Associative Property of Addition

9. Simplify the expression $(3x^3)(5x^4)$.

Solution

$(3x^3)(5x^4) = 15x^{3+4} = 15x^7$

10. Simplify the expression $(a^3b^2)(ab)^5$.

Solution

$$(a^3b^2)(ab)^5 = (a^3b^2)(a^5b^5)$$
$$= a^{3+5}b^{2+5}$$
$$= a^8b^7$$

11. Simplify the expression $2x^2 - 3x + 5x^2 - (2 + 3x)$.

Solution

$2x^2 - 3x + 5x^2 - (2 + 3x)$

$= 2x^2 - 3x + 5x^2 - 2 - 3x$

$= 7x^2 - 6x - 2$

12. Solve the equation $12x - 3 = 7x + 27$ and check your solution.

Solution

$12x - 3 = 7x + 27$

$5x - 3 = 27$

$5x = 30$

$x = \frac{30}{5}$

$x = 6$

13. Solve the following equation and check your solution.

$2x - \frac{5x}{4} = 13$

Solution

$2x - \frac{5x}{4} = 13$

$4\left(2x - \frac{5x}{4}\right) = 4(13)$

$8x - 5x = 52$

$3x = 52$

$x = \frac{52}{3}$

14. Solve the equation $2(x - 3) + 3 = 12 - x$ and check your solution.

Solution

$2(x - 3) + 3 = 12 - x$

$2x - 6 + 3 = 12 - x$

$2x - 3 = 12 - x$

$3x - 3 = 12$

$3x = 15$

$x = 5$

15. Solve and graph the following inequality.

$-1 \le \frac{x + 3}{2} < 2$

Solution

$-1 \le \frac{x + 3}{2} < 2$

$2(-1) \le 2\left(\frac{x + 3}{2}\right) < 2(2)$

$-2 \le x + 3 < 4$

$-2 - 3 \le x < 4 - 3$

$-5 \le x < 1$

16. The sticker on a new car gives the fuel efficiency as 28.3 miles per gallon. In your own words, explain how to estimate the annual fuel cost for the buyer if the car will be driven approximately 15,000 miles per year and the fuel cost is $1.179 per gallon.

Solution

The annual fuel cost can be estimated by (a) dividing the 15,000 miles by the 28.3 miles per gallon to determine how many gallons of fuel will be needed, and then (b) multiplying that result by the $1.179 cost per gallon of fuel.

Verbal model: $\boxed{\begin{array}{c}\text{Annual}\\\text{cost}\end{array}} = \left(\boxed{\begin{array}{c}\text{Miles driven}\\\text{per year}\end{array}} \div \boxed{\begin{array}{c}\text{Miles per}\\\text{gallon}\end{array}} \right) \cdot \boxed{\begin{array}{c}\text{Cost per}\\\text{gallon}\end{array}}$

Labels: Annual cost = x (dollars)
Miles driven per year = 15,000 (miles)
Miles per gallon = 28.3 (miles per gallon)
Cost per gallon = 1.179 (dollars per gallon)

Equation: $x = \dfrac{15,000}{28.3}(1.179)$

$x \approx 624.91$

Therefore, the annual fuel cost is approximately $624.91.

17. Express the ratio "24 ounces to 2 pounds" as a fraction in reduced form.

Solution

$$\frac{24 \text{ ounces}}{2 \text{ pounds}} = \frac{24 \text{ \sout{ounces}}}{2(16) \text{ \sout{ounces}}}$$

$$= \frac{24}{32}$$

$$= \frac{3}{4}$$

18. The sum of two consecutive even integers is 494. Find the two numbers.

Solution

Verbal model: $\boxed{\begin{array}{c}\text{First consecutive}\\\text{even integer}\end{array}} + \boxed{\begin{array}{c}\text{Second consecutive}\\\text{even integer}\end{array}} = \boxed{494}$

Labels: First consecutive even integer = $2n$
Second consecutive even integer = $2n + 2$

Equation: $2n + (2n + 2) = 494$

$4n + 2 = 494$

$4n = 492$

$n = \dfrac{492}{4}$

$n = 123$

$2n = 246$

$2n + 2 = 248$

The two consecutive even integers are 246 and 248.

19. The suggested retail price of a camcorder is $1150. The camcorder is on sale for "20% off" the list price. Find the sale price.

Solution

Verbal model: | Sale price | = | List price | − | Discount |

Labels: Sale price $= x$ (dollars)
                    List price $= \$1150$
                    Discount rate $= 0.20$ (percent in decimal form)
                    Discount $= 0.20(1150) = \$230$

Equation: $x = 1150 - 230$

                $x = 920$

The sale price of the camcorder is $920.

20. The figure in the textbook shows two pieces of property. The assessed values of the properties are proportional to their areas. The value of the larger piece is $95,000. What is the value of the smaller piece?

Solution

Verbal model:

$$\frac{\text{Assessed value of larger piece}}{\text{Area of larger piece}} = \frac{\text{Assessed value of smaller piece}}{\text{Area of smaller piece}}$$

Labels: Assessed value of larger piece $= 95,000$ (dollars)
                    Area of larger piece $= (100)(80) = 8000$ (square units)
                    Assessed value of smaller piece $= x$ (dollars)
                    Area of smaller piece $= (60)(80) = 4800$ (square units)

Equation: $\dfrac{95,000}{8000} = \dfrac{x}{4800}$

               $8000x = 95,000(4800)$

               $\dfrac{8000x}{8000} = \dfrac{95,000(4800)}{8000}$

                   $x = 57,000$

Therefore, the assessed value of the smaller piece is $57,000.

CHAPTER FOUR
Graphs and Linear Applications

4.1 | Ordered Pairs and Graphs

7. Plot the points $(3, 2)$, $(-4, 2)$, and $(2, -4)$ on a rectangular coordinate system.

Solution

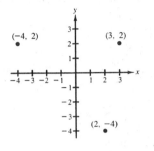

9. Plot the points $\left(\frac{3}{2}, -1\right)$, $\left(-3, \frac{3}{4}\right)$, and $\left(\frac{1}{2}, -\frac{1}{2}\right)$ on a rectangular coordinate system.

Solution

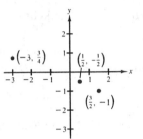

11. *Graphical Estimation* Estimate the coordinates of the points shown on the graph in the textbook.

Solution

A. $(5, 2)$
B. $(-3, 4)$
C. $(2, -5)$
D. $(-2, -2)$

13. Plot the points $(-1, 1)$, $(2, -1)$, and $(3, 4)$ and connect them with line segments to form a triangle.

Solution

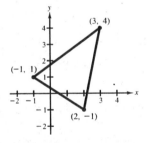

15. Plot the points $(5, 2)$, $(7, 0)$, $(1, -2)$, and $(-1, 0)$ and connect them with line segments to form a parallelogram.

Solution

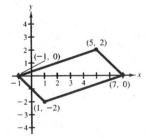

17. Determine the quadrant in which the point $\left(-3, \frac{1}{8}\right)$ is located.

Solution

The point $\left(-3, \frac{1}{8}\right)$ is located 3 units to the left of the vertical axis and $\frac{1}{8}$ above the horizontal axis; it is in Quadrant II.

19. Determine the quadrant in which the point $(-100, -365.6)$ is located.

Solution

The point $(-100, -365.6)$ is located 100 units to the left of the vertical axis and 365.6 units below the horizontal axis; it is in Quadrant III.

21. Determine the quadrant or quadrants in which the point $(-5, y)$ must be located; y is a real number.

Solution

Quadrants II or III. This point is located 5 units to the *left* of the vertical axis. If y is positive, the point would be *above* the horizontal axis and in the second quadrant. If y is negative, the point would be *below* the horizontal axis and in the third quadrant. (If y is zero, the point would be on the horizontal axis *between* the second and third quadrants.)

23. *Organizing Data*   The table in the textbook gives the normal temperature y (Fahrenheit) in Anchorage, Alaska for each month of the year. The months are numbered 1 through 12, with 1 corresponding to January. (*Source:* National Oceanic and Atmospheric Administration)

(a) Plot the data given in the table.

(b) Did you use the same scale on both axes? Explain.

(c) Using the graph, find the three consecutive months when the normal temperature changes the least.

Solution

(a)

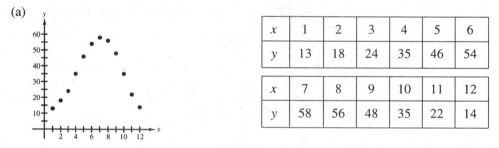

x	1	2	3	4	5	6
y	13	18	24	35	46	54

x	7	8	9	10	11	12
y	58	56	48	35	22	14

(b) No. On the graph shown above, each mark on the horizontal axis represents 1 unit and each mark on the vertical axis represents 5 units. *Note:* Answers may vary on this question.

(c) Normal temperatures change the least during the months of June, July, and August.

25. Decide whether the following ordered pairs are solutions of the equation $y = 2x + 4$.

(a) $(3, 10)$ (b) $(-1, 3)$ (c) $(0, 0)$ (d) $(-2, 0)$

Solution

(a) The ordered pair $(3, 10)$ *is* a solution because $10 = 2(3) + 4$.

(b) The ordered pair $(-1, 3)$ is *not* a solution because $3 \neq 2(-1) + 4$.

(c) The ordered pair $(0, 0)$ is *not* a solution because $0 \neq 2(0) + 4$.

(d) The ordered pair $(-2, 0)$ *is* a solution because $0 = 2(-2) + 4$.

27. Determine whether the following ordered pairs are solution points of $y = \frac{2}{3}x$.

(a) $(6, 6)$ (b) $(-9, -6)$ (c) $(0, 0)$ (d) $\left(-1, \frac{2}{3}\right)$

Solution

(a) The ordered pair $(6, 6)$ is *not* a solution because $6 \neq \frac{2}{3}(6)$.

(b) The ordered pair $(-9, -6)$ *is* a solution because $-6 = \frac{2}{3}(-9)$.

(c) The ordered pair $(0, 0)$ *is* a solution because $0 = \frac{2}{3}(0)$.

(c) The ordered pair $\left(-1, \frac{2}{3}\right)$ is *not* a solution because $\frac{2}{3} \neq \frac{2}{3}(-1)$.

29. Complete the table shown in the textbook. Plot the resulting data on a rectangular coordinate system.

Solution

x	-2	0	2	4	6
$y = 3x - 4$	-10	-4	2	8	14

When $x = -2$, $y = 3(-2) - 4 = -6 - 4 = -10$.
When $x = 0$, $y = 3(0) - 4 = 0 - 4 = -4$.
When $x = 2$, $y = 3(2) - 4 = 6 - 4 = 2$.
When $x = 4$, $y = 3(4) - 4 = 12 - 4 = 8$.
When $x = 6$, $y = 3(6) - 4 = 18 - 4 = 14$.

31. Complete the table shown in the textbook. Plot the resulting data in a rectangular coordinate system.

Solution

x	-5	$\frac{3}{2}$	5	10	20
$y = -\frac{3}{2}x + 5$	$\frac{25}{2}$	$\frac{11}{4}$	$-\frac{5}{2}$	-10	-25

When $x = -5$, $y = -\frac{3}{2}(-5) + 5 = \frac{15}{2} + \frac{10}{2} = \frac{25}{2}$.

When $x = \frac{3}{2}$, $y = -\frac{3}{2}\left(\frac{3}{2}\right) + 5 = -\frac{9}{4} + \frac{20}{4} = \frac{11}{4}$.

When $x = 5$, $y = -\frac{3}{2}(5) + 5 = -\frac{15}{2} + \frac{10}{2} = -\frac{5}{2}$.

When $x = 10$, $y = -\frac{3}{2}(10) + 5 = -15 + 5 = -10$.

When $x = 20$, $y = -\frac{3}{2}(20) + 5 = -30 + 5 = -25$.

33. *Think About It*   Review the tables in Exercises 29–32 and observe that in some cases the y-coordinates of the solution points increase and in others the y-coordinates decrease. What factor in the equation causes this? Explain.

Solution

The y-coordinates increase when the coefficient of x is positive; the y-coordinates decrease when the coefficient of x is negative.

35. *Graphical Estimation*   Use the bar graph in the textbook which shows new housing starts (in thousands) in the United States from 1980 through 1992 (*Source:* U.S. Bureau of the Census). Estimate the number of new housing starts in 1986.

Solution

The number of new housing starts in 1986 was approximately 1,800,000.

37. *Graphical Estimation*   Use the bar graph in the textbook which shows new housing starts (in thousands) in the United States from 1980 through 1992 (*Source:* U.S. Bureau of the Census). Estimate the increase and the percent increase in housing starts from 1982 to 1983.

Solution

From the graph, it appears that the estimated increase in housing starts from 1982 to 1983 was between 600,000 and 650,000. This is an approximate increase of between 60% and 65%.

39. Solve the equation $125(r-1) = 625$.

Solution

$$125(r-1) = 625$$
$$125r - 125 = 625$$
$$125r - 125 + 125 = 625 + 125$$
$$125r = 750$$
$$r = 6$$

41. Solve the equation $20 - \frac{1}{9}x = 4$.

Solution

$$20 - \frac{1}{9}x = 4$$
$$9\left(20 - \frac{1}{9}x\right) = 9(4)$$
$$180 - x = 36$$
$$180 - 180 - x = 36 - 180$$
$$-x = -144$$
$$x = 144$$

43. *Housing Costs*   The total cost of a lot and house is $154,000. The cost of constructing the house is seven times the cost of the lot. What is the cost of the lot?

Solution

Verbal model: $\boxed{\text{Cost of lot}} + \boxed{\text{Cost of house}} = \boxed{\text{Total cost}}$

Labels: Cost of lot $= x$ (dollars)
Cost of house $= 7x$ (dollars)
Total cost $= 154,000$ (dollars)

Equation: $x + 7x = 154,000$

$$8x = 154,000$$
$$x = 19,250$$

The cost of the lot is $19,250.

45. Plot the points $(-10, -4)$, $(4, -4)$, and $(4, 3)$ on a rectangular coordinate system.

Solution

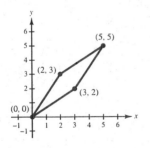

47. Plot the points $(3, -4)$, $(5, 0)$, and $(0, 3)$ on a rectangular coordinate system.

Solution

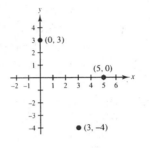

49. *Graphical Estimation* Estimate the coordinates of the points shown on the graph in the textbook.

Solution

A. $(-1, 3)$
B. $(5, -3)$
C. $(2, 1)$
D. $(-1, -2)$

51. *Geometry* Plot the points $(2, 4)$, $(5, 1)$, $(2, -2)$, and $(-1, 1)$ and connect them with line segments to form a square.

Solution

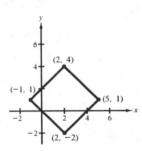

53. *Geometry* Plot the points $(0, 0)$, $(3, 2)$, $(2, 3)$, and $(5, 5)$ and connect them with line segments to form a rhombus.

Solution

55. *Graphical Interpretation* The table in the textbook gives the hours x that a student studied for five different algebra exams and the resulting scores y.

(a) Plot the data given in the table.

(b) Use the graph to describe the relationship between the number of hours studied and the resulting exam score.

Solution

(a)

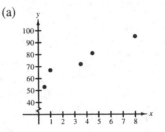

(b) As the number of hours of study increases, the exam score increases.

57. Decide whether the following ordered pairs are solutions of the equation $y = 3 - 4x$.

(a) $(-1, 7)$ (b) $(2, 5)$ (c) $(0, 0)$ (d) $(2, -5)$

Solution

(a) The ordered pair $(-1, 7)$ *is* a solution because $7 = 3 - 4(-1)$.
(b) The ordered pair $(2, 5)$ is *not* a solution because $5 \neq 3 - 4(2)$.
(c) The ordered pair $(0, 0)$ is *not* a solution because $0 \neq 3 - 4(0)$.
(d) The ordered pair $(2, -5)$ *is* a solution because $-5 = 3 - 4(2)$.

59. Decide whether the following ordered pairs are solutions of the equation $2y - 3x + 1 = 0$.

(a) $(1, 1)$ (b) $(5, 7)$ (c) $(-3, -1)$ (d) $\left(-2, -\frac{7}{2}\right)$

Solution

(a) The ordered pair $(1, 1)$ *is* a solution because $2(1) - 3(1) + 1 = 2 - 3 + 1 = 0$.
(b) The ordered pair $(5, 7)$ *is* a solution because $2(7) - 3(5) + 1 = 14 - 15 + 1 = 0$.
(c) The ordered pair $(-3, -1)$ is *not* a solution because $2(-1) - 3(-3) + 1 = -2 + 9 + 1 = 6 \neq 0$.
(d) The ordered pair $\left(-2, -\frac{7}{2}\right)$ *is* a solution because $2\left(-\frac{7}{2}\right) - 3(-2) + 1 = -7 + 6 + 1 = 0$.

61. Complete the table shown in the textbook. Plot the results on a rectangular coordinate system.

Solution

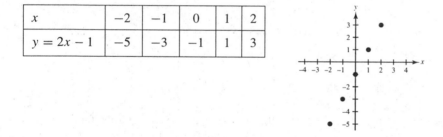

x	-2	-1	0	1	2
$y = 2x - 1$	-5	-3	-1	1	3

63. Complete the table shown in the textbook. Plot the results on a rectangular coordinate system.

Solution

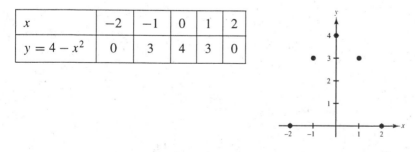

x	-2	-1	0	1	2
$y = 4 - x^2$	0	3	4	3	0

65. *Organizing Data*   The cost y of producing x units of a product is given by $y = 35x + 5000$. Complete the table shown in the textbook to determine the cost for producing the specified number of units. Plot the results on a rectangular coordinate system.

Solution

x	100	150	200	250	300
$y = 35x + 5000$	8500	10,250	12,000	13,750	15,500

When $x = 100$, $y = 35(100) + 5000 = 8500$.
When $x = 150$, $y = 35(150) + 5000 = 10{,}250$.
When $x = 200$, $y = 35(200) + 5000 = 12{,}000$.
When $x = 250$, $y = 35(250) + 5000 = 13{,}750$.
When $x = 300$, $y = 35(300) + 5000 = 15{,}500$.

67. *Graphical Comparisons*   Which kind of graph is shown in the textbook: a *straight-line depreciation* graph in which the value depreciates by the same amount each year, or a *declining balances* graph in which the value depreciates by the same percent each year?

Solution

This is a graph for the type of depreciation called declining balances.

69. *Graphical Interpretation*   In Exercises 67 and 68, what is the original cost of the equipment that is being depreciated?

Solution

The original cost of the equipment is $1000.

71. *Graphical Estimation*   Use the bar graph in the textbook showing the per capita personal income in the United States from 1984 through 1992 (*Source:* U.S. Bureau of Economic Analysis). Estimate the increase in per capita personal income from 1984 to 1989.

Solution

From the graph, it appears that the approximate increase in per capita personal income from 1984 to 1989 was between $4000 and $5000.

73. *Graphical Estimation*   Use the bar graph in the textbook which compares the percents of gross domestic product spent on health care in selected countries in 1991 (*Source:* Organization for Economic Cooperation and Development). Estimate the percent of gross domestic product spent on health care in Sweden.

Solution

The percent of gross domestic product spent on health care in Sweden is approximately 8.5%.

75. (a) Plot the points $(3, 2)$, $(-5, 4)$, and $(6, -4)$ on a rectangular coordinate system.

(b) Change the sign of the x-coordinate of each point and plot the three new points on the same axes.

(c) What can you infer about the location of a point when the sign of the x-coordinate is changed?

Solution

(a) & (b)

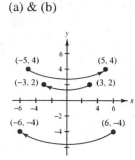

Original Points	New Points
$(3, 2)$	$(-3, 2)$
$(-5, 4)$	$(5, 4)$
$(6, -4)$	$(-6, -4)$

(c) When the sign of the x-coordinate is changed, the location of the point is reflected about the y-axis.

4.2   Graphs of Equations

7. Match the equation $y = 3 - x$ with its graph. [The graphs are labeled (a), (b), (c), and (d).]

Solution

(d)

9. Match the equation $y = -x^2 + 1$ with its graph. [The graphs are labeled (a), (b), (c), and (d).]

Solution

(a)

11. Solve the equation $3x + 4y = 12$ for y.

Solution

$$3x + 4y = 12$$

$$3x - 3x + 4y = 12 - 3x$$

$$4y = 12 - 3x$$

$$\frac{4y}{4} = \frac{12 - 3x}{4}$$

$$y = \frac{12 - 3x}{4} \text{ or } y = \frac{1}{4}(12 - 3x)$$

$$\text{or } y = 3 - \frac{3}{4}x \text{ or } y = -\frac{3}{4}x + 3$$

13. Solve the equation $x - 2y = 8$ for y.

Solution

$$x - 2y = 8$$

$$x - x - 2y = -x + 8$$

$$-2y = -x + 8$$

$$\frac{-2y}{-2} = \frac{-x + 8}{-2}$$

$$y = \frac{(-x + 8)(-1)}{(-2)(-1)}$$

$$y = \frac{x - 8}{2} \text{ or }$$

$$y = \frac{1}{2}(x - 8) \text{ or } y = \frac{1}{2}x - 4$$

15. Complete the table shown in the textbook and use the results to sketch the graph of the equation $y = 9 - x$.

Solution

x	-2	-1	0	1	2
y	11	10	9	8	7
(x, y)	$(-2, 11)$	$(-1, 0)$	$(0, 9)$	$(1, 8)$	$(2, 7)$

When $x = -2$, $y = 9 - (-2) = 11$.
When $x = -1$, $y = 9 - (-1) = 10$.
When $x = 0$, $y = 9 - 0 = 9$.
When $x = 1$, $y = 9 - 1 = 8$.
When $x = 2$, $y = 9 - 2 = 7$.

17. *Graphical Solution/Algebraic Check*   Given $4x - 2y = -8$, graphically estimate the x- and y-intercepts of the graph. Then check your results algebraically.

Solution

Graphical solution: It appears that the x-intercept is $(-2, 0)$ and the y-intercept is $(0, 4)$.

Algebraic check:

$$4x - 2y = -8 \qquad\qquad 4x - 2y = -8$$
$$\text{Let } y = 0. \qquad\qquad \text{Let } x = 0.$$
$$4x - 2(0) = -8 \qquad\qquad 4(0) - 2y = -8$$
$$4x = -8 \qquad\qquad\qquad -2y = -8$$
$$x = -2 \qquad\qquad\qquad y = 4$$
$$x\text{-intercept: } (-2, 0) \qquad y\text{-intercept: } (0, 4)$$

19. *Graphical Solution/Algebraic Check*   Given $y = |x| - 3$, graphically estimate the x- and y-intercepts of the graph. Then check your results algebraically.

Solution

Graphical solution: It appears that the x-intercepts are $(-3, 0)$ and $(3, 0)$; it appears that the y-intercept is $(0, -3)$.

Algebraic check:

$$y = |x| - 3 \qquad\qquad y = |x| - 3$$
$$\text{Let } y = 0. \qquad\qquad \text{Let } x = 0.$$
$$0 = |x| - 3 \qquad\qquad y = |0| - 3$$
$$3 = |x| \qquad\qquad\qquad y = -3$$
$$x = -3 \text{ or } x = 3$$
$$x\text{-intercepts: } (-3, 0), (3, 0) \qquad y\text{-intercept: } (0, -3)$$

21. Find the x- and y-intercepts of the graph of the equation $x - y = 1$.

Solution

x-intercept	y-intercept
Let $y = 0$.	Let $x = 0$.
$x - 0 = 1$	$0 - y = 1$
$x = 1$	$-y = 1$
$(1, 0)$	$y = -1$
	$(0, -1)$

The graph has one x-intercept at $(1, 0)$ and one y-intercept at $(0, -1)$.

25. Find the x- and y-intercepts of the graph of the equation $y = \frac{1}{2}x - 1$.

Solution

x-intercept	y-intercept
Let $y = 0$.	Let $x = 0$.
$0 = \frac{1}{2}x - 1$	$y = \frac{1}{2}(0) - 1$
$2(0) = 2\left(\frac{1}{2}x - 1\right)$	$y = 0 - 1$
$0 = x - 2$	$y = -1$
$2 = 0$	$(0, -1)$
$(2, 0)$	

The graph has one x-intercept at $(2, 0)$ and one y-intercept at $(0, -1)$.

23. Find the x- and y-intercepts of the graph of the equation $2x + y - 4 = 0$.

Solution

x-intercept	y-intercept
Let $y = 0$.	Let $x = 0$.
$2x + 0 - 4 = 0$	$2(0) + y - 4 = 0$
$2x - 4 = 0$	$0 + y - 4 = 0$
$2x = 4$	$y - 4 = 0$
$x = 2$	$y = 4$
$(2, 0)$	$(0, 4)$

The graph has one x-intercept at $(2, 0)$ and one y-intercept at $(0, 4)$.

27. Sketch the graph of the equation $y = 2 - x$ and label the coordinates of three solution points.

Solution

x	0	2	3
$y = 2 - x$	2	0	-1

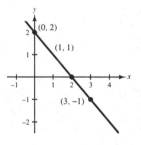

29. Sketch the graph of the equation $2x - 3y = 12$ and label the coordinates of three solution points.

Solution

$$2x - 3y = 12$$
$$-3y = -2x + 12$$
$$y = \frac{2}{3}x - 4$$

x	0	6	3
$y = \frac{2}{3}x - 4$	-4	0	-2

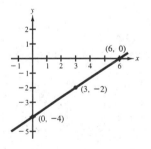

31. Sketch the graph of the equation $4x + y = 2$ and label the coordinates of three solution points.

Solution

$4x + y = 2$

$\qquad y = -4x + 2$

x	0	$\frac{1}{2}$	1
$y = -4x + 2$	2	0	-2

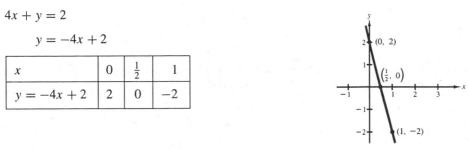

33. Sketch the graph of the equation $y = x^2$ and label the coordinates of three solution points.

Solution

x	-2	-1	0	1	2
$y = x^2$	4	1	0	1	4

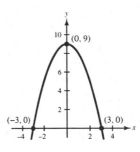

35. Sketch the graph of the equation $y = -x^2 + 9$ and label the coordinates of three solution points.

Solution

x	-3	-2	0	1	3
$y = -x^2 + 9$	0	5	9	6	0

37. Sketch the graph of the equation $y = |x| - 3$ and label the coordinates of three solution points.

Solution

x	-4	-3	0	2	3
$y = |x| - 3$	1	0	-3	-1	0

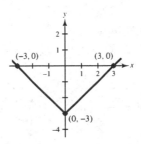

39. *Research Project*   Use a weekly news magazine or newspaper to find examples of misleading graphs and explain why they are misleading.

Solution

Answers will vary.

41. *Creating a Model*   The cost of printing a book is $500, plus $5 per book. Let C represent the total cost and let x represent the number of books. Write an equation that relates C and x and sketch its graph.

Solution

$C = 500 + 5x$

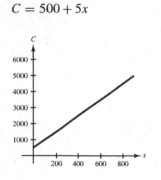

43. Solve the equation $\frac{5}{6}x - 7 = 0$.

Solution

$$\frac{5}{6}x - 7 = 0$$

$$6\left(\frac{5}{6}x - 7\right) = 6(0)$$

$$5x - 42 = 0$$

$$5x - 42 + 42 = 0 + 42$$

$$5x = 42$$

$$\frac{5x}{5} = \frac{42}{5}$$

$$x = \frac{42}{5}$$

45. Solve the following equation.

$$\frac{t}{2} + \frac{t}{4} = 30$$

Solution

$$\frac{t}{2} + \frac{t}{4} = 30$$

$$8\left(\frac{t}{2} + \frac{t}{4}\right) = 8(30)$$

$$4t + 2t = 240$$

$$6t = 240$$

$$\frac{6t}{6} = \frac{240}{6}$$

$$t = 40$$

47. *Geometry*   The width of a rectangular mirror is $\frac{3}{5}$ its length. The perimeter of the mirror is 80 inches. What are the dimensions of the mirror?

Solution

Verbal model: $2\;\boxed{\text{Length of mirror}}+2\;\boxed{\text{Width of mirror}}=\boxed{\text{Perimeter of mirror}}$

Labels: Length of mirror $= x$ (inches)
Width of mirror $=\frac{3}{5}x$ (inches)
Perimeter $= 80$ (inches)

Equation: $2x + 2\left(\frac{3}{5}x\right) = 80$

$2x + \frac{6}{5}x = 80$

$5\left(2x + \frac{6}{5}x\right) = 5(80)$

$10x + 6x = 400$

$16x = 400$

$x = 25$ and $\frac{3}{5}x = 15$

The dimensions of the mirror are 25 inches by 15 inches.

49. Complete the table shown in the textbook and use the results to sketch the graph of the equation $y = 4x - 2$.

Solution

x	-2	-1	0	1	2
y	-10	-6	-2	2	6

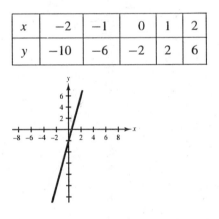

51. Complete the table shown in the textbook and use the results to sketch the graph of the equation $y = |x + 1|$.

Solution

x	-3	-2	-1	0	1
y	2	1	0	1	2

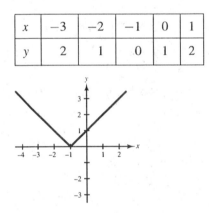

53. *Graphical Solution/Algebraic Check*   Given $y = 4 - |x|$, graphically estimate the x- and y-intercepts of the graph. Then check your results algebraically.

Solution

Graphical solution: It appears that the x-intercepts are $(-4, 0)$ and $(4, 0)$; it appears that the y-intercept is $(0, 4)$.

Algebraic check:

$y = 4 - |x|$
Let $y = 0$.
$0 = 4 - |x|$

$|x| = 4$

$x = -4$ or $x = 4$

x-intercepts: $(-4, 0), (4, 0)$

$y = 4 - |x|$
Let $x = 0$.
$y = 4 - |0|$

$y = 4$

y-intercept: $(0, 4)$

55. Find the x- and y-intercepts of the graph of the equation $y = 6x + 2$.

Solution

x-intercept	y-intercept
Let $y = 0$.	Let $x = 0$.
$0 = 6x + 2$	$y = 6(0) + 2$
$-2 = 6x$	$y = 0 + 2$
$-\frac{1}{3} = x$	$y = 2$
$\left(-\frac{1}{3}, 0\right)$	$(0, 2)$

The graph has one x-intercept at $\left(-\frac{1}{3}, 0\right)$ and one y-intercept at $(0, 2)$.

57. Find the x- and y-intercepts of the graph of the equation $2x + 6y = 9$.

Solution

x-intercept	y-intercept
Let $y = 0$.	Let $x = 0$.
$2x + 6(0) = 9$	$2(0) + 6y = 9$
$2x = 9$	$6y = 9$
$x = \frac{9}{2}$	$y = \frac{9}{6}$
$\left(\frac{9}{2}, 0\right)$	$y = \frac{3}{2}$
	$\left(0, \frac{3}{2}\right)$

The graph has one x-intercept at $\left(\frac{9}{2}, 0\right)$ and one y-intercept at $\left(0, \frac{3}{2}\right)$.

59. Find the x- and y-intercepts of the graph of the equation $\frac{3}{4}x - \frac{1}{2}y = 3$.

Solution

x-intercept	y-intercept
Let $y = 0$.	Let $x = 0$.
$\frac{3}{4}x - \frac{1}{2}(0) = 3$	$\frac{3}{4}(0) - \frac{1}{2}y = 3$
$\frac{3}{4}x = 3$	$-\frac{1}{2}y = 3$
$4\left(\frac{3}{4}x\right) = 4(3)$	$2\left(-\frac{1}{2}y\right) = 2(3)$
$3x = 12$	$-y = 6$
$x = 4$	$y = -6$
$(4, 0)$	$(0, -6)$

The graph has one x-intercept at $(4, 0)$ and one y-intercept at $(0, -6)$.

61. Sketch the graph of the equation $y = x - 1$ and label the coordinates of three solution points.

Solution

x	0	1	3
$y = x - 1$	-1	0	2

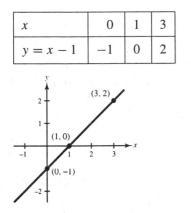

63. Sketch the graph of the equation $4x + 5y = 20$ and label the coordinates of three solution points.

Solution

$$4x + 5y = 20$$
$$5y = -4x + 20$$
$$\frac{5y}{5} = \frac{-4x}{5} + \frac{20}{5}$$
$$y = -\frac{4}{5}x + 4$$

x	0	$\frac{5}{2}$	5
$y = -\frac{4}{5}x + 4$	4	2	0

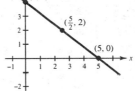

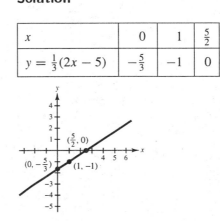

65. Sketch the graph of the equation $y = \frac{3}{8}x + 15$ and label the coordinates of three solution points.

Solution

x	-40	0	8
$y = \frac{3}{8}x + 15$	0	15	18

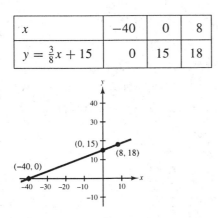

67. Sketch the graph of the equation $y = \frac{1}{3}(2x - 5)$ and label the coordinates of three solution points.

Solution

x	0	1	$\frac{5}{2}$
$y = \frac{1}{3}(2x - 5)$	$-\frac{5}{3}$	-1	0

69. Sketch the graph of the equation $y = 3x$ and label the coordinates of three solution points.

Solution

x	-1	0	2
$y = 3x$	-3	0	6

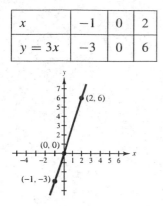

71. Sketch the graph of the equation $y = -x^2$ and label the coordinates of three solution points.

Solution

x	-2	0	1	2
$y = -x^2$	-4	0	-1	-4

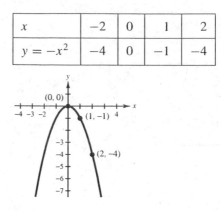

73. Sketch the graph of the equation $y = x^2 + 3$ and label the coordinates of three solution points.

Solution

x	-1	0	2
$y = x^2 + 3$	4	3	7

75. Sketch the graph of the equation $y = |x - 5|$ and label the coordinates of three solution points.

Solution

x	0	4	5	6
$y = |x - 5|$	5	1	0	1

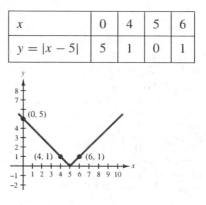

77. *Interpreting Intercepts* The model $5F - 9C = 160$ relates the temperature in degrees Celsius C and degrees Fahrenheit F.

(a) Graph the equation where F is measured on the horizontal axis.

(b) Explain what the intercepts represent.

Solution

(a) $5F - 9C = 160$

$$-9C = -5F + 160$$

$$\frac{-9C}{-9} = \frac{-5F + 160}{-9}$$

$$C = \frac{5F - 160}{9}$$

F	-4	0	32	68	212
$C = \dfrac{5F - 160}{9}$	-20	$-17\frac{7}{9}$	0	20	100

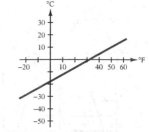

(b) The x-intercept $(32, 0)$ represents the Fahrenheit temperature of 32 degrees which corresponds to the Celsius temperature of 0 degrees. The y-intercept $\left(0, -17\frac{7}{9}\right)$ represents the Celsius temperature of $-17\frac{7}{9}$ degrees which corresponds to the Fahrenheit temperature of 0 degrees.

4.3 Graphs and Graphing Utilities

7. Use a graphing utility to graph the equation $y = x + 1$. (Use a standard setting.)

Solution

Keystrokes: $\boxed{Y=}$ $\boxed{X|T}$ $\boxed{+}$ 1 $\boxed{GRAPH}$

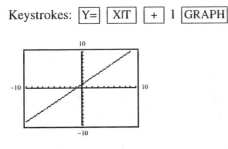

9. Use a graphing utility to graph the equation $y = -\frac{1}{3}x$. (Use a standard setting.)

Solution

Keystrokes: $\boxed{Y=}$ $\boxed{(-)}$ $\boxed{X|T}$ $\boxed{\div}$ 3 $\boxed{GRAPH}$

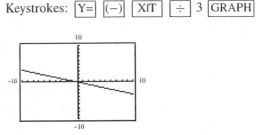

11. Use a graphing utility to graph the equation $y = -2x^2 + 5$. (Use a standard setting.)

Solution

Keystrokes: $\boxed{Y-}$ $\boxed{(-)}$ 2 $\boxed{X|T}$ $\boxed{x^2}$ $\boxed{+}$ 5 $\boxed{GRAPH}$

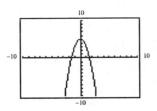

13. Use a graphing utility to graph the equation $y = |x + 1| - 2$. (Use a standard setting.)

Solution

Keystrokes:

$\boxed{Y=}$ $\boxed{ABS}$ $\boxed{(}$ $\boxed{X|T}$ $\boxed{+}$ 1 $\boxed{)}$ $\boxed{-}$ 2 $\boxed{GRAPH}$

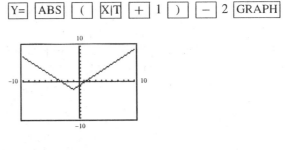

15. Use a graphing utility to graph the equation $y = 25 - 3x^2$.

```
RANGE
Xmin=-4
Xmax=4
Xscl=1
Ymin=-5
Ymax=25
Yscl=5
```

Solution

Keystrokes: $\boxed{Y=}$ 25 $\boxed{-}$ 3 $\boxed{X|T}$ $\boxed{x^2}$ $\boxed{GRAPH}$

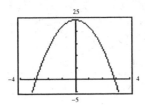

17. Use a graphing utility to graph the equation $y = 3x^2 - 9x$.

```
RANGE
Xmin=-2
Xmax=4
Xscl=1
Ymin=-8
Ymax=8
Yscl=2
```

Solution

Keystrokes: $\boxed{Y=}$ 3 $\boxed{X|T}$ $\boxed{x^2}$ $\boxed{-}$ 9 $\boxed{X|T}$ $\boxed{GRAPH}$

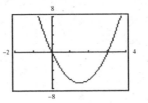

19. Use a graphing utility to find a viewing window that matches the one shown in the textbook for the equation $y = \frac{1}{2}x + 2$.

Solution

```
RANGE
Xmin=-15
Xmax=15
Xscl=1
Ymin=-10
Ymax=10
Yscl=1
```

21. Use a graphing utility to find a viewing window that matches the one shown in the textbook for the equation $y = \frac{1}{4}x^2 - 4x + 12$.

Solution

```
RANGE
Xmin=-5
Xmax=20
Xscl=5
Ymin=-5
Ymax=20
Yscl=5
```

23. Use a graphing utility to find a viewing window that matches the one shown in the textbook for the equation $y = |2x - 1|$.

Solution

```
RANGE
Xmin=-8
Xmax=8
Xscl=4
Ymin=-10
Ymax=10
Yscl=1
```

25. Match the equation $y = x^2$ with its graph. [The graphs are labeled (a), (b), (c), and (d) in the textbook.]

Solution

(b)

27. Match the equation $y = \frac{1}{4}x^2$ with its graph. [The graphs are labeled (a), (b), (c), and (d) in the textbook.]

Solution

(d)

29. Graph both $y_1 = \frac{1}{3}x - 1$ and $y_2 = -1 + \frac{1}{3}x$ on the same screen. Are the graphs identical? If so, what Rule of Algebra is being illustrated?

Solution

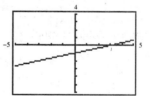

Yes, the graphs are identical. By the Commutative Property of Addition, $\frac{1}{3}x - 1 = -1 + \frac{1}{3}x$.

31. Graph both $y_1 = x(x - 2)$ and $y_2 = x^2 - 2x$ on the same screen. Are the graphs identical? If so, what Rule of Algebra is being illustrated?

Solution

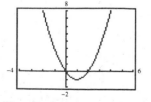

Yes, the graphs are identical. By the Distributive Property $x(x - 2) = x^2 - 2x$.

33. Given $y = 2x - 5$, use the TRACE feature of a graphing utility to approximate the x- and y-intercepts of the graph.

Solution

x-intercept: $\left(\frac{5}{2}, 0\right)$

y-intercept: $(0, -5)$

35. Given $y = x^2 + = 1.5x - 1$, use the TRACE feature of a graphing utility to approximate the x- and y-intercepts of the graph.

Solution

x-intercepts: $\left(\frac{1}{2}, 0\right)$, $(-2, 0)$

y-intercept: $(0, -1)$

37. *Modeling Data* Use the following models, which give the revenues from stamps and postal cards and from postage paid under permit and meter.

Stamps and postal cards: $y = 0.06x^2 + 1.59x + 18.08,\ 4 \le x \le 11$

Permit and metered mail: $y = 0.03x^2 + 1.19x + 10.91,\ 4 \le x \le 11$

In these models, y is the revenue (in billions of dollars) and x is the year, with $x = 0$ corresponding to 1980 (*Source:* U.S. Postal Service). Use the following setting to graph both models on the same display of a graphing utility.

```
RANGE
Xmin=4
Xmax=11
Xscl=1
Ymin=10
Ymax=45
Yscl=5
```

Solution

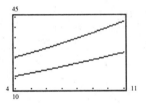

39. Given $y = 15 + |x - 12|$, find a viewing window that shows the important characteristics of the graph.

Solution

```
RANGE
Xmin=-5
Xmax=25
Xscl=5
Ymin=-2
Ymax=40
Yscl=5
```

Note: Answers may vary.

41. Given $y = -15 + |x + 12|$, find a viewing window that shows the important characteristics of the graph.

Solution

```
RANGE
Xmin=-40
Xmax=10
Xscl=5
Ymin=-20
Ymax=10
Yscl=5
```

Note: Answers may vary.

43. Write the coordinates of three points that are solutions of the equation $y = \frac{4}{5}x + 2$.

Solution

Note: Answers may vary.

x	-5	0	5
$y = \frac{4}{5}x + 2$	-2	2	6
Solution	$(-5, -2)$	$(0, 2)$	$(5, 6)$

45. Write the coordinates of three points that are solutions of the equation $y = 8 - 0.75x$.

Solution

Note: Answers may vary.

x	-4	0	4
$y = 8 - 0.75x$	11	8	5
Solution	$(-4, 11)$	$(0, 8)$	$(4, 5)$

47. Simplify the expression $v^2 \cdot v^3$.

Solution

$v^2 \cdot v^3 = v^5$

49. Simplify the expression $x(x - 3) - (x^2 + 6x)$

Solution

$$x(x - 3) - (x^2 + 6x) = x^2 - 3x - x^2 - 6x$$
$$= -9x$$

51. Use a graphing utility to graph the equation $y = -3x$. (Use the standard setting.)

Solution

Keystrokes:

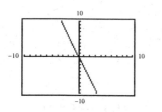

53. Use a graphing utility to graph the equation $y = \frac{3}{4}x - 6$. (Use the standard setting.)

Solution

Keystrokes:

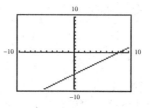

55. Use a graphing utility to graph the equation $y = \frac{1}{2}x^2$. (Use the standard setting.)

Solution

Keystrokes: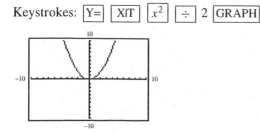

57. Use a graphing utility to graph the equation $y = x^2 - 4x + 2$. (Use the standard setting.)

Solution

Keystrokes:

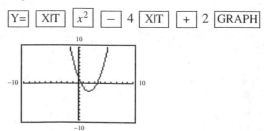

59. Use a graphing utility to graph the equation $y = |x - 3|$. (Use the standard setting.)

Solution

Keystrokes:

$\boxed{\text{Y=}}$ $\boxed{\text{ABS}}$ $\boxed{(}$ $\boxed{\text{X}|\text{T}}$ $\boxed{-}$ 3 $\boxed{)}$ $\boxed{\text{GRAPH}}$

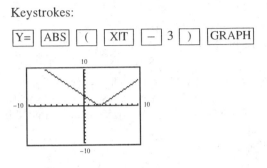

61. Use a graphing utility to graph the equation $y = |x^2 - 4|$. (Use the standard setting.)

Solution

Keystrokes:

$\boxed{\text{Y=}}$ $\boxed{\text{ABS}}$ $\boxed{(}$ $\boxed{\text{X}|\text{T}}$ $\boxed{x^2}$ $\boxed{-}$ 4 $\boxed{)}$ $\boxed{\text{GRAPH}}$

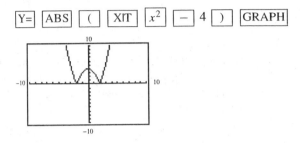

63. Use a graphing utility to graph the equation $y = 27x + 100$.

```
RANGE
Xmin=0
Xmax=5
Xscl=.5
Ymin=75
Ymax=250
Yscl=25
```

Solution

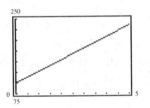

65. Use a graphing utility to graph the equation $y = 0.001x^2 + 0.5x$.

```
RANGE
Xmin=-500
Xmax=200
Xscl=50
Ymin=-100
Ymax=100
Yscl=20
```

Solution

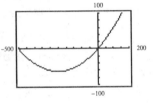

67. Graph both $y_1 = 2x + (x + 1)$ and $y_2 = (2x + x) + 1$ on the same screen. Are the graphs identical? If so, what Rule of Algebra is being illustrated?

Solution

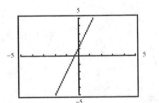

Yes, the graphs are identical. By the Associative Property of Addition $2x + (x + 1) = (2x + x) + 1$.

69. Graph both $y_1 = 2\left(\frac{1}{2}\right)$ and $y_2 = 1$ on the same screen. Are the graphs identical? If so, what Rule of Algebra is being illustrated?

Solution

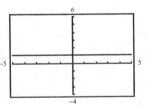

Yes, the graphs are identical. By the Multiplicative Inverse Property $2\left(\frac{1}{2}\right) = 1$.

71. Given $y = 9 - x^2$, use the TRACE feature of a graphing utility to approximate the x- and y-intercepts of the graph.

Solution

x-intercepts: $(3, 0)$, $(-3, 0)$
y-intercept: $(0, 9)$

73. Given $y = 6 - |x + 2|$, use the TRACE feature of a graphing utility to approximate the x- and y-intercepts of the graph.

Solution

x-intercepts: $(-8, 0)$, $(4, 0)$
y-intercept: $(0, 4)$

75. *Geometry* Graph the equations $y = -4$ and $y = -|x|$ on the same display. Using a "square setting," determine the geometrical shape bounded by the graphs.

Solution

Triangle

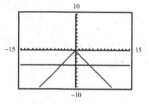

77. *Geometry* Graph the equations $y = |x| - 8$ and $y = -|x| + 8$ on the same display. Using a "square setting," determine the geometrical shape bounded by the graphs.

Solution

Square

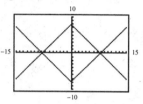

79. *Modeling Data* Use the following models, which give the number of births and the number of deaths in the United States from 1980 through 1991.

$y = 0.004x^2 + 0.010x + 3.602$, $0 \le x \le 11$ Births
$y = -0.001x^2 + 0.027x + 1.959$, $0 \le x \le 11$ Deaths

In these models, y is the number of births and deaths (in millions), and x is the year, with $x = 0$ corresponding to 1980 (*Source:* National Center for Health Statistics). Graph both models on the same screen.

```
RANGE
Xmin=0
Xmax=11
Xscl=.5
Ymin=1.5
Ymax=5
Yscl=.5
```

Solution

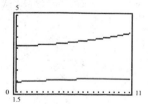

Mid-Chapter Quiz for Chapter 4

1. Plot the points $(4, -2)$ and $\left(-1, -\frac{5}{2}\right)$ on a rectangular coordinate system.

Solution

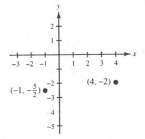

2. Determine the quadrants in which the points $(x, 5)$ must be located. (x is a real number.)

Solution

Quadrants I or II. This point is located 5 units *above* the horizontal axis. If x is positive, the point would be to the *right* of the vertical axis and in the first quadrant. If x is negative, the point would be to the *left* of the vertical axis and in the second quadrant. (If x is zero, the point would be on the vertical axis *between* the second and third quadrants.)

3. Decide whether the ordered pairs are solutions of the equation $y = 9 - |x|$.

(a) $(2, 7)$
(b) $(-3, 12)$
(c) $(-9, 0)$
(d) $(0, -9)$

Solution

(a) Yes, the ordered pair $(2, 7)$ *is* a solution because $7 = 9 - |2|$.
(b) No, the ordered pair $(-3, 12)$ is *not* a solution because $12 \neq 9 - |-3|$.
(c) Yes, the ordered pair $(-9, 0)$ *is* a solution because $0 = 9 - |-9|$.
(d) No, the ordered pair $(0, -9)$ is *not* a solution because $-9 \neq 9 - |0|$.

4. Complete the table shown in the textbook for $y = -x + 3$. Plot the resulting data.

Solution

x	-2	0	2	4	6
y	5	3	1	-1	-3

5. The bar graph in the textbook shows the average number (in millions) of shares traded per day on the New York Stock Exchange for the years 1988 through 1992. Estimate the percent increase in the average number of shares traded per day from 1990 to 1992. (*Source:* The New York Stock Exchange)

Solution

The increase in the average number of shares traded per day from 1990 to 1992 is approximately 30%.

6. What is the y-coordinate of any point on the x-axis?

Solution

Any point on the x-axis has a y-coordinate of 0.

7. Find the intercepts of the graph of the equation $x - 3y = 12$.

Solution

x-intercept	*y-intercept*
Let $y = 0$.	Let $x = 0$.
$x - 3(0) = 12$	$0 - 3y = 12$
$x = 12$	$-3y = 12$
$(12, 0)$	$\dfrac{-3y}{-3} = \dfrac{12}{-3}$
	$y = -4$
	$(0, -4)$

The graph has one x-intercept at $(12, 0)$ and one y-intercept at $(0, -4)$.

8. Find the intercepts of the graph of the equation $y = 6 - 4x$.

Solution

x-intercept	*y-intercept*
Let $y = 0$.	Let $x = 0$.
$0 = 6 - 4x$	$y = 6 - 4(0)$
$4x = 6$	$y = 6$
$x = \frac{6}{4}$	$(0, 6)$
$x = \frac{3}{2}$	
$\left(\frac{3}{2}, 0\right)$	

The graph has one x-intercept at $\left(\frac{3}{2}, 0\right)$ and one y-intercept at $(0, 6)$.

9. Sketch the graph of $y = x - 1$ and label the intercepts.

Solution

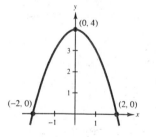

10. Sketch the graph of $y = 5 - 2x$ and label the intercepts.

Solution

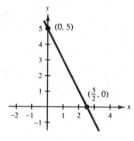

11. Sketch the graph of $y = 4 - x^2$ and label the intercepts.

Solution

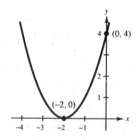

12. Sketch the graph of $y = (x + 2)^2$ and label the intercepts.

Solution

13. Sketch the graph of $y = x^3$ and label the intercepts.

Solution

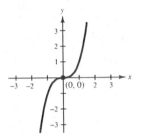

14. Sketch the graph of $y = 1 - |x|$ and label the intercepts.

Solution

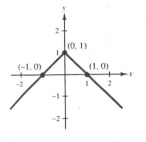

15. Use a graphing utility to graph $y = \frac{1}{2}x^2 - 12$. Find a viewing window that shows the important characteristics of the graph. In your own words, describe the graph's important characteristics.

Solution

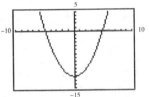

Note: The viewing window could vary.

16. Use a graphing utility to graph $y = x^3 - x$. Find a viewing window that shows the important characteristics of the graph. In your own words, describe the graph's important characteristics.

Solution

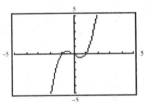

Note: The viewing window could vary.

17. Use the TRACE feature of a graphing utility to approximate the x- and y-intercepts of the graph of $y = x^2 - 6x - 7$.

Solution

The x-intercepts are $(-1, 0)$ and $(7, 0)$, and the y-intercept is $(0, -7)$.

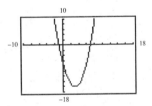

18. Use a graphing utility to graph the equations $y_1 = x(x - 2)$ and $y_2 = x^2 - 2x$ on the same screen. Are the graphs identical? If so, what Rule of Algebra is illustrated?

Solution

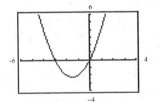

Yes, the graphs are identical. By the Distributive Property, $x(x + 3) = x^2 + 3x$.

19. Use a graphing utility to graph the equations $y_1 = \sqrt{25 - x^2}$ and $y_2 = -\sqrt{25 - x^2}$ on the same screen. The resulting graphs should form a circle. What change to the calculator settings is required if the graph does not appear circular?

Solution

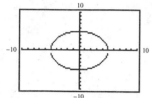

Use the square setting to make the graph appear circular.

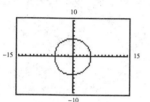

20. The cost of a new computer system is approximately $3000 and depreciates at the rate of $500 per year for 4 years. Let y represent the value of the system after x years. Write an equation that relates y and x and sketch its graph.

Solution

$y = 3000 - 500x$

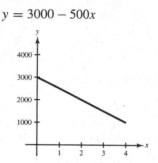

4.4   Business Applications and Graphs

7. Find the markup and the markup rate when the cost is $26.97 and the selling price is $49.95. The markup rate is a percent of the cost.

Solution

Verbal model: $\boxed{\text{Selling price}} = \boxed{\text{Cost}} + \boxed{\text{Markup}}$

Labels: Selling price = $49.95
Cost = $26.97
Markup = x (dollars)

Equation: $49.95 = 26.97 + x$

$49.95 - 26.97 = x$

$22.98 = x$

The markup is $22.98.

Verbal model: $\boxed{\text{Markup}} = \boxed{\text{Markup rate}} \cdot \boxed{\text{Cost}}$

Labels: Markup = $22.98
Markup rate = p (percent in decimal form)
Cost = $26.97

Equation: $22.98 = p(26.97)$

$\dfrac{22.98}{26.97} = p$

$0.852 \approx p$

The markup rate is approximately 85.2%.

9. Find the cost and the markup when the selling price is $74.38 and the markup rate is 81.5%. The markup rate is a percent of the cost.

Solution

Verbal model: $\boxed{\text{Selling price}} = \boxed{\text{Cost}} + \boxed{\text{Markup}}$

Labels: Selling price = $74.38
Cost = c (dollars)
Markup rate = 0.815
Markup = $0.815c$

Equation: $74.38 = c + 0.815c$

$74.38 = 1.815c$

$\dfrac{74.38}{1.815} = c$

$40.98 \approx c$

$0.815c \approx 0.815(40.98) \approx 33.40$

The cost is approximately $40.98 and the markup is $33.40.

11. Find the discount and the discount rate when the list price is $39.95 and the sale price is $29.95. The discount rate is a percent of the list price.

Solution

Verbal model: $\boxed{\text{Sale price}} = \boxed{\text{List price}} - \boxed{\text{Discount}}$

Labels: Sale price = $29.95
List price = $39.95
Discount rate = p (percent in decimal form)
Discount = $p(39.95)$ (dollars)

Equation: $29.95 = 39.95 - p(39.95)$

$29.95 - 39.95 = -p(39.95)$

$-10.00 = -p(39.95)$

$\dfrac{-10.00}{-39.95} = p$

$0.2503 \approx p$

The discount is $39.95 - $29.95 = $10.00; the discount rate is approximately 25.03% (approximately 25%).

13. Find the list price and the discount when the sale price is $18.95 and the discount rate is 20%. The discount rate is a percent of the list price.

Solution

Verbal model: $\boxed{\text{Sale price}} = \boxed{\text{List price}} - \boxed{\text{Discount}}$

Labels: Sale price = $18.95
List price = x (dollars)
Discount rate = 0.2 (percent in decimal form)
Discount = $0.2x$ (dollars)

Equation: $18.95 = x - 0.2x$ *Note:* $x - 0.2x = (1 - 0.2)x = 0.8x$

$$18.95 = 0.8x$$

$$\frac{18.95}{0.8} = x$$

$$23.69 \approx x$$

The list price is $23.69 and the discount is $23.69 - $18.95 = $4.74.

15. *Labor Charge*   An appliance repair shop charges $30 for a service call and $\frac{1}{2}$ hour of service. For each additional half hour of labor there is a charge of $16. Find the length of a service call if the bill is $78.

Solution

Verbal model: $\boxed{\text{Total bill}} = \boxed{\text{Basic cost}} + \boxed{\text{Additional labor cost}}$

Labels: Total bill = $78
Basic cost = $30
Additional half-hours of labor = x
Additional labor cost = $16x$ (dollars)

Equation: $78 = 30 + 16x$

$$78 - 30 = 16x$$

$$48 = 16x$$

$$\frac{48}{16} = x$$

$$3 = x$$

There were 3 **additional** half-hours of labor, so the length of the service call was 4 half-hours, or 2 hours.

17. *Telephone Charge*   The weekday rate for a telephone call is $0.55 for the first minute plus $0.40 for each additional minute, as shown in the graph in the textbook. Determine the length of a call that costs $2.95. What would a call of the same length have cost if it had been made during the weekend when there is a 60% discount?

Solution

Verbal model: | Total price | = | Cost of first minute | + | Cost of additional minutes |

Labels: Total price = $2.95
Cost of first minute = $0.55
Additional minutes = x
Cost for additional minutes = $0.40x$ (dollars)

Equation: $2.95 = 0.55 + 0.40x$

$2.95 - 0.55 = 0.40x$

$2.40 = 0.40x$

$\dfrac{2.40}{0.40} = x$

$6 = x$

There were 6 **additional** minutes. Therefore, the call lasted 7 minutes in all.

Verbal model: | Weekend cost | = | Weekday cost | − | Discount |

Labels: Weekend cost = x
Weekday cost = $2.95
Discount rate = 0.60 (percent in decimal form)
Discount = 0.60(2.95) = $1.77

Equation: $x = 2.95 - 1.77 = 1.18$

A call of the same length would cost $1.18 on the weekend.

19. *Amount Financed*   You buy a motorbike that costs $1450 plus 6% sales tax.

(a) Find the amount of the sales tax and the total bill.

(b) How much of the total bill must you finance if you make a down payment of $500?

Solution

(a) *Verbal model:* | Total cost | = | List price | + | Tax |

Labels: Total cost = x (dollars)
List price = $1450
Tax rate = 0.06 (percent in decimal form)
Tax = 0.06(1450) = $87

Equation: $x = 1450 + 87$

$x = 1537$

The sales tax is $87 and the total bill is $1537.

(b) *Verbal model:* | Total cost | = | Down payment | + | Amount financed |

Labels: Total cost = $1537
Down payment = $500
Amount financed = x

Equation: $1537 = 500 + x$

$1537 - 500 = x$

$1037 = x$

Thus, $1037 must be financed if a $500 down payment is made.

21. *Commission Rate* Determine the commission rate for an employee who earned $500 in commissions on sales of $4000.

Solution

Verbal model: $\boxed{\text{Commission}} = \boxed{\text{Commission rate}} \cdot \boxed{\text{Sales}}$

Labels: Commission = $500
 Commission rate = p (percent in decimal form)
 Sales = $4000

Equation: $500 = p(4000)$

$$\frac{500}{4000} = p$$

$$0.125 = p$$

The commission rate is 12.5%.

23. *Comparing Prices* A mail-order catalog lists automobile shock absorbers for $48.99 a pair, plus a shipping charge of $4.69. A local store has a special sale with 25% off a list price of $63.99. Which is the better bargain?

Solution

Verbal model (catalog): $\boxed{\text{Total cost}} = \boxed{\text{List price}} + \boxed{\text{Shipping charge}}$

Labels: Total cost = x (dollars)
 List price = $48.99
 Shipping charge = $4.69

Equation: $x = 48.99 + 4.69 = 53.68$

Verbal model (store): $\boxed{\text{Sale price}} = \boxed{\text{List price}} - \boxed{\text{Discount}}$

Labels: Sale price = x (dollars)
 List price = $63.99
 Discount rate = 0.25 (percent in decimal form)
 Discount = $0.25(63.99) \approx 16.00$

Equation: $x = 63.99 - 16.00 = 47.99$

The shock absorbers would cost $53.68 from the mail order catalog and $47.99 from the local store. The local store offers the better bargain.

25. *Graphical Estimation* As a sales representative, you receive a weekly salary of $300 plus a 3% commission on all sales.

(a) Write a linear equation giving your weekly salary y in terms of your sales x.

(b) Use a graphing utility to graph the equation in part (a). Use the following settings.

 Xmin = 0 Ymin = 200
 Xmax = 10000 Ymax = 600
 Xscl = 1000 Yscl = 50

(c) Use the graph to estimate your weekly salary if your sales are $5500.

Solution

(a) $y = 300 + 0.03x$

(b)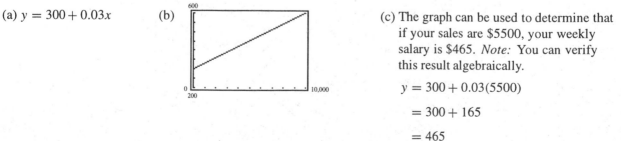

(c) The graph can be used to determine that if your sales are $5500, your weekly salary is $465. *Note:* You can verify this result algebraically.

$$y = 300 + 0.03(5500)$$

$$= 300 + 165$$

$$= 465$$

27. Solve the following equation.

$$\frac{x}{3} = \frac{4}{9}$$

Solution

$$\frac{x}{3} = \frac{4}{9}$$

$$9x = 12$$

$$x = \frac{12}{9}$$

$$x = \frac{4}{3}$$

29. Solve the following equation.

$$0.3x + 4.5 = 20.7$$

Solution

$$0.3x + 4.5 = 20.7$$

$$0.3x = 16.2$$

$$x = \frac{16.2}{0.3}$$

$$x = 54$$

31. *Modeling*   Write an algebraic expression that represents the the distance traveled by a train traveling at 65 miles per hour for t hours.

Solution

Distance $= 65t$

Note: Distance $=$ (Rate)(Time)

33. Find the cost and markup rate when the selling price is \$125.98 and the markup is \$56.69. The markup rate is a percent of the cost.

Solution

Verbal model: | Selling price | $=$ | Cost | $+$ | Markup |

Labels: Selling price $=$ \$125.98
Cost $= c$
Markup $=$ \$56.69
Markup rate $= p$ (percent in decimal form)

Equation: $125.98 = c + 56.69$

$$125.98 - 56.69 = c$$

$$69.29 = c$$

Verbal model: | Markup | $=$ | Markup rate | $\cdot$ | Cost |

Equation: $56.69 = p(69.29)$

$$\frac{56.69}{69.29} = p$$

$$0.818 \approx p$$

The cost is \$69.29 and the markup rate is approximately 81.8%.

35. Find the cost and the markup rate when the selling price is $15,900.00 and the markup is $2650.00. The markup rate is a percent of the cost.

Solution

Verbal model: $\boxed{\text{Selling price}} = \boxed{\text{Cost}} + \boxed{\text{Markup}}$

Labels: Selling price = $15,900
Cost = c (dollars)
Markup rate = p (percent in decimal form)
Markup = $2650

Equation:
$$15,900 = c + 2650$$
$$15,900 - 2650 = c$$
$$13,250 = c$$

Verbal model: $\boxed{\text{Markup}} = \boxed{\text{Markup rate}} \cdot \boxed{\text{Cost}}$

Equation:
$$2650 = p(13,250)$$
$$\frac{2650}{13,250} = p$$
$$0.2 = p$$

The cost is $13,250 and the markup rate is 20%.

37. Find the selling price and the markup when the cost is $107.97 and the markup rate is 85.2%. The markup rate is a percent of the cost.

Solution

Verbal model: $\boxed{\text{Selling price}} = \boxed{\text{Cost}} + \boxed{\text{Markup}}$

Labels: Selling price = x (dollars)
Cost = $107.97
Markup rate = 0.852 (percent in decimal form)
Markup = 0.852(107.97) ≈ $91.99

Equation: $x = 107.97 + 91.99 = 199.96$

The markup is $91.99 and the selling price is $199.96.

39. Find the sale price and the discount rate when the list price is $189.99 and the discount is $30.00. The discount rate is a percent of the list price.

Solution

Verbal model: $\boxed{\text{Sale price}} = \boxed{\text{List price}} - \boxed{\text{Discount}}$

Labels: Sale price = x (dollars)
List price = $189.99
Discount rate = p (percent in decimal form)
Discount = $30.00

Equation: $x = 189.99 - 30.00 = 159.99$

Verbal model: $\boxed{\text{Discount}} = \boxed{\text{Discount rate}} \cdot \boxed{\text{List price}}$

Equation:
$$30.00 = p(189.99)$$
$$\frac{30.00}{189.99} = p$$
$$0.158 \approx p$$

The sale price is $159.99 and the discount rate is approximately 15.8%.

41. Find the sale price and the discount when the list price is $119.96 and the discount rate is 50%. The discount rate is a percent of the list price.

Solution

Verbal model: | Sale price | = | List price | − | Discount |

Labels: Sale price = x (dollars)
List price = $119.96
Discount rate = 0.50 (percent in decimal form)
Discount = 0.50(119.96) ≈ 59.98 (dollars)

Equation: $x = 119.96 - 59.98 = 59.98$

The sale price is $59.98 and the discount is $59.98.

43. Find the list price and the discount rate when the sale price is $695.00 and the discount is $300.00. The discount rate is a percent of the list price.

Solution

Verbal model: | Sale price | = | List price | − | Discount |

Labels: Sale price = $695.00
List price = x (dollars)
Discount rate = p (percent in decimal form)
Discount = $300

Equation: $695.00 = x - 300$

$$695.00 + 300 = x$$

$$995 = x$$

Verbal model: | Discount | = | Discount rate | · | List price |

Equation: $300 = p(995)$

$$\frac{300}{995} = p$$

$$0.302 \approx p$$

The list price is $995.00 and the discount rate is approximately 30.2%.

45. *Labor Charge* A computer company charges $50 for a service call and $\frac{1}{2}$ hour of service. Each additional half hour of labor costs $19. Find the length of a service call if the bill is $145.

Solution

Verbal model: | Basic charge | + | Additional labor charge | = | Total bill |

Labels: Basic charge for service call = 50 (dollars)
Additional $\frac{1}{2}$ hours of labor = x (half-hours)
Additional labor charge = $19x$ (dollars per half-hour)
Total bill = 145 (dollars)

Equation: $50 + 19x = 145$

$$19x = 95$$

$$x = \frac{95}{19}$$

$$x = 5$$

There were 5 additional half-hours of labor, so the length of the service call was 6 half-hours, or 3 hours.

47. *Tip Rate* A customer left $20 for a meal that cost $16.95. Determine the tip rate.

Solution

Verbal model: $\boxed{\text{Tip}} = \boxed{\text{Percent}} \cdot \boxed{\text{Price of meal}}$

Labels: Tip $= 20 - 16.95 = \$3.05$
Percent $= p$ (percent in decimal form)
Price of meal $= \$16.95$

Equation: $3.05 = p(16.95)$

$$\frac{3.05}{16.95} = p$$

$$0.18 \approx p$$

The tip rate is approximately 18%.

49. *Telephone Charge* The weekday rate for a telephone call is $0.85 for the first minute plus $0.70 for each additional minute. How long is a call that costs $16.25? What would such a call cost during the weekend when there is a 60% discount?

Solution

Verbal model: $\boxed{\text{Cost of first minute}} + \boxed{\text{Cost of additional minutes}} = \boxed{\text{Total cost}}$

Labels: Cost of first minute $= 0.85$ (dollars)
Additional minutes $= x$ (minutes)
Cost of additional minutes $= 0.70x$ (dollars)
Total cost $= 16.25$ (dollars)

Equation: $0.85 + 0.70x = 16.25$

$$0.70x = 15.40$$

$$x = \frac{15.40}{0.70}$$

$$x = 22$$

There were 22 *additional* minutes. Therefore, the call lasted 23 minutes in all.

Verbal model: $\boxed{\text{Weekday cost}} - \boxed{\text{Discount}} = \boxed{\text{Weekend cost}}$

Labels: Weekday cost $= 16.25$ (dollars)
Discount rate $= 0.60$ (percent in decimal form)
Discount $= 0.60(16.25) = 9.75$ (dollars)
Weekend cost $= x$ (dollars)

Equation: $16.25 - 9.75 = x$

$$6.50 = x$$

A call of the same length would cost $6.50 on the weekend.

51. *Insurance Premium*   The annual insurance premium for a policyholder is normally $739. However, after having an automobile accident, the policyholder was charged an additional 30%. What is the new annual premium?

Solution

Verbal model: $\boxed{\text{New premium}} = \boxed{\text{Normal premium}} + \boxed{\text{Increase}}$

Labels: New premium $= x$ (dollars)
Normal premium $= \$739$
Rate of increase $= 0.3$ (percent in decimal form)
Increase $= 0.3(739) = \$221.70$

Equation: $x = 739 + 221.70$

$x = 960.70$

The new annual premium is $960.70.

53. *Amount of Sales*   The monthly salary of an employee is $1000 plus a 7% commission on the total sales. How much must the employee sell in order to obtain a monthly salary of $3500?

Solution

Verbal model: $\boxed{\text{Total income}} = \boxed{\text{Base salary}} + \boxed{\text{Commission}}$

Labels: Total income $= \$3500$
Base salary $= \$1000$
Commission rate $= 0.07$ (percent in decimal form)
Sales $= x$ (dollars)
Commission $= 0.07x$ (dollars)

Equation: $3500 = 1000 + 0.07x$

$3500 - 1000 = 0.07x$

$2500 = 0.07x$

$\dfrac{2500}{0.07} = x$

$35{,}714 \approx x$

The employee must have sales of approximately $35,714.

55. *Hours of Overtime*   An employee is paid $11.25 per hour for the first 40 hours and $16 for each additional hour (see figure in textbook). During the first week on the job, the employee's gross pay was $622. How many hours of overtime did the employee work?

Solution

Verbal model: $\boxed{\text{Total pay}} = \boxed{\text{Regular pay}} + \boxed{\text{Overtime pay}}$

Labels: Total pay $= \$622$
Regular pay $= 11.25(40)$ (dollars)
Hours of overtime $= x$
Overtime pay $= 16x$ (dollars)

Equation: $622 = 11.25(40) + 16x$

$622 = 450 + 16x$

$622 - 450 = 16x$

$172 = 16x$

$\frac{172}{16} = x$

$10.75 = x$

The employee worked 10.75 hours of overtime.

57. *Wholesale Cost*   The list price of an automobile tire is $88. During a promotional sale, the fourth tire is free with the purchase of three at the list price. Counting the free tire, the markup rate on cost is 10%. Find the cost of each tire.

Solution

Verbal model:	$\boxed{\text{Selling price for 4 tires}} = \boxed{\text{Cost for 4 tires}} + \boxed{\text{Markup}}$
Labels:	Selling price $= 3(88) + 1(0) = \$264$
	Cost per tire $= x$ (dollars)
	Cost for 4 tires $= 4x$ (dollars)
	Markup rate $= 0.10$ (percent in decimal form)
	Markup $= 0.10(4x)$ (dollars)

Equation:

$$264 = 4x + 0.10(4x)$$

$$264 = 4x + 0.4x$$

$$264 = 4.4x$$

$$\frac{264}{4.4} = x$$

$$60 \approx x$$

The cost of each tire is $60.00.

4.5   Formulas and Scientific Applications

7. *Geometry*   Each room in the floor plan of a house is square (see figure). The perimeter of the bathroom is 32 feet. The perimeter of the kitchen is 80 square feet. Find the area of the living room.

Solution

The living room is a square. The area of a square is given by the formula $A = s^2$, where s is the side of the square. To find the area of the living room, you need to find the side of the living room.

Verbal model:	$\boxed{\text{Side of living room}} = \boxed{\text{Side of bathroom}} + \boxed{\text{Side of kitchen}}$
Labels:	Side of living room $= s$ (feet)
	Side of bathroom $= b$ (feet)
	Side of kitchen $= k$ (feet)

The perimeter of the bathroom is 32 feet.

Common formula:	$P = 4b$
Label:	$P = 32$ (feet)
Equation:	$32 = 4b$
	$\frac{32}{4} = b$
	$8 = b$

The side of the bathroom is 8 feet.

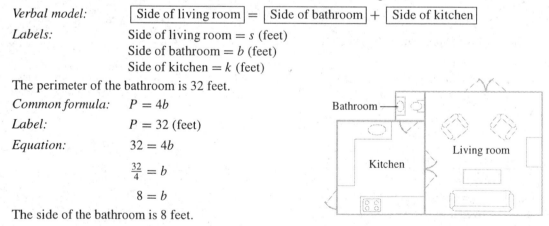

7. **—CONTINUED—**

The perimeter of the kitchen is 80 feet.

Common formula: $P = 4k$

Label: $P = 80$ (feet)

Equation: $80 = 4k$

$$\frac{80}{4} = k$$

$$20 = k$$

The side of the kitchen is 20 feet.

Equation: $s = b + k$

$$s = 8 + 20$$

$$s = 28$$

The side of the living room is 28 feet. Therefore, the area of the living room $A = s^2 = 28^2$.

$$A = (28)(28) = 784$$

The living room has an area of 784 square feet.

9. *Geometry* A circle has a circumference of 15 meters. What is the radius of the circle? Round your result to two decimal places.

Solution

$$C = 2\pi r$$

$$15 = 2\pi r$$

$$\frac{15}{2\pi} = \frac{2\pi r}{2\pi}$$

$$\frac{15}{2(3.14)} \approx r$$

$$2.39 \approx r$$

Therefore, the radius of the circle is approximately 2.39 meters.

11. *Simple Interest* Find the interest on a $1000 bond paying an annual rate of 9% for 6 years.

Solution

Common formula: $I = Prt$

Labels: $I =$ interest (dollars)
$P = \$1000$
$r = 0.09$ (annual interest rate)
$t = 6$ (years)

Equation: $I = Prt$

$$I = (1000)(0.09)(6)$$

$$I = 540$$

The interest is $540.

13. *Simple Interest* You borrow $15,000 for $\frac{1}{2}$ year. You promise to pay back the principal and the interest in one lump sum. The annual interest rate is 13%. What is your payment?

Solution

Common formula: $A = P + Prt$

Labels: $A =$ amount of payment (dollars)
$P = \$15,000$
$r = 0.13$
$t = \frac{1}{2}$ (year)

Equation: $A = 15,000 + 15,000(0.13)\left(\frac{1}{2}\right)$

$$A = 15,000 + 975$$

$$A = 15,975$$

The amount of the payment is $15,975.

15. *Geometry* Evaluate the following formula. List the units of measure for your result.

The volume of a right circular cylinder is $V = \pi r^2 h$. Find the volume of a right circular cylinder that has a radius of 2 meters and a height of 3 meters.

Solution

$V = \pi r^2 h$

$V = \pi (2)^2 \cdot 3$

$V = 12\pi$

$V \approx 12(3.14)$

$V \approx 37.7$ cubic meters

17. Solve $A = \frac{1}{2}bh$ for h.

Solution

$A = \frac{1}{2}bh$

$2A = 2 \cdot \frac{1}{2}bh$

$2A = bh$

$\dfrac{2A}{b} = \dfrac{bh}{b}$

$\dfrac{2A}{b} = h$

19. Solve $S = C + RC$ for C.

Solution

$S = C + RC$

$S = C(1 + R)$

$\dfrac{S}{1 + R} = \dfrac{C(1 + R)}{1 + R}$

$\dfrac{S}{1 + R} = C$

21. Find the distance, d, when the rate is 55 miles per hour and the time is 3 hours.

Solution

Verbal model: $\boxed{\text{Distance}} = \boxed{\text{Rate}} \cdot \boxed{\text{Time}}$

Labels: Distance = d (miles)
Rate = 55 (miles per hour)
Time = 3 (hours)

Equation: $d = 55(3)$

$d = 165$

The distance is 165 miles.

23. Find the time, t, when the distance is 500 kilometers and the rate is 90 kilometers per hour.

Solution

Verbal model: $\boxed{\text{Distance}} = \boxed{\text{Rate}} \cdot \boxed{\text{Time}}$

Labels: Distance = 500 (kilometers)
Rate = 90 (kilometers per hour)
Time = t (hours)

Equation: $500 = 90t$

$\frac{500}{90} = t$

$5.6 \approx t$

The time is approximately 5.6 hours (or five hours and approximately 36 minutes).

25. Find the rate, r, when the distance is 5280 feet and the time is $\frac{5}{2}$ seconds.

Solution

Verbal model: $\boxed{\text{Distance}} = \boxed{\text{Rate}} \cdot \boxed{\text{Time}}$

Labels: Distance = 5280 (feet)
Rate = r (feet per second)
Time = $\frac{5}{2}$ (seconds)

Equation:
$$5280 = r\left(\frac{5}{2}\right)$$

$$5280(2) = r\left(\frac{5}{2}\right)(2)$$

$$10{,}560 = 5r$$

$$\frac{10{,}560}{5} = r$$

$$2112 = r$$

The rate is 2112 feet per second.

27. *Space Shuttle Time*   The speed of a space shuttle is 17,000 miles per hour (see figure). How long will it take the shuttle to travel a distance of 3000 miles?

Solution

Verbal model: $\boxed{\text{Distance}} = \boxed{\text{Rate}} \cdot \boxed{\text{Time}}$

Labels: Distance = 3000 (miles)
Rate = 17,000 (miles per hour)
Time = t (hours)

Equation:
$$3000 = 17{,}000t$$

$$\frac{3000}{17{,}000} = t$$

$$0.176 \approx t$$

The time is approximately 0.176 hours (or approximately 10.6 minutes).

3000 miles

29. *Speed*   Determine the average speed of an Olympic runner who completes the 10,000-meter race in 27 minutes and 45 seconds.

Solution

Verbal model: $\boxed{\text{Distance}} = \boxed{\text{Rate}} \cdot \boxed{\text{Time}}$

Labels: Distance = 10,000 (meters)
Rate = r (meters per minute)
Time = 27 minutes, 45 seconds or 27.75 (minutes)

Equation:
$$10{,}000 = r(27.75)$$

$$\frac{10{,}000}{27.75} = r$$

$$360 \approx r$$

The rate is approximately 360 meters per minute or 6 meters per second.

31. *Interpreting a Table*   An agricultural corporation places an order for 100 tons of cattle feed. The feed is to be a mixture of soybeans, which cost $200 per ton, and corn, which costs $125 per ton. Complete the table shown in the textbook, where x is the number of tons of corn in the mixture.

(a) How does the increase in the corn weight affect the soybean weight?

(b) How does the increase in the number of tons of corn affect the price per ton of the mixture?

(c) If there were an equal number of tons of corn and soybeans in the mixture, how would the resulting price of the mixture relate to the price of each component?

Solution

Verbal model: $\boxed{\text{Value of corn}} + \boxed{\text{Value of soybeans}} = \boxed{\text{Value of mixture}}$

Labels: Corn: price per ton = $125, number of tons = x, value of corn = $125x$
Soybeans: price per ton = $200, number of tons = $100 - x$, value of soybeans = $200(100 - x)$
Mixture: price per ton = m (dollars), number of tons = 100, value of mixture = $100m$

Equation: $125x + 200(100 - x) = 100m$

$$\frac{125x + 200(100 - x)}{100} = m$$

Corn weight, x	Soybean weight, $100 - x$	Price/ton of the mixture, m
0	100	$\frac{125(0) + 200(100)}{100} = \frac{20{,}000}{100} = \$200$
20	80	$\frac{125(20) + 200(80)}{100} = \frac{18{,}500}{100} = \$185$
40	60	$\frac{125(40) + 200(60)}{100} = \frac{17{,}000}{100} = \$170$
60	40	$\frac{125(60) + 200(40)}{100} = \frac{15{,}500}{100} = \$155$
80	20	$\frac{125(80) + 200(20)}{100} = \frac{14{,}000}{100} = \$140$
100	0	$\frac{125(100) + 200(0)}{100} = \frac{12{,}500}{100} = \$125$

(a) As the number of tons of corn increases, the number of tons of soybeans *decreases* by the same amount.

(b) As the number of tons of corn increases, the price per ton of the mixture *decreases* and gets closer to the price per ton of the corn.

(c) If there were equal numbers of tons of corn and soybeans in the mixture, the price per ton of the mixture would be halfway between the prices of the two components. In other words, the price per ton of the mixture would be the *average of the two prices,* 125 and 200, which is

$$\frac{125 + 200}{2} = \frac{325}{2} = 162.50.$$

Note: This result can be verified with the equation above, using $x = 50$ and $100 - x = 50$.

$$m = \frac{125(50) + 200(50)}{100} = \$162.50.$$

33. *Number of Coins*   A person has 50 coins in dimes and quarters with a combined value of $7.70. Determine the number of coins of each type.

Solution

Verbal model: | Value of dimes | + | Value of quarters | = | Total value |

Labels: Mixed coins: total value = $7.70, number of coins = 50
Dimes: value per coin = $0.10, number of coins = x
Quarters: value per coin = $0.25, number of coins = $50 - x$

Equation:
$$0.10x + 0.25(50 - x) = 7.70$$
$$0.10x + 12.50 - 0.25x = 7.70$$
$$-0.15x + 12.50 = 7.70$$
$$-0.15x = -4.80$$
$$x = \frac{-4.80}{-0.15}$$
$$x = 32$$
$$50 - x = 50 - 32 = 18$$

There are 32 dimes and 18 quarters.

35. *Flower Order*   A floral shop receives an order for flowers that totals $384. The price per dozen for the roses and carnations are $18 and $12, respectively. The order contains twice as many roses as carnations. How many of each type of flower are in the order?

Solution

Verbal model: | Cost of carnations per dozen | · | Number of dozens of carnations | + | Cost of roses per dozen | · | Number of dozens of roses | = | Total cost of flowers |

Labels: Cost of carnations per dozen = 12 (dollars per dozen)
Number of dozens of carnations = x (dozens)
Cost of dozens of roses = 18 (dollars per dozen)
Number of dozens of roses = $2x$ (dollars)
Total cost of flowers = 384 (dollars)

Equation:
$$12x + 18(2x) = 384$$
$$12x + 36x = 384$$
$$48x = 384$$
$$x = 8 \text{ and } 2x = 16$$

Therefore, 8 dozen carnations and 16 dozen roses were ordered.

37. Evaluate the expression $-50 - 4(3 - 8)$.

Solution

$$-50 - 4(3 - 8) = -50 - 4(-5)$$
$$= -50 + 20$$
$$= -30$$

39. Evaluate the following expression.

$$\frac{-|7 + 3^2|}{4}$$

Solution

$$\frac{-|7 + 3^2|}{4} = \frac{-|7 + 9|}{4}$$
$$= \frac{-|16|}{4}$$
$$= \frac{-16}{4}$$
$$= -4$$

41. Evaluate the expression $|-3 + 5(6 - 8)|$.

Solution

$$|-3 + 5(6 - 8)| = |-3 + 5(-2)|$$
$$= |-3 - 10|$$
$$= |-13|$$
$$= 13$$

43. *Sales Tax*   You buy a computer for $2750 plus 6% sales tax. What is your total bill?

Solution

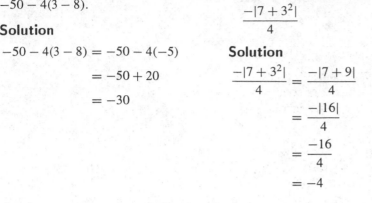

Verbal model: [Computer price] + [Sales tax] = [Total price]

Labels: Computer price: $= 2750$ (dollars)
                 Sales tax rate $= 0.06$ (percent in decimal form)
                 Sales tax $= 0.06(2750)$ (dollars)
                 Total price $= x$

Equation: $2750 + 0.06(2750) = x$
                 $2750 + 165 = x$
                 $2915 = x$

Therefore, your total bill is $2915.

45. *Geometry*   A triangle has an area of 48 square meters and a height of 12 meters. Find the length of the base.

Solution

$$A = \tfrac{1}{2}bh$$
$$48 = \tfrac{1}{2}b(12)$$
$$48 = 6b$$
$$8 = b$$

47. *Geometry*   The perimeter of a square is 48 feet. Find its area.

Solution

$$P = 4s \text{ and } A = s^2$$
$$48 = 4s$$
$$12 = s \text{ and } s^2 = 144$$

Therefore, the length of the side is 12 feet, so the area of the square is $(12)^2$, or 144 square feet.

49. Use the closed rectangular box shown in the figure in the textbook to find the area of the base.

Solution

Common formula: $A = lw$

Labels: A = area in rectangular base (square inches)
$l = 8$ (inches)
$w = 3$ (inches)

Equation: $A = lw$

$A = (8)(3)$

$A = 24$

Thus, the area is 24 square inches.

51. Use the closed rectangular box shown in the figure in the textbook to find the volume of the box.

Solution

Common formula: $V = lwh$

Labels: V = volume of box (cubic inches)
$l = 8$ (inches)
$w = 3$ (inches)
$h = 4$ (inches)

Equation: $V = lwh$

$V = 8(3)(4)$

$V = 96$

Thus, the volume of the box is 96 cubic inches.

53. *Simple Interest* Find the annual interest rate on a savings account that earns $110 interest in 1 year on a principal of $1000. Use the formula for simple interest.

Solution

Common formula: $I = Prt$

Labels: $I = \$110$
$P = \$1000$
r = annual interest rate
$t = 1$ (year)

Equation: $I = Prt$

$110 = (1000)(r)(1)$

$\frac{110}{1000} = r$

$0.11 = r$

The annual interest rate is 11%.

55. *Simple Interest* Find the principal required to earn $408 interest in 4 years, if the annual interest rate is $8\frac{1}{2}\%$. Use the formula for simple interest.

Solution

Common formula: $I = Prt$

Labels: $I = \$408$
P = principal (dollars)
$r = 0.085$
$t = 4$ (years)

Equation: $I = Prt$

$408 = P(0.085)(4)$

$\frac{408}{(0.085)(4)} = P$

$\frac{408}{0.34} = P$

$1200 = P$

The required principal is $1200.

57. Solve $E = IR$ for R.

Solution

$E = IR$

$\frac{E}{I} = \frac{IR}{I}$

$\frac{E}{I} = R$

59. Solve $V = lwh$ for l.

Solution

$V = lwh$

$\frac{V}{wh} = \frac{lwh}{wh}$

$\frac{V}{wh} = l$

61. Solve $A = P + Prt$ for r.

Solution

$$A = P + Prt$$
$$A - P = P - P + Prt$$
$$A - P = Prt$$
$$\frac{A - P}{Pt} = \frac{Prt}{Pt}$$
$$\frac{A - P}{Pt} = r$$

63. Solve $V = \frac{1}{3}\pi h^2(3r - h)$ for r.

Solution

$$V = \frac{1}{3}\pi h^2(3r - h)$$
$$3V = \pi h^2(3r - h)$$
$$3V = 3\pi h^2 r - \pi h^3$$
$$3V + \pi h^3 = 3\pi h^2 r$$
$$\frac{3V + \pi h^3}{3\pi h^2} = \frac{3\pi h^2 r}{3\pi h^2}$$
$$\frac{3V + \pi h^3}{3\pi h^2} = r$$

Note: This answer could also be written as
$$r = \frac{V}{\pi h^2} + \frac{h}{3}.$$

65. *Distance*   Two cars start at a given point and travel in the same direction at average speeds of 45 miles per hour and 52 miles per hour (see figure). How far apart will they be in 4 hours?

Solution

Verbal model: | Faster car's distance | − | Slower car's distance | = | Distance between cars |

Labels: Faster car's rate = 52 (miles per hour)
                 Slower car's rate = 45 (miles per hour)
                 Time = 4 hours
                 Distance between cars = x (miles)

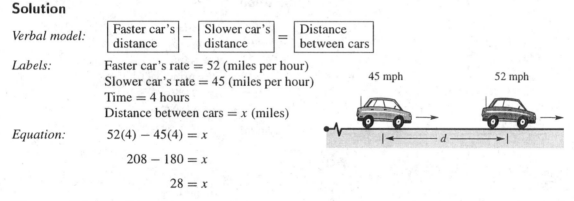
45 mph   52 mph

Equation: $52(4) - 45(4) = x$
$$208 - 180 = x$$
$$28 = x$$

The cars will be 28 miles apart.

67. *Speed*   Determine the average speed of an experimental plane that can travel a distance of 3000 miles in 2.6 hours.

Solution

Verbal model: | Distance | = | Rate | · | Time |

Labels: Distance = 3000 (miles)
                 Rate = r (miles per hour)
                 Time = 2.6 (hours)

Equation: $3000 = r(2.6)$
$$\frac{3000}{2.6} = r$$
$$1154 \approx r$$

The rate is approximately 1154 miles per hour.

69. Determine the numbers of units of alcohol solutions 1 and 2 needed to obtain the desired amount and the alcohol concentration of the final solution.

Concentration Solution 1	Concentration Solution 2	Concentration Final Solution	Amount of Final Solution
10%	30%	25%	100 gal

Solution

Verbal model: | Amount of alcohol in solution 1 | $+$ | Amount of alcohol in solution 2 | $=$ | Amount of alcohol in final solution |

Labels: Solution 1: percent alcohol = 0.10, amount = x (gallons)
Solution 2: percent alcohol = 0.30, amount = $100 - x$ (gallons)
Final solution: percent alcohol = 0.25, amount = 100 (gallons)

Equation: $\quad 0.10x + 0.30(100 - x) = 0.25(100)$

$$0.10x + 30 - 0.30x = 25$$

$$-0.20x + 30 = 25$$

$$-0.20x = -5$$

$$x = \frac{-5}{-0.20} = 25$$

$$100 - x = 100 - 25 = 75$$

The final solution will contain 25 gallons of solution 1 and 75 gallons of solution 2.

71. Determine the numbers of units of alcohol solutions 1 and 2 needed to obtain the desired amount and the alcohol concentration of the final solution.

Concentration Solution 1	Concentration Solution 2	Concentration Final Solution	Amount of Final Solution
15%	45%	30%	10 qt

Solution

Verbal model: | Amount of alcohol in solution 1 | $+$ | Amount of alcohol in solution 2 | $=$ | Amount of alcohol in final solution |

Labels: Solution 1: percent alcohol = 0.15, amount = x (quarts)
Solution 2: percent alcohol = 0.45, amount = $10 - x$ (quarts)
Final solution: percent alcohol = 0.30, amount = 10 (quarts)

Equation: $\quad 0.15x + 0.45(10 - x) = 0.30(10)$

$$0.15x + 4.5 - 0.45x = 3$$

$$-0.3x + 4.5 = 3$$

$$-0.3x = -1.5$$

$$x = \frac{-1.5}{-0.3} = 5$$

$$10 - x = 5$$

The final solution will contain 5 quarts of solution 1 and 5 quarts of solution 2.

73. *Antifreeze* The cooling system in a truck contains 4 gallons of coolant that is 30% antifreeze. How much must be withdrawn and replaced with 100% antifreeze to bring the coolant in the system to 50% antifreeze?

Solution

Verbal model: $\boxed{\text{Amount of antifreeze in first solution}} + \boxed{\text{Amount of antifreeze in pure solution}} = \boxed{\text{Amount of antifreeze in final solution}}$

Labels: Original solution: percent antifreeze $= 0.30$, amount $= 4 - x$ (gallons)
Pure antifreeze: percent antifreeze $= 1.00$, amount $= x$ (gallons)
Final solution: percent antifreeze $= 0.50$, amount $= 4$ (gallons)

Equation: $0.30(4 - x) + 1.00(x) = 0.50(4)$

$$1.2 - 0.3x + x = 2$$
$$1.2 + 0.7x = 2$$
$$0.7x = 0.8$$
$$x = \frac{0.8}{0.7} \approx 1.14$$

Approximately 1.14 gallons (or $1\frac{1}{7}$ gallons) must be withdrawn and replaced.

75. *Nut Mixture* A grocer mixes two kinds of nuts that cost $2.49 and $3.89 per pound to make 100 pounds of a mixture that costs $3.47 per pound. How many pounds of each kind of nut were put into the mixture?

Solution

Verbal model: $\boxed{\text{Total cost of first nuts}} + \boxed{\text{Total cost of second nuts}} = \boxed{\text{Total cost of mixture}}$

Labels: First nuts: price per pound $= \$2.49$, number of pounds $= x$
Second nuts: price per pound $= \$3.89$, number of pounds $= 100 - x$
Mixture: price per pound $= \$3.47$, number of pounds $= 100$

Equation: $2.49(x) + 3.89(100 - x) = 3.47(100)$

$$2.49x + 389 - 3.89x = 347$$
$$-1.40x + 389 = 347$$
$$-1.40x = -42$$
$$x = \frac{-42}{-1.40}$$
$$x = 30$$
$$100 - x = 100 - 30 = 70$$

The mixture contained 30 pounds of the nuts costing $2.49 per pound and 70 pounds of the nuts costing $3.89 per pound.

77. *Number of Coins* A person has 20 coins in nickels and dimes with a combined value of $1.60. Determine the number of coins of each type.

Solution

Verbal model: $\boxed{\text{Value of nickels}} + \boxed{\text{Value of dimes}} = \boxed{\text{Total value}}$

Labels: Mixed coins: total value = $1.60, number of coins = 20
Nickels: value per coin = $0.05, number of coins = x
Dimes: value per coin = $0.10, number of coins = $20 - x$

Equation: $0.05x + 0.10(20 - x) = 1.60$

$$0.05x + 2 - 0.10x = 1.60$$

$$-0.05x + 2 = 1.60$$

$$-0.05x = -0.40$$

$$x = \frac{-0.40}{-0.05}$$

$$x = 8$$

$$20 - x = 20 - 8 = 12$$

There are 8 nickels and 12 dimes.

79. *Poll Results* One thousand people were surveyed in an opinion poll. Candidates A and B received approximately the same number of votes. Candidate C received twice as many votes as each of the other two candidates. How many votes did each candidate receive?

Solution

Verbal model: $\boxed{\begin{array}{c}\text{Votes for}\\\text{candidate A}\end{array}} + \boxed{\begin{array}{c}\text{Votes for}\\\text{candidate B}\end{array}} + \boxed{\begin{array}{c}\text{Votes for}\\\text{candidate C}\end{array}} = \boxed{\begin{array}{c}\text{Total}\\\text{votes}\end{array}}$

Labels: Votes for candidate A = x
Votes for candidate B = x
Votes for candidate C = $2x$
Total votes = 1000

Equation: $x + x + 2x = 1000$

$$4x = 1000$$

$$x = \frac{1000}{4}$$

$$x = 250$$

$$2x = 2(250) = 500$$

Candidates A and B received 250 votes each, and candidate C received 500 votes.

81. *Work Rate*   One person can complete a typing project in 6 hours, and another can complete the same project in 8 hours. If they both work on the project, in how many hours can it be completed?

Solution

Verbal model: $\boxed{\text{Work done}} = \boxed{\begin{array}{l}\text{Portion done by}\\ \text{first person}\end{array}} + \boxed{\begin{array}{l}\text{Portion done by}\\ \text{second person}\end{array}}$

Labels: Both persons: work done = 1 project, time = t (hours)

First person: rate = $\frac{1}{6}$ project per hour, time = t (hours)

Second person: rate = $\frac{1}{8}$ project per hour, time = t (hours)

Equation: $1 = \frac{1}{6}(t) + \frac{1}{8}(t)$

$1 = \left(\frac{1}{6} + \frac{1}{8}\right)t$

$1 = \left(\frac{4}{24} + \frac{3}{24}\right)t$

$1 = \frac{7}{24}t$

$\frac{24}{7}(1) = \frac{24}{7}\left(\frac{7}{24}t\right)$

$\frac{24}{7} = t$

It would take $\frac{24}{7}$ hours (or $3\frac{3}{7}$ hours) to complete the project.

83. *Ages*   A mother was 30 years old when her son was born. How old will the son be when his age is $\frac{1}{3}$ his mother's age?

Solution

Verbal model: $\boxed{\text{Son's age}} = \frac{1}{3} \cdot \boxed{\text{Mother's age}}$

Labels: Son's age = x (years)

Mother's age = $x + 30$ (years)

Equation: $x = \frac{1}{3}(x + 30)$

$3 \cdot x = 3 \cdot \frac{1}{3}(x + 30)$

$3x = x + 30$

$2x = 30$

$x = \frac{30}{2}$

$x = 15$

The son will be 15 years old. (His mother will be 15 + 30 or 45 years old.)

Review Exercises for Chapter 4

1. Plot the points $(-2, 0)$, $\left(\frac{3}{2}, 4\right)$, and $(-1, -3)$ on a rectangular coordinate system.

Solution

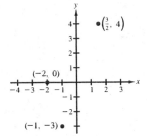

3. Determine the quadrant or quadrants in which the point $(-5, 3)$ must be located.

Solution

The point $(-5, 3)$ is located 5 units to the *left* of the vertical axis and 3 units *above* the horizontal axis in Quadrant II.

5. Determine the quadrant or quadrants in which the points $(x, 5)$, $x < 0$, must be located.

Solution

The point $(x, 5)$, $x < 0$, is located to the left of the vertical axis because x is negative; it is 5 units above the horizontal axis. Therefore, the point is in Quadrant II.

7. Decide whether each of the following ordered pairs is a solution of the equation $y = \frac{2}{3}x + 3$.

(a) $(3, 5)$ (b) $(-2, 0)$

Solution

(a) Yes, $(3, 5)$ is a solution because $5 = \frac{2}{3}(3) + 3$, or $5 = 2 + 3$.

(b) No, $(-2, 0)$ is not a solution because $0 \neq \frac{2}{3}(-2) + 3$.

9. *Graphical Interpretation*   The line graph in the textbook shows the receipts (in billions of dollars) for businesses associated with lodging, food, and recreation in the travel industry in the United States for the years 1981 through 1990. (*Source:* U.S. Travel Data Center)

(a) Approximate the receipts for eating places in the travel industry in 1985.

(b) In what year was approximately 40 billion dollars spent on recreation in the travel industry?

(c) Approximate the percent increase in receipts of lodging establishments from 1988 to 1989.

Solution

(a) Approximately $125 billion

(b) 1989

(c) Approximately 12%

11. Sketch the graph of $y = 4 - \frac{1}{2}x$ and determine any intercepts of the graph.

Solution

x	0	8	4
y	4	0	2

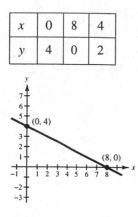

13. Sketch the graph of $y - 2x - 3 = 0$ and determine any intercepts of the graph.

Solution

x	0	$-\frac{3}{2}$	2	-3
y	3	0	7	-3

Note: You could rewrite the equation as $y = 2x + 3$.

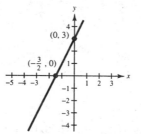

15. Sketch the graph of $y = 3 - |x|$ and determine any intercepts of the graph.

Solution

x	0	3	-3	6	-6
y	3	0	0	-3	-3

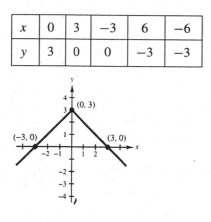

17. Sketch the graph of $y = (x - 4)^2$ and determine any intercepts of the graph.

Solution

x	0	4	2	6	5	1	7
y	16	0	4	4	1	9	9

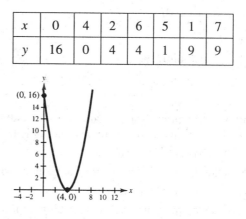

19. Use a graphing utility to graph $y = \frac{7}{8}x + 1$. (Use a standard setting.)

Solution

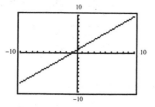

21. Use a graphing utility to graph $y = -\frac{1}{4}x^2 + x$. (Use a standard setting.)

Solution

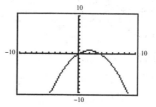

23. Graph $y = 250 - 50x$ using a graphing utility.

```
RANGE
Xmin=-5
Xmax=5
Xscl=1
Ymin=0
Ymax=500
Yscl=25
```

Solution

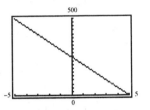

25. Use the TRACE feature of a graphing utility to approximate the x- and y-intercepts of the graph of $y = 5.2 - |x|$.

Solution

x-intercepts: $(5.2, 0), (-5.2, 0)$

y-intercept: $(0, 5.2)$

27. *Business Expense* A company reimburses its sales representatives \$125 per day for lodging and meals plus 27¢ per mile driven. Write an equation giving the daily cost y to the company in terms of x, the number of miles driven. Graph the equation.

Solution

$y = 125 + 0.27x$

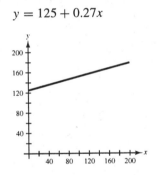

29. *Price* A food processor that costs a retailer \$85 is marked up by 35%. Find the price for the consumer.

Solution

Verbal model: | Selling price | $=$ | Cost | $+$ | Markup |

Labels:
Selling price $= x$ (dollars)
Cost $= \$85$
Markup rate $= 0.35$ (percent in decimal form)
Markup $= (0.35)(85) = \$29.75$

Equation: $x = 85 + 29.75$

$x = 114.75$

The selling price is \$114.75.

31. *Sale Price* While shopping for a new suit you find one you like with a list price of $279. The list price has been reduced by 35% to form the sale price. Find the sale price of the suit.

Solution

Verbal model: ⎡Sale price⎤ = ⎡List price⎤ − ⎡Discount⎤

Labels: Sale price = x (dollars)
               List price = $279
               Discount rate = 0.35 (percent in decimal form)
               Discount = 0.35(279) = $97.65

Equation: $x = 279 - 97.65$

              $x = 181.35$

The sale price of the suit is $181.35.

33. *Comparing Two Prices* A mail-order catalog lists the price of luggage at $89 plus $4 for shipping and handling. A local department store has the same luggage for $112.98. The department store has a special 20%-off sale. Which is the better price?

Solution

Verbal model (catalog): ⎡Total cost⎤ = ⎡List price⎤ + ⎡Added charge⎤

Labels: Total cost = x (dollars)
               List price = $89
               Added charge = $4

Equation: $x = 89 + 4$

              $x = 93$

Verbal model (store): ⎡Sale price⎤ = ⎡List price⎤ − ⎡Discount⎤

Labels: Sale price = x (dollars)
               List price = $112.98
               Discount rate = 0.20 (percent in decimal form)
               Discount = 0.20(112.98) ≈ $22.60

Equation: $x = 112.98 - 22.60$

              $x = 90.38$

The luggage would cost $93 from the catalog and $90.38 from the department store. The store offers the better price.

35. *Time* A train's average speed is 60 miles per hour. How long will it take the train to travel 562 miles?

Solution

Verbal model: ⎡Distance⎤ = ⎡Rate⎤ · ⎡Time⎤

Labels: Distance = 562 (miles)
               Rate = 60 (miles per hour)
               Time = x (hours)

Equation: $562 = 60x$

             $\frac{562}{60} = x$

           $9.37 \approx x$

It would take the train approximately 9.37 hours (or approximately 9 hours and 22 minutes).

37. *Dimensions of a Swimming Pool*   The width of a rectangular swimming pool is 4 feet less than its length. The perimeter of the pool is 112 feet. Find the dimensions of the pool.

Solution

Common formula: $P = 2l + 2w$

Labels: $P = 112$ (feet)
$l = $ length (feet)
$w = $ width $= l - 4$ (feet)

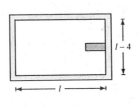

Equation: $112 = 2l + 2(l - 4)$

$112 = 2l + 2l - 8$

$112 = 4l - 8$

$120 = 4l$

$\frac{120}{4} = l$

$30 = l$

$26 = l - 4$

The length of the pool is 30 feet and the width is 26 feet.

39. *Simple Interest*   Find the principal required to have an annual interest income of $25,000 if the annual interest rate on the principal is 8.75%.

Solution

Common formula: $I = Prt$

Labels: $I = \$25,000$
$P = $ principal (dollars)
$r = 0.0875$ (percent in decimal form)
$t = 1$

Equation: $25,000 = P(0.0875)1$

$25,000 = P(0.0875)$

$\dfrac{25,000}{0.0875} = P$

$285,714 \approx P$

The required principal is approximately $285,714.

41. Solve the following formula for θ.

$$A = \frac{r^2\theta}{2}$$

Solution

$$A = \frac{r^2\theta}{2}$$

$$2A = 2\left(\frac{r^2\theta}{2}\right)$$

$$2A = r^2\theta$$

$$\frac{2A}{r^2} = \frac{r^2\theta}{r^2}$$

$$\frac{2A}{r^2} = \theta$$

43. *Work Rate* Find the time for two people working together to complete a task that, if they work individually, takes them 5 hours and 6 hours, respectively.

Solution

Verbal model: $\boxed{\text{Work done}} = \boxed{\begin{array}{l}\text{Portion done by}\\ \text{first person}\end{array}} + \boxed{\begin{array}{l}\text{Portion done by}\\ \text{second person}\end{array}}$

Labels: Both persons: work done = 1 complete task, time = t (hours)

First person: rate = $\frac{1}{5}$ task per hour, time = t (hours)

Second person: rate = $\frac{1}{6}$ task per hour, time = t (hours)

Equation: $1 = \left(\frac{1}{5}\right)t + \left(\frac{1}{6}\right)t$

$1 = \left(\frac{1}{5} + \frac{1}{6}\right)t$

$1 = \left(\frac{6}{30} + \frac{5}{30}\right)t$

$1 = \left(\frac{11}{30}\right)t$

$\frac{30}{11}(1) = \frac{30}{11}\left(\frac{11}{30}\right)t$

$\frac{30}{11} = t$

It would take $\frac{30}{11}$ hours (or approximately 2.7 hours) for both people to complete the task.

Test for Chapter 4

1. Plot the points $(-1, 2)$, $(1, 4)$, and $(2, -1)$ on a rectangular coordinate system. Connect the points with line segments to form a right triangle.

Solution

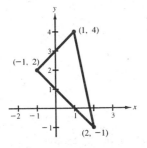

2. When an employee produces x units per hour, the hourly wage is $y = 0.75x + 4$. Complete the table shown in the textbook to determine the hourly wages for producing the specified numbers of units. Plot the results.

Solution

x	2	4	6	8	10	12
$y = 0.75x + 4$	5.5	7	8.5	10	11.5	13

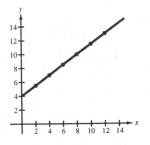

3. Which ordered pairs are solutions of $y = |x| + |x - 2|$?

(a) $(0, -2)$ (b) $(0, 2)$ (c) $(-4, 10)$ (d) $(-2, 2)$

Solution

(a) No, $(0, -2)$ is not a solution because $-2 \neq |0| + |0 - 2|$ or $0 + 2$.

(b) Yes, $(0, 2)$ is a solution because $2 = |0| + |0 - 2|$ or $0 + 2$.

(c) Yes, $(-4, 10)$ is a solution because $10 = |-4| + |-4 - 2|$ or $4 + 6$.

(d) No, $(-2, 2)$ is not a solution because $2 \neq |-2| + |-2 - 2|$ or $2 + 4$.

4. Find the x- and y-intercepts of the graph of $3x - 4y + 12 = 0$.

Solution

x-intercept

Let $y = 0$.

$3x - 4 \cdot 0 + 12 = 0$

$\quad 3x + 12 = 0$

$\qquad 3x = -12$

$\qquad\quad x = -4$

$\quad (-4, 0)$

y-intercept

Let $x = 0$.

$3 \cdot 0 - 4y + 12 = 0$

$\quad -4y + 12 = 0$

$\qquad -4y = -12$

$\qquad\quad y = 3$

$\quad (0, 3)$

The graph has *one* x-intercept at $(-4, 0)$ and *one* y-intercept at $(0, 3)$.

5. Use a graphing utility to sketch the graph of $x - 2y = 6$.

Solution

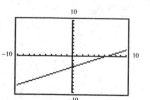

6. Use a graphing utility to sketch the graph of $y = |x + 2|$.

Solution

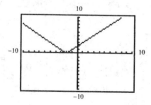

7. Use a graphing utility to sketch the graph of $y = 0.6(x - 2)$.

Solution

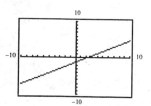

8. Use a graphing utility to sketch the graph of $y = 9 - (x - 3)^2$.

Solution

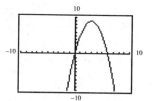

9. A calculator is marked up from \$80 to \$112. What is the markup rate?

Solution

Verbal model: | List price | = | Cost | + | Markup |

Labels: List price = \$112
                   Cost = \$80
                   Markup rate = p (percent in decimal form)
                   Markup = $p(80)$ (dollars)

Equation: $112 = 80 + p(80)$

                   $32 = p(80)$

                   $\frac{32}{80} = p$

                   $0.4 = p$

Thus, the markup rate is 40%.

10. A person has 20 coins in dimes and quarters with a combined value of \$3.80. Determine the number of coins of each type.

Solution

Verbal model: | Value of dimes | + | Value of quarters | = | Total value |

Labels: Mixed coins: total value = \$3.80, number of coins = 20
                   Dimes: value per coin = \$0.10, number of coins = x
                   Quarters: value per coin = \$0.25, number of coins = $20 - x$

Equation: $0.10x + 0.25(20 - x) = 3.80$

                   $0.10x + 5 - 0.25x = 3.80$

                   $-0.15x + 5 = 3.80$

                   $-0.15x = -1.20$

                   $x = \dfrac{-1.20}{-0.15}$

                   $x = 8$

                   $20 - x = 20 - 8 = 12$

There are 8 dimes and 12 quarters.

11. Roses cost $18 per dozen and carnations cost $9 per dozen. How many roses are in a $12 arrangement of one dozen roses and carnations?

Solution

Verbal model: $\boxed{\text{Value of roses}} + \boxed{\text{Value of carnations}} = \boxed{\text{Total value}}$

Labels: Arrangement: price per dozen $= \$12$, number of flowers $= 12$

Roses: price per flower $= \frac{18}{12} = \$1.50$, number of flowers $= x$

Carnations: price per flower $= \frac{9}{12} = 0.75$, number of flowers $= 12 - x$

Equation: $1.50x + 0.75(12 - x) = 12$

$$1.50x + 9 - 0.75x = 12$$

$$0.75x + 9 = 12$$

$$0.75x = 3$$

$$x = \frac{3}{0.75}$$

$$x = 4$$

There are 4 roses in the arrangement.

12. You traveled 264 miles in $5\frac{1}{2}$ hours. What was your average speed?

Solution

Verbal model: $\boxed{\text{Distance}} = \boxed{\text{Rate}} \cdot \boxed{\text{Time}}$

Labels: Distance $= 264$ (miles)

Rate $= x$ (miles per hour)

Time $= 5\frac{1}{2}$ (hours)

Equation: $264 = x\left(5\frac{1}{2}\right)$

$$\frac{264}{5.5} = x$$

$$48 = x$$

The average speed was 48 miles per hour.

13. You can paint a building in 9 hours. Your friend would take 12 hours. Working together, how long will it take the two of you to paint the building?

Solution

Verbal model: $\boxed{\text{Work done}} = \boxed{\begin{array}{c}\text{Portion done}\\\text{by you}\end{array}} + \boxed{\begin{array}{c}\text{Portion done}\\\text{by your friend}\end{array}}$

Labels: Both persons: work done $= 1$ task, time $= t$ (hours)

Your work: rate $= \frac{1}{9}$ task per hour, time $= t$ (hours)

Friend's work: rate $= \frac{1}{12}$ task per hour, time $= t$ (hours)

Equation: $1 = \left(\frac{1}{9}\right)t + \left(\frac{1}{12}\right)t$

$1 = \left(\frac{1}{9} + \frac{1}{12}\right)t$

$1 = \left(\frac{4}{36} + \frac{3}{36}\right)t$

$1 = \frac{7}{36}t$

$\frac{36}{7}(1) = \frac{36}{7}\left(\frac{7}{36}\right)t$

$\frac{36}{7} = t$

It would take $\frac{36}{7}$ hours (or approximately 5.1 hours) to paint the building.

14. Find three consecutive integers whose sum is 93.

Solution

Verbal model: $\boxed{\begin{array}{c}\text{First}\\\text{consecutive}\\\text{integer}\end{array}} + \boxed{\begin{array}{c}\text{Second}\\\text{consecutive}\\\text{integer}\end{array}} + \boxed{\begin{array}{c}\text{Third}\\\text{consecutive}\\\text{integer}\end{array}} = \boxed{93}$

Labels: First consecutive integer $= n$
Second consecutive integer $= n + 1$
Third consecutive integer $= n + 2$

Equation: $n + (n + 1) + (n + 2) = 93$

$3n + 3 = 93$

$3n = 90$

$n = \frac{90}{3}$

$n = 30$

$n + 1 = 31$

$n + 2 = 32$

Thus, the three consecutive integers are 30, 31, and 32.

15. Solve for R in the formula: $S = C + RC$.

Solution

$$S = C + RC$$

$$S - C = C - C + RC$$

$$S - C = RC$$

$$\frac{S - C}{C} = \frac{RC}{C}$$

$$\frac{S - C}{C} = R$$

Note: This answer could also be written as

$$\frac{S}{C} - 1 = R.$$

16. Find the total interest for a 6-month $1000 bond paying an annual interest rate of 8%. (Use the formula for simple interest.)

Solution

Common formula:	$I = Prt$

Labels:	I = interest (dollars)
	$P = \$1000$
	$r = 0.08$ (percent in decimal form)
	$t = \frac{1}{2}$ (years) *Note:* 6 months $= \frac{1}{2}$ year

Equation:	$I = (1000)(0.08)\left(\frac{1}{2}\right)$
	$I = 40$

The interest is $40.

17. How much must you deposit to earn $500 per year at 8% simple interest?

Solution

Common formula:	$I = Prt$

Labels:	$I = \$500$
	P = principal (dollars)
	$r = 0.08$ (percent in decimal form)
	$t = 1$ (year)

Equation:	$500 = P(0.08)(1)$
	$500 = P(0.08)$
	$\dfrac{500}{0.08} = P$
	$6250 = P$

The required principal is $6250.

18. Use the figure in the textbook to find the area of the blue region.

Solution

Length = 5 units
Width = 2 units
Area = 5(2) or 10 square units

19. Use the figure in the textbook to find the perimeter of the green region. Describe the method you used.

Solution

Perimeter $= 5 + 2 + 4 + 2 + 1 + 4$
or 18 units

20. Use the figure in the textbook to find the area of the green region. Describe the method you used.

Solution

$$\text{Area} = 5(2) + 1(2)$$

$$= 10 + 2$$

$$= 12 \text{ square units}$$

CHAPTER FIVE
Exponents and Polynomials

5.1 | Adding and Subtracting Polynomials

7. Write the polynomial $5 - 32x$ in standard form. Then find its degree and leading coefficient.

Solution

Polynomial:	$5 - 32x$
Standard Form:	$-32x + 5$
Degree:	1
Leading Coefficient:	-32

9. Write the polynomial $8x + 2x^5 - x^2 - 1$ in standard form. Then find its degree and leading coefficient.

Solution

Polynomial:	$8x + 2x^5 - x^2 - 1$
Standard Form:	$2x^5 - x^2 + 8x - 1$
Degree:	5
Leading Coefficient:	2

11. Determine whether the polynomial $x^3 - 4$ is a monomial, binomial, or trinomial.

Solution

The polynomial $x^3 - 4$ has two terms; it is a binomial.

13. Determine whether the polynomial 5 is a monomial, binomial, or trinomial.

Solution

The polynomial 5 has one term; it is a monomial.

15. Determine whether the expression $6/x$ is a polynomial. If it is not, explain why.

Solution

No, the expression $6/x$ is not a polynomial; this term is not of the form ax^k.

17. Determine whether the expression $9 - z$ is a polynomial. If it is not, explain why.

Solution

Yes, the expression $9 - z$ is a polynomial.

19. Give an example of a polynomial that fits the discription, "A binomial in one variable of degree 3." (*Note:* There are many correct answers.)

Solution

$8x^3 + 5x$ or $-x^3 + 2$

21. Give an example of a polynomial that fits the discription, "A monomial in one variable of degree 2." (*Note:* There are many correct answers.)

Solution

$10x^2$ or $-2x^2$

23. Use a vertical arrangement to perform the polynomial addition in

$$\begin{array}{r} -x^3 \qquad + 3 \\ 3x^3 + 2x^2 + 5 \end{array}$$

Solution

$$\begin{array}{r} -x^3 \qquad + 3 \\ 3x^3 + 2x^2 + 5 \\ \hline 2x^3 + 2x^2 + 8 \end{array}$$

25. Use a vertical arrangement to perform the polynomial addition in $(2 - 3y) + (y^4 + 3y + 2)$.

Solution

$$\begin{array}{r} -3y + 2 \\ y^4 + \quad 3y + 2 \\ \hline y^4 \qquad\quad + 4 \end{array}$$

27. Use a horizontal arrangement to perform the polynomial addition in $(3z^2 - z + 2) + (z^2 - 4)$.

Solution

$(3z^2 - z + 2) + (z^2 - 4)$

$= (3z^2 + z^2) + (-z) + (2 - 4)$

$= 4z^2 - z - 2$

29. Use a horizontal arrangement to perform the polynomial addition in $(2a - 3) + (a^2 - 2a) + (4 - a^2)$.

Solution

$(2a - 3) + (a^2 - 2a) + (4 - a^2)$

$= (a^2 - a^2) + (2a - 2a) + (-3 + 4)$

$= 1$

31. *Comparing Two Formats* Add the two polynomials $6x^2 + 5$ and $3 - 2x^2$.

(a) Use a horizontal arrangement.

(b) Use a vertical arrangement. Which format do you prefer? Explain

Solution

(a) $(6x^2 + 5) + (3 - 2x^2) = (6x^2 - 2x^2) + (5 + 3)$

$= 4x^2 + 8$

(b) $\begin{array}{r} 6x^2 + 5 \\ -2x^2 + 3 \\ \hline 4x^2 + 8 \end{array}$

Answers regarding format preference will vary.

33. Use a vertical arrangement to perform the polynomial subtraction in

$\begin{array}{r} 2x^2 - x + 2 \\ -(3x^2 + x - 1) \end{array}$

Solution

$\begin{array}{r} 2x^2 - x + 2 \\ -(3x^2 + x - 1) \end{array} \qquad \begin{array}{r} 2x^2 - x + 2 \\ -3x^2 - x + 1 \\ \hline -x^2 - 2x + 3 \end{array}$

35. Use a vertical arrangement to perform the polynomial subtraction in $(4t^3 - 3t + 5) - (3t^2 - 3t - 10)$.

Solution

$\begin{array}{r} 4t^3 - 3t + 5 \\ -(3t^2 - 3t + 5) \end{array} \qquad \begin{array}{r} 4t^3 \phantom{{}- 3t^2} - 3t + 5 \\ - 3t^2 + 3t + 10 \\ \hline 4t^3 - 3t^2 + 15 \end{array}$

37. Use a vertical arrangement to subtract $7x^3 - 4x + 5$ from $10x^3 + 15$.

Solution

$\begin{array}{r} 10x^3 + 15 \\ -(7x^3 - 4x + 5) \end{array} \qquad \begin{array}{r} 10x^3 + 15 \\ -7x^3 + 4x - 5 \\ \hline 3x^3 + 4x + 10 \end{array}$

39. Use a horizontal arrangement to perform the polynomial subtraction in $(4 - 2x - x^3) - (3 - 2x + 2x^3)$.

Solution

$(4 - 2x - x^3) - (3 - 2x + 2x^3)$

$= 4 - 2x - x^3 - 3 + 2x - 2x^3$

$= (-x^3 - 2x^3) + (-2x + 2x) + (4 - 3)$

$= -3x^3 + 1$

41. Perform the operations in $2(x^4 + 2x) + (5x + 2)$.

Solution

$2(x^4 + 2x) + (5x + 2)$

$= 2x^4 + 4x + 5x + 2$

$= 2x^4 + (4x + 5x) + 2$

$= 2x^4 + 9x + 2$

43. Perform the operations in $5z - [3z - (10z + 8)]$.

Solution

$5z - [3z - (10z + 8)] = 5z - [3z - 10z - 8]$

$= 5z - [-7z - 8] = 5z + 7z + 8 = 12z + 8$

45. *Comparing Models*   From 1970 through 1991, the average number of pounds of beef, B, and poultry, P, consumed by Americans can be modeled by

$$B = -0.04t^2 - 0.79t + 76.86, \quad -10 \leq t \leq 11$$

$$P = 0.06t^2 + 1.14t + 39.04, \quad -10 \leq t \leq 11$$

where $t = 0$ represents 1980.

(a) Add the polynomials to find a model for the total, T, amount of beef and poultry consumed.

(b) Use a graphing utility to graph the models B, P, and T.

(c) Use the graphs of part (b) to determine whether Americans are increasing of decreasing their consumption of each type of meat. Is the model for the sum increasing or decreasing?

Solution

(a) $B + P = (-0.04t^2 - 0.79t + 76.86) + (0.06t^2 + 1.14t + 39.04)$

$$= 0.02t^2 + 0.35t + 115.9$$

(b)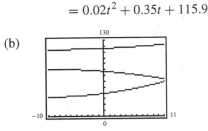

(c) Beef: Decreasing
    Pork: Increasing
    Total: Increasing

47. *Geometry*   Find the perimeter of the figure in the textbook.

Solution

Perimeter $= 2z + 4z + 2z + z + 1 + 2 + 1 + z$

$$= 10z + 4$$

49. *Geometry*   Write a polynomial that represents the area of the shaded portion of the figure in the textbook.

Solution

Area of Shaded Portion = Area of Larger Rectangle − Area of Smaller Rectangle

$$= (6x)\left(\tfrac{7}{2}x\right) - (10)\left(\tfrac{4}{5}x\right)$$

$$= 21x^2 - 8x$$

51. Use the exponential form to rewrite the expression $3z \cdot 3z \cdot 3z \cdot 3z$. Then state the degree and leading coefficient.

Solution

$3z \cdot 3z \cdot 3z \cdot 3z = (3z)^4$; Degree: 4

53. Use the exponential form to rewrite the expression $4 \cdot 4 \cdot m \cdot m \cdot m^2$. Then state the degree and leading coefficient.

Solution

$4 \cdot 4 \cdot m \cdot m \cdot m^2 = 4^2 m^4$; Degree: 4

55. Expand the expression $3^2 x^4 y^3$ as a product of factors.

Solution

$3^2 x^4 y^3 = 3 \cdot 3 \cdot x \cdot x \cdot x \cdot x \cdot y \cdot y \cdot y$

57. Use the Distributive Property to expand the expression $-2(3t - 4)$.

Solution

$-2(3t - 4) = -6t + 8$

59. Write the polynomial $x^3 - 4x^2 + 9$ in standard form. Then find its degree and leading coefficient.

Solution

Polynomial:	$x^3 - 4x^2 + 9$
Standard Form:	$x^3 - 4x^2 + 9$
Degree:	3
Leading Coefficient:	1

61. Write the polynomial 10 in standard form. Then find its degree and leading coefficient.

Solution

Polynomial:	10
Standard Form:	10
Degree:	0
Leading Coefficient:	10

63. Write the polynomial $3r + \pi r^2$ (π is a constant) in standard form. Then find its degree and leading coefficient.

Solution

Polynomial:	$3r + \pi r^2$
Standard form:	$\pi r^2 + 3r$
Degree:	2
Leading Coefficient:	π

65. Determine whether the expression $t^3 - 3t + 4$ is a polynomial. If it is not, explain why.

Solution

Yes, the expression $t^3 - 3t + 4$ is a polynomial.

67. Determine whether the expression $z^{-1} + z^2 - 2$ is a polynomial. If it is not, explain why.

Solution

No, the expression $z^{-1} + z^2 - 2$ is not a polynomial because the exponent in the first term is a *negative* integer.

69. Use a vertical arrangement to perform the polynomial addition in:

$$3x^4 - 2x^3 - 4x^2 + 2x - 5$$
$$\underline{ x^2 - 7x + 5}$$

Solution

$$3x^4 - 2x^3 - 4x^2 + 2x - 5$$
$$\underline{ x^2 - 7x + 5}$$
$$3x^4 - 2x^3 - 3x^2 - 5x$$

71. Use a vertical arrangement to perform the polynomial addition in $(n^2 + 1) + (2n^2 - 3)$.

Solution

$$n^2 + 1$$
$$\underline{2n^2 - 3}$$
$$3n^2 - 2$$

73. Use a vertical arrangement to perform the polynomial addition in $(x^2 - 4) + (2x^2 + 6)$.

Solution

$$x^2 - 4$$
$$\underline{2x^2 + 6}$$
$$3x^2 + 2$$

75. Use a vertical arrangement to perform the polynomial addition in $(x^2 - 2x + 2) + (x^2 + 4x) + 2x^2$.

Solution

$$x^2 - 2x + 2$$
$$x^2 + 4x$$
$$\underline{2x^2 }$$
$$4x^2 + 2x + 2$$

77. Use a horizontal arrangement to perform the polynomial addition in $b^2 + (b^3 - 2b^2 + 3) + (b^3 - 3)$.

Solution

$$b^2 + (b^3 - 2b^2 + 3) + (b^3 - 3) = (b^3 + b^3) + (b^2 - 2b^2) + (3 - 3)$$
$$= 2b^3 - b^2$$

79. Use a horizontal arrangement to perform the polynomial addition in $\left(\frac{2}{3}y^2 - \frac{3}{4}\right) + \left(\frac{5}{6}y^2 + 2\right)$.

Solution

$$\left(\frac{2}{3}y^2 - \frac{3}{4}\right) + \left(\frac{5}{6}y^2 + 2\right) = \left(\frac{2}{3}y^2 + \frac{5}{6}y^2\right) + \left(-\frac{3}{4} + 2\right)$$
$$= \left(\frac{4}{6}y^2 + \frac{5}{6}y^2\right) + \left(-\frac{3}{4} + \frac{8}{4}\right)$$
$$= \left(\frac{9}{6}y^2\right) + \left(\frac{5}{4}\right)$$
$$= \frac{3}{2}y^2 + \frac{5}{4}$$

81. Use a vertical arrangement to perform the polynomial subtraction in:

$$\begin{array}{r} -3x^3 - 4x^2 + 2x - 5 \\ -(\,2x^4 + 2x^3 \qquad - 4x + 5) \\ \hline \end{array}$$

Solution

$$\begin{array}{r} -3x^3 - 4x^2 + 2x - 5 \\ -(\,2x^4 + 2x^3 \qquad - 4x + 5) \\ \hline \end{array} \qquad \begin{array}{r} -3x^3 - 4x^2 + 2x - 5 \\ - 2x^4 - 2x^3 + 4x - 5 \\ \hline -2x^4 - 5x^3 - 4x^2 + 6x - 10 \end{array}$$

83. Use a vertical arrangement to perform the polynomial subtraction in $(2 - x^3) - (2 + x^3)$.

Solution

$$\begin{array}{r} -x^3 + 2 \\ - (x^3 + 2\,) \\ \hline \end{array} \qquad \begin{array}{r} -x^3 + 2 \\ - x^3 - 2 \\ \hline -2x^3 \end{array}$$

85. Use a vertical arrangement to perform the polynomial subtraction in $(6x^3 - 3x^2 + x) - [(x^3 + 3x^2 + 3 + (x - 3)]$.

Solution

$$(6x^3 - 3x^2 + x) - [(x^3 + 3x^2 + 3) + (x - 3)] = (6x^3 - 3x^2 + x) - (x^3 + 3x^2 + 3 + x - 3)$$
$$= (6x^3 - 3x^2 + x) - (x^3 + 3x^2 + x)$$

$$\begin{array}{r} 6x^3 - 3x^2 + x \\ - (x^3 + 3x^2 + x\,) \\ \hline \end{array} \qquad \begin{array}{r} 6x^3 - 3x^2 + x \\ - x^3 - 3x^2 - x \\ \hline 5x^3 - 6x^2 \end{array}$$

87. Use a horizontal arrangement to perform the polynomial subtraction in $(x^2 - x) - (x - 2)$.

Solution

$$(x^2 - x) - (x - 2) = x^2 - x - x + 2$$
$$= (x^2) + (-x - x) + 2$$
$$= x^2 - 2x + 2$$

91. Perform the indicated operations.

$$(6x - 5) - (8x + 15)$$

Solution

$$(6x - 5) - (8x + 15)$$
$$= 6x - 5 - 8x - 15$$
$$= (6x - 8x) + (-5 - 15)$$
$$= -2x - 20$$

95. Perform the indicated operations.

$$(15x^2 - 6) - (-8x^3 - 14x^2 - 17)$$

Solution

$$(15x^2 - 6) - (-8x^3 - 14x^2 - 17) = 15x^2 - 6 + 8x^3 + 14x^2 + 17$$
$$= (8x^3) + (15x^2 + 14x^2) + (-6 + 17)$$
$$= 8x^3 + 29x^2 + 11$$

97. Perform the indicated operations.

$$2(t^2 + 5) - 3(t^2 + 5) + 5(t^2 + 5)$$

Solution

$$2(t^2 + 5) - 3(t^2 + 5) + 5(t^2 + 5) = 2t^2 + 10 - 3t^2 - 15 + 5t^2 + 25$$
$$= (2t^2 - 3t^2 + 5t^2) + (10 - 15 + 25)$$
$$= 4t^2 + 20$$

99. Perform the indicated operations.

$$8v - 6(3v - v^2) + 10(10v + 3)$$

Solution

$$8v - 6(3v - v^2) + 10(10v + 3) = 8v - 18v + 6v^2 + 100v + 30$$
$$= (6v^2) + (8v - 18v + 100v) + 30$$
$$= 6v^2 + 90v + 30$$

89. Use a horizontal arrangement to perform the polynomial subtraction in $10 - (u^2 + 5)$.

Solution

$$10 - (u^2 + 5)$$
$$= 10 - u^2 - 5$$
$$= (-u^2) + (10 - 5)$$
$$= -u^2 + 5$$

93. Perform the indicated operations.

$$-(x^3 - 2) + (4x^3 - 2x)$$

Solution

$$-(x^3 - 2) + (4x^3 - 2x)$$
$$= -x^3 + 2 + 4x^3 - 2x$$
$$= (-x^3 + 4x^3) + (-2x) + 2$$
$$= 3x^3 - 2x + 2$$

101. *Geometry* Write a polynomial that represents the area of the shaded portion of the figure in the textbook.

Solution

Verbal model: $\boxed{\text{Area of shaded region}} = \boxed{\text{Area of larger rectangle}} - \boxed{\text{Area of smaller rectangle}}$

Labels:
Area of shaded region $= A$
Length of larger rectangle $= 2x$
Width of larger rectangle $= x$
Area of larger rectangle $= 2x(x) = 2x^2$

Length of smaller rectangle $= 4$
Width of smaller rectangle $= x/2$
Area of smaller rectangle $= 4(x/2) = 2x$

Solution: $A = 2x^2 - 2x$

The area of the shaded region is $2x^2 - 2x$.

Note: To find the area of each rectangle, we use the formula $A = lw$ or Area $=$ (length)(width).

103. Write a polynomial that represents the area of the shaded portion of the figure in the textbook.

Solution

Verbal Model: $\boxed{\text{Area of shaded region}} = \boxed{\text{Area of larger triangle}} - \boxed{\text{Area of smaller triangle}}$

Labels:
Area of shaded region $= A$
Base of larger triangle $= 4x$
Height of larger triangle $= 5$
Area of larger triangle $= \frac{1}{2}(4x)(5)$
Base of smaller triangle $= 4x$
Height of smaller triangle $= 2$
Area of smaller trangle $= \frac{1}{2}(4x)(2)$

Solution: $A = \frac{1}{2}(4x)(5) - \frac{1}{2}(4x)(2)$

$= 10x - 4x$

$= 6x$

Therefore, the area of the shaded region is $6x$. Note: To find the area of a triangle, use the formula:

Area $= \frac{1}{2}$(Base)(Height)

105. *Comparing Business Models*   The cost of producing x units is $C = 100 + 30x$. The revenue for selling x units is $R = 90x - x^2$, where $0 \le x \le 40$. The profit is given by the revenue minus the cost.

(a) Perform the subtraction required to find the polynomial representing profit.

(b) Use a graphing utility to graph the polynomial representing profit.

(c) Determine the profit when $x = 30$ units are produced and sold. Use the graph of part (b) to predict the change in profit if x is some value other than 30.

Solution

(a) Profit = Revenue $-$ Cost

$$C = 100 + 30x$$

$$R = 90x - x^2$$

$$\text{Profit} = (90x - x^2) - (100 + 30x)$$

$$= 90x - x^2 - 100 - 30x$$

$$= -x^2 + 60x - 100$$

(b)

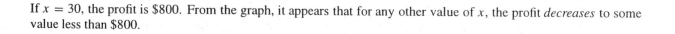

(c) Profit $= -x^2 + 60x - 100$

If $x = 30$,

$$-x^2 + 60x - 100 = -(30)^2 + 60(30) - 100$$

$$= -900 + 1800 - 100$$

$$= 800.$$

If $x = 30$, the profit is \$800. From the graph, it appears that for any other value of x, the profit *decreases* to some value less than \$800.

5.2   Multiplying Polynomials: Special Products

7. Multiply $x(-2x)$ and simplify.

Solution

$$x(-2x) = -2x^2$$

9. Multiply $(-2b^2)(-3b)$ and simplify.

Solution

$$(-2b^2)(-3b) = 6b^3$$

Note: $(-2b^2)(-3b) = (-2)(-3)(b^{2+1})$

$$= 6b^3$$

11. Multiply $2x(3x)^2$ and simplify.

Solution

$$2x(3x)^2 = 2x(3)^2(x)^2 = 2x(9x^2) = 18x^3$$

13. Multiply $(x + 3)(x + 4)$ and simplify.

Solution

$$(x + 3)(x + 4) = x(x + 4) + 3(x + 4)$$

$$= (x)(x) + (x)(4) + (3)(x) + (3)(4)$$

$$= x^2 + 4x + 3x + 12$$

$$= x^2 + 7x + 12$$

Using the FOIL Method:

$$\begin{array}{cccc} \text{F} & \text{O} & \text{I} & \text{L} \end{array}$$

$$(x + 3)(x + 4) = x^2 + 4x + 3x + 12 = x^2 + 7x + 12$$

15. Multiply $(5x + 7)(2x - 3)$ and simplify.

Solution

$$\begin{array}{cccc} \text{F} & \text{O} & \text{I} & \text{L} \end{array}$$
$$(5x + 7)(2x - 3) = 10x^2 - 15x + 14x - 21 = 10x^2 - x - 21$$

17. Multiply $-4x(3 + 3x^2 - 6x^3)$ and simplify.

Solution

$$-4x(3 + 3x^2 - 6x^3) = (-4x)(3) + (-4x)(3x^2) - (-4x)(6x^3)$$
$$= -12x - 12x^3 + 24x^4 \text{ or } 24x^4 - 12x^3 - 12x$$

19. Multiply $(s - 2t)(s + t) - (s - 2t)(s - t)$ using a horizontal format.

Solution

$$\begin{array}{cccccccc} \text{F} & \text{O} & \text{I} & \text{L} & \text{F} & \text{O} & \text{I} & \text{L} \end{array}$$
$$(s - 2t)(s + t) - (s - 2t)(s - t) = (s^2 + st - 2st - 2t^2) - (s^2 - st - 2st + 2t^2)$$
$$= (s^2 - st - 2t^2) - (s^2 - 3st + 2t^2)$$
$$= s^2 - st - 2t^2 - s^2 + 3st - 2t^2$$
$$= 2st - 4t^2$$

21. Multiply $(x^3 - 2x + 1)(x - 5)$ using a horizontal format.

Solution

$$(x^3 - 2x + 1)(x - 5) = x^3(x - 5) - 2x(x - 5) + 1(x - 5)$$
$$= x^4 - 5x^3 - 2x^2 + 10x + x - 5$$
$$= x^4 - 5x^3 - 2x^2 + 11x - 5$$

23. Multiply $(x - 2)(x^2 + 2x + 4)$ using a horizontal format.

Solution

$$(x - 2)(x^2 + 2x + 4) = x(x^2 + 2x + 4) - 2(x^2 + 2x + 4)$$
$$= x^3 + 2x^2 + 4x - 2x^2 - 4x - 8$$
$$= x^3 - 8$$

25. Multiply $(x^2 - 3x + 9) \times (x + 3)$ using a vertical format.

Solution

$$\begin{array}{r} x^2 - 3x + 9 \\ \times \quad\quad x + 3 \\ \hline 3x^2 - 9x + 27 \\ x^3 - 3x^2 + 9x \quad\quad \\ \hline x^3 \quad\quad\quad\quad + 27 \end{array}$$

Note: Here is an alternate vertical arrangement.

$$\begin{array}{r} x + 3 \\ \times \quad x^2 - 3x + 9 \\ \hline 9x + 27 \\ -3x^2 - 9x \quad\quad \\ x^3 + 3x^2 \quad\quad\quad \\ \hline x^3 \quad\quad\quad\quad + 27 \end{array}$$

27. Multiply $(x^2 - x + 2)(x^2 + x - 2)$ using a vertical format.

Solution

$$
\begin{array}{r}
x^2 + x - 2 \\
\times \quad x^2 - x + 2 \\
\hline
2x^2 + 2x - 4 \\
-x^3 - x^2 + 2x \\
x^4 + x^3 - 2x^2 \\
\hline
x^4 \qquad - x^2 + 4x - 4
\end{array}
$$

29. Multiply $(x^3 + x + 3)(x^2 + 5x - 4)$ using a vertical format.

Solution

$$
\begin{array}{r}
x^3 + x + 3 \\
\times \quad x^2 + 5x - 4 \\
\hline
-4x^3 \qquad - 4x - 12 \\
5x^4 \qquad + 5x^2 + 15x \\
x^5 \qquad + x^3 + 3x^2 \\
\hline
x^5 + 5x^4 - 3x^3 + 8x^2 + 11x - 12
\end{array}
$$

31. Use a special product pattern to find the product of $(x + 2)(x - 2)$.

Solution

Multiply

$(x + 2)(x - 2)$

Solution

$(x + 2)(x - 2) = (x)^2 - (2)^2$

$\qquad \qquad = x^2 - 4$

Pattern

$(a + b)(a - b) = a^2 - b^2$

33. Use a special product pattern to find the product of $(x + 6)^2$.

Solution

Multiply

$(x + 6)^2$

Solution

$(x + 6)^2 = (x)^2 + 2(x)(6) + (6)^2$

$\qquad \quad = x^2 + 12x + 36$

Pattern

$(a + b)^2 = a^2 + 2ab + b^2$

35. Use a special product pattern to find the product of $(2x - 5y)^2$.

Solution

Multiply

$(2x - 5y)^2$

Solution

$(2x - 5y)^2 = (2x)^2 - 2(2x)(5y) + (5y)^2$

$\qquad \qquad = 4x^2 - 20xy + 25y^2$

Pattern

$(a - b)^2 = a^2 - 2ab + b^2$

37. Use a special product pattern to find the product of $[u - (v - 3)]^2$.

Solution

Multiply:

$[u - (v - 3)]^2$

Solution:

$[u - (v - 3)]^2 = (u)^2 - 2(u)(v - 3) + (v - 3)^2$

$\qquad \qquad = u^2 - 2u(v - 3) + (v - 3)^2$

$\qquad \qquad = u^2 - 2u(v - 3) + v^2 - 6v + 9$

$\qquad \qquad = u^2 - 2uv + 6u + v^2 - 6v + 9$

$\qquad$ or $u^2 + v^2 - 2uv + 6u - 6v + 9$

Pattern:

$(a - b)^2 = a^2 - 2ab + b^2$

Use same pattern for $(v - 3)^2$.

39. *Geometry*   The base of a triangular sail is $2x$ feet and its height is $x + 10$ feet (see figure in textbook). Find the area A of the sail.

Solution

$$\text{Area } A = \tfrac{1}{2}(\text{Base})(\text{Height})$$
$$= \tfrac{1}{2}(2x)(x + 10)$$
$$= x(x + 10)$$
$$= x^2 + 10x$$

41. *Using Mathematical Models*   From 1980 through 1992, each American's share, S, of the debt of the federal government can be modeled by $S = 32.26t^2 + 594.49t + 3798.18$, $0 \le t \le 12$. where $t = 0$ represents 1980. The population, P (in millions), during the same time can be modeled by $P = 2.26t + 227.42$, $0 \le t \le 12$.

(a) Use a graphing utility to graph the model of the per capita debt S.

(b) Multiply the polynomials representing the population P and the per capita debt S.

(c) Use the product of part (b) to estimate the total federal debt for 1990. (*Note:* The answer will be in millions of dollars.)

Solution

(a)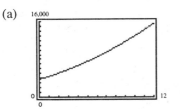

(b) $(P)(S) = (32.26t^2 + 594.49t + 3798.18)(2.26t + 227.42)$

$= 72.9076t^3 + 7336.5692t^2 + 1343.5474t^2 + 135,198.9158t + 8583.8868t + 863,782.0956$

$= 72.9076t^3 + 8680.1166t^2 + 143,782.8026t + 863,782.0956$

(c) In this model, $t = 0$ represents 1980. Therefore, for 1990, $t = 10$. Substituting $t = 10$ in the expression for the product PS in part (b) yields

$72.9076(10)^3 + 8680.1166(10)^2 + 143,782.8026(10) + 863,782.0956 \approx \$3,242,529.38$ million.

43. What polynomial product is represented by the figure? Explain.

Solution

$2x(x + 2) = 2x^2 + 4x$

45. Simplify the expression $2(x - 4) + 5x$.

Solution

$(x - 4) + 5x = 2x - 8 + 5x$
$= 7x - 8$

47. Simplify the expression $-3(z - 2) - (z - 6)$.

Solution

$-3(z - 2) - (z - 6) = -3z + 6 - z + 6$
$= -4z + 12$

49. *Amount of Sales* You earn a sales commission rate of 5.5% . Your commission is $1600. How much did you sell?

Solution

Verbal model: | Commission | $=$ | Commission rate | $\cdot$ | Sales | $(a = pb)$

Labels:
Commission $=$ \$1600 (dollars)
Commission $=$ 0.055 (percent in decimal form)
Sales $= x$

Equation:
$$1600 = 0.055x$$

$$\frac{1600}{0.055} = x$$

$$\$29{,}090.91 = x$$

Therefore, the sales were \$29,090.91.

51. Multiply $\left(\dfrac{x}{4}\right)(10x)$ and simplify.

Solution

$$\left(\frac{x}{4}\right)(10x) = \frac{5}{2}x^2 \quad \text{or} \quad \frac{5x^2}{2}$$

Note: $\left(\dfrac{x}{4}\right)(10x) = \left(\dfrac{x}{4}\right)\left(\dfrac{10x}{1}\right) = \dfrac{10x^2}{4} = \dfrac{\cancel{2}(5x^2)}{\cancel{2}(2)} = \dfrac{5x^2}{2}$

53. Multiply $t^2(4t)$ and simplify.

Solution

$$t^2(4t) = 4t^3$$

55. Multiply $(-3t)^3$ and simplify.

Solution

$$(-3t)^3 = (-3)^3 t^3 = -27t^3$$

57. Multiply $y(3 - y)$ and simplify.

Solution

$$y(3 - y) = (y)(3) - (y)(y)$$
$$= 3y - y^2 \quad \text{or} \quad -y^2 + 3y$$

59. Multiply $-x(x^2 - 4)$ and simplify.

Solution

$$-x(x^2 - 4) = (-x)(x^2) - (-x)(4)$$
$$= -x^3 + 4x$$

61. Multiply $3t(2t - 5)$ and simplify.

Solution

$$3t(2t - 5) = (3t)(2t) - (3t)(5) = 6t^2 - 15t$$

63. Multiply $3x(x^2 - 2x + 1)$ and simplify.

Solution

$$3x(x^2 - 2x + 1) = (3x)(x^2) - (3x)(2x) + (3x)(1)$$
$$= 3x^3 - 6x^2 + 3x$$

65. Multiply $2x(x^2 - 2x + 8)$ and simplify.

Solution

$$2x(x^2 - 2x + 8) = (2x)(x^2) - (2x)(2x) + (2x)(8) = 2x^3 - 4x^2 + 16x$$

67. Multiply $-2x(-3x)(5x + 2)$ and simplify.

Solution

$$-2x(-3x)(5x + 2) = [(-2x)(-3x)](5x + 2)$$
$$= (6x^2)(5x + 2)$$
$$= (6x^2)(5x) + (6x^2)(2)$$
$$= 30x^3 + 12x^2$$

Note: This problem could also be done by first multiplying $(-3x)$ and $(5x + 2)$ as follows.

$$-2x(-3x)(5x + 2) = -2x[(-3x)(5x + 2)]$$
$$= -2x[(-3x)(5x) + (-3x)(2)]$$
$$= -2x[-15x^2 - 6x]$$
$$= (-2x)(-15x^2) - (-2x)(6x)$$
$$= 30x^3 + 12x^2$$

69. Multiply $(x - 7)(x - 9)$ and simplify.

Solution

$$\qquad\qquad \text{F} \quad \text{O} \quad \text{1} \quad \text{L}$$
$$(x - 7)(x - 9) = x^2 - 9x - 7x + 63 = x^2 - 16x + 63$$

71. Multiply $(3x - 5)(2x + 1)$ and simplify.

Solution

$$\qquad\qquad \text{F} \quad \text{O} \quad \text{I} \quad \text{L}$$
$$(3x - 5)(2x + 1) = 6x^2 + 3x - 10x - 5 = 6x^2 - 7x - 5$$

73. Multiply $(x + y)(x + 2y)$ and simplify.

Solution

$$\qquad\qquad \text{F} \quad \text{O} \quad \text{I} \quad \text{L}$$
$$(x + y)(x + 2y) = x^2 + 2xy + xy + 2y^2 = x^2 + 3xy + 2y^2$$

75. Multiply $2x(6x^4) - 3x^2(2x^2)$ and simplify.

Solution

$$(2x)(6x^4) - 3x^2(2x^2) = 12x^5 - 6x^4$$

77. Multiply $5x(x + 1) - 3x(2x - 4)$ and simplify.

Solution

$$5x(x + 1) - 3x(2x - 4) = 5x^2 + 5x - 6x^2 + 12x$$
$$= -x^2 + 17x$$

79. Multiply $(u - 1)(2u + 3)(2u + 1)$ and simplify.

Solution

$$(u - 1)(2u + 3)(2u + 1) = (2u^2 + 3u - 2u - 3)(2u + 1)$$
$$= (2u^2 + u - 3)(2u + 1)$$
$$= (4u^3 + 2u^2 + 2u^2 + u - 6u - 3)$$
$$= 4u^3 + 4u^2 - 5u - 3$$

81. Use a special product pattern to find the product of $(x + 5)(x - 5)$.

Solution

Multiply

$(x + 5)(x - 5)$

Solution

$(x + 5)(x - 5) = x^2 - 5^2$

$= x^2 - 25$

Pattern

$(a + b)(a - b) = a^2 - b^2$

83. Use a special product pattern to find the product of $(y + 9)(y - 9)$.

Solution

Multiply

$(y + 9)(y - 9)$

Solution

$(y + 9)(y - 9) = y^2 - 9^2$

$= y^2 - 81$

Pattern

$(a + b)(a - b) = a^2 - b^2$

85. Use a special product pattern to find the product of $(2x + 3y)(2x - 3y)$.

Solution

Multiply

$(2x + 3y)(2x - 3y)$

Solution

$(2x + 3y)(2x - 3y) = (2x)^2 - (3y)^2$

$= 4x^2 - 9y^2$

Pattern

$(a + b)(a - b) = a^2 - b^2$

87. Use a special product pattern to find the product of $(a - 2)^2$.

Solution

Multiply

$(a - 2)^2$

Solution

$(a - 2)^2 = (a)^2 - 2(a)(2) + (2)^2$

$= a^2 - 4a + 4$

Pattern

$(a - b)^2 = a^2 - 2ab + b^2$

89. Use a special product pattern to find the product of $(3x + 2)^2$.

Solution

Multiply

$(3x + 2)^2$

Solution

$(3x + 2)^2 = (3x)^2 + 2(3x)(2) + 2^2$

$= 9x^2 + 12x + 4$

Pattern

$(a + b)^2 = a^2 + 2ab + b^2$

91. Use a special product pattern to find the product of $(8 - 3z)^2$.

Solution

Multiply

$(8 - 3z)^2$

Solution

$(8 - 3z)^2 = (8)^2 - 2(8)(3z) + (3z)^2$

$= 64 - 48z + 9z^2$

or $9z^2 - 48z + 64$

Pattern

$(a - b)^2 = a^2 - 2ab + b^2$

93. Use a special product pattern to find the product of $[(x + 1) + y]^2$.

Solution

Multiply:
$[(x + 1) + y]^2$

Solution:
$$[(x + 1) + y]^2 = (x + 1)^2 + 2(x + 1)(y) + (y)^2$$
$$= (x + 1)^2 + 2y(x + 1) + y^2$$
$$= x^2 + 2x + 1 + 2y(x + 1) + y^2$$
$$= x^2 + 2x + 1 + 2xy + 2y + y^2$$
$$\text{or } x^2 + y^2 + 2xy + 2x + 2y + 1$$

Pattern:
$$(a + b)^2 = a^2 + 2ab + b^2$$

Use pattern again for $(x + 1)^2$.

95. Multiply $(x + 2)^2 - (x - 2)^2$. Then simplify.

Solution

$$(x + 2)^2 - (x - 2)^2 = [(x)^2 + 2(x)(2) + (2)^2] - [(x)^2 - 2(x)(2) + (2)^2]$$
$$= [x^2 + 4x + 4] - [x^2 - 4x + 4]$$
$$= x^2 + 4x + 4 - x^2 + 4x - 4 = 8x$$

97. Is the equation $(x + y)^3 = x^3 + 3x^2y + 3xy^2 + y^3$ an identity? Explain your reasoning.

Solution

Yes, this is an identity.

$$(x + y)^3 = (x + y)(x + y)(x + y)$$
$$= (x^2 + 2xy + y^2)(x + y)$$
$$= x^3 + x^2y + 2x^2y + 2xy^2 + xy^2 + y^3$$
$$= x^3 + 3x^2y + 3xy^2 + y^3$$

99. Use the results of Exercise 97 to find the product of $(x + 2)^3$.

Solution

$$(x + 2)^3 = (x)^3 + 3(x)^2(2) + 3(x)(2)^2 + (2)^3 = x^3 + 6x^2 + 12x + 8$$

Pattern: $(a + b)^3 = a^3 + 3a^2b + 3ab^2 + b^3$

101. *Geometry*   The height of a rectangular sign is twice its width w (see figure). Find (a) the perimeter and (b) the area of the rectangle.

Solution

(a)   *Common formula:* $P = 2l + 2w$

      Labels: Width $= w$

                      Length (height) $= 2w$

      Perimeter: $P = 2(2w) + 2(w)$

$$= 4w + 2w$$

$$= 6w$$

(b)   *Common formula:* $A = l \cdot w$

      Labels: Width $= w$

                      Length (height) $= 2w$

      Area: $A = (2w)(w)$

$$= 2w^2$$

The area of the sign is $2w^2$.

103. Find a polynomial product the represents the area of the region shown in the textbook.

Solution

Area $=$ (Width)(Length)

The width is $x + x$, or $2x$, and the length is $(x + 1) + (x + 1)$ or $2(x + 1)$.

$$2x[2(x + 1)] = 2x(2x + 2) = 4x^2 + 4x$$

105. Find two different expressions that represent the area of the shaded portion of the figure in the textbook.

Solution

Area $=$ (Length)(Width)

$$= x(4)$$

$$= 4x$$

Area $=$ Area of largest rectangle $-$ Area of 3 smaller rectangles

$$= (x + 3)(x + 4) - (x)(x) - (3)(x) - (3)(4)$$

Therefore, $4x = (x + 3)(x + 4) - (x)(x) - (3)(x) - (3)(4)$

$$= (x + 3)(x + 4) - x^2 - 3x - 12$$

Note: You can verify algebraically that the two expressions are the same.

107. *Interpreting Graphs*   When x units are sold, the revenue R is given by $R = x(900 - 0.5x)$.

(a) Use a graphing utility to graph the expression.

(b) Multiply the expression for revenue and use a graphing utility to graph the product. Verify that the graph is the same as in part (a).

(c) Find the revenue when $x = 500$. Use the graph to determine whether the revenue would increase or decrease if more units were sold.

Solution

(a)
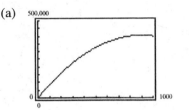

(b) $R = x(900 - 0.5x)$

$\qquad = 900x - 0.5x^2$

The graph of the product is identical to the graph in (a).

(c) When $x = 500$, $R = 900(500) - 0.5(500)^2$

$$= 450,000 - 0.5(250,000)$$

$$= 325,000$$

When $x = 500$, the revenue R is \$325,000. From the graph, you can determine that the revenue would *increase* if more units were sold.

Mid-Chapter Quiz for Chapter 5

1. Explain why $x^2 + 2x - (3/x)$ is not a polynomial.

Solution

Each term in a polynomial must be in the form ax^k with a nonnegative k; the last term in this expression, $3/x$, is not in this form.

2. Determine the degree and the leading coefficient of the polynomial $-3x^4 + 2x^2 - x$.

Solution

Degree: 4

Leading Coefficient: -3

3. Give an example of a trinomial in one variable of degree 5.

Solution

$6x^5 + x^2 - 1$

Note: There are many correct answers.

4. *True or False?*   The product of two binomials is a binomial. If false, give an example to show it is false.

Solution

False. For example, $(x - 2)(x + 7) = x^2 + 5x - 14$.

5. Perform the operations for the expression $y + (4 + 3y)$ and simplify.

Solution

$$y + (4 + 3y) = (y + 3y) + 4$$

$$= 4y + 4$$

6. Perform the operations for the expression $(3v^2 - 5) - (v^3 + 2v^2 - 6v)$ and simplify.

Solution

$$(3v^2 - 5) - (v^3 + 2v^2 - 6v) = 3v^2 - 5 - v^3 - 2v^2 + 6v$$

$$= -v^3 + v^2 + 6v - 5$$

7. Perform the operations for the expression $9s - [6 - (s - 5) + 7s]$ and simplify.

Solution

$$9s - [6 - (s - 5) + 7s] = 9s - [6 - s + 5 + 7s]$$
$$= 9s - 6 + s - 5 - 7s$$
$$= (9s + s - 7s) + (-6 - 5)$$
$$= 3s - 11$$

8. Perform the operations for the expression $-3(4 - x) + 4(x^2 + 2) - (x^2 - 2x)$ and simplify.

Solution

$$-3(4 - x) + 4(x^2 + 2) - (x^2 - 2x) = -12 + 3x + 4x^2 + 8 - x^2 + 2x$$
$$= (4x^2 - x^2) + (3x + 2x) + (-12 + 8)$$
$$= 3x^2 + 5x - 4$$

9. Perform the operations for the expression $2r(5r)^2$ and simplify.

Solution

$$2r(5r)^2 = 2r(25r^2)$$
$$= 50r^3$$

10. Perform the operations for the expression $m(-2m)^3$ and simplify.

Solution

$$m(-2m)^3 = m(-8m^3)$$
$$= -8m^4$$

11. Perform the operations for the expression $6a(2a/3)^3$ and simplify.

Solution

$$6a\left(\frac{2a}{3}\right)^3 = \frac{6a}{1}\left(\frac{8a^3}{27}\right)$$
$$= \frac{48a^4}{27}$$
$$= \frac{16a^4}{9}$$

12. Perform the operations for the expression $(2y - 3)(y + 5)$ and simplify.

Solution

$$(2y - 3)(y + 5) = 2y^2 + 10y - 3y - 15$$
$$= 2y^2 + 7y - 15$$

13. Perform the operations for the expression $(4 - 3x)^2$ and simplify.

Solution

$$(4 - 3x)^2 = 4^2 - 2(4)(3x) + (3x)^2$$
$$= 16 - 24x + 9x^2$$

Note: This problem can also be done by using FOIL to multiply $(4 - 3x)(4 - 3x)$.

14. Perform the operations for the expression $(2u - 3)(2u + 3)$ and simplify.

Solution

$$(2u - 3)(2u + 3) = (2u)^2 - 3^2$$
$$= 4u^2 - 9$$

15. Add $5x^4 + 2x^2 + x - 3$ and $3x^3 - 2x^2 - 3x + 5$ using a vertical format.

Solution

$$
\begin{array}{r}
5x^4 \quad\quad + 2x^2 + \ x - 3 \\
+ \quad\quad 3x^3 - 2x^2 - 3x + 5 \\
\hline
5x^4 + 3x^3 \quad\quad\quad - 2x + 2
\end{array}
$$

16. Subtract $5x^2 - 3x - 9$ from $2x^3 + x^2 - 8$ using a vertical format.

Solution

$$
\begin{array}{r}
2x^3 + \ x^2 \quad\quad - 8 \\
- \quad\quad (5x^2 - 3x - 9) \\
\hline
\end{array}
\qquad
\begin{array}{r}
2x^3 + \ x^2 \quad\quad - 8 \\
- \quad\quad 5x^2 + 3x + 9 \\
\hline
2x^3 - 4x^2 + 3x + 1
\end{array}
$$

17. Multiply $3x^2 + 7x + 1$ and $2x - 5$ using a vertical format.

Solution

$$
\begin{array}{r}
3x^2 + \ 7x + 1 \\
\times \quad\quad\quad 2x - 5 \\
\hline
-15x^2 - 35x - 5 \\
6x^3 + \ 14x^2 + \ 2x \quad\quad\quad \\
\hline
6x^3 - \quad x^2 - 33x - 5
\end{array}
$$

18. Multiply $5x^3 - 6x^2 + 3$ and $x^2 - 3x$ using a vertical format.

Solution

$$
\begin{array}{r}
5x^3 - 6x^2 + \ 3 \\
\times \quad\quad\quad\quad x^2 - 3x \\
\hline
- \ 15x^4 + 18x^3 \quad\quad\quad - 9x \\
5x^5 - \ 6x^4 \quad\quad\quad + 3x^2 \quad\quad \\
\hline
5x^5 - 21x^4 + 18x^3 + 3x^2 - 9x
\end{array}
$$

19. Find the perimeter of the figure in the textbook.

Solution

$$
\begin{aligned}
\text{Perimeter} &= 5x + 18 + 2x + 2x + 3x + (18 - 2x) \\
&= (5x + 2x + 2x + 3x - 2x) + (18 + 18) \\
&= 10x + 36
\end{aligned}
$$

20. Find the area of the figure in the textbook.

Solution

Area = Area of rectangle on left + Area of rectangle on right

$$
\begin{aligned}
&= x(x + 1) + 3(x) \\
&= x^2 + x + 3x \\
&= x^2 + 4x
\end{aligned}
$$

or

Area = Area of upper rectangle + Area of lower rectangle

$$
\begin{aligned}
&= x(1) + (x + 3)x \\
&= x + x^2 + 3x \\
&= x^2 + 4x
\end{aligned}
$$

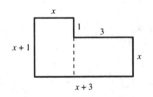

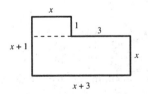

5.3   Dividing Polynomials

7. Perform the following division by cancellation *and* by subtracting exponents. (Assume the denominator is not zero.)

$$\frac{x^5}{x^2}$$

Solution

By Cancellation

$$\frac{x^5}{x^2} = \frac{x \cdot x \cdot x \cdot \cancel{x} \cdot \cancel{x}}{\cancel{x} \cdot \cancel{x}} = x^3$$

By Subtracting Exponents

$$\frac{x^5}{x^2} = x^{5-2} = x^3$$

9. Perform the following division by cancellation *and* by subtracting exponents. (Assume the denominator is not zero.)

$$\frac{4^5 x^3}{4 x^5}$$

Solution

By Cancellation

$$\frac{4^5 x^3}{4 x^5} = \frac{\cancel{4} \cdot 4 \cdot 4 \cdot 4 \cdot 4 \cdot \cancel{x} \cdot \cancel{x} \cdot \cancel{x}}{\cancel{4} \cdot \cancel{x} \cdot \cancel{x} \cdot \cancel{x} \cdot x \cdot x} = \frac{4^4}{x^2} \quad \text{or} \quad \frac{256}{x^2}$$

By Subtracting Exponents

$$\frac{4^5 x^3}{4 x^5} = 4^{5-1} \cdot \frac{1}{x^{5-3}}$$

$$= 4^4 \cdot \frac{1}{x^2} = \frac{4^4}{x^2} \quad \text{or} \quad \frac{256}{x^2}$$

11. Simplify the following expression. (Assume the denominator is not zero.)

$$\frac{-3x^2}{6x}$$

Solution

$$\frac{-3x^2}{6x} = \left(-\frac{3}{6}\right)(x^{2-1}) = -\frac{1}{2}x \text{ or } -\frac{x}{2}$$

13. Simplify the following expression. (Assume the denominator is not zero.)

$$\frac{-18s^4}{-12r^2 s}$$

Solution

$$\frac{-18s^4}{-12r^2 s} = \left(\frac{-18}{-12}\right) \cdot \frac{s^{4-1}}{r^2} = \frac{3}{2} \cdot \frac{s^3}{r^2} = \frac{3s^3}{2r^2}$$

15. Simplify the following expression. (Assume the denominator is not zero.)

$$\frac{(-3z)^2}{18z^3}$$

Solution

$$\frac{(-3z)^2}{18z^3} = \frac{9z^2}{18z^3}$$

$$= \left(\frac{9}{18}\right)\left(\frac{1}{z^{3-2}}\right)$$

$$= \left(\frac{1}{2}\right)\left(\frac{1}{z}\right)$$

$$= \frac{1}{2z}$$

17. Simplify the following expression. (Assume the denominator is not zero.)

$$\frac{24(u^2 v)^4}{18u^2 v^6}$$

Solution

$$\frac{24(u^2 v)^4}{18u^2 v^6} = \frac{24u^8 v^4}{18u^2 v^6} = \left(\frac{24}{18}\right) \cdot \frac{u^{8-2}}{v^{6-4}} = \frac{4u^6}{3v^2}$$

19. Perform the following division and simplify.

$$\frac{3z + 3}{3}$$

Solution

$$\frac{3z + 3}{3} = \frac{3z}{3} + \frac{3}{3} = z + 1$$

21. Perform the following division and simplify. (Assume the denominator is not zero.)

$$\frac{8z^3 + 3z^2 - 2z}{z}$$

Solution

$$\frac{8z^3 + 3z^2 - 2z}{z} = \frac{8z^3}{z} + \frac{3z^2}{z} - \frac{2z}{z}$$
$$= 8z^2 + 3z - 2$$

23. Perform the following division and simplify. (Assume the denominator is not zero.)

$$\frac{x^2 - x - 2}{x + 1}$$

Solution

$$
\begin{array}{r}
x - 2 \\
x + 1 \overline{)\, x^2 - x - 2} \\
\underline{x^2 + x} \\
-2x - 2 \\
\underline{-2x - 2} \\
0
\end{array}
$$

Thus, $\dfrac{x^2 - x - 2}{x + 1} = x - 2$.

25. Perform the following division and simplify. (Assume the denominator is not zero.)

$$\frac{3y^2 + 4y - 4}{3y - 2}$$

Solution

$$
\begin{array}{r}
y + 2 \\
3y - 2 \overline{)\, 3y^2 + 4y - 4} \\
\underline{3y^2 - 2y} \\
6y - 4 \\
\underline{6y - 4}
\end{array}
$$

Thus, $\dfrac{3y^2 + 4y - 4}{3y - 2} = y + 2$.

27. Divide $4x^2 + 3x + 1$ by $x + 1$ and simplify. (Assume the denominator is not zero.)

Solution

$$
\begin{array}{r}
4x - 1 + \dfrac{2}{x + 1} \\
x + 1 \overline{)\, 4x^2 + 3x + 1} \\
\underline{4x^2 + 4x} \\
-x + 1 \\
\underline{-x - 1} \\
2
\end{array}
$$

Thus, $\dfrac{4x^2 + 3x + 1}{x + 1} = 4x - 1 + \dfrac{2}{x + 1}$.

29. Divide $(x^3 - 4x^2 + 9x - 7) \div (x - 2)$ and simplify. (Assume the denominator is not zero.)

Solution

$$
\begin{array}{r}
x^2 - 2x + 5 + \dfrac{3}{x - 2} \\
x - 2 \overline{)\, x^3 - 4x^2 + 9x - 7} \\
\underline{x^3 - 2x^2} \\
-2x^2 + 9x \\
\underline{-2x^2 + 4x} \\
5x - 7 \\
\underline{5x - 10} \\
3
\end{array}
$$

Thus, $\dfrac{x^3 - 4x^2 + 9x - 7}{x - 2} = x^2 - 2x + 5 + \dfrac{3}{x - 2}$.

31. Simplify the following expression. (Assume the denominator is not zero.)

$$\frac{4x^3}{x^2} - 2x$$

Solution

$$\frac{4x^3}{x^2} - 2x = 4x - 2x = 2x$$

33. Determine whether the cancellation is valid in

$$\frac{3 + 4}{3} = \frac{\cancel{3} + 4}{\cancel{3}} = 4.$$

Solution

No, the cancellation is *not* valid because a term (rather than a factor) was cancelled from the numerator. Only common *factors* can be cancelled from the numerator and denominator.

35. Determine whether the cancellation is valid in
$$\frac{7 \cdot 12}{19 \cdot 7} = \frac{\cancel{7} \cdot 12}{19 \cdot \cancel{7}} = \frac{12}{19}.$$

Solution

Yes, this cancellation *is* valid.

37. What is 62% of 25?

Solution

Verbal model: $\boxed{\begin{array}{c}\text{What} \\ \text{number}\end{array}} = \boxed{62\% \text{ of } 25}$ $(a = pb)$

Label: Unknown number $= a$
            $p = 0.62$ (percent in decimal form)
            $b = 25$

Equation: $a = 0.62(25)$

            $a = 15.5$

Therefore, 15.5 is 62% of 25.

39. 32% of what number is 145.6?

Solution

Verbal model: $\boxed{145.6} = \boxed{\begin{array}{c}32\% \text{ of an} \\ \text{unknown} \\ \text{number}\end{array}}$ $(a = pb)$

Labels: $a = 145.6$
            $p = 0.32$ (percent in decimal form)
            Unknown number $= b$

Equation: $145.6 = 0.32b$

            $\dfrac{145.6}{0.32} = b$

            $455 = b$

Therefore, 145.6 is 32% of 455.

41. Write an algebraic expression that represents the product of two consecutive odd integers, the first of which is $2n + 1$.

Solution

First consecutive odd integer $= 2n + 1$

Second consecutive odd integer $= 2n + 3$

$\begin{aligned} \text{Product} &= (2n + 1)(2n + 3) \\ &= 4n^2 + 6n + 2n + 3 \\ &= 4n^2 + 8n + 3 \end{aligned}$

43. Perform the following division by cancellation *and* by subtracting exponents. (Assume the denominator is not zero.)
$$\frac{2^3 y^4}{2^2 y^2}$$

Solution

By Cancellation

$$\frac{2^3 y^4}{2^2 y^2} = \frac{2 \cdot \cancel{2} \cdot \cancel{2} \cdot y \cdot y \cdot \cancel{y} \cdot \cancel{y}}{\cancel{2} \cdot \cancel{2} \cdot \cancel{y} \cdot \cancel{y}} = 2y^2$$

By Subtracting Exponents

$$\frac{2^3 y^4}{2^2 y^2} = 2^{3-2} y^{4-2} = 2y^2$$

45. Perform the following division by cancellation *and* by subtracting exponents. (Assume the denominator is not zero.)
$$\frac{z^4}{z^7}$$

Solution

By Cancellation

$$\frac{z^4}{z^7} = \frac{\cancel{z} \cdot \cancel{z} \cdot \cancel{z} \cdot \cancel{z}}{z \cdot z \cdot z \cdot \cancel{z} \cdot \cancel{z} \cdot \cancel{z} \cdot \cancel{z}} = \frac{1}{z^3}$$

By Subtracting Exponents

$$\frac{z^4}{z^7} = \frac{1}{z^{7-4}} = \frac{1}{z^3}$$

47. Perform the following division by cancellation *and* by subtracting exponents. (Assume the denominator is not zero.)

$$\frac{x^5}{x^5}$$

Solution

By Cancellation

$$\frac{x^5}{x^5} = \frac{\cancel{x} \cdot \cancel{x} \cdot \cancel{x} \cdot \cancel{x} \cdot \cancel{x}}{\cancel{x} \cdot \cancel{x} \cdot \cancel{x} \cdot \cancel{x} \cdot \cancel{x}} = 1$$

By Subtracting Exponents

$$\frac{x^5}{x^5} = x^{5-5} = x^0 = 1$$

49. Perform the following division by cancellation *and* by subtracting exponents. (Assume the denominator is not zero.)

$$\frac{3^4(ab)^2}{3(ab)^3}$$

Solution

$$\frac{3^4(ab)^2}{3(ab)^3} = \frac{3^4 a^2 b^2}{3 a^3 b^3} = \frac{3^{4-1}}{a^{3-2} b^{3-2}} = \frac{3^3}{ab} \text{ or } \frac{27}{ab}$$

51. Simplify the expression $\dfrac{-12z^3}{-3z}$.

Solution

$$\frac{-12z^3}{-3z} = \left(\frac{-12}{-3}\right)(z^{3-1}) = 4z^2$$

53. Simplify the expression $\dfrac{32b^4}{12b^3}$.

Solution

$$\frac{32b^4}{12b^3} = \left(\frac{32}{12}\right)(b^{4-3}) = \frac{8}{3}b \text{ or } \frac{8b}{3}$$

55. Simplify the expression $\dfrac{-22y^2}{4y}$.

Solution

$$\frac{-22y^2}{4y} = \left(\frac{-22}{4}\right)y^{2-1} = -\frac{11}{2}y \quad \text{or} \quad -\frac{11y}{2}$$

57. Simplify the expression $\dfrac{(-3xy)^3}{-9x^6y^2}$.

Solution

$$\frac{(-3xy)^3}{-9x^6y^2} = \frac{(-3)^3 x^3 y^3}{-9x^6y^2}$$

$$= \frac{-27x^3 y^3}{-9x^6 y^2}$$

$$= \left(\frac{-27}{-9}\right)\frac{y^{3-2}}{x^{6-3}}$$

$$= \frac{3y}{x^3}$$

59. Perform the following division and simplify.

$$\frac{4z - 12}{4}$$

Solution

$$\frac{4z - 12}{4} = \frac{4z}{4} - \frac{12}{4} = z - 3$$

61. Divide $(5x^2 - 2x) \div x$ and simplify. (Assume the denominator is not zero.)

Solution

$$(5x^2 - 2x) \div x = \frac{5x^2}{x} - \frac{2x}{x} = 5x - 2$$

63. Perform the following division and simplify. (Assume the denominator is not zero.)

$$\frac{25z^3 + 10z^2}{-5z}$$

Solution

$$\frac{25z^3 + 10z^2}{-5z} = \frac{25z^3}{-5z} + \frac{10z^2}{-5z} = -5z^2 - 2z$$

65. Perform the following division and simplify. (Assume the denominator is not zero.)

$$-\frac{4x^2 - 3x}{x}$$

Solution

$$-\frac{4x^2 - 3x}{x} = \frac{-4x^2 + 3x}{x} = \frac{-4x^2}{x} + \frac{3x}{x} = -4x + 3$$

Note: Here is an alternate method for this exercise.

$$-\frac{4x^2 - 3x}{x} = -\left(\frac{4x^2}{x} - \frac{3x}{x}\right)$$

$$= -(4x - 3)$$

$$= -4x + 3$$

67. Perform the following division and simplify. (Assume the denominator is not zero.)

$$\frac{m^3 + 3m - 4}{m}$$

Solution

$$\frac{m^3 + 3m - 4}{m} = \frac{m^3}{m} + \frac{3m}{m} - \frac{4}{m} = m^2 + 3 - \frac{4}{m}$$

69. Perform the following division and simplify. (Assume the denominator is not zero.)

$$\frac{x^2 + 9x + 20}{x + 4}$$

Solution

$$\require{enclose}
\begin{array}{r}
x + 5 \\
x + 4 \enclose{longdiv}{x^2 + 9x + 20} \\
\underline{x^2 + 4x} \\
5x + 20 \\
\underline{5x + 20} \\
0
\end{array}$$

Thus, $(x^2 + 9x + 20) \div (x + 4) = x + 5$.

71. Divide $9x^2 - 1$ by $3x + 1$ and simplify. (Assume that the denominator is not zero.)

Solution

$$\begin{array}{r}
3x - 1 \\
3x + 1 \enclose{longdiv}{9x^2 + 0x - 1} \\
\underline{9x^2 + 3x} \\
-3x - 1 \\
\underline{-3x - 1} \\
0
\end{array}$$

Thus, $\dfrac{9x^2 - 1}{3x + 1} = 3x - 1$.

73. Divide $2z^2 + 5z - 3$ by $z + 3$ and simplify. (Assume the denominator is not zero.)

Solution

$$\begin{array}{r}
2z - 1 \\
z + 3 \enclose{longdiv}{2z^2 + 5z - 3} \\
\underline{2z^2 + 6z} \\
-z - 3 \\
\underline{-z - 3} \\
0
\end{array}$$

Thus, $\dfrac{2z^2 + 5z - 3}{z + 3} = 2z - 1$.

75. Divide $(18t^2 - 21t - 4) \div (3t - 4)$ and simplify. (Assume the denominator is not zero.)

Solution

$$\begin{array}{r}
6t + 1 \\
3t - 4 \enclose{longdiv}{18t^2 - 21t - 4} \\
\underline{18t^2 - 24t} \\
3t - 4 \\
\underline{3t - 4} \\
0
\end{array}$$

Thus, $\dfrac{18t^2 - 21t - 4}{3t - 4} = 6t + 1$.

77. Perform the following division and simplify. (Assume the denominator is not zero.)

$$\frac{x^3 - 8}{x - 2}$$

Solution

$$
\begin{array}{r}
x^2 + 2x + 4 \\
x - 2 \overline{)\, x^3 + 0x^2 + 0x - 8} \\
\underline{x^3 - 2x^2} \\
2x^2 \\
\underline{2x^2 - 4x} \\
4x - 8 \\
\underline{4x - 8} \\
0
\end{array}
$$

Thus, $\dfrac{x^3 - 8}{x - 2} = x^2 + 2x + 4.$

81. Divide $(7x + 3) \div (x + 2)$ and simplify. (Assume the denominator is not zero.)

Solution

$$
\begin{array}{r}
7 - \dfrac{11}{x + 2} \\
x + 2 \overline{)\, 7x + \quad 3} \\
\underline{7x + \ 14} \\
-11
\end{array}
$$

Thus, $\dfrac{7x + 3}{x + 2} = 7 - \dfrac{11}{x + 2}.$

85. Perform the following division and simplify. (Assume the denominator is not zero.)

$$\frac{2x^2 + 7x + 8}{2x + 3}$$

Solution

$$
\begin{array}{r}
x + 2 + \dfrac{2}{2x + 3} \\
2x + 3 \overline{)\, 2x^2 + 7x + 8} \\
\underline{2x^2 + 3x} \\
4x + 8 \\
\underline{4x + 6} \\
2
\end{array}
$$

Thus, $\dfrac{2x^2 + 7x + 8}{2x + 3} = x + 2 + \dfrac{2}{2x + 3}.$

79. Perform the following division and simplify. (Assume the denominator is not zero.)

$$\frac{x^3 - 7x + 6}{x - 2}$$

Solution

$$
\begin{array}{r}
x^2 + \ 2x - 3 \\
x - 2 \overline{)\, x^3 + 0x^2 - \ 7x + 6} \\
\underline{x^3 - 2x^2} \\
2x^2 - \ 7x \\
\underline{2x^2 - \ 4x} \\
-3x + 6 \\
\underline{-3x + 6} \\
0
\end{array}
$$

Thus, $\dfrac{x^3 - 7x + 6}{x - 2} = x^2 + 2x - 3.$

83. Perform the following division and simplify. (Assume the denominator is not zero.)

$$\frac{x^2 + 9}{x + 3}$$

Solution

$$
\begin{array}{r}
x - \ 3 + \dfrac{18}{x + 3} \\
x + 3 \overline{)\, x^2 + \quad 0x + \ 9} \\
\underline{x^2 + \ 3x} \\
-3x + \ 9 \\
\underline{-3x - \ 9} \\
18
\end{array}
$$

Thus, $\dfrac{x^2 + 9}{x + 3} = x - 3 + \dfrac{18}{x + 3}.$

87. Perform the following division and simplify. (Assume the denominator is not zero.)

$$\frac{4z^2 + 3z}{2z - 1}$$

Solution

$$
\begin{array}{r}
2z + \frac{5}{2} \\
2z - 1 \overline{)\, 4z^2 + 3z} \\
\underline{4z^2 - 2z} \\
5z \\
\underline{5z - \frac{5}{2}} \\
\frac{5}{2}
\end{array}
$$

Thus, $\dfrac{4z^2 + 3z}{2z - 1} = 2z + \dfrac{5}{2} + \dfrac{5}{2(2z - 1)}.$

89. Perform the following division and simplify. (Assume the denominator is not zero.)

$$\frac{3t^3 + 7t^2 + 3t - 2}{t + 2}$$

Solution

$$
\begin{array}{r}
3t^2 + t + 1 - \dfrac{4}{t+2} \\
t+2\,)\overline{3t^3 + 7t^2 + 3t - 2} \\
\underline{3t^3 + 6t^2} \\
t^2 + 3t \\
\underline{t^2 + 2t} \\
t - 2 \\
\underline{t + 2} \\
-4
\end{array}
$$

Thus, $\dfrac{3t^3 + 7t^2 + 3t - 2}{t + 2} = 3t^2 + t + 1 - \dfrac{4}{t+2}$.

91. Divide $x^4 - 1$ by $x - 1$ and simplify. (Assume the denominator is not zero.)

Solution

$$
\begin{array}{r}
x^3 + x^2 + x + 1 \\
x-1\,)\overline{x^4 + 0x^3 + 0x^2 + 0x - 1} \\
\underline{x^4 - x^3} \\
x^3 \\
\underline{x^3 - x^2} \\
x^2 \\
\underline{x^2 - x} \\
x - 1 \\
\underline{x - 1} \\
0
\end{array}
$$

Thus, $\dfrac{x^4 - 1}{x - 1} = x^3 + x^2 + x + 1$.

93. Simplify the following expression. (Assume the denominator is not zero.)

$$\frac{8u^2 v}{2u} + \frac{(uv)^2}{uv}$$

Solution

$$\frac{8u^2 v}{2u} + \frac{(uv)^2}{uv} = \frac{8u^2 v}{2u} + \frac{u^2 v^2}{uv} = 4uv + uv = 5uv$$

95. *Comparing Ages*   You have two children: one is 18 years old and the other is 8 years old. In t years, their ages will be $t + 18$ and $t + 8$.

(a) Use long division to rewrite the ratio of your older child's age to your younger child's age.

(b) Complete the table shown in the textbook.

(c) What happens to the values of the ratio as t increases? Use the result of part (a) to explain your conclusion.

Solution

(a)
$$
\begin{array}{r}
1 \\
t+8\,)\overline{t + 18} \\
\underline{t + 8} \\
10
\end{array}
$$

Thus, $\dfrac{t + 18}{t + 8} = 1 + \dfrac{10}{t + 8}$.

(b)

t	0	10	20	30	40	50	60
$\dfrac{t + 18}{t + 8}$	$\dfrac{18}{8} = 2.25$	$\dfrac{28}{18} \approx 1.56$	$\dfrac{38}{28} \approx 1.36$	$\dfrac{48}{38} \approx 1.26$	$\dfrac{58}{48} \approx 1.21$	$\dfrac{68}{58} \approx 1.17$	$\dfrac{78}{68} \approx 1.15$

(c) As the value of t increases, the value of the ratio approaches 1. The ratio

$$\frac{t + 18}{t + 8} = 1 + \frac{10}{t + 8}$$

approaches 1 as t increases because the fraction

$$\frac{10}{t + 8}$$

approaches 0.

97. *Exploration* Use a graphing utility to compare the graphs of the following equations. On most graphing utilities, the graphs appear the same. There is however, a slight difference in the graphs. What is it?

$$y = \frac{x^2 - 3x + 2}{x - 1} \quad \text{and} \quad y = x - 2$$

Solution

The value of

$$y = \frac{x^2 - 3x + 2}{x - 1}$$

is undefined for $x = 1$ because the denominator would be 0, and division by zero is undefined. The two graphs are identical except that there is a "hole" in the first graph at $x = 1$.

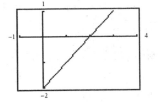

99. Match the graph in the textbook with two of the following expressions.

(a) $x - 1$ (b) $1 - x$ (c) $\dfrac{x^2 - 1}{x + 1}$ (d) $\dfrac{x^2 + 2x + 1}{x + 1}$ (e) $\dfrac{1 - 2x + x^2}{1 - x}$ (f) $\dfrac{x^2 - 1}{x - 1}$

Solution

Expressions (b) and (e)

5.4 | Negative Exponents and Scientific Notation

7. Rewrite $5x^{-4}$ with positive exponents.

Solution

$$5x^{-4} = 5\left(\frac{1}{x^4}\right) = \frac{5}{x^4}$$

9. Rewrite the following expression with positive exponents.

$$\frac{1}{2z^{-4}}$$

Solution

$$\frac{1}{2z^{-4}} = \frac{1z^4}{2} = \frac{z^4}{2}$$

11. Rewrite 3^{-2} with positive exponents. Then evaluate the expression.

Solution

$$3^{-2} = \frac{1}{3^2} = \frac{1}{9}$$

13. Rewrite the following expression with positive exponents. Then evaluate the expression.

$$\frac{2^{-4}}{3^{-2}}$$

Solution

$$\frac{2^{-4}}{3^{-2}} = \frac{3^2}{2^4} = \frac{9}{16}$$

15. Use a calculator to evaluate the expression 3.8^{-4}.

Solution

$$3.8^{-4} \approx 0.0048$$

17. Use a calculator to evaluate the expression $\dfrac{100}{1.06^{-15}}$.

Solution

$$\frac{100}{1.06^{-15}} = 239.66$$

19. Rewrite the following expression with positive exponents. (Assume that no variable is equal to zero.)

$$\frac{x^2}{x^{-3}}$$

Solution

$$\frac{x^2}{x^{-3}} = x^2 \cdot x^3 = x^{2+3} = x^5$$

Note: We could also use the rule $\frac{a^m}{a^n} = a^{m-n}$.

$$\frac{x^2}{x^{-3}} = x^{2-(-3)} = x^{2+3} = x^5$$

21. Rewrite $(2x^2)^{-2}$ with positive exponents. (Assume that no variable is equal to zero.)

Solution

$$(2x^2)^{-2} = 2^{-2}x^{-4} = \frac{1}{2^2 x^4} = \frac{1}{4x^4}$$

23. Rewrite $(4a^{-2}b^3)^{-3}$ with positive exponents. (Assume that no variable is equal to zero.)

Solution

$$(4a^{-2}b^3)^{-3} = 4^{-3}(a^{-2})^{-3}(b^3)^{-3}$$
$$= 4^{-3}a^6 b^{-9}$$
$$= \frac{a^6}{4^3 b^9}$$
$$= \frac{a^6}{64b^9}$$

25. Rewrite $(-2x^2)^3(4x^3)^{-1}$ with positive exponents. (Assume that no variable is equal to zero.)

Solution

$$(-2x^2)^3(4x^3)^{-1} = (-2)^3(x^2)^3(4)^{-1}(x^3)^{-1}$$
$$= -8x^6(4)^{-1}x^{-3}$$
$$= -\frac{8x^6}{4x^3}$$
$$= -2x^3$$

27. Rewrite the following expression with positive exponents. (Assume that no variable is equal to zero.)

$$\left(\frac{x}{10}\right)^{-1}$$

Solution

$$\left(\frac{x}{10}\right)^{-1} = \frac{x^{-1}}{10^{-1}} = \frac{10}{x}$$

29. Rewrite the following expression with positive exponents. (Assume that no variable is equal to zero.)

$$\frac{3}{2} \cdot \left(\frac{-2}{3}\right)^{-3}$$

Solution

$$\frac{3}{2} \cdot \left(\frac{-2}{3}\right)^{-3} = \frac{3}{2} \cdot \frac{(-2)^{-3}}{3^{-3}} = \frac{3}{2} \cdot \frac{3^3}{(-2)^3}$$
$$= \frac{3}{2} \cdot \frac{27}{-8} = \frac{3(27)}{2(-8)}$$
$$= \frac{81}{-16} = -\frac{81}{16}$$

31. Write 1.09×10^6 in decimal form.

Solution

$$1.09 \times 10^6 = 1,090,000$$

33. Write 8.52×10^{-3} in decimal form.

Solution

$$8.52 \times 10^{-3} = 0.00852$$

35. Write 1,637,000,000 in scientific notation.

Solution

$$1,637,000,000 = 1.637 \times 10^9$$

37. Write 0.000435 in scientific notation.

Solution

$$0.000435 = 4.35 \times 10^{-4}$$

39. Use a calculator to evaluate the expression $8,000,000 \times 623,000$.

Solution

$8,000,000 \times 623,000 = (8 \times 10^6)(6.23 \times 10^5) = 4.984 \times 10^{12}$

Use the following calculator steps.

8 $\boxed{\text{EE}}$ 6 $\boxed{\times}$ 6.23 $\boxed{\text{EE}}$ 5 $\boxed{=}$ (Scientific)

8 $\boxed{\text{EE}}$ 6 $\boxed{\times}$ 6.23 $\boxed{\text{EE}}$ 5 $\boxed{\text{ENTER}}$ (Graphing)

41. Use a calculator to evaluate the expression
$$\frac{67,000,000}{0.0052}.$$

Solution

$$\frac{67,000,000}{0.0052} = \frac{6.7 \times 10^7}{5.2 \times 10^{-3}}$$

$$\approx 1.2885 \times 10^{10}$$

43. *Time for Light to Travel* Light travels from the sun to the earth in approximately
$$\frac{9.3 \times 10^7}{1.1 \times 10^7}.$$

Solution

$$\frac{9.3 \times 10^7}{1.1 \times 10^7} = 8.\overline{45} \text{ minutes}$$

45. Evaluate the expression $25 - 3^2 \cdot 2$.

Solution

$$25 - 3^2 \cdot 2 = 25 - 9 \cdot 2$$

$$= 25 - 18$$

$$= 7$$

47. Graph the equation $y = x^2 - 2x + 1$ using a graphing utility.

Solution

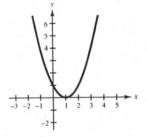

49. Evaluate the expression $10 + 3x - 2x^2$ for the values (a) $x = -2$ and (b) $x = 0$. (If it is not possible, state the reason.)

Solution

(a) If $x = -2$, the value of the expression is
$$10 + 3(-2) - 2(-2)^2 = 10 - 6 - 2(4)$$
$$= 10 - 6 - 8 = -4.$$

(b) If $x = 0$, the value of the expression is
$$10 - 3(0) - 2(0)^2 = 10.$$

51. Rewrite $(-4)^{-3}$ with positive exponents. Then evaluate the result.

Solution

$$(-4)^{-3} = \frac{1}{(-4)^3} = \frac{1}{-64} = -\frac{1}{64}$$

53. Rewrite the following expression with positive exponents. Then evaluate the result.
$$\frac{1}{16^{-1}}$$

Solution

$$\frac{1}{16^{-1}} = 1 \cdot 16 = 16$$

55. Rewrite the following expression with positive exponents. Then evaluate the result.

$$\frac{4^{-3}}{2}$$

Solution

$$\frac{4^{-3}}{2} = \frac{1}{2 \cdot 4^3}$$

$$= \frac{1}{2 \cdot 64}$$

$$= \frac{1}{128}$$

57. Rewrite $\left(\frac{2}{3}\right)^{-2}$ with positive exponents. Then evaluate the result.

Solution

$$\left(\frac{2}{3}\right)^{-2} = \frac{2^{-2}}{3^{-2}}$$

$$= \frac{3^2}{2^2}$$

$$= \frac{9}{4}$$

59. Rewrite the following expression with positive exponents and simplify. (Assume that no variable equals zero.)

$$\frac{y^{-5}}{y}$$

Solution

$$\frac{y^{-5}}{y} = \frac{1}{y \cdot y^5}$$

$$= \frac{1}{y^{1+5}}$$

$$= \frac{1}{y^6}$$

Note: We could also use the rule $\dfrac{a^m}{a^n} = a^{m-n}$.

$$\frac{y^{-5}}{y} = y^{-5-1}$$

$$= y^{-6}$$

$$= \frac{1}{y^6}$$

61. Rewrite the following expression with positive exponents and simplify. (Assume that no variable equals zero.)

$$\frac{x^{-4}}{x^{-2}}$$

Solution

$$\frac{x^{-4}}{x^{-2}} = \frac{x^2}{x^4} = \frac{1}{x^2}$$

Note: There are several ways to work this problem.

$$\frac{x^{-4}}{x^{-2}} = x^{-4-(-2)} \qquad \text{or} \qquad \frac{x^{-4}}{x^{-2}} = \frac{1}{x^{-2} \cdot x^4} \qquad \text{or} \qquad \frac{x^{-4}}{x^{-2}} = x^{-4}x^2$$

$$= x^{-4+2} \qquad\qquad\qquad = \frac{1}{x^{-2+4}} \qquad\qquad\qquad = x^{-4+2}$$

$$= x^{-2} \qquad\qquad\qquad = \frac{1}{x^2} \qquad\qquad\qquad = x^{-2}$$

$$= \frac{1}{x^2} \qquad\qquad\qquad\qquad\qquad\qquad\qquad = \frac{1}{x^2}$$

63. Rewrite $(y^{-3})^2$ with positive exponents and simplify. (Assume that no variable equals zero.)

Solution

$$(y^{-3})^2 = y^{-6} = \frac{1}{y^6}$$

65. Rewrite $(s^2)^{-1}$ with positive exponents and simplify. (Assume that no variable equals zero.)

Solution

$$(s^2)^{-1} = s^{-2} = \frac{1}{s^2}$$

67. Rewrite the following expression with positive exponents and simplify. (Assume that no variable equals zero.)

$$\frac{b^2 \cdot b^{-3}}{b^4}$$

Solution

$$\frac{b^2 b^{-3}}{b^4} = \frac{b^2}{b^4 b^3}$$

$$= \frac{b^2}{b^{4+3}}$$

$$= \frac{b^2}{b^7}$$

$$= \frac{1}{b^5}$$

Note: Here is another way to work this problem.

$$\frac{b^2 b^{-3}}{b^4} = \frac{b^{2+(-3)}}{b^4}$$

$$= \frac{b^{-1}}{b^4}$$

$$= \frac{1}{b^4 b}$$

$$= \frac{1}{b^{4+1}}$$

$$= \frac{1}{b^5}$$

69. Rewrite $(3x^2 y)^{-2}$ with positive exponents and simplify. (Assume that no variable equals zero.)

Solution

$$(3x^2 y)^{-2} = 3^{-2}(x^2)^{-2} y^{-2}$$

$$= 3^{-2} x^{-4} y^{-2}$$

$$= \frac{1}{3^2 x^4 y^2}$$

$$= \frac{1}{9x^4 y^2}$$

71. Rewrite $(5x^2 y^4)^3 (5x^2 y^4)^{-3}$ with positive exponents and simplify. (Assume that no variable equals zero.)

Solution

$$(5x^2 y^4)^3 (5x^2 y^4)^{-3} = (5x^2 y^4)^{3+(-3)}$$

$$= (5x^2 y^4)^0$$

$$= 1$$

73. Rewrite the following expression with positive exponents and simplify. (Assume that no variable equals zero.)

$$\left(\frac{3z^2}{x}\right)^{-2}$$

Solution

$$\left(\frac{3z^2}{x}\right)^{-2} = \frac{3^{-2}(z^2)^{-2}}{x^{-2}}$$

$$= \frac{3^{-2} z^{-4}}{x^{-2}}$$

$$= \frac{x^2}{3^2 z^4}$$

$$= \frac{x^2}{9z^4}$$

75. Rewrite $(2x^2 y)^0$ with positive exponents and simplify. (Assume that no variable equals zero.)

Solution

$$(2x^2 y)^0 = 1$$

Note: Any number (other than zero) when raised to the zero power is 1.

77. Rewrite the following expression with positive exponents and simplify.

$$\frac{3}{8} \cdot \left(-\frac{5}{2}\right)^3$$

Solution

$$\frac{3}{8} \cdot \left(-\frac{5}{2}\right)^3 = \frac{3}{8} \cdot \frac{(-5)^3}{2^3}$$

$$= \frac{3}{8} \cdot \frac{-125}{8}$$

$$= \frac{3(-125)}{8(8)}$$

$$= -\frac{375}{64}$$

79. Write 6.21×10^0 in decimal form.

Solution

$6.21 \times 10^0 = 6.21$

Note: $10^0 = 1$

81. Write 8.67×10^{-2} in decimal form.

Solution

$8.67 \times 10^{-2} = 0.0867$

83. Write $(8 \times 10^3) + (3 \times 10^0) + (5 \times 10^{-2})$ in decimal form.

Solution

$$(8 \times 10^3) + (3 \times 10^0) + (5 \times 10^{-2})$$

$$= 8000 + 3 + 0.05$$

$$= 8003.05$$

85. Write 93,000,000 in scientific notation.

Solution

$93,000,000 = 9.3 \times 10^7$

87. Write 0.004392 in scientific notation.

Solution

$0.004392 = 4.392 \times 10^{-3}$

89. Use a calculator to evaluate the expression $0.000345 \times 8,980,000,000$.

Solution

$$0.000345 \times 8,980,000,000 = (3.45 \times 10^{-4})(8.98 \times 10^9)$$

$$= 3,098,100 \quad \text{or} \quad 3.0981 \times 10^6$$

Use the following calculator steps.

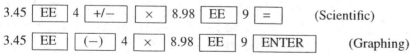

3.45 [EE] 4 [+/−] [×] 8.98 [EE] 9 [=]   (Scientific)

3.45 [EE] [(−)] 4 [×] 8.98 [EE] 9 [ENTER]   (Graphing)

91. Use a calculator to evaluate the expression $3,200,000^5$.

Solution

$3,200,000^5 = (3.2 \times 10^6)^5 \approx 3.3554 \times 10^{32}$

Use the following calculator steps.

3.2 [EE] 6 [y^x] 5 [=]   (Scientific)

3.2 [EE] 6 [∧] 5 [ENTER]   (Graphing)

93. Use a calculator to evaluate the expression $(3.28 \times 10^{-6})^4$.

Solution

$(3.28 \times 10^{-6})^4 \approx 1.1574 \times 10^{-22}$

Use the following calculator keystrokes.

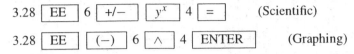

3.28 [EE] 6 [+/−] [y^x] 4 [=]   (Scientific)

3.28 [EE] [(−)] 6 [∧] 4 [ENTER]   (Graphing)

95. *Boltzman's Constant*   The study of the kinetic energy of an ideal gas uses Boltzman's Constant. This constant k is given by

$$k = \frac{8.31 \times 10^7}{6.02 \times 10^{23}}.$$

Perform the division, leaving your result in scientific notation.

Solution

$$k = \frac{8.31 \times 10^7}{6.01 \times 10^{23}} = \frac{8.31}{6.01} \times 10^{7-23} \approx 1.38 \times 10^{-16}$$

97. *Numerical and Graphical Approach*   A new car is purchased for $24,000. Its value after t years is $V = 24,000(1.2)^{-t}$.

(a) Use the model to complete the table in the textbook.

(b) Represent the data in the table graphically.

(c) *Guess, Check, and Revise*   When will the car be valued as less than $100?

Solution

(a)

t	0	2	4	6	8
$24,000(1.2)^{-t}$	$24,000	$16,667	$11,574	$8038	$5582

(b)

V
24,000 ● (0, 24,000)
20,000
16,000 ● (2, 16,667)
12,000 ● (4, 11,574)
8,000 ● (6, 8,038)
        ● (8, 5,581)
4,000
1 2 3 4 5 6 7 8 9   t

(c) The car will be valued at less than $100 after 31 years.

99. *An Infinite Sum* Successively remove from the unit square in the accompanying figure the fractional parts given in the table of Exercise 98 in the textbook. If this process were continued, what do you think would happen? Use this to estimate the sum

$$\frac{1}{2} + \frac{1}{4} + \frac{1}{8} + \frac{1}{16} + \cdots.$$

Solution

$$\frac{1}{2} + \frac{1}{4} + \frac{1}{8} + \frac{1}{16} + \cdots = 1$$

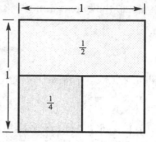

Review Exercises for Chapter 5

1. Write the polynomial $10x - 4 - 5x^3$ in standard form. Then find its degree and leading coefficient.

Solution

Standard form: $-5x^3 + 10x - 4$

Degree: 3

Leading coefficient: -5

3. Write the polynomial $2(x - 5) + 10$ in standard form. Then find its degree and leading coefficient.

Solution

$$2(x - 5) + 10 = 2x - 10 + 10$$
$$= 2x$$

Standard from: $2x$

Degree: 1

Leading coefficient: 2

5. Give an example of a polynomial that satisfies the condition "A trinomial of degree 4." (*Note:* There are many correct answers.)

Solution

$$8x^4 + 3x - 2$$

7. Determine whether the expressions $\frac{3}{8} + \frac{1}{8} \overset{?}{=} \frac{4}{16}$ are equal. If not, rewrite the right side so that it is a simplification of the left side.

Solution

$$\frac{3}{8} + \frac{1}{8} \neq \frac{4}{16}$$

$$\frac{3}{8} + \frac{1}{8} = \frac{4}{8} = \frac{1}{2}$$

9. Determine whether the expressions $(2x)^4 \overset{?}{=} 2x^4$ are equal. If not, rewrite the right side so that it is a simplification of the left side.

Solution

$$(2x)^4 \neq 2x^4$$

$$(2x)^4 = 2^4 x^4 = 16x^4$$

11. Determine whether the expressions $3^{-2} \overset{?}{=} -9$ are equal. If not, rewrite the right side so that it is a simplification of the left side.

Solution

$$3^{-2} \neq -9$$

$$3^{-2} = \frac{1}{3^2} = \frac{1}{9}$$

13. Determine whether the following expressions are equal. If not, rewrite the right side so that it is a simplification of the left side.

$$\frac{7-x}{7} \overset{?}{=} 1 - x$$

Solution

$$\frac{7-x}{7} \ne 1 - x$$

$$\frac{7-x}{7} = \frac{7}{7} - \frac{x}{7} = 1 - \frac{x}{7}$$

15. Determine whether the expressions $-3(x-2) \overset{?}{=} -3x-6$ are equal. If not, rewrite the right side so that it is a simplification of the left side.

Solution

$$-3(x-2) \ne -3x - 6$$

$$-3(x-2) = -3x + 6$$

17. Determine whether the expressions $(x+2)^2 \overset{?}{=} x^2 + 4$ are equal. If not, rewrite the right side so that it is a simplification of the left side.

Solution

$$(x+2)^2 \ne x^2 + 4$$

$$(x+2)^2 = (x+2)(x+2)$$

$$= x^2 + 4x + 4$$

19. Perform the following operations and simplify.

$$(2x + 3) + (x - 4)$$

Solution

$$(2x+3) + (x-4) = (2x+x) + (3-4)$$

$$= 3x - 1$$

21. Perform the following operations and simplify.

$$(t - 5) - (t^2 - t - 5)$$

Solution

$$(t-5) - (t^2 - t - 5) = t - 5 - t^2 + t + 5$$

$$= -t^2 + 2t$$

23. Perform the following operations and simplify.

$$(4 - x^2) + 2(x - 2)$$

Solution

$$(4 - x^2) + 2(x - 2) = 4 - x^2 + 2x - 4$$

$$= -x^2 + 2x$$

25. Perform the following operations and simplify.

$$(-x^3 - 3x) - 2(2x^3 + x + 1)$$

Solution

$$(-x^3 - 3x) - 2(2x^3 + x + 1) = -x^3 - 3x - 4x^3 - 2x - 2$$

$$= -5x^3 - 5x - 2$$

27. Perform the following operations and simplify.

$$4y^2 - [y - 3(y^2 + 2)]$$

Solution

$$4y^2 - [y - 3(y^2 + 2)] = 4y^2 - [y - 3y^2 - 6]$$

$$= 4y^2 - y + 3y^2 + 6$$

$$= 7y^2 - y + 6$$

29. Perform the following operations and simplify.

$$2x(x + 4)$$

Solution

$$2x(x + 4) = 2x^2 + 8x$$

31. Perform the following operations and simplify.

$$(x + 3)(2x - 4)$$

Solution

$$\overset{\text{F} \quad\quad \text{O} \quad\quad \text{I} \quad\quad \text{L}}{(x + 3)(2x - 4) = 2x^2 - 4x + 6x - 12}$$

$$= 2x^2 + 2x - 12$$

33. Perform the following operations and simplify.

$$(x^2 + 5x + 2)(2x + 3)$$

Solution

$$(x^2 + 5x + 2)(2x + 3) = x^2(2x + 3) + 5x(2x + 3) + 2(2x + 3)$$

$$= 2x^3 + 3x^2 + 10x^2 + 15x + 4x + 6$$

$$= 2x^3 + 13x^2 + 19x + 6$$

35. Perform the following operations and simplify.

$$2u(u - 5) - (u + 1)(u - 5)$$

Solution

$$2u(u - 5) - (u + 1)(u - 5) = 2u^2 - 10u - (u^2 - 4u - 5)$$

$$= 2u^2 - 10u - u^2 + 4u + 5$$

$$= u^2 - 6u + 5$$

37. Perform the following operations and simplify. (Assume that the denominator is not zero.)

$$(5x^2 + 15x) \div 5x$$

Solution

$$(5x^2 + 15x) \div 5x = \frac{5x^2}{5x} + \frac{15x}{5x}$$

$$= x + 3$$

39. Perform the following operations and simplify. (Assume that the denominator is not zero.)

$$7(x + 4)^3 \div 3(x + 4)^2$$

Solution

$$7(x + 4)^3 \div 3(x + 4)^2 = \frac{7(x + 4)^3}{3(x + 4)^2}$$

$$= \frac{7(x + 4)^{3-2}}{3}$$

$$= \frac{7(x + 4)}{3} \text{ or } \frac{7}{3}(x + 4)$$

41. Perform the following operations and simplify. (Assume that the denominator is not zero.)

$$\frac{x^2 - x - 6}{x - 3}$$

Solution

$$\begin{array}{r} x + 2 \\ x - 3 \overline{)\, x^2 - x - 6} \\ \underline{x^2 - 3x} \\ 2x - 6 \\ \underline{2x - 6} \\ 0 \end{array}$$

Thus, $\dfrac{x^2 - x - 6}{x - 3} = x + 2.$

43. Perform the following operations and simplify. (Assume that the denominator is not zero.)

$$\frac{24x^2 - x - 8}{3x - 2}$$

Solution

$$\begin{array}{r} 8x + 5 + \dfrac{2}{3x - 2} \\ 3x - 2 \overline{)\, 24x^2 - x - 8} \\ \underline{24x^2 - 16x} \\ 15x - 8 \\ \underline{15x - 10} \\ 2 \end{array}$$

Thus, $\dfrac{24x^2 - x - 8}{3x - 2} = 8x + 5 + \dfrac{2}{3x - 2}.$

45. Perform the following operations and simplify. (Assume that the denominator is not zero.)

$$\frac{2x^3 + 2x^2 - x + 2}{x - 1}$$

Solution

$$
\begin{array}{r}
2x^2 + 4x + 3 + \dfrac{5}{x-1} \\
x - 1 \overline{)\, 2x^3 + 2x^2 - x + 2 } \\
\underline{2x^3 - 2x^2 } \\
4x^2 - x \\
\underline{4x^2 - 4x } \\
3x + 2 \\
\underline{3x - 3} \\
5
\end{array}
$$

Thus,

$$\frac{2x^3 + 2x^2 - x + 2}{x - 1} = 2x^2 + 4x + 3 + \frac{5}{x - 1}.$$

47. Perform the following operations and simplify. (Assume that the denominator is not zero.)

$$\frac{x^4 - 3x^2 + 2}{x^2 - 1}$$

Solution

$$
\begin{array}{r}
x^2 - 2 \\
x^2 - 1 \overline{)\, x^4 + 0x^3 - 3x^2 + 0x + 2} \\
\underline{x^4 - x^2 } \\
-2x^2 \\
\underline{-2x^2 + 2} \\
0
\end{array}
$$

Thus, $\dfrac{x^4 - 3x^2 + 2}{x^2 - 1} = x^2 - 2.$

49. Use a special product pattern to expand the expression $(x + 3)^2$.

Solution

$$(x + 3)^2 = (x)^2 + 2(x)(3) + (3)^2$$
$$= x^2 + 6x + 9$$

51. Use a special product pattern to expand the expression $(4x - 7)^2$.

Solution

$$(4x - 7)^2 = (4x)^2 - 2(4x)(7) + (7)^2$$
$$= 16x^2 - 56x + 49$$

53. Use a special product pattern to expand the expression $(u - 6)(u + 6)$.

Solution

$$(u - 6)(u + 6) = (u)^2 - (6)^2$$
$$= u^2 - 36$$

55. Use a special product pattern to expand the expression $(3t - 1)(3t + 1)$.

Solution

$$(3t - 1)(3t + 1) = (3t)^2 - (1)^2$$
$$= 9t^2 - 1$$

57. Use a special product pattern to expand the expression $[(a - 1) + b][(a - 1) - b]$.

Solution

$$[(a - 1) + b][(a - 1) - b] = (a - 1)^2 - (b)^2$$
$$= (a)^2 - 2(a)(1) + (1)^2 - b^2$$
$$= a^2 - 2a + 1 - b^2$$
$$\text{or } a^2 - b^2 - 2a + 1$$

59. Evaluate the expression 4^{-2}.

Solution

$$4^{-2} = \frac{1}{4^2} = \frac{1}{16}$$

61. Evaluate the expression $6^{-4}6^2$.

Solution

$$6^{-4}6^2 = 6^{-4+2} = 6^{-2} = \frac{1}{6^2} = \frac{1}{36}$$

63. Evaluate the expression $\left(\frac{3}{5}\right)^{-3}$.

Solution

$$\left(\frac{3}{5}\right)^{-3} = \frac{3^{-3}}{5^{-3}} = \frac{5^3}{3^3} = \frac{125}{27}$$

65. Evaluate the expression $\left(-\frac{2}{5}\right)^3\left(\frac{5}{2}\right)^2$.

Solution

$$\left(-\frac{2}{5}\right)^3\left(\frac{5}{2}\right)^2 = -\frac{2^3}{5^3}\cdot\frac{5^2}{2^2} = -\frac{2^{3-2}}{5^{3-2}} = -\frac{2}{5}$$

67. Evaluate the expression $(3\times10^3)^2$.

Solution

$$(3\times10^3)^2 = 3^2\times10^6 = 9\times10^6$$

69. Evaluate the expression $\dfrac{1.85\times10^9}{5\times10^4}$.

Solution

$$\frac{1.85\times10^9}{5\times10^4} = 3.7\times10^4 = 37,000$$

Use the following calculator keystrokes.

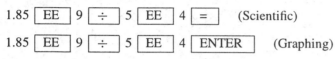

1.85 $\boxed{\text{EE}}$ 9 $\boxed{\div}$ 5 $\boxed{\text{EE}}$ 4 $\boxed{=}$ (Scientific)

1.85 $\boxed{\text{EE}}$ 9 $\boxed{\div}$ 5 $\boxed{\text{EE}}$ 4 $\boxed{\text{ENTER}}$ (Graphing)

71. Rewrite $(-3a^2)^{-2}$ with positive exponents and simplify. (Assume that no variable equals zero.)

Solution

$$(-3a^2)^{-2} = (-3)^{-2}(a^2)^{-2}$$
$$= (-3)^{-2}a^{-4} = \frac{1}{(-3)^2a^4} = \frac{1}{9a^4}$$

73. Rewrite $(x^2y^{-3})^2$ with positive exponents and simplify. (Assume that no variable equals zero.)

Solution

$$(x^2y^{-3})^2 = (x^2)^2(y^{-3})^2 = x^4y^{-6} = \frac{x^4}{y^6}$$

75. Rewrite the following expression with positive exponents and simplify. (Assume that no variable equals zero.)

$$\frac{t^{-4}}{t^{-1}}$$

Solution

$$\frac{t^{-4}}{t^{-1}} = \frac{t}{t^4} = \frac{1}{t^3}$$

77. Rewrite the following expression with positive exponents and simplify. (Assume that no variable equals zero.)

$$\left(\frac{y}{5}\right)^{-2}$$

Solution

$$\left(\frac{y}{5}\right)^{-2} = \frac{y^{-2}}{5^{-2}} = \frac{5^2}{y^2} = \frac{25}{y^2}$$

79. Find the polynomial that represents the area of the shaded portion of the figure in the textbook.

Solution

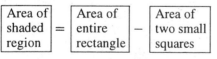

| Area of shaded region | = | Area of entire rectangle | − | Area of two small squares |

Area of entire rectangle $= 10(8) = 80$

Area of two small squares $= 2(x\cdot x) = 2x^2$

Area of shaded region $= 80 - 2x^2$

81. Find the polynomial that represents the area of the shaded portion of the figure in the textbook.

Solution

$\boxed{\text{Area of shaded region}} = \frac{1}{2}\boxed{\text{Area of square}}$

Area of square $= (2x+4)^2$

Area of shaded region $= \frac{1}{2}(2x+4)^2$
$$= \frac{1}{2}(4x^2+16x+16)$$
$$= 2x^2+8x+8$$

83. *Geometry*   The length of a rectangular wall is x units, and its height is $x - 3$ units (see figure in textbook). Find (a) the perimeter and (b) the area of the rectangle.

Solution

(a) *Common formula:* $P = 2l + 2w$

    Labels: $P = $ perimeter of wall

                        $l = x$

                        $w = x - 3$

    Equation: $P = 2x + 2(x - 3)$

                         $= 2x + 2x - 6$

                         $= 4x - 6$

    The perimeter of the wall is $4x - 6$ units.

(b) *Common formula:* $A = lw$

    Labels: $A = $ area of wall

                        $l = x$

                        $w = x - 3$

    Equation: $A = x(x - 3)$

                       $A = x^2 - 3x$

    The area of the wall is $x^2 - 3x$ square units.

85. *Probability*   The probability of three successes in five trials of an experiment is $10p^3(1 - p)^2$. Find this product.

Solution

$$10p^3(1 - p)^2 = 10p^3[(1)^2 - 2(1)(p) + (p)^2]$$
$$= 10p^3(1 - 2p + p^2)$$
$$= 10p^3 - 20p^4 + 10p^5$$

87. *Special Product*   What special product does the figure illustrate? Explain your reasoning.

Solution

(a) The area of the larger square is x^2, and the area of the smaller square is y^2. If the smaller square is removed, the area of the remaining figure is $x^2 - y^2$.

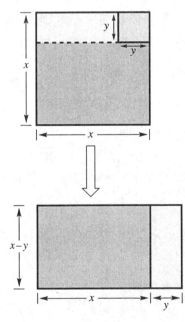

(b) If the remaining figure is rearranged as shown, the resulting rectangle has length $x + y$ and width $x - y$. The area of this resulting rectangle is the product of its length and width or $(x + y)(x - y)$.

The area of the remaining figure in part (a) and the area of the resulting rectangle in part (b) are equal. Thus, $x^2 - y^2 = (x + y)(x - y)$. This demonstrates geometrically the special product formula $(x + y)(x - y) = x^2 - y^2$.

Test for Chapter 5

1. Explain how to determine the degree and identify the leading coefficient of $-3x^4 - 5x^2 + 2x - 10$.

Solution

Degree: 4

Leading Coefficient: -3

2. Give an example of a trinomial in one variable of degree 4.

Solution

$-9^4 + 8x^3 + 5$ or $y^4 - y^2 - y$

Note: There are many correct answers.

3. Perform the indicated operations in $(3z^2 - 3z + 7) + (8 - z^2)$ and simplify.

Solution

$$(3z^2 - 3z + 7) + (8 - z^2) = 3z^2 - 3z + 7 + 8 - z^2$$
$$= 2z^2 - 3z + 15$$

4. Perform the indicated operations in $(8u^3 + 3u^2 - 2u - 1) - (u^3 + 3u^2 - 2u)$ and simplify.

Solution

$$(8u^3 + 3u^2 - 2u - 1) - (u^3 + 3u^2 - 2u) = 8u^3 + 3u^2 - 2u - 1 - u^3 - 3u^2 + 2u$$
$$= 7u^3 - 1$$

5. Perform the indicated operations in $6y - [2y - (3 + 4y - y^2)]$ and simplify.

Solution

$$6y - [2y - (3 + 4y - y^2)] = 6y - [2y - 3 - 4y + y^2]$$
$$= 6y - [y^2 - 2y - 3]$$
$$= 6y - y^2 + 2y + 3$$
$$= -y^2 + 8y + 3$$

6. Perform the indicated operations in $-5(x^2 - 1) + 3(4x + 7) - (x^2 + 26)$ and simplify.

Solution

$$-5(x^2 - 1) + 3(4x + 7) - (x^2 + 26) = -5x^2 + 5 + 12x + 21 - x^2 - 26$$
$$= -6x^2 + 12x$$

7. Perform the indicated operations in $(5b + 3)(2b - 1)$ and simplify.

Solution

$$\overset{\text{F}\quad\text{O}\quad\text{I}\quad\text{L}}{(5b + 3)(2b - 1) = 10b^2 - 5b + 6b - 3} = 10b^2 + b - 3$$

8. Perform the indicated operations and simplify.

$$4x\left(\frac{3x}{2}\right)^2$$

Solution

$$4x\left(\frac{3x}{2}\right)^2 = 4x\left(\frac{3^2x^2}{2^2}\right) = \frac{4x}{1} \cdot \frac{9x^2}{4} = \frac{\cancel{4} \cdot 9x^3}{\cancel{4}} = 9x^3$$

9. Perform the indicated operations in $(z + 2)(2z^2 - 3z + 5)$ and simplify.

Solution

$$(z + 2)(2z^2 - 3z + 5) = 2z^3 - 3z^2 + 5z + 4z^2 - 6z + 10$$
$$= 2z^3 + z^2 - z + 10$$

10. Perform the indicated operations in $(x - 5)^2$ and simplify.

Solution

$$(x - 5)^2 = (x)^2 - 2(x)(5) + (5)^2 = x^2 - 10x + 25$$

11. Perform the indicated operations in $(2x - 3)(2x + 3)$ and simplify.

Solution

$$(2x - 3)(2x + 3) = (2x)^2 - (3)^2 = 4x^2 - 9$$

12. Perform the indicated operations and simplify.

$$\frac{15x + 25}{5}$$

Solution

$$\frac{15x + 25}{5} = \frac{15x}{5} + \frac{25}{5} = 3x + 5$$

13. Perform the indicated operations and simplify. (Assume that no variable or denominator equals zero.)

$$\frac{x^3 - x - 6}{x - 2}$$

Solution

$$
\begin{array}{r}
x^2 + 2x + 3 \\
x - 2 \overline{\smash{\big)}\ x^3 + 0x^2 - x - 6} \\
\underline{x^3 - 2x^2} \\
2x^2 - x \\
\underline{2x^2 - 4x} \\
3x - 6 \\
\underline{3x - 6} \\
0
\end{array}
$$

Thus, $\dfrac{x^2 - x - 6}{x - 2} = x^2 + 2x + 3.$

14. Perform the indicated operations and simplify. (Assume that no variable or denominator equals zero.)

$$\frac{4x^3 + 10x^2 - 2x - 5}{2x + 1}$$

Solution

$$
\begin{array}{r}
2x^2 + 4x - 3 - \dfrac{2}{2x+1} \\
2x + 1 \overline{\smash{\big)}\ 4x^3 + 10x^2 - 2x - 5} \\
\underline{4x^3 + 2x^2} \\
8x^2 - 2x \\
\underline{8x^2 + 4x} \\
-6x - 5 \\
\underline{-6x - 3} \\
-2
\end{array}
$$

Thus, $\dfrac{4x^3 + 10x^2 - 2x - 5}{2x + 1} = 2x^2 + 4x - 3 - \dfrac{2}{2x + 1}.$

15. Simplify the following expression> (Assume that no variable equals zero.)

$$\frac{-6a^2b}{-9ab}$$

Solution

$$\frac{-6a^2b}{-9ab} = \frac{2a}{3}$$

16. Simplify the expression $(3x^{-2}y^3)^{-2}$. (Assume that no variable equals zero.)

Solution

$$(3x^{-2}y^3)^{-2} = 3^{-2}(x^{-2})^{-2}(y^3)^{-2}$$
$$= 3^{-2}x^4y^{-6}$$
$$= \frac{x^4}{3^2y^6}$$
$$= \frac{x^4}{9y^6}$$

17. Evaluate the expression *without* using a calculator. Show your work.

(a) $\dfrac{2^{-3}}{3^{-1}}$

(b) $(1.5 \times 10^5)^2$

Solution

(a) $\dfrac{2^{-3}}{3^{-1}} = \dfrac{3^1}{2^3} = \dfrac{3}{8}$

(b) $(1.5 \times 10^5)^2 = (1.5)^2 \times (10^5)^2$

$$= 2.25 \times 10^{10}$$

$$= 22{,}500{,}000{,}000$$

18. Write an expression that represents the area of the triangle in the textbook. Explain your reasoning.

Solution

Area of triangle $= \frac{1}{2}$(Base)(Height)

$$= \tfrac{1}{2}(4x - 2)(x + 6)$$

$$= (2x - 1)(x + 6)$$

$$= 2x^2 + 12x - x - 6$$

$$= 2x^2 + 11x - 6$$

19. The mean distance from earth to the moon is 3.84×10^8 meters. Write the distance in decimal form.

Solution

$3.84 \times 10^8 = 384{,}000{,}000$

20. The standard atmospheric pressure is 101,300 newtons per square meter. Write this pressure in scientific notation.

Solution

$101{,}300 = 1.013 \times 10^5$

CHAPTER SIX
Factoring and Solving Equations

6.1 Factoring Polynomials with Common Factors

7. Find the greatest common factor of 24 and 90.

Solution

$24 = 2 \cdot 2 \cdot 2 \cdot 3 = 6(4)$

$90 = 2 \cdot 3 \cdot 3 \cdot 5 = 6(15)$

The greatest common factor is 6.

9. Find the greatest common factor of $2x^2$ and $12x$.

Solution

$2x^2 = 2 \cdot x \cdot x = 2x(x)$

$12x = 2 \cdot 2 \cdot 3 \cdot x = 2x(6)$

The greatest common factor is $2x$.

11. Find the greatest common factor of $9yz^2$ and $-12y^2z^3$.

Solution

$9yz^2 = 3 \cdot 3 \cdot y \cdot z \cdot z = 3yz^2(3)$

$-12y^2z^3 = -(2 \cdot 2 \cdot 3 \cdot y \cdot y \cdot z \cdot z \cdot z)$

$\qquad = 3yz^2(-4yz)$

The greatest common factor is $3yz^2$.

13. Find the greatest common factor of $14x^2$, 1, and $7x^4$.

Solution

$14x^2 = 2 \cdot 7 \cdot x \cdot x = 1(14x^2)$

$\quad 1 = 1(1)$

$7x^4 = 7 \cdot x \cdot x \cdot x \cdot x = 1(7x^4)$

The greatest common factor is 1.

15. Factor $3x + 3$.

Solution

$3x + 3 = 3(x + 1)$

17. Factor $8t - 16$.

Solution

$8t - 16 = 8(t - 2)$

19. Factor $-25x - 10$.

Solution

$-25x - 10 = -5(5x + 2)$

or

$5(-5x - 2)$

21. Factor $12x^2 - 2x$.

Solution

$12x^2 - 2x = 2x(6x - 1)$

23. Factor $10ab + 10a^2b$.

Solution

$10ab + 10a^2b = 10ab(1 + a)$

25. Factor $100 + 75z - 50z^2$.

Solution

$100 + 75z - 50z^2$

$\qquad = 25(4 + 3z - 2z^2)$

27. Factor $x(x - 3) + 5(x - 3)$.

Solution

$x(x - 3) + 5(x - 3)$

$\qquad = (x - 3)(x + 5)$

29. Factor $z^2(z + 5) + (z + 5)$.

Solution

$z^2(z + 5) + (z + 5)$

$\qquad = (z + 5)(z^2 + 1)$

31. Factor $x^2 - 5x + xy - 5y$ by grouping.

Solution

$$x^2 - 5x + xy - 5y = x(x - 5) + y(x - 5)$$
$$= (x - 5)(x + y)$$

33. Factor $t^3 - 3t^2 + 2t - 6$ by grouping.

Solution

$$t^3 - 3t^2 + 2t - 6 = t^2(t - 3) + 2(t - 3)$$
$$= (t - 3)(t^2 + 2)$$

35. Factor a negative real number from $5 - 10x$ and then use the Commutative Property to write the polynomial factor with a positive leading coefficient.

Solution

$$5 - 10x = -5(-1 + 2x) = -5(2x - 1)$$

37. Factor a negative real number from $4 + 2x - x^2$ and then use the Commutative Property to write the polynomial factor with a positive leading coefficient.

Solution

$$4 + 2x - x^2 = -1(-4 - 2x + x^2)$$
$$= -1(x^2 - 2x - 4) \text{ or } -(x^2 - 2x - 4)$$

39. Complete the factorization $\frac{1}{2}x + \frac{3}{4} = \frac{1}{4}(\boxed{})$.

Solution

$$\frac{1}{2}x + \frac{3}{4} = \frac{2}{4}x + \frac{3}{4}$$
$$= \frac{1}{4}(2x + 3)$$

41. *Geometry*   The front of a microwave oven has an area of $44h - h^2$. Factor this expression to determine the length of the microwave in terms of h.

Solution

Area $= 44h - h^2 = h(44 - h)$

The area is the product of h and $(44 - h)$. Thus, the length is $44 - h$.

43. Use a graphing utility to graph $y = 8 - 4x$. Then identify the coordinates of at least three solution points.

Solution

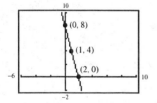

45. Use a graphing utility to graph $y = -\frac{1}{2}x^2$. Then identify the coordinates of at least three solution points.

Solution

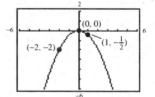

47. *Commission Rate*   Determine the commission rate for an employee who earned $1620 in commissions on sales of $54,000.

Solution

Verbal model: $\boxed{\text{Sales}} \cdot \boxed{\text{Commission rate}} = \boxed{\text{Commission}}$

Labels: Sales $= 54,000$ (dollars)
Commission rate $= r$ (percent in decimal form)
Commission $= 1620$ (dollars)

Equation: $54,000r = 1620$

$$r = \frac{1620}{54,000}$$

$$r = 0.03$$

Therefore, the commission rate is 3%.

49. Find the greatest common factor of 18, 150, and 100.

Solution

$18 = 2 \cdot 3 \cdot 3 = 2(9)$

$150 = 2 \cdot 3 \cdot 5 \cdot 5 = 2(75)$

$100 = 2 \cdot 2 \cdot 5 \cdot 5 = 2(50)$

The greatest common factor is 2.

51. Find the greatest common factor of z^2 and $-z^6$.

Solution

$z^2 = z \cdot z = z^2(1)$

$-z^6 = -(z \cdot z \cdot z \cdot z \cdot z \cdot z) = z^2(-z^4)$

The greatest common factor is z^2.

53. Find the greatest common factor of u^2v and u^3v^2.

Solution

$u^2v = u \cdot u \cdot v$

$\quad\ = u^2v(1)$

$u^3v^2 = u \cdot u \cdot u \cdot v \cdot v$

$\quad\quad = u^2v(uv)$

The greatest common factor is u^2v.

55. Find the greatest common factor of $28a^4b^2$, $14a^3b^3$, and $42a^2b^5$.

Solution

$28a^4b^2 = 2 \cdot 2 \cdot 7 \cdot a \cdot a \cdot a \cdot a \cdot b \cdot b$

$\quad\quad\quad = 14a^2b^2(2a^2)$

$14a^3b^3 = 2 \cdot 7 \cdot a \cdot a \cdot a \cdot b \cdot b \cdot b$

$\quad\quad\quad = 14a^2b^2(ab)$

$42a^2b^5 = 2 \cdot 3 \cdot 7 \cdot a \cdot a \cdot b \cdot b \cdot b \cdot b \cdot b$

$\quad\quad\quad = 14a^2b^2(3b^3)$

The greatest common factor is $14a^2b^2$.

57. Factor $6z - 6$.

Solution

$6z - 6 = 6(z - 1)$

59. Factor $24y^2 - 18$.

Solution

$24y^2 - 18 = 6(4y^2 - 3)$

61. Factor $x^2 + x$.

Solution

$x^2 + x = x(x + 1)$

63. Factor $25u^2 - 14u$.

Solution

$25u^2 - 14u = u(25u - 14)$

65. Factor $2x^4 + 6x^3$.

Solution

$2x^4 + 6x^3 = 2x^3(x + 3)$

67. Factor $7s^2 + 9t^2$.

Solution

$7s^2 + 9t^2$ (No common factor)

69. Factor $-10r^3 - 35r$.

Solution

$-10r^3 - 35r = -5r(2r^2 + 7)$

71. Factor $16a^3b^3 + 24a^4b^3$.

Solution

$16a^3b^3 + 24a^4b^3 = 8a^3b^3(2 + 3a)$

73. Factor $12x^2 + 16x - 8$.

Solution

$12x^2 + 16x - 8 = 4(3x^2 + 4x - 2)$

75. Factor $9x^4 + 6x^3 + 18x^2$.

Solution

$9x^4 + 6x^3 + 18x^2$

$\quad\quad\quad = 3x^2(3x^2 + 2x + 6)$

77. Factor $5u^2 + 5u^2 + 5u$.

Solution

$$5u^2 + 5u^2 + 5u = 10u^2 + 5u$$

$$= 5u(2u + 1)$$

79. Factor $t(s + 10) - 8(s + 10)$.

Solution

$$t(s + 10) - 8(s + 10) = (s + 10)(t - 8)$$

81. Factor $a^2(b + 2) - a(b + 2)$.

Solution

$$a^2(b + 2) - a(b + 2) = a(b + 2)(a - 1)$$

Note: The greatest common factor is $a(b + 2)$.

83. Factor $(a + b)(c + 7) + (a + b)(c + 7)^2$.

Solution

$$(a + b)(c + 7) + (a + b)(c + 7)^2$$

$$= (a + b)(c + 7)[1 + (c + 7)]$$

$$= (a + b)(c + 7)(c + 8)$$

85. Factor $a^2 - 4a + ab - 4b$ by grouping.

Solution

$$a^2 - 4a + ab - 4b = a(a - 4) + b(a - 4)$$

$$= (a - 4)(a + b)$$

87. Factor $xy^2 - 4xy + 2y - 8$ by grouping.

Solution

$$xy^2 - 4xy + 2y - 8 = xy(y - 4) + 2(y - 4)$$

$$= (y - 4)(xy + 2)$$

89. Factor $x^3 + 2x^2 + x + 2$ by grouping.

Solution

$$x^3 + 2x^2 + x + 2 = x^2(x + 2) + 1(x + 2)$$

$$= (x + 2)(x^2 + 1)$$

91. Factor $z^3 + 3z^2 - 2z - 6$ by grouping.

Solution

$$z^3 + 3z^2 - 2z - 6 = z^2(z + 3) - 2(z + 3)$$

$$= (z + 3)(z^2 - 2)$$

93. Factor a negative real number from $3000 - 3x$ and then use the Commutative Property to write the polynomial factor with a positive leading coefficient.

Solution

$$3000 - 3x = -3(-1000 + x)$$

$$= -3(x - 1000)$$

95. Factor a negative real number from $4 + 12x - 2x^2$ and then use the Commutative Property to write the polynomial factor with a positive leading coefficient.

Solution

$$4 + 12x - 2x^2 = -2(-2 - 6x + x^2)$$

$$= -2(x^2 - 6x - 2)$$

97. Complete the factorization of $2y - \frac{1}{5} = \frac{1}{5}(\boxed{})$.

Solution

$$2y - \frac{1}{5} = \frac{10}{5}y - \frac{1}{5}$$

$$= \frac{1}{5}(10y - 1)$$

99. Complete the factorization of $\frac{7}{8}x + \frac{5}{16}y = \frac{1}{16}(\boxed{})$.

Solution

$$\frac{7}{8}x + \frac{5}{16}y = \frac{14}{16}x + \frac{5}{16}y$$

$$= \frac{1}{16}(14x + 5y)$$

101. Use a graphing utility to graph the equations $y_1 = 9 - 3x$ and $y_2 = -3(x - 3)$ on the same screen. Use the graphs to verify the factorization.

Solution

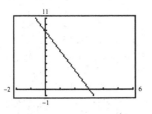

103. Use a graphing utility to graph the equations $y_1 = 6x - x^2$ and $y_2 = x(6 - x)$ on the same screen. Use the graphs to verify the factorization.

Solution

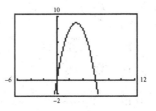

105. *Geometry* Find the length of the rectangle that has an area of $2x^2 + 2x$ and a width of $2x$.

Solution

Area = (Length)(Width)

Area $= 2x^2 + 2x$ and Width $= 2x$

$2x^2 + 2x = 2x(x + 1)$

Therefore, the length of the rectangle must be $x + 1$.

107. *Geometry* The surface area of a right circular cylinder is $S = 2\pi r^2 + 2\pi rh$. Factor the expression for the surface area.

Solution

$S = 2\pi r^2 + 2\pi rh$

$\quad = 2\pi r(r + h)$

109. *Unit Price* The revenue R for selling x units of a product at a price of p dollars per unit is given by $R = xp$. For a particular commodity the revenue is $900x - 0.1x^2$. Factor the revenue model and determine an expression that represents the price p in terms of x.

Solution

$R = xp$

$R = 900x - 0.1x^2$

$R = x(900 - 0.1x)$

Therefore, $p = 900 - 0.1x$.

6.2 Factoring Trinomials

7. Find the missing factor. Then check your answer by multiplying the two factors.

$$x^2 + 4x + 3 = (x + 3)(\boxed{})$$

Solution

$x^2 + 4x + 3 = (x + 3)(x + 1)$

Check:

$$\qquad\qquad\quad \text{F}\quad\;\text{O}\quad\;\text{I}\quad\;\text{L}$$
$$(x + 3)(x + 1) = x^2 + x + 3x + 3 = x^2 + 4x + 3$$

9. Find the missing factor. Then check your answer by multiplying the two factors.

$$y^2 - 2y - 15 = (y + 3)(\boxed{})$$

Solution

$$y^2 - 2y - 15 = (y + 3)(y - 5)$$

Check:

$$\text{F} \quad \text{O} \quad \text{I} \quad \text{L}$$
$$(y + 3)(y - 5) = y^2 - 5y + 3y - 15 = y^2 - 2y - 15$$

11. Find all possible products of the form $(x + m)(x + n)$ where $mn = 11$. (Assume m and n are integers.)

Solution

$(x + 11)(x + 1)$

$(x - 11)(x - 1)$

13. Find all possible products of the form $(x + m)(x + n)$ where $mn = 12$. (Assume m and n are integers.)

Solution

$(x + 12)(x + 1) \qquad (x - 12)(x - 1)$

$(x + 6)(x + 2) \qquad (x - 6)(x - 2)$

$(x + 4)(x + 3) \qquad (x - 4)(x - 3)$

15. Factor $x^2 + 6x + 8$.

Solution

$$x^2 + 6x + 8 = (x + 4)(x + 2)$$

17. Factor $x^2 - x - 6$.

Solution

$$x^2 - x - 6 = (x - 3)(x + 2)$$

19. Factor $x^2 - 17x + 72$.

Solution

$$x^2 - 17x + 72 = (x - 8)(x - 9)$$

21. Factor $y^2 + 5y + 11$.

Solution

The trinomial is prime.

23. Factor $u^2 - 22u - 48$.

Solution

$$u^2 - 22u - 48 = (u - 24)(u + 2)$$

25. Factor $a^2 + 2ab - 15b^2$.

Solution

$$a^2 + 2ab - 15b^2 = (a + 5b)(a - 3b)$$

27. Find all integer values of b such that $x^2 + bx + 15$ can be factored.

Solution

$b = 8 \qquad (x + 3)(x + 5)$

$b = -8 \qquad (x - 3)(x - 5)$

$b = 16 \qquad (x + 15)(x + 1)$

$b = -16 \quad (x - 15)(x - 1)$

29. Find *two* values of c such that $x^2 - 6x + c$ can be factored.

Solution

$c = 5 \qquad (x - 5)(x - 1)$

$c = 8 \qquad (x - 4)(x - 2)$

$c = 9 \qquad (x - 3)(x - 3)$

$c = -7 \qquad (x - 7)(x + 1)$

$c = -16 \quad (x - 8)(x + 2)$

(There are many other correct answers.)

31. Factor $3x^2 + 21x + 30$ completely.

Solution

$$3x^2 + 21x + 30 = 3(x^2 + 7x + 10)$$
$$= 3(x + 5)(x + 2)$$

33. Factor $3z^2 + 5z + 6$ completely.

Solution

This trinomial is prime.

35. Factor $2x^3y + 4x^2y^2 - 6xy^3$ completely.

Solution

$$2x^3y + 4x^2y^2 - 6xy^3 = 2xy(x^2 + 2xy - 3y^2)$$
$$= 2xy(x + 3y)(x - y)$$

37. *Geometric Modeling of Factoring* Factor $x^2 + 4x + 3$ and draw a geometric model of the result. [The sample given in the textbook shows a geometric model for factoring $x^2 + 3x + 2$ as $(x + 1)(x + 2)$.]

Solution

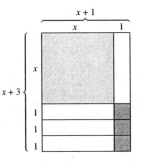

39. *Geometric Modeling of Factoring* Factor $x^2 + 5x + 6$ and draw a geometric model of the result. [The sample given in the textbook shows a geometric model for factoring $x^2 + 3x + 2$ as $(x + 1)(x + 2)$.]

Solution

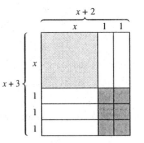

41. Find $-a^2(a - 1)$.

Solution

$$-a^2(a - 1) = -a^3 + a^2$$

43. Find $(u - 8)(u + 3)$.

Solution

$$(u - 8)(u + 3) = u^2 + 3u - 8u - 24$$
$$= u^2 - 5u - 24$$

45. Graph $y = x^2 - 2x - 8$.

Solution

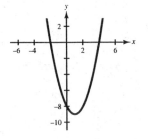

47. *Profit* A company had a loss of $2,500,000 during the first 6 months of the year. The company ended the year with an overall profit of $1,475,000. What was the profit during the second 6 months of the year?

Solution

Verbal model: | Result of first six months | + | Result of next six months | = | Results from entire year |

Labels: Results of first six months = $-2,500,000$ (dollars)
Results of next six months = x
Results from entire year = $1,475,000$

Equation: $-2,500,000 + x = 1,475,000$

$$x = 1,475,000 + 2,500,000$$

$$x = 3,975,000$$

Therefore, the profit during the second six months of the year was $3,975,000.

49. Find the missing factor. Then check your answer by multiplying the two factors.

$$a^2 + a - 6 = (a + 3)(\boxed{})$$

Solution

$a^2 + a - 6 = (a + 3)(a - 2)$

Check:

$$\text{F}\quad\text{O}\quad\text{I}\quad\text{L}$$
$$(a + 3)(a - 2) = a^2 - 2a + 3a - 6 = a^2 + a - 6$$

51. Find the missing factor. Then check your answer by multiplying the two factors.

$$z^2 - 5z + 6 = (z - 3)(\boxed{})$$

Solution

$z^2 - 5z + 6 = (z - 3)(z - 2)$

Check:

$$\text{F}\quad\text{O}\quad\text{I}\quad\text{L}$$
$$(z - 3)(z - 2) = z^2 - 2z - 3z + 6 = z^2 - 5z + 6$$

53. Factor $x^2 - 13x + 40$.

Solution

$x^2 - 13x + 40 = (x - 5)(x - 8)$

55. Factor $z^2 - 7z + 12$.

Solution

$z^2 - 7z + 12 = (z - 3)(z - 4)$

57. Factor $x^2 + 2x - 15$.

Solution

$x^2 + 2x - 15 = (x + 5)(x - 3)$

59. Factor $x^2 + 3x - 70$.

Solution

$x^2 + 3x - 70 = (x + 10)(x - 7)$

61. Factor $y^2 - 6y + 10$.

Solution

This trinomial is prime.

63. Factor $x^2 + 19x + 60$.

Solution

$x^2 + 19x + 60 = (x + 15)(x + 4)$

65. Factor $x^2 + xy - 2y^2$.

Solution

$x^2 + xy - 2y^2 = (x + 2y)(x - y)$

67. Factor $x^2 + 8xy + 15y^2$.

Solution

$x^2 + 8xy + 15y^2 = (x + 3y)(x + 5y)$

69. Factor $x^2 - 7xz - 18z^2$.

Solution

$x^2 - 7xz - 18z^2 = (x - 9z)(x + 2z)$

71. Factor $x^3 - 13x^2 + 30x$ completely.

Solution

$$x^3 - 13x^2 + 30x = x(x^2 - 13x + 30)$$
$$= x(x - 3)(x - 10)$$

73. Factor $4y^2 - 8y - 12$ completely.

Solution

$$4y^2 - 8y - 12 = 4(y^2 - 2y - 3)$$
$$= 4(y - 3)(y + 1)$$

75. Factor $10x^3 + 50x^2y + 60xy^2$.

Solution

$$10x^3 + 50x^2y + 60xy^2 = 10x(x^2 + 5xy + 6y^2)$$
$$= 10x(x + 3y)(x + 2y)$$

77. Use a graphing utility to graph $y_1 = x^2 - x - 6$ and $y_2 = (x + 2)(x - 3)$ on the same viewing rectangle. What can you conclude?

Solution

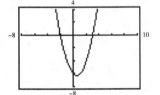

The graphs are identical, and the two expressions are equivalent.

79. Use a graphing utility to graph $y_1 = x^3 + x^2 - 20x$ and $y_2 = x(x - 4)(x + 5)$ on the same viewing rectangle. What can you conclude?

Solution

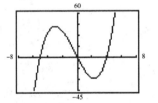

The graphs are identical, and the two expressions are equivalent.

81. *Exploration* An open box is to be made from a 4-foot by 6-foot sheet of metal by cutting equal squares from each corner and turning up the sides (see the figure shown in the textbook). The volume of the box can be modeled by

$$V = 4x^3 - 20x^2 + 24x, \quad 0 < x < 2.$$

(a) Factor the trinomial modeling the volume of the box. Use the factored form to explain how the model was found.

(b) Use a graphing utility to graph the trinomial over the specified interval. Use the graph to approximate the size of the squares to be cut from each corner so the volume of the box is greatest.

Solution

(a) $V = 4x^3 - 20x^2 + 24x, \quad 0 < x < 2$

$\qquad = 4x(x^2 - 5x + 6)$

$\qquad = 4x(x - 3)(x - 2) \quad$ or $\ x(2)(x - 3)(2)(x - 2)$ $\qquad\qquad$ Note: $4 = 2(2)$

$\qquad\qquad\qquad\qquad\qquad$ or $\ x(2x - 6)(2x - 4)$

$\qquad\qquad\qquad\qquad\qquad$ or $\ x(2x - 6)(-1)(-1)(2x - 4)$ $\qquad\qquad$ Note: $(-1)(-1)=1$

$\qquad\qquad\qquad\qquad\qquad$ or $\ x[(2x - 6)(-1)][(-1)(2x - 4)]$

$\qquad\qquad\qquad\qquad\qquad$ or $\ x(-2x + 6)(-2x + 4)$

$\qquad\qquad\qquad\qquad\qquad$ or $\ x(6 - 2x)(4 - 2x)$

The volume is the product of the length, width, and height. The length is $6 - 2x$ feet, the width is $4 - 2x$ feet, and the height is x feet.

Note: Here is an alternate factoring approach which also points out the dimensions of the box.

$\quad V = 4x^3 - 20x^2 + 24x$

$\qquad = 4x(x^2 - 5x + 6)$

$\qquad = 4x(6 - 5x + x^2)$

$\qquad = 4x(3 - x)(2 - x)$

$\qquad = x(2)(3 - x)(2)(2 - x)$

$\qquad = x[2(3 - x)][2(2 - x)]$

$\qquad = x(6 - 2x)(4 - 2x)$

(b)

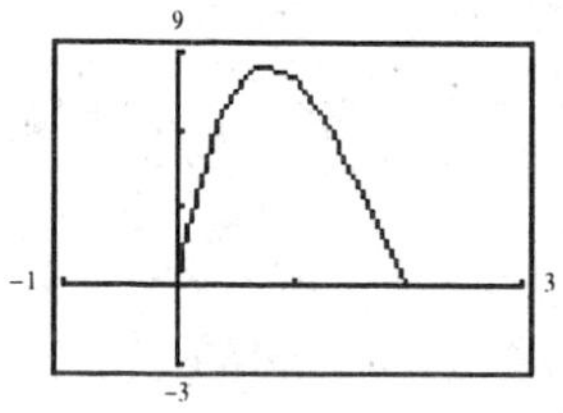

The volume is greatest when x is approximately 0.785 feet.

6.3 More About Factoring Trinomials

7. Find the missing factor.

$$5x^2 + 18x + 9 = (x + 3)(\boxed{})$$

Solution

$$5x^2 + 18x + 9 = (x + 3)(5x + 3)$$

9. Find the missing factor.

$$4z^2 - 13z + 3 = (z - 3)(\boxed{})$$

Solution

$$4z^2 - 13z + 3 = (z - 3)(4z - 1)$$

11. Find all possible products of the form $(5x + m)(x + n)$ where $mn = 3$. (Assume m and n are integers.)

Solution

$(5x + 3)(x + 1)$

$(5x - 3)(x - 1)$

$(5x + 1)(x + 3)$

$(5x - 1)(x - 3)$

13. Find all possible products of the form $(5x + m)(x + n)$, where $mn = 12$. (Assume m and n are integers.)

Solution

$(5x + 12)(x + 1)$ $(5x - 12)(x - 1)$

$(5x + 1)(x + 12)$ $(5x - 1)(x - 12)$

$(5x + 6)(x + 2)$ $(5x - 6)(x - 2)$

$(5x + 2)(x + 6)$ $(5x - 2)(x - 6)$

$(5x + 4)(x + 3)$ $(5x - 4)(x - 3)$

$(5x + 3)(x + 4)$ $(5x - 3)(x - 4)$

15. Factor $2x^2 + 5x + 3$.

Solution

$$2x^2 + 5x + 3 = (2x + 3)(x + 1)$$

17. Factor $2y^2 - 3y + 1$.

Solution

$$2y^2 - 3y + 1 = (2y - 1)(y - 1)$$

19. Factor $2x^2 + x + 3$.

Solution

This trinomial is prime.

21. Factor $16z^2 - 34z + 15$.

Solution

$$16z^2 - 34z + 15 = (8x - 5)(2z - 3)$$

23. Factor $-2x^2 + x + 3$.

Solution

$$\begin{aligned} -2x^2 + x + 3 &= (-1)(2x^2 - x - 3) \\ &= -(2x - 3)(x + 1) \\ &\text{or } (-2x + 3)(x + 1) \\ &\text{or } (2x - 3)(-x - 1) \end{aligned}$$

25. Factor $1 - 4x - 60x^2$.

Solution

$$1 - 4x - 60x^2 = (1 - 10x)(1 + 6x)$$

Alternate Method:

$$\begin{aligned} 1 - 4x - 60x^2 &= (-1)(60x^2 + 4x - 1) \\ &= -(10x - 1)(6x + 1) \\ &\text{or } (-10x + 1)(6x + 1) \\ &\text{or } (10x - 1)(-6x - 1) \end{aligned}$$

27. Factor $x^2 - 3x$ completely.

Solution

$$x^2 - 3x = x(x - 3)$$

29. Factor $v^2 + v - 42$ completely.

Solution

$$v^2 + v - 42 = (v + 7)(v - 6)$$

31. Factor $2x^2 + 2x + 1$ completely.

Solution

This trinomial is prime.

33. Find all integers b such that $3x^2 + bx + 10$ can be factored.

Solution

$$
\begin{array}{ll}
b = 13 & (3x + 10)(x + 1) \\
b = -13 & (3x - 10)(x - 1) \\
b = 31 & (3x + 1)(x + 10) \\
b = -31 & (3x - 1)(x - 10) \\
b = 11 & (3x + 5)(x + 2) \\
b = -11 & (3x - 5)(x - 2) \\
b = 17 & (3x + 2)(x + 5) \\
b = -17 & (3x - 2)(x - 5)
\end{array}
$$

35. Find all integers b such that $2x^2 + bx - 6$ can be factored.

Solution

$$
\begin{array}{ll}
b = 4 & 2(x + 3)(x - 1) \text{ or } (2x + 6)(x - 1) \text{ or } (2x - 2)(x + 3) \\
b = -4 & 2(x - 3)(x + 1) \text{ or } (2x - 6)(x + 1) \text{ or } (2x + 2)(x - 3) \\
b = -11 & (2x + 1)(x - 6) \\
b = 11 & (2x - 1)(x + 6) \\
b = -1 & (2x + 3)(x - 2) \\
b = 1 & (2x - 3)(x + 2)
\end{array}
$$

37. Factor $3x^2 + 7x + 2$ by grouping.

Solution

$ac = 3 \cdot 2 = 6$

$b = 7$

The two numbers with a product of 6 and a sum of 7 are 6 and 1.

$$
\begin{aligned}
3x^2 + 7x + 2 &= 3x^2 + 6x + x + 2 \\
&= (3x^2 + 6x) + (x + 2) \\
&= 3x(x + 2) + (x + 2) \\
&= (x + 2)(3x + 1)
\end{aligned}
$$

Note: When the middle term $7x$ is rewritten as the sum of $6x$ and x, these two terms may also be written in the opposite order.

$$
\begin{aligned}
3x^2 + 7x + 2 &= 3x^2 + x + 6x + 2 \\
&= (3x^2 + x) + (6x + 2) \\
&= x(3x + 1) + 2(3x + 1) \\
&= (3x + 1)(x + 2)
\end{aligned}
$$

39. Factor $15x^2 - 11x + 2$ by grouping.

Solution

$ac = 15(2) = 30$

$b = -11$

The two numbers with a product of 30 and a sum of -11 are -6 and -5.

$$15x^2 - 11x + 2 = 15x^2 - 6x - 5x + 2$$
$$= (15x^2 - 6x) - (5x - 2)$$
$$= 3x(5x - 2) - (5x - 2)$$
$$= (5x - 2)(3x - 1)$$

41. Consider the following equations.

$$y_1 = 2x^3 + 3x^2 - 5x$$
$$y_2 = x(2x + 5)(x - 1)$$

(a) Factor the trinomial represented by y_1. What is the relationship between y_1 and y_2?

(b) Demonstrate your answer to part (a) graphically by using a graphing utility to graph y_1 and y_2.

(c) Identify the x- and y-intercepts of the graphs of y_1 and y_2.

Solution

(a) $y_1 = 2x^3 + 3x^2 - 5x$

$\qquad = x(2x^2 + 3x - 5)$

$\qquad = x(2x + 5)(x - 1)$

$y_1 = y_2$

(b)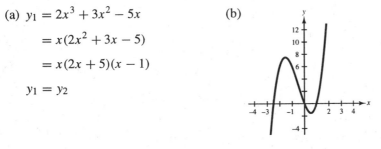

(c) The x-intercepts are $(0, 0)$, $(1, 0)$, and $\left(-\frac{5}{2}, 0\right)$. The y-intercept is $(0, 0)$.

43. *Geometric Model of Factoring* Factor $2x^2 + 5x + 2$ and draw a geometric model of the result. (The sample in the textbook shows a geometric model for factoring.)

Solution

$2x^2 + 5x + 2 = (2x + 1)(x + 2)$

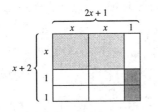

45. Find $-3t(t + 2)$.

Solution

$-3t(t + 2) = -3t^2 - 6t$

47. Find $(x + 4)^2$.

Solution

$(x + 4)^2 = x^2 + 8x + 16$

49. *Distance* The minimum and maximum speeds on an interstate highway are 40 miles per hour and 55 miles per hour. You travel nonstop for 3 hours on the highway. Give an interval that describes the distance you could have legally traveled.

Solution

Distance = (Rate)(Time)

The rate is in the interval $40 \le r \le 55$, and the time is $t = 3$. Therefore, the distance $3r$ is in the interval

$$40(3) \le 3r \le 55(3)$$

$$120 \le 3r \le 165$$

$$120 \le d \le 165$$

The distance d must be between 120 miles and 165 miles.

51. Find the missing factor.

$$5a^2 + 12a - 9 = (a + 3)(\boxed{})$$

Solution

$$5a^2 + 12a - 9 = (a + 3)(5a - 3)$$

53. Find the missing factor.

$$5x^2 + 19x + 12 = (x + 3)(\boxed{})$$

Solution

$$5x^2 + 19x + 12 = (x + 3)(5x + 4)$$

55. Factor $4y^2 + 5y + 1$.

Solution

$$4y^2 + 5y + 1 = (4y + 1)(y + 1)$$

57. Factor $2x^2 - x - 3$.

Solution

$$2x^2 - x - 3 = (2x - 3)(x + 1)$$

59. Factor $5x^2 - 2x + 1$.

Solution

This trinomial is prime.

61. Factor $15a^2 + 14a - 8$.

Solution

$$15a^2 + 14a - 8 = (5a - 2)(3a + 4)$$

63. Factor $18u^2 - 9u - 2$.

Solution

$$18u^2 - 9u - 2 = (6u + 1)(3u - 2)$$

65. Factor $10t^2 - 3t - 18$.

Solution

$$10t^2 - 3t - 18 = (5t + 6)(2t - 3)$$

67. Factor $15m^2 + 16m - 15$.

Solution

$$15m^2 + 16m - 15 = (5m - 3)(3m + 5)$$

69. Factor $5s^2 - 10s + 6$.

Solution

This trinomial is prime.

71. Factor $4 - 4x - 3x^2$.

Solution

$$4 - 4x - 3x^2 = (2 - 3x)(2 + x) \text{ or}$$

$$4 - 4x - 3x^2 = -1(-4 + 4x + 3x^2)$$

$$= -1(3x^2 + 4x - 4)$$

$$= -(3x - 2)(x + 2)$$

$$\text{or } (-3x + 2)(x + 2)$$

$$\text{or } (3x - 2)(-x - 2)$$

73. Factor $-6x^2 + 7x + 10$.

Solution

$$-6x^2 + 7x + 10 = (-1)(6x^2 - 7x - 10)$$

$$= -(6x + 5)(x - 2)$$

$$\text{or } (-6x - 5)(x - 2)$$

$$\text{or } (6x + 5)(-x + 2)$$

75. Factor $15y^2 + 18y$ completely.

Solution

$15y^2 + 18y = 3y(5y + 6)$

77. Factor $u(u - 3) + 9(u - 3)$ completely.

Solution

$u(u - 3) + 9(u - 3) = (u - 3)(u + 9)$

79. Factor $x^2 + 6x - 40$ completely.

Solution

$x^2 + 6x - 40 = (x + 10)(x - 4)$

81. Factor $6x^2 + 8x - 8$ completely.

Solution

$$6x^2 + 8x - 8 = 2(3x^2 + 4x - 4)$$
$$= 2(3x - 2)(x + 2)$$

83. Factor $15y^2 - 7y^3 - 2y^4$ completely.

Solution

$$15y^2 - 7y^3 - 2y^4 = y^2(15 - 7y - 2y^2)$$
$$= y^2(5 + y)(3 - 2y)$$
$$\text{or } -y^2(y + 5)(2y - 3)$$

85. Factor $9u^2 + 18u - 27$ completely.

Solution

$$9u^2 + 18u - 27 = 9(u^2 + 2u - 3)$$
$$= 9(u + 3)(u - 1)$$

87. Factor $2x^2 + x - 3$ by grouping.

Solution

$ac = 2(-3) = -6$

$b = 1$

The two numbers with a product of -6 and a sum of 1 are 3 and -2.

$$2x^2 + x - 3 = 2x^2 + 3x - 2x - 3$$
$$= (2x^2 + 3x) - (2x + 3)$$
$$= x(2x + 3) - (2x + 3)$$
$$= (2x + 3)(x - 1)$$

89. Factor $6x^2 + 5x - 4$ by grouping.

Solution

$ac = 6(-4) = -24$

$b = 5$

The two numbers with a product of -24 and a sum of 5 are -3 and 8.

$$6x^2 + 5x - 4 = 6x^2 - 3x + 8x - 4$$
$$= (6x^2 - 3x) + (8x - 4)$$
$$= 3x(2x - 1) + 4(2x - 1)$$
$$= (2x - 1)(3x + 4)$$

91. Factor $3a^2 + 11a + 10$ by grouping.

Solution

$ac = 3(10) = 30$

$b = 11$

The two numbers with a product of 30 and a sum of 11 are 6 and 5.

$$3a^2 + 11a + 10 = 3a^2 + 6a + 5a + 10$$
$$= 3a(a + 2) + 5(a + 2)$$
$$= (a + 2)(3a + 5)$$

93. Factor $16x^2 + 2x - 3$ by grouping.

Solution

$ac = 16(-3) = -48$

$b = 2$

The two numbers with a product of -48 and a sum of 2 are 8 and -6.

$$16x^2 + 2x - 3 = 16x^2 + 8x - 6x - 3$$
$$= (16x^2 + 8x) - (6x + 3)$$
$$= 8x(2x + 1) - 3(2x + 1)$$
$$= (2x + 1)(8x - 3)$$

95. Find two values of c such that $4x^2 + 3x + c$ can be factored.

Solution

There are *many* correct answers. These are some examples.

$c = -1$	$(4x - 1)(x + 1)$
$c = -10$	$(4x - 5)(x + 2)$
$c = -27$	$(4x - 9)(x + 3)$
$c = -7$	$(4x + 7)(x - 1)$
$c = -22$	$(4x + 11)(x - 2)$
$c = -45$	$(4x + 15)(x - 3)$

97. Use a graphing utility to graph $y_1 = 4x^2 - 11x - 45$ and $y_2 = (x - 5)(4x + 9)$ on the same screen. What can you conclude?

Solution

$y_1 = y_2$

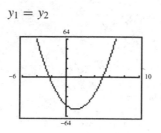

99. Use a graphing utility to graph $y_1 = 2x^3 + 3x^2 - 2x$ and $y_2 = x(2x - 1)(x + 2)$ on the same screen. What can you conclude?

Solution

$y_1 = y_2$

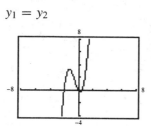

101. *Geometry*   The cake box in the figure shown in the textbook has a height of x and a width of $x + 1$. The volume of the box is $3x^3 + 4x^2 + x$. Find the length of the box.

Solution

$$\begin{aligned} \text{Volume} &= 3x^3 + 4x^2 + x \\ &= x(3x^2 + 4x + 1) \\ &= x(3x + 1)(x + 1) \end{aligned}$$

The volume is the product of the length, width, and height. The height is x, and the width is $x + 1$. Therefore, the third factor, $3x + 1$, must be the length.

Mid-Chapter Quiz for Chapter 6

1. Find the missing factor.

$$\tfrac{2}{3}x - 1 = \tfrac{1}{3}(\boxed{})$$

Solution

$$\begin{aligned} \tfrac{2}{3}x - 1 &= \tfrac{2}{3}x - \tfrac{3}{3} \\ &= \tfrac{1}{3}(2x - 3) \end{aligned}$$

The missing factor is $2x - 3$.

2. Find the missing factor.

$$x^2y - xy^2 = xy(\boxed{})$$

Solution

$$x^2y - xy^2 = xy(x - y)$$

The missing factor is $x - y$.

3. Find the missing factor.

$$y^2 + y - 42 = (y + 7)(\boxed{})$$

Solution

$$y^2 + y - 42 = (y + 7)(y - 6)$$

The missing factor is $y - 6$.

4. Find the missing factor.

$$2x^2 - x - 1 = (x - 1)(\boxed{})$$

Solution

$$2x^2 - x - 1 = (x - 1)(2x + 1)$$

The missing factor is $2x + 1$.

5. Factor $10x^2 + 70$.

Solution

$10x^2 + 70 = 10(x^2 + 7)$

6. Factor $2a^3b - 4a^2b^2$.

Solution

$2a^3b - 4a^2b^2 = 2a^2b(a - 2b)$

7. Factor $x(x + 2) - 3(x + 2)$.

Solution

$x(x + 2) - 3(x + 2) = (x + 2)(x - 3)$

8. Factor $t^3 - 3t^2 + t - 3$.

Solution

$$t^3 - 3t^2 + t - 3 = t^2(t - 3) + (t - 3)$$
$$= t^2(t - 3) + 1(t - 3)$$
$$= (t - 3)(t^2 + 1)$$

9. Factor $y^2 + 11y + 30$.

Solution

$y^2 + 11y + 30 = (y + 6)(y + 5)$

10. Factor $u^2 + u - 30$.

Solution

$u^2 + u - 30 = (u + 6)(u - 5)$

11. Factor $x^3 - x^2 - 30x$.

Solution

$$x^3 - x^2 - 30x = x(x^2 - x - 30)$$
$$= x(x - 6)(x + 5)$$

12. Factor $2x^2y + 8xy - 64y$.

Solution

$$2x^2y + 8xy - 64y = 2y(x^2 + 4x - 32)$$
$$= 2y(x + 8)(x - 4)$$

13. Factor $3v^2 - 4v - 2$.

Solution

This trinomial is prime.

14. Factor $6 - 13z - 5z^2$.

Solution

$$6 - 13z - 5z^2 = (3 + z)(2 - 5z) \text{ or}$$
$$6 - 13z - 5z^2 = -1(5z^2 + 13z - 6)$$
$$= -(5z - 2)(z + 3)$$
$$\text{or } (-5z + 2)(z + 3)$$
$$\text{or } (5z - 2)(-z - 3)$$

15. Factor $6x^2 - x - 2$.

Solution

$6x^2 - x - 2 = (3x - 2)(2x + 1)$

16. Factor $10s^4 - 14s^3 + 2s^2$.

Solution

$10s^4 - 14s^3 + 2s^2 = 2s^2(5s^2 - 7s + 1)$

17. Find all integer values of b such that the polynomial $x^2 + bx + 12$ can be factored. Describe the method you used.

Solution

$(x + 3)(x + 4)$	$b = 7$
$(x - 3)(x - 4)$	$b = -7$
$(x + 6)(x + 2)$	$b = 8$
$(x - 6)(x - 2)$	$b = -8$
$(x + 12)(x + 1)$	$b = 13$
$(x - 12)(x - 1)$	$b = -13$

18. Find two values of c such that the polynomial $x^2 - 10x + c$ can be factored. Describe the method you used.

Solution

$(x - 6)(x - 4)$	$c = 24$
$(x - 2)(x - 8)$	$c = 16$
$(x + 2)(x - 12)$	$c = -24$
$(x - 11)(x + 1)$	$c = -11$
$(x - 5)(x - 5)$	$c = 25$

These are just examples of some of the possible values of c. There are many correct answers.

19. Find all possible products of the form $(3x + m)(x + n)$ such that $mn = 6$. Describe the method you used.

Solution

$(3x + 2)(x + 3)$	$(3x - 2)(x - 3)$
$(3x + 3)(x + 2)$	$(3x - 3)(x - 2)$
$(3x + 6)(x + 1)$	$(3x - 6)(x - 1)$
$(3x + 1)(x + 6)$	$(3x - 1)(x - 6)$

20. The area of the rectangle shown in the figure in the textbook is $x^2 + 4x - 5$. Given that the width is $x - 1$, find its length.

Solution

Area = (Length)(Width)

The area of the rectangle is $x^2 + 4x - 5$, or $(x + 5)(x - 1)$, and the width is $x - 1$. Therefore, the other factor, $x + 5$, represents the length of the rectangle.

6.4 Factoring Polynomials with Special Forms

7. Factor the difference of two squares $x^2 - 36$.

Solution
$$x^2 - 36 = x^2 - 6^2$$
$$= (x + 6)(x - 6)$$

9. Factor the difference of two squares $u^2 - \frac{1}{4}$.

Solution
$$u^2 - \frac{1}{4} = u^2 - \left(\frac{1}{2}\right)^2$$
$$= \left(u + \frac{1}{2}\right)\left(u - \frac{1}{2}\right)$$

11. Factor $2x^2 - 72$ completely.

Solution
$$2x^2 - 72 = 2(x^2 - 36)$$
$$= 2(x^2 - 6^2)$$
$$= 2(x + 6)(x - 6)$$

13. Factor $x^4 - 1$ completely.

Solution
$$x^4 - 1 = (x^2)^2 - 1^2$$
$$= (x^2 + 1)(x^2 - 1)$$
$$= (x^2 + 1)(x^2 - 1^2)$$
$$= (x^2 + 1)(x + 1)(x - 1)$$

15. Factor the perfect square trinomial $x^2 - 4x + 4$.

Solution
$$x^2 - 4x + 4 = x^2 - 2(2x) + 2^2$$
$$= (x - 2)^2$$

17. Factor the perfect square trinomial $25y^2 - 10y + 1$.

Solution
$$25y^2 - 10y + 1 = (5y)^2 - 2(5y) + 1^2$$
$$= (5y - 1)^2$$

19. Factor the perfect square trinomial
$$b^2 + b + \tfrac{1}{4}.$$

Solution
$$b^2 + b + \tfrac{1}{4} = b^2 + 2\left(\tfrac{1}{2}b\right) + \left(\tfrac{1}{2}\right)^2$$
$$= \left(b + \tfrac{1}{2}\right)^2$$

23. Find two values of b so that the expression $x^2 + bx + 100$ is a perfect square trinomial.

Solution
$$x^2 + bx + 100 = x^2 + bx + 10^2$$
$$(x + 10)^2 = x^2 + 2(x)(10) + 10^2$$
$$= x^2 + 20x + 100$$
$$(x - 10)^2 = x^2 - 2(x)(10) + 10^2$$
$$= x^2 - 20x + 100$$

Thus, $b = 20$ or $b = -20$.

27. Factor the difference of two cubes $x^3 - 8$.

Solution
$$x^3 - 8 = x^3 - 2^3$$
$$= (x - 2)(x^2 + (x)(2) + 2^2)$$
$$= (x - 2)(x^2 + 2x + 4)$$

31. Factor $x^3 - 4x^2$ completely.

Solution
$$x^3 - 4x^2 = x^2(x - 4)$$

35. Factor $x(x - 1) + (x - 1)^2$ completely.

Solution
$$x(x - 1) + (x - 1)^2 = (x - 1)[x + (x - 1)]$$
$$= (x - 1)(2x - 1)$$

21. Factor the perfect square trinomial
$$x^2 - 6xy + 9y^2.$$

Solution
$$x^2 - 6xy + 9y^2 = x^2 - 2(3xy) + (3y)^2$$
$$= (x - 3y)^2$$

25. Find two values of b so that the expression $4x^2 + bx + \tfrac{1}{4}$ is a perfect square trinomial.

Solution
$$4x^2 + bx + \tfrac{1}{4} = (2x)^2 + bx + \left(\tfrac{1}{2}\right)^2$$
$$\left(2x + \tfrac{1}{2}\right)^2 = (2x)^2 + 2(2x)\left(\tfrac{1}{2}\right) + \left(\tfrac{1}{2}\right)^2$$
$$= 4x^2 + 2x + \tfrac{1}{4}$$
$$\left(2x - \tfrac{1}{2}\right)^2 = (2x)^2 - 2(2x)\left(\tfrac{1}{2}\right) + \left(\tfrac{1}{2}\right)^2$$
$$= 4x^2 - 2x + \tfrac{1}{4}$$

Thus, $b = 2$ or $b = -2$.

29. Factor the sum of two cubes $1 + 8t^3$.

Solution
$$1 + 8t^3 = 1^3 + (2t)^3$$
$$= (1 + 2t)[1^2 - (1)(2t) + (2t)^2]$$
$$= (1 + 2t)(1 - 2t + 4t^2)$$

33. Factor $1 - 4x + 4x^2$ completely.

Solution
$$1 - 4x + 4x^2 = 1^2 - 2(2x) + (2x)^2$$
$$= (1 - 2x)^2$$

Alternate Method:
$$1 - 4x + 4x^2 = 4x^2 - 4x + 1$$
$$= (2x)^2 - 2(2x) + 1^2$$
$$= (2x - 1)^2$$

37. Factor $9t^2 - 16$ completely.

Solution
$$9t^2 - 16 = (3t)^2 - 4^2$$
$$= (3t + 4)(3t - 4)$$

39. Factor $2y^2 - 3y - 5$ completely.

Solution

$$2y^2 - 3y - 5 = (2y - 5)(y + 1)$$

41. Factor $2t^3 - 16$ completely.

Solution

$$2t^3 - 16 = 2(t^3 - 8)$$
$$= 2(t^3 - 2^3)$$
$$= 2(t - 2)(t^2 + (t)(2) + 2^2)$$
$$= 2(t - 2)(t^2 + 2t + 4)$$

43. *Geometric Factoring Models*   Write the factoring problem represented by the geometric factoring model shown in the textbook.

Solution

$$x^2 + 2x + 1 = (x + 1)(x + 1)$$
$$= (x + 1)^2$$

45. *Geometry*   An annulus is the region between two concentric circles. The area of an annulus in the figure shown in the textbook is $\pi R^2 - \pi r^2$. Give the complete factorization of the expression for the area.

Solution

$$\pi R^2 - \pi r^2 = \pi(R^2 - r^2)$$
$$= \pi(R + r)(R - r)$$

47. Solve $2(x + 1) = 0$ and check your result.

Solution

$$2(x + 1) = 0$$
$$2x + 2 = 0$$
$$2x + 2 - 2 = 0 - 2$$
$$2x = -2$$
$$\frac{2x}{2} = \frac{-2}{2}$$
$$x = -1$$

49. Solve $\frac{3}{4}(12x - 8) = 10$ and check your result.

Solution

$$\frac{3}{4}(12x - 8) = 10$$
$$\frac{3}{4}(12x) - \frac{3}{4}(8) = 10$$
$$9x - 6 = 10$$
$$9x - 6 + 6 = 10 + 6$$
$$9x = 16$$
$$\frac{9x}{9} = \frac{16}{9}$$
$$x = \frac{16}{9}$$

51. *Membership*   The current membership for a public television station is 120% of what it was a year ago. The current number is 8346. How many members did the station have last year?

Solution

Verbal model: $\boxed{8346} = \boxed{120\% \text{ of what number}}$ $(a = pb)$

Labels: $a = 8346$
$p = 1.2$ (percent in decimal form)
$b =$ unknown number

Equation: $8346 = 1.2b$
$$\frac{8346}{1.2} = \frac{1.2b}{1.2}$$
$$6955 = b$$

The station had 6955 members last year.

53. Factor the difference of squares $81 - x^2$.

Solution

$$81 - x^2 = 9^2 - x^2$$
$$= (9 + x)(9 - x)$$

55. Factor the difference of squares $t^2 - \frac{1}{16}$.

Solution

$$t^2 - \frac{1}{16} = t^2 - \left(\frac{1}{4}\right)^2$$
$$= \left(t + \frac{1}{4}\right)\left(t - \frac{1}{4}\right)$$

57. Factor the difference of squares $16y^2 - 9$.

Solution

$$16y^2 - 9 = (4y)^2 - 3^2$$
$$= (4y + 3)(4y - 3)$$

59. Factor the difference of squares $25 - (z + 5)^2$.

Solution

$$25 - (z + 5)^2 = 5^2 - (z + 5)^2$$
$$= [5 + (z + 5)][5 - (z + 5)]$$
$$= [5 + z + 5][5 - z - 5]$$
$$= (z + 10)(-z)$$
$$= -z(z + 10)$$

61. Factor $y^4 - 81$ completely.

Solution

$$y^4 - 81 = (y^2)^2 - 9^2$$
$$= (y^2 + 9)(y^2 - 9)$$
$$= (y^2 + 9)(y^2 - 3^2)$$
$$= (y^2 + 9)(y + 3)(y - 3)$$

63. Factor $8 - 50x^2$ completely.

Solution

$$8 - 50x^2 = 2(4 - 25x^2)$$
$$= 2(2^2 - (5x)^2)$$
$$= 2(2 + 5x)(2 - 5x)$$

65. Factor the perfect square trinomial
$$z^2 + 6z + 9.$$

Solution

$$z^2 + 6z + 9 = z^2 + 2(3z) + 3^2$$
$$= (z + 3)^2$$

67. Factor the perfect square trinomial
$$4t^2 + 4t + 1.$$

Solution

$$4t^2 + 4t + 1 = (2t)^2 + 2(2t) + 1^2$$
$$= (2t + 1)^2$$

69. Factor the perfect square trinomial
$$4x^2 - x + \frac{1}{16}.$$

Solution

$$4x^2 - x + \frac{1}{16} = (2x)^2 - (2)(2x)\left(\frac{1}{4}\right) + \left(\frac{1}{4}\right)^2$$
$$= \left(2x - \frac{1}{4}\right)^2$$

71. Factor the perfect square trinomial.
$$4y^2 + 20yz + 25z^2.$$

Solution

$$4y^2 + 20yz + 25z^2 = (2y)^2 + 2(10yz) + (5z)^2$$
$$= (2y)^2 + 2(2y)(5z) + (5z)^2$$
$$= (2y + 5z)^2$$

73. Factor the perfect square trinomial.
$$9a^2 - 12ab + 4b^2.$$

Solution

$$9a^2 - 12ab + 4b^2 = (3a)^2 - 2(3a)(2b) + (2b)^2$$
$$= (3a - 2b)^2$$

75. *Think About It*   Find two values of b so that $x^2 + bx + 1$ is a perfect square trinomial.

Solution

$$x^2 + bx + 1 = x^2 + bx + 1^2$$
$$(x + 1)(x + 1) = x^2 + 2(x)(1) + 1 \qquad b = 2$$
$$(x - 1)(x - 1) = x^2 - 2(x)(1) + 1 \qquad b = -2$$

77. *Think About It*   Find two values of b so that $4x^2 + bx + 9$ is a perfect square trinomial.

Solution

$$4x^2 + bx + 9 = (2x)^2 + bx + 3^2$$

$$(2x + 3)(2x + 3) = (2x)^2 + 2(2x)(3) + 3^2$$

$$= 4x^2 + 12x + 9 \qquad b = 12$$

$$(2x - 3)(2x - 3) = (2x)^2 - 2(2x)(3) + 3^2$$

$$= 4x^2 - 12x + 9 \qquad b = -12$$

79. *Think About It*   Find a number c so that $x^2 + 6x + c$ is a perfect square trinomial.

Solution

$$x^2 + 6x + c = x^2 + 2(x)(3) + c$$

$$(x + 3)^2 = x^2 + 2(x)(3) + 3^2$$

$$= x^2 + 6x + 9$$

Thus, $c = 9$.

81. *Think About It*   Find a number c so that $y^2 - 4y + c$ is a perfect square trinomial.

Solution

$$y^2 - 4y + c = (y)^2 - 2(y)(2) + c$$

$$(y - 2)^2 = y^2 - 2(y)(2) + 2^2$$

$$= y^2 - 4y + 4$$

Thus, $c = 4$.

83. Factor the sum of cubes $y^3 + 64$.

Solution

$$y^3 + 64 = y^3 + 4^3$$

$$= (y + 4)(y^2 - (y)(4) + 4^2)$$

$$= (y + 4)(y^2 - 4y + 16)$$

85. Factor the sum of cubes $27u^3 + 8$.

Solution

$$27u^3 + 8 = (3u)^2 + 2^3$$

$$= (3u + 2)[(3u)^2 - (3u)(2) + 2^2]$$

$$= (3u + 2)(9u^2 - 6u + 4)$$

87. Factor $y^4 - 25y^2$ completely.

Solution

$$y^4 - 25y^2 = y^2(y^2 - 25)$$

$$= y^2(y^2 - 5^2)$$

$$= y^2(y + 5)(y - 5)$$

89. Factor $z^4 - \frac{4}{9}z^2$ completely.

Solution

$$z^4 - \frac{4}{9}z^2 = z^2\left(z^2 - \frac{4}{9}\right)$$

$$= z^2\left[z^2 - \left(\frac{2}{3}\right)^2\right]$$

$$= z^2\left(z + \frac{2}{3}\right)\left(z - \frac{2}{3}\right)$$

91. Factor $x^2 - 2x + 1$ completely.

Solution

$$x^2 - 2x + 1 = x^2 - 2(x) + 1^2$$

$$= (x - 1)^2$$

93. Factor $4v^2 + 4v + 1$ completely.

Solution

$$4v^2 + 4v + 1 = (2v)^2 + 2(2v)(1) + 1^2$$
$$= (2v + 1)^2 \text{ or } (2v + 1)(2v + 1)$$

95. Factor $2x^2 + 4x - 2x^3$ completely.

Solution

$$2x^2 + 4x - 2x^3 = 2x(x + 2 - x^2)$$
$$= 2x(-x^2 + x + 2)$$
$$= -2x(x^2 - x - 2)$$
$$= -2x(x - 2)(x + 1)$$

Note: This answer could also be written in other equivalent forms, such as $2x(-x + 2)(x + 1)$, $2x(x - 2)(-x - 1)$, $2x(2 - x)(x + 1)$, or $2x(2 - x)(1 + x)$.

97. Factor $9x^2 + 10x + 1$ completely.

Solution

$$9x^2 + 10x + 1 = (9x + 1)(x + 1)$$

99. Factor $(x - 1)^2 - 2(x - 1)$ completely.

Solution

$$(x - 1)^2 - 2(x - 1) = (x - 1)[(x - 1) - 2]$$
$$= (x - 1)(x - 3)$$

101. Factor $5 - x + 5x^2 - x^3$ completely.

Solution

$$5 - x + 5x^2 - x^3 = (5 - x) + (5x^2 - x^3)$$
$$= (5 - x) + x^2(5 - x)$$
$$= (5 - x)(1 + x^2)$$

103. Factor $x^4 - 4x^3 + x^2 - 4x$ completely.

Solution

$$x^4 - 4x^3 + x^2 - 4x = x[x^3 - 4x^2 + x - 4]$$
$$= x[x^2(x - 4) + (x - 4)]$$
$$= x[(x - 4)(x^2 + 1)]$$
$$= x(x - 4)(x^2 + 1)$$

105. Factor $(y + 2)^2 - 1$ completely.

Solution

$$(y + 2)^2 - 1 = (y + 2)^2 - 1^2$$
$$= [(y + 2) + 1][(y + 2) - 1]$$
$$= (y + 3)(y + 1)$$

This expression can also be factored by first squaring the binomial.

$$(y + 2)^2 - 1 = y^2 + 4y + 4 - 1$$
$$= y^2 + 4y + 3$$
$$= (y + 3)(y + 1)$$

107. Factor $64 - (z + 8)^2$ completely.

Solution

$$64 - (z + 8)^2 = 8^2 - (z + 8)^2$$
$$= [8 + (z + 8)][8 - (z + 8)]$$
$$= (z + 16)(-z)$$

This expression can also be factored by first squaring the binomial.

$$64 - (z + 8)^2 = 64 - (z^2 + 16z + 64)$$
$$= 64 - z^2 - 16z - 64$$
$$= -z^2 - 16z$$
$$= -z(z + 16)$$

109. Factor $3u^3 - 27u$ completely.

Solution

$$3u^3 - 27u = 3u(u^2 - 9)$$
$$= 3u(u^2 - 3^2)$$
$$= 3u(u + 3)(u - 3)$$

111. Factor $8x^2 + 2$ completely.

Solution

$$8x^2 + 2 = 2(4x^2 + 1)$$

113. Factor $y^3 + \frac{1}{8}$ completely.

Solution

$$y^3 + \frac{1}{8} = y^3 + \left(\frac{1}{2}\right)^3$$

$$= \left(y + \frac{1}{2}\right)\left[y^2 - (y)\left(\frac{1}{2}\right) + \left(\frac{1}{2}\right)^2\right]$$

$$= \left(y + \frac{1}{2}\right)\left(y^2 - \frac{1}{2}y + \frac{1}{4}\right)$$

117. Factor $(x - 3)^3 - 1$ completely.

Solution

$$(x - 3)^3 - 1 = (x - 3)^3 - 1^3$$

$$= [(x - 3) - 1][(x - 3)^2 + (x - 3)(1) + 1^2]$$

$$= (x - 4)(x^2 - 6x + 9 + x - 3 + 1)$$

$$= (x - 4)(x^2 - 5x + 7)$$

119. Factor $u^3 + 2u^2 + 3u$ completely.

Solution

$$u^3 + 2u^2 + 3u = u(u^2 + 2u + 3)$$

123. Factor $1 - x^4$ completely.

Solution

$$1 - x^4 = 1^2 - (x^2)^2$$

$$= (1 + x^2)(1 - x^2)$$

$$= (1 + x^2)(1^2 - x^2)$$

$$= (1 + x^2)(1 + x)(1 - x)$$

Note: This answer could be written in other forms, such as $-(x^2 + 1)(x + 1)(x - 1)$.

115. Factor $16x^3 - 2$ completely.

Solution

$$16x^3 - 2 = 2(8x^3 - 1)$$

$$= 2[(2x)^3 - 1^3]$$

$$= 2(2x - 1)[(2x)^2 + (2x)(1) + 1^2]$$

$$= 2(2x - 1)(4x^2 + 2x + 1)$$

121. Factor $x^4 - 81$ completely.

Solution

$$x^4 - 81 = (x^2)^2 - 9^2$$

$$= (x^2 + 9)(x^2 - 9)$$

$$= (x^2 + 9)(x^2 - 3^2)$$

$$= (x^2 + 9)(x + 3)(x - 3)$$

125. Use a graphing utility to graph $y_1 = x^2 - 36$ and $y_2 = (x + 6)(x - 6)$ on the same screen. What can you conclude?

Solution

The two graphs are the same because

$$1x^2 - 36 = (x + 6)(x - 6)$$

$$y_1 = y_2.$$

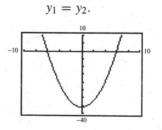

127. Use a graphing utility to graph $y_1 = x^3 - 6x^2 + 9x$ and $y_2 = x(x - 3)^2$ on the same screen. What can you conclude?

Solution

The two graphs are the same because

$$x^3 - 6x^2 + 9x = x(x^2 - 6x + 9) \ .$$

$$= x(x - 3)^2$$

$$y_1 = y_2$$

129. *Mental Math*   Evaluate 21^2 using the two samples as models.

Samples: $29^2 = (30 - 1)^2$

$$= 30^2 - 2 \cdot 30 \cdot 1 + 1^2$$

$$= 900 - 60 + 1$$

$$= 841$$

$$48 \cdot 52 = (50 - 2)(50 + 2)$$

$$= 50^2 - 2^2$$

$$= 2496$$

Solution

$21^2 = (20 + 1)^2$

$$= 20^2 + 2(20)(1) + 1^2$$

$$= 400 + 40 + 1$$

$$= 441$$

131. *Mental Math*   Evaluate $59 \cdot 61$ using the two samples as models.

Samples: $29^2 = (30 - 1)^2$

$$= 30^2 - 2 \cdot 30 \cdot 1 + 1^2$$

$$= 900 - 60 + 1$$

$$= 841$$

$$48 \cdot 52 = (50 - 2)(50 + 2)$$

$$= 50^2 - 2^2$$

$$= 2496$$

Solution

$59 \cdot 61 = (60 - 1)(60 + 1)$

$$= 60^2 - 1^2$$

$$= 3600 - 1$$

$$= 3599$$

133. Write $x^2 + 6x + 10$ as the sum of two squares.

$$x^2 + 6x + 10 = (x^2 + 6x + 9) + 1$$

$$= \boxed{}^2 + \boxed{}^2$$

Solution

$x^2 + 6x + 10 = x^2 + 6x + 9 + 1$

$$= (x + 3)^2 + 1^2$$

135. *Geometric Factoring Models*   Write the factoring problem represented by the geometric factoring model shown in the textbook.

Solution

$x^2 + 4x = x(x + 4)$

6.5 Solving Equations and Problem Solving

7. Use the Zero-Factor Property to solve $x(x - 5) = 0$.

Solution

$$x(x - 5) = 0$$

$$x = 0$$

$$x - 5 = 0 \rightarrow x = 5$$

Note: There are *two* solutions, 0 and 5.

9. Use the Zero-Factor Property to solve $y(y - 1)(y + 3) = 0$.

Solution

$$y(y - 1)(y + 3) = 0$$

$$y = 0$$

$$y - 1 = 0 \rightarrow y = 1$$

$$y + 3 = 0 \rightarrow y = -3$$

11. Solve $x^2 - 16 = 0$.

Solution

$$x^2 - 16 = 0$$

$$(x + 4)(x - 4) = 0$$

$$x + 4 = 0 \rightarrow x = -4$$

$$x - 4 = 0 \rightarrow x = 4$$

13. Solve $3y^2 - 27 = 0$.

Solution

$$3y^2 - 27 = 0$$

$$3(y^2 - 9) = 0$$

$$3(y + 3)(y - 3) = 0$$

$$3 \neq 0$$

$$y + 3 = 0 \rightarrow y = -3$$

$$y - 3 = 0 \rightarrow y = 3$$

15. Solve $6x^2 + 3x = 0$.

Solution

$$6x^2 + 3x = 0$$

$$3x(2x + 1) = 0$$

$$3x = 0 \rightarrow x = 0$$

$$2x + 1 = 0 \rightarrow 2x = -1 \rightarrow x = -\tfrac{1}{2}$$

17. Solve $x^2 - 2x - 8 = 0$.

Solution

$$x^2 - 2x - 8 = 0$$

$$(x - 4)(x + 2) = 0$$

$$x - 4 = 0 \rightarrow x = 4$$

$$x + 2 = 0 \rightarrow x = -2$$

19. Solve $3 + 5x - 2x^2 = 0$.

Solution

$$3 + 5x - 2x^2 = 0$$

$$(3 - x)(1 + 2x) = 0$$

$$3 - x = 0 \rightarrow -x = -3 \rightarrow x = 3$$

$$1 + 2x = 0 \rightarrow 2x = -1 \rightarrow x = -\tfrac{1}{2}$$

21. Solve $x(x - 5) = 14$.

Solution

$$x(x - 5) = 14$$

$$x^2 - 5x = 14$$

$$x^2 - 5x - 14 = 0$$

$$(x - 7)(x + 2) = 0$$

$$x - 7 = 0 \rightarrow x = 7$$

$$x + 2 = 0 \rightarrow x = -2$$

23. Solve $u(u + 2) - 3(u + 2) = 0$.

Solution

$$u(u + 2) - 3(u + 2) = 0$$

$$(u + 2)(u - 3) = 0$$

$$u + 2 = 0 \rightarrow u = -2$$

$$u - 3 = 0 \rightarrow u = 3$$

25. Solve $x^2(x - 2) - 9(x - 2) = 0$.

Solution

$$x^2(x - 2) - 9(x - 2) = 0$$

$$(x - 2)(x^2 - 9) = 0$$

$$(x - 2)(x + 3)(x - 3) = 0$$

$$x - 2 = 0 \rightarrow x = 2$$

$$x + 3 = 0 \rightarrow x = -3$$

$$x - 3 = 0 \rightarrow x = 3$$

27. *Height of an Object*   An object is dropped from a weather balloon 1600 feet above the ground (see figure). Find the time t for the object to reach the ground. The height (above ground) of the object is modeled by

$$\text{Height} = -16t^2 + 1600$$

where the height is measured in feet and the time t is measured in seconds.

Solution

$$\text{Height} = -16t^2 + 1600$$

$$0 = -16t^2 + 1600 \quad \text{(Set height equal to 0)}$$

$$16t^2 - 1600 = 0$$

$$16(t^2 - 100) = 0$$

$$16(t + 10)(t - 10) = 0$$

$$16 \neq 0$$

$$t + 10 = 0 \rightarrow t = -10$$

$$t - 10 = 0 \rightarrow t = 10$$

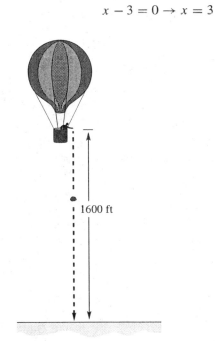

1600 ft

The time could not be -10 seconds, so we discard this negative answer. Thus, $t = 10$; the object reaches the ground in 10 seconds.

29. *Graphical Estimation*   Estimate the *x*-intercepts visually by referring to the graph shown in the textbook. Then check your estimates algebraically.

Solution

The *x*-intercepts are $(1, 0)$ and $(-3, 0)$.

Algebraic check:
$$y = x^2 + 2x - 3$$
$$0 = x^2 + 2x - 3$$
$$0 = (x + 3)(x - 1)$$
$$x + 3 = 0 \rightarrow x = -3$$
$$x - 1 = 0 \rightarrow x = 1$$
$$x\text{-intercepts } (-3, 0), (1, 0)$$

31. *Graphical Estimation*   Estimate the *x*-intercepts visually by referring to the graph shown in the textbook. Then check your estimates algebraically.

Solution

The *x*-intercepts are $(0, 0)$ and $(3, 0)$.

Algebraic check:
$$y = x^3 - 6x^2 + 9x$$
$$0 = x^3 - 6x^2 + 9x$$
$$0 = x(x^2 - 6x + 9)$$
$$0 = x(x - 3)^2$$
$$x = 0$$
$$x - 3 = 0 \rightarrow x = 3$$
$$x\text{-intercepts } (0, 0), (3, 0)$$

33. Use a graphing utility to graph $y = x^2 - 4x$. Use the graph to estimate the *x*-intercepts. Then check your estimates by substituting into the equation.

Solution

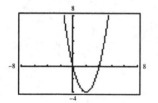

The *x*-intercepts are $(0, 0)$ and $(4, 0)$.

35. Use a graphing utility to graph $y = x^3 - 4x^2$. Use the graph to estimate the *x*-intercepts. Then check your estimates by substituting into the equation.

Solution

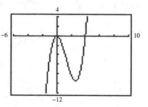

The *x*-intercepts are $(0, 0)$ and $(4, 0)$.

37. *Problem Solving*   Find two consecutive positive integers whose product is 72.

Solution

Verbal model: $\boxed{\begin{array}{c}\text{First consecutive} \\ \text{positive integer}\end{array}} \cdot \boxed{\begin{array}{c}\text{Second consecutive} \\ \text{positive integer}\end{array}} = \boxed{72}$

Labels: First integer $= n$
Second integer $= n + 1$

Equation:
$$n(n + 1) = 72$$
$$n^2 + n = 72$$
$$n^2 + n - 72 = 0$$
$$(n + 9)(n - 8) = 0$$
$$n + 9 = 0 \rightarrow n = -9$$
$$n - 8 = 0 \rightarrow n = 8$$

The original problem required that the integers be positive, so we discard -9 as a solution. Thus, $n = 8$ and $n + 1 = 9$; the two consecutive integers are 8 and 9.

39. Evaluate $2(-3) + 9$.

Solution

$2(-3) + 9 = -6 + 9 = 3$

41. Evaluate $4 - \frac{5}{2}$.

Solution

$4 - \frac{5}{2} = \frac{4}{1} - \frac{5}{2}$

$\qquad = \frac{8}{2} - \frac{5}{2}$

$\qquad = \frac{3}{2}$

43. *Simple Interest*   Find the interest on a $1000 bond paying an annual percentage rate of 7.5% for 10 years.

Solution

$I = Prt$

$\quad = (1000)(.075)(10)$

$\quad = 750$

The interest is $750.

45. Use the Zero-Factor Property to solve $z(z - 3) = 0$.

Solution

$z(z - 3) = 0$

$z = 0$

$z - 3 = 0 \to z = 3$

47. Use the Zero-Factor Property to solve
$x(x - 3)(x + 25) = 0$.

Solution

$x(x - 3)(x + 25) = 0$

$x = 0$

$x - 3 = 0 \to x = 3$

$x + 25 = 0 \to x = -25$

49. Solve $(a + 1)(a - 2) = 0$.

Solution

$(a + 1)(a - 2) = 0$

$a + 1 = 0 \to a = -1$

$a - 2 = 0 \to a = 2$

51. Solve $v^2 - 100 = 0$.

Solution

$v^2 - 100 = 0$

$v^2 - 10^2 = 0$

$(v + 10)(v - 10) = 0$

$v + 10 = 0 \to v = -10$

$v - 10 = 0 \to v = 10$

53. Solve $4x^2 - 9 = 0$.

Solution

$4x^2 - 9 = 0$

$(2x)^2 - 3^2 = 0$

$(2x + 3)(2x - 3) = 0$

$2x + 3 = 0 \to 2x = -3 \to x = -\frac{3}{2}$

$2x - 3 = 0 \to 2x = 3 \to x = \frac{3}{2}$

55. Solve $(t - 3)^2 - 25 = 0$.

Solution

$$(t - 3)^2 - 25 = 0$$
$$[(t - 3) + 5][(t - 3) - 5] = 0$$
$$[t + 2][t - 8] = 0$$
$$t + 2 = 0 \rightarrow t = -2$$
$$t - 8 = 0 \rightarrow t = 8$$

Note: Here is an alternative solution.

$$(t - 3)^2 - 25 = 0$$
$$t^2 - 6t + 9 - 25 = 0$$
$$t^2 - 6t - 16 = 0$$
$$(t + 2)(t - 8) = 0$$
$$t + 2 = 0 \rightarrow t = -2$$
$$t - 8 = 0 \rightarrow t = 8$$

57. Solve $x^2 - 2x = 0$.

Solution

$$x^2 - 2x = 0$$
$$x(x - 2) = 0$$
$$x = 0$$
$$x - 2 = 0 \rightarrow x = 2$$

59. Solve $x(x - 8) + 2(x - 8) = 0$.

Solution

$$x(x - 8) + 2(x - 8) = 0$$
$$(x - 8)(x + 2) = 0$$
$$x - 8 = 0 \rightarrow x = 8$$
$$x + 2 = 0 \rightarrow x = -2$$

Note: Here is an alternative solution.

$$x(x - 8) + 2(x - 8) = 0$$
$$x^2 - 8x + 2x - 16 = 0$$
$$x^2 - 6x - 16 = 0$$
$$(x - 8)(x + 2) = 0$$
$$x - 8 = 0 \rightarrow x = 8$$
$$x + 2 = 0 \rightarrow x = -2$$

61. Solve $m^2 - 2m + 1 = 0$.

Solution

$$m^2 - 2m + 1 = 0$$
$$(m - 1)^2 = 0$$
$$m - 1 = 0 \rightarrow m = 1$$

63. Solve $x^2 + 14x + 49 = 0$.

Solution

$$x^2 + 14x + 49 = 0$$
$$(x + 7)^2 = 0$$
$$x + 7 = 0 \rightarrow x = -7$$

65. Solve $4t^2 - 12t + 9 = 0$.

Solution

$$4t^2 - 12t + 9 = 0$$
$$(2t - 3)^2 = 0$$
$$2t - 3 = 0 \rightarrow 2t = 3 \rightarrow t = \tfrac{3}{2}$$

67. Solve $6x^2 + 4x - 10 = 0$.

Solution

$$6x^2 + 4x - 10 = 0$$

$$2(3x^2 + 2x - 5) = 0$$

$$2(3x + 5)(x - 1) = 0$$

$$2 \neq 0$$

$$3x + 5 = 0 \rightarrow 3x = -5 \rightarrow x = -\tfrac{5}{3}$$

$$x - 1 = 0 \rightarrow x = 1$$

69. Solve $z(z + 2) = 15$.

Solution

$$z(z + 2) = 15$$

$$z^2 + 2z = 15$$

$$z^2 + 2z - 15 = 0$$

$$(z + 5)(z - 3) = 0$$

$$z + 5 = 0 \rightarrow z = -5$$

$$z - 3 = 0 \rightarrow z = 3$$

71. Solve $y(2y + 1) = 3$.

Solution

$$y(2y + 1) = 3$$

$$2y^2 + y = 3$$

$$2y^2 + y - 3 = 0$$

$$(2y + 3)(y - 1) = 0$$

$$2y + 3 = 0 \rightarrow 2y = -3 \rightarrow y = -\tfrac{3}{2}$$

$$y - 1 = 0 \rightarrow y = 1$$

73. Solve $(x + 1)(x + 4) = 4$.

Solution

$$(x + 1)(x + 4) = 4$$

$$x^2 + 5x + 4 = 4$$

$$x^2 + 5x = 0$$

$$x(x + 5) = 0$$

$$x = 0$$

$$x + 5 = 0 \rightarrow x = -5$$

75. Solve $2t^3 + 5t^2 - 12t = 0$.

Solution

$$2t^3 + 5t^2 - 12t = 0$$

$$t(2t^2 + 5t - 12) = 0$$

$$t(2t - 3)(t + 4) = 0$$

$$t = 0$$

$$2t - 3 = 0 \rightarrow 2t = 3 \rightarrow t = \tfrac{3}{2}$$

$$t + 4 = 0 \rightarrow t = -4$$

77. Solve $y^2(y + 3) - (y + 3) = 0$.

Solution

$$y^2(y + 3) - (y + 3) = 0$$

$$(y + 3)(y^2 - 1) = 0$$

$$(y + 3)(y + 1)(y - 1) = 0$$

$$y + 3 = 0 \rightarrow y = -3$$

$$y + 1 = 0 \rightarrow y = -1$$

$$y - 1 = 0 \rightarrow y = 1$$

79. *Graphical Estimation*   Estimate the x-intercepts visually by referring to the figure in the textbook. Check your estimates algebraically.

Solution

The x-intercepts are $(-3, 0)$ and $(4, 0)$.

Algebraic check: $y = x^2 - x - 12$

$$0 = x^2 - x - 12$$

$$0 = (x - 4)(x + 3)$$

$$x - 4 = 0 \rightarrow x = 4$$

$$x + 3 = 0 \rightarrow x = -3$$

x-intercepts $(4, 0)$ and $(-3, 0)$

81. *Graphical Estimation* Estimate the x-intercepts visually by referring to the figure in the textbook. Check your estimates algebraically.

Solution

The x-intercepts are $(0, 0)$, $(1, 0)$ and $\left(-\frac{5}{2}, 0\right)$.

Algebraic check: $\quad y = 5x - 3x^2 - 2x^3$

$$0 = 5x - 3x^2 - 2x^3$$

$$0 = x(5 - 3x - 2x^2)$$

$$0 = x(5 + 2x)(1 - x)$$

$$x = 0$$

$$5 + 2x = 0 \rightarrow 2x = -5$$

$$x = -\frac{5}{2}$$

$$1 - x = 0 \rightarrow -x = -1$$

$$x = 1$$

x-intercepts $(0, 0)$, $(1, 0)$ and $\left(-\frac{5}{2}, 0\right)$

83. Use a graphing utility to graph $y = x^2 - 4$. Use the graph to estimate the x-intercepts. Then check your estimates by substituting into the equation.

Solution

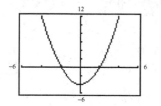

The x-intercepts are $(2, 0)$ and $(-2, 0)$.

85. Use a graphing utility to graph $y = x^2(x + 2) - 9(x + 2)$. Use the graph to estimate the x-intercepts. Then check your estimates by substituting into the equation.

Solution

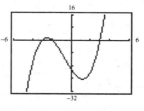

The x-intercepts are $(-2, 0)$, $(-3, 0)$, and $(3, 0)$.

87. Find two consecutive positive integers whose product is 240.

Solution

Verbal model: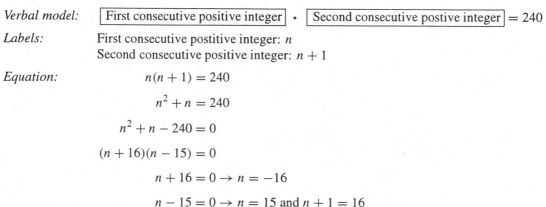

Labels: First consecutive positive integer: n
Second consecutive positive integer: $n + 1$

Equation: $\quad n(n + 1) = 240$

$$n^2 + n = 240$$

$$n^2 + n - 240 = 0$$

$$(n + 16)(n - 15) = 0$$

$$n + 16 = 0 \rightarrow n = -16$$

$$n - 15 = 0 \rightarrow n = 15 \text{ and } n + 1 = 16$$

We discard the negative solution. The two consecutive positive integers are 15 and 16.

89. *Geometry*   The length of a rectangle is 3 inches greater than the width. The area of the rectangle is 108 square inches. Find the dimensions of the rectangle.

Solution

Verbal model: $\boxed{\text{Length of rectangle}} \cdot \boxed{\text{Width of rectangle}} = \boxed{\text{Area of rectangle}}$

Labels: Width $= x$ (inches)
Length $= x + 3$ (inches)
Area $= 108$ (square inches)

Equation: $$(x + 3)x = 108$$

$$x^2 + 3x = 108$$

$$x^2 + 3x - 108 = 0$$

$$(x + 12)(x - 9) = 0$$

$$x + 12 = 0 \rightarrow x = -12$$

$$x - 9 = 0 \rightarrow x = 9$$

The width of the rectangle could not be -12, so we discard this answer. Thus, $x = 9$ and $x + 3 = 12$; the width of the rectangle is 9 inches and the length is 12 inches.

91. *Profit*   The profit for selling x units is $P = -0.4x^2 + 8x - 10$.

(a) Use a graphing utility to graph the expression for revenue.
(b) Use the graph to estimate any values of x that yield a profit of $P = \$20$.
(c) Use factorization to find any values of x that yield a profit of $P = \$20$.

Solution

(a)

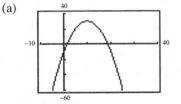

(b) The profit is $20 for two x-values, $x = 5$ and $x = 15$.

(c) $$P = -0.4x^2 + 8x - 10$$

$$20 = -0.4x^2 + 8x - 10$$

$$0 = -0.4x^2 + 8x - 30$$

$$10(0) = 10(-0.4x^2 + 8x - 30)$$

$$0 = -4x^2 + 80x - 300$$

$$0 = -4(x^2 + 20x + 75)$$

$$0 = -4(x - 5)(x - 15)$$

$$-4 \neq 0$$

$$x - 5 = 0 \rightarrow x = 5$$

$$x - 15 = 0 \rightarrow x = 15$$

93. *Exploration* An open box is to be made from a square piece of cardboard by cutting 2-inch squares from each corner and turning up the sides (see figure). (a) Show that the volume of the box is given by $V = 2x^2$. (b) Complete the table shown in the textbook. (c) Find the dimensions of the original piece of cardboard if $V = 200$ cubic inches.

Solution

(a) Volume = (length)(width)(height)

Length $= x$

Width $= x$

Height $= 2$

Volume $= (x)(x)(2) = 2x^2$

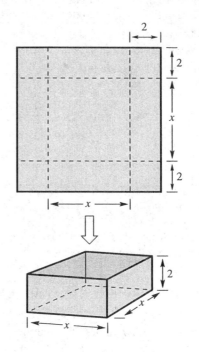

(b)

x	2	4	6	8
V	8	32	72	128

$x = 2, \quad V = 2(2)^2 = 2 \cdot 4 = 8$

$x = 4, \quad V = 2(4)^2 = 2 \cdot 16 = 32$

$x = 6, \quad V = 2(6)^2 = 2 \cdot 36 = 72$

$x = 8, \quad V = 2(8)^2 = 2 \cdot 64 = 128$

(c)
$$V = 200 \rightarrow 2x^2 = 200$$
$$2x^2 = 200$$
$$2x^2 - 200 = 0$$
$$2(x^2 - 100) = 0$$
$$2(x + 10)(x - 10) = 0$$
$$2 \neq 0$$
$$x + 10 = 0 \rightarrow x = -10$$
$$x - 10 = 0 \rightarrow x = 10$$

The length of the box could not be -10, so we discard this negative solution. Thus, $x = 10$. The original piece of cardboard is a square with each side $2 + x + 2$ or $x + 4$ inches long. We know $x = 10$, so $x + 4 = 14$. Therefore, the original piece of cardboard is a 14 inch $\times$ 14 inch square.

95. *Exploration* If a is a nonzero real number, find two solutions of the equation $ax^2 - ax = 0$.

Solution

$$ax^2 - ax = 0$$
$$ax(x - 1) = 0$$
$$ax = 0 \rightarrow x = 0$$
$$x - 1 = 0 \rightarrow x = 1$$

The two solutions are 0 and 1.

REVIEW EXERCISES for Chapter 6

Review Exercises for Chapter 6

1. Find the greatest common factor of 20, 60, and 150.

Solution

$20 = (2)(2)(5) = 10(2)$

$60 = (2)(2)(3)(5) = 10(6)$

$150 = (2)(3)(5)(5) = 10(15)$

The greatest common factor is $2(5)$, or 10.

3. Find the greatest common factor of $18ab^2$ and $27a^2b$.

Solution

$18ab^2 = (2)(3)(3)(a)(b)(b) = 9ab(2b)$

$27a^2b = (3)(3)(3)(a)(a)(b) = 9ab(3a)$

The greatest common factor is $(3)(3)(a)(b)$, or $9ab$.

5. Factor $5x^2 + 10x^3$.

Solution

$5x^2 + 10x^3 = 5x^2(1 + 2x)$

7. Factor $8a - 12a^3$.

Solution

$8a - 12a^3 = 4a(2 - 3a^2)$

9. Factor $24(x + 1) - 18(x + 1)^2$.

Solution

$$24(x + 1) - 18(x + 1)^2 = 6(x + 1)[4 - 3(x + 1)]$$
$$= 6(x + 1)[4 - 3x - 3]$$
$$= 6(x + 1)(-3x + 1)$$
$$\text{or } -6(x + 1)(3x - 1)$$

11. Factor $y^3 + 3y^2 + 2y + 6$.

Solution

$$y^3 + 3y^2 + 2y + 6 = y^2(y + 3) + 2(y + 3)$$
$$= (y + 3)(y^2 + 2)$$

13. Factor $a^2 - 100$.

Solution

$$a^2 - 100 = a^2 - 10^2$$
$$= (a + 10)(a - 10)$$

15. Factor $(u + 1)^2 - 4$.

Solution

$$(u + 1)^2 - 4 = (u + 1)^2 - 2^2$$
$$= [(u + 1) + 2][(u + 1) - 2]$$
$$= (u + 3)(u - 1)$$

This expression can also be factored by first squaring the binomial.

17. Factor $x^2 - 8x + 16$.

Solution

$$x^2 - 8x + 16 = x^2 - 2(x)(4) + 4^2$$
$$= (x - 4)^2$$

19. Factor $9s^2 + 12s + 4$.

Solution

$$9s^2 + 12s + 4 = (3s)^2 + 2(3s)(2) + 2^2$$
$$= (3s + 2)^2$$

21. Factor $4x^2 + 8x + 3$.

Solution

$$4x^2 + 8x + 3 = (2x + 3)(2x + 1)$$

23. Factor $50 - 5x - x^2$.

Solution

$$50 - 5x - x^2 = (10 + x)(5 - x) \text{ or}$$
$$50 - 5x - x^2 = -1(-50 + 5x + x^2)$$
$$= -1(x^2 + 5x - 50)$$
$$= -1(x + 10)(x - 5)$$

25. Insert the missing factor.

$$\tfrac{1}{3}x + \tfrac{5}{6} = \tfrac{1}{6}(\boxed{})$$

Solution

$$\tfrac{1}{3}x + \tfrac{5}{6} = \tfrac{2}{6}x + \tfrac{5}{6}$$
$$= \tfrac{1}{6}(2x + 5)$$

27. Insert the missing factor.

$$3x^2 + 14x + 8 = (x + 4)(\boxed{})$$

Solution

$$3x^2 + 14x + 8 = (x + 4)(3x + 2)$$

29. Insert the missing factor.

$$x^4 - 2x^2 + 1 = (x + 1)^2(\boxed{})^2$$

Solution

$$x^4 - 2x^2 + 1 = (x^2)^2 - 2(x^2)(1) + 1^2$$
$$= (x^2 - 1)^2$$
$$= (x^2 - 1^2)^2$$
$$= [(x + 1)(x - 1)]^2$$
$$= (x + 1)^2(x - 1)^2$$

31. Factor $x^2 - 3x - 28$ completely.

Solution

$$x^2 - 3x - 28 = (x - 7)(x + 4)$$

33. Factor $6x^2 + 7x + 2$ completely.

Solution

$$6x^2 + 7x + 2 = (3x + 2)(2x + 1)$$

35. Factor $6u^3 + 3u^2 - 30u$ completely.

Solution

$$6u^3 + 3u^2 - 30u = 3u(2u^2 + u - 10)$$
$$= 3u(2u + 5)(u - 2)$$

37. Factor $10x^2 + 9xy + 2y^2$ completely.

Solution

$$10x^2 + 9xy + 2y^2 = (5x + 2y)(2x + y)$$

39. Factor $s^3t - st^3$ completely.

Solution

$$s^3t - st^3 = st(s^2 - t^2) = st(s + t)(s - t)$$

41. Factor $2x^2 - 3x + 1$ completely.

Solution

$$2x^2 - 3x + 1 = (2x - 1)(x - 1)$$

43. Factor $27 - 8t^3$ completely.

Solution

$$27 - 8t^3 = 3^3 - (2t)^3$$
$$= (3 - 2t)(3^2 + (3)(2t) + (2t)^2)$$
$$= (3 - 2t)(9 + 6t + 4t^2)$$

45. Factor $-16a^3 - 16a^2 - 4a$ completely.

Solution

$$-16a^3 - 16a^2 - 4a = -4a(4a^2 + 4a + 1)$$
$$= -4a[(2a)^2 + 2(2a)(1) + 1^2]$$
$$= -4a(2a + 1)^2$$

47. Factor $x^3 + 2x^2 + x + 2$ completely.

Solution

$$x^3 + 2x^2 + x + 2 = (x^3 + 2x^2) + (x + 2)$$
$$= x^2(x + 2) + 1(x + 2)$$
$$= (x + 2)(x^2 + 1)$$

49. Factor $x^3 - 4x^2 - 4x + 16$ completely.

Solution

$$x^3 - 4x^2 - 4x + 16 = (x^3 - 4x^2) - (4x - 16)$$
$$= x^2(x - 4) - 4(x - 4)$$
$$= (x - 4)(x^2 - 4)$$
$$= (x - 4)(x^2 - 2^2)$$
$$= (x - 4)(x + 2)(x - 2)$$

53. Find all values of b such that $z^2 + bz + 11$ is factorable.

Solution

$b = 12 \quad (z + 11)(z + 1)$
$b = -12 \quad (z - 11)(z - 1)$

51. Find all values of b such that $x^2 + bx + 9$ is factorable.

Solution

$b = 6 \quad (x + 3)(x + 3)$
$b = -6 \quad (x - 3)(x - 3)$
$b = 10 \quad (x + 9)(x + 1)$
$b = -10 \quad (x - 9)(x - 1)$

55. Find all values of b such that $x^2 + bx - 24$ is factorable.

Solution

$b = 23 \quad (b + 24)(b - 1)$
$b = -23 \quad (b - 24)(b + 1)$
$b = 10 \quad (b + 12)(b - 2)$
$b = -10 \quad (b - 12)(b + 2)$
$b = 5 \quad (b + 8)(b - 3)$
$b = -5 \quad (b - 8)(b + 3)$
$b = 2 \quad (b + 6)(b - 4)$
$b = -2 \quad (b - 6)(b + 4)$

57. Find all values of b such that $3x^2 + bx - 20$ is factorable.

Solution

$b = 59 \quad (3x - 1)(x + 20)$
$b = -59 \quad (3x + 1)(x - 20)$
$b = 17 \quad (3x + 20)(x - 1)$
$b = -17 \quad (3x - 20)(x + 1)$
$b = 28 \quad (3x - 2)(x + 10)$
$b = -28 \quad (3x + 2)(x - 10)$
$b = 4 \quad (3x + 10)(x - 2)$
$b = -4 \quad (3x - 10)(x + 2)$
$b = 11 \quad (3x - 4)(x + 5)$
$b = -11 \quad (3x + 4)(x - 5)$
$b = 7 \quad (3x - 5)(x + 4)$
$b = -7 \quad (3x + 5)(x - 4)$

59. Find *two* values of c such that $x^2 + 6x + c$ is factorable.

Solution

$c = 9 \quad (x + 3)^2$
$c = 8 \quad (x + 4)(x + 2)$
$c = 5 \quad (x + 5)(x + 1)$
$c = -7 \quad (x + 7)(x - 1)$
$c = -16 \quad (x + 8)(x - 2)$
$c = -27 \quad (x + 9)(x - 3)$

(There are *many* other correct answers.)

61. Find *two* values of c such that $2x^2 - 4x + c$ is factorable.

Solution

$c = 2 \quad 2(x - 1)^2$
$c = -6 \quad 2(x + 1)(x - 3)$
$c = -16 \quad 2(x + 2)(x - 4)$
$c = -30 \quad 2(x + 3)(x - 5)$
$c = -48 \quad 2(x + 4)(x - 6)$

(There are *many* other correct answers.)

63. Use a graphing utility to graph $y_1 = x^2 - 2x - 3$ and $y_2 = (x+1)(x-3)$ on the same screen. What can you conclude?

Solution

The two graphs are the same.

$y_1 = y_2$

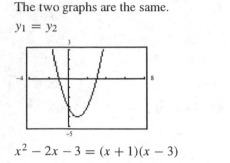

$x^2 - 2x - 3 = (x+1)(x-3)$

65. Use a graphing utility to graph $y_1 = x^3 - 4x^2 + 4x$ and $y_2 = x(x-2)^2$ on the same screen. What can you conclude?

Solution

The two graphs are the same.

$y_1 = y_2$

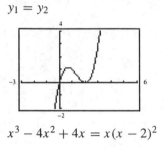

$x^3 - 4x^2 + 4x = x(x-2)^2$

67. *Geometry* A rectangular sheet of metal has dimensions 2 feet by 3 feet. An open box is to be made from the metal by cutting equal squares from each corner and turning up the sides. The volume of the box is

$V = 4x^3 - 10x^2 + 6x, \quad 0 < x < 1.$

(a) Sketch the rectangular sheet and the open box. Label the height of the box as x.

(b) Factor the expression for the volume and use the result to label the length and width of the box.

(c) Use a graphing utility to graph the volume over the specified interval. Use the graph to approximate the size of the squares to be cut from the corners so that the volume of the box is greatest.

Solution

(a)

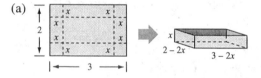

(b) $V = 4x^3 - 10x^2 + 6x$

$\quad = 2x(2x^2 - 5x + 3)$

$\quad = 2x(x-1)(2x-3)$

$\quad = x[2(x-1)](2x-3)$

$\quad = x(2x-2)(2x-3)$

$\quad = x(2x-2)(-1)(-1)(2x-3)$

$\quad = x[(2x-2)(-1)][(-1)(2x-3)]$

$\quad = x(-2x+2)(-2x+3)$

$\quad = x(2-2x)(3-2x)$

or $V = 4x^3 - 10x^2 + 6x$

$\quad = 2x(2x^2 - 5x + 3)$

$\quad = 2x(3 - 5x + 2x^2)$

$\quad = 2x(1-x)(3-2x)$

$\quad = x(2-2x)(3-2x)$

The height of the box is x feet, the length is $3 - 2x$ feet, and the width is $2 - 2x$ feet.

(c)

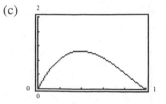

The volume appears to be greatest when x is approximately 0.4 feet.

69. Solve $x(2x - 3) = 0$.

Solution

$$x(2x - 3) = 0$$

$$x = 0$$

$$2x - 3 = 0 \rightarrow 2x = 3 \rightarrow x = \tfrac{3}{2}$$

71. Solve $x^2 - 81 = 0$.

Solution

$$x^2 - 81 = 0$$

$$(x + 9)(x - 9) = 0$$

$$x + 9 = 0 \rightarrow x = -9$$

$$x - 9 = 0 \rightarrow x = 9$$

73. Solve $x^2 - 12x + 36 = 0$.

Solution

$$x^2 - 12x + 36 = 0$$

$$(x - 6)^2 = 0$$

$$x - 6 = 0 \rightarrow x = 6$$

75. Solve $4s^2 + s - 3 = 0$.

Solution

$$4s^2 + s - 3 = 0$$

$$(4s - 3)(s + 1) = 0$$

$$4s - 3 = 0 \rightarrow 4s = 3 \rightarrow s = \tfrac{3}{4}$$

$$s + 1 = 0 \rightarrow s = -1$$

77. Solve $x(7 - x) = 12$.

Solution

$$x(7 - x) = 12$$

$$7x - x^2 = 12$$

$$-x^2 + 7x - 12 = 0$$

$$x^2 - 7x + 12 = 0 \quad \text{(Multiply both sides of equation by } -1)$$

$$(x - 3)(x - 4) = 0$$

$$x - 3 = 0 \rightarrow x = 3$$

$$x - 4 = 0 \rightarrow x = 4$$

79. Solve $u^3 + 5u^2 - u = 5$.

Solution

$$u^3 + 5u^2 - u = 5$$

$$u^3 + 5u^2 - u - 5 = 0$$

$$u^2(u + 5) - 1(u + 5) = 0$$

$$(u + 5)(u^2 - 1) = 0$$

$$(u + 5)(u^2 - 1^2) = 0$$

$$(u + 5)(u + 1)(u - 1) = 0$$

$$u + 5 = 0 \rightarrow u = -5$$

$$u + 1 = 0 \rightarrow u = -1$$

$$u - 1 = 0 \rightarrow u = 1$$

81. *Revenue*   The revenue for selling x units is $R = 12x - 0.3x^2$.

    (a) Use a graphing utility to graph the revenue.

    (b) Use the graph to estimate the value of x that yields a revenue of $R = \$120$.

    (c) Use factorization to find the value of x that yield a revenue of $R = \$120$.

Solution

(a)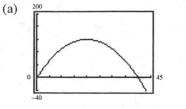

(b) From the graph, it appears that $R = \$120$ when $x = 20$ units.

(c)
$$R = 12x - 0.3x^2$$
$$120 = 12x - 0.3x^2$$
$$0.3x^2 - 12x + 120 = 0$$
$$\frac{0.3x^2 - 12x + 120}{0.3} = \frac{0}{0.3}$$
$$x^2 - 40x + 400 = 0$$
$$(x - 20)(x - 20) = 0$$
$$x - 20 = 0 \rightarrow x = 20$$

The revenue is $\$120$ when $x = 20$ units.

83. *Geometry*   The height of a rectangular window is one and one-half times the width. The area of the window is 2400 square inches. Find the dimensions of the window.

Solution

Common formula: $lw = A$

Labels: $A = 2400$ (square inches)
                    $w = $ width (inches)
                    $l = $ height $= \frac{3}{2}w$

Equation:
$$\left(\tfrac{3}{2}w\right)(w) = 2400$$
$$\tfrac{3}{2}w^2 = 2400$$
$$2\left(\tfrac{3}{2}w^2\right) = 2(2400)$$
$$3w^2 = 4800$$
$$3w^2 - 4800 = 0$$
$$3(w^2 - 1600) = 0$$
$$3(w + 40)(w - 40) = 0$$
$$3 \neq 0$$
$$w + 40 = 0 \rightarrow w = -40$$
$$w - 40 = 0 \rightarrow w - 40$$

The width of the window could not be a negative number, so we discard the negative answer. Thus, $w = 40$ and $\frac{3}{2}w = 60$. Thus, the window is 60 inches by 40 inches.

85. *Geometry* A box with a square base has a surface area of 400 square inches (see figure shown in the textbook). The height of the box is 5 inches. Find the dimensions of the box. (*Hint*: The surface area is given by $S = 2x^2 + 4xh$.)

Solution

$$S = x^2 + 4xh$$

$$300 = x^2 + 4x(5) \qquad \text{Set } S \text{ equal to 300 and } h \text{ equal to 5.}$$

$$300 = x^2 + 20x$$

$$0 = x^2 + 20x - 300$$

$$0 = (x + 30)(x - 10)$$

$$x + 30 = 0 \rightarrow x = -30$$

$$x - 10 = 0 \rightarrow x = 10$$

The negative answer is discarded. Thus $x = 10$, and the dimensions of the box are 10 inches by 10 inches by 5 inches.

Test for Chapter 6

1. Completely factor $7x^2 - 14x^3$.

Solution

$7x^2 - 14x^3 = 7x^2(1 - 2x)$

2. Completely factor $z(z + 7) - 3(z + 7)$.

Solution

$z(z + 7) - 3(z + 7) = (z + 7)(z - 3)$

3. Completely factor $t^2 - 4t - 5$.

Solution

$t^2 - 4t - 5 = (t - 5)(t + 1)$

4. Completely factor $6x^2 - 11x + 4$.

Solution

$6x^2 - 11x + 4 = (3x - 4)(2x - 1)$

5. Completely factor $6y^3 + 45y^2 + 75y$.

Solution

$$6y^3 + 45y^2 + 75y = 3y(2y^2 + 15y + 25)$$

$$= 3y(2y + 5)(y + 5)$$

6. Completely factor $4 - 25v^2$.

Solution

$$4 - 25v^2 = 2^2 - (5v)^2$$

$$= (2 + 5v)(2 - 5v)$$

7. Completely factor $4x^2 - 20x + 25$.

Solution

$$4x^2 - 20x + 25 = (2x)^2 - 2(2x)(5) + 5^2$$

$$= (2x - 5)^2$$

8. Completely factor $16 - (z + 9)^2$.

Solution

$$16 - (z + 9)^2 = 4^2 - (z + 9)^2$$

$$= [4 - (z + 9)][4 + (z + 9)]$$

$$= [4 - z - 9][4 + z + 9]$$

$$= (-z - 5)(z + 13)$$

$$\text{or } -(z + 5)(z + 13)$$

9. Completely factor $x^3 + 2x^2 - 9x - 18$.

Solution

$$x^3 + 2x^2 - 9x - 18 = (x^3 + 2x^2) - (9x + 18)$$
$$= x^2(x + 2) - 9(x + 2)$$
$$= (x + 2)(x^2 - 9)$$
$$= (x + 2)(x + 3)(x - 3)$$

10. Completely factor $16 - z^4$.

Solution

$$16 - z^4 = [4^2 - (z^2)^2]$$
$$= (4 + z^2)(4 - z^2)$$
$$= (4 + z^2)(2^2 - z^2)$$
$$= (4 + z^2)(2 + z)(2 - z)$$

or

$$16 - z^4 = -1(z^2 + 4)(z + 2)(z - 2)$$

11. Find the missing factor: $\frac{2}{3}x - \frac{3}{4} = \frac{1}{12}(\quad)$

Solution

$$\frac{2}{3}x - \frac{3}{4} = \frac{8}{12}x - \frac{9}{12} = \frac{1}{12}(8x - 9)$$

12. Find all values of b such that $x^2 + bx + 5$ can be factored.

Solution

$$b = 6 \qquad (x + 5)(x + 1)$$
$$b = -6 \qquad (x - 5)(x - 1)$$

13. Find a number c such that $x^2 + 12x + c$ is a perfect square trinomial.

Solution

$$x^2 + 12x + c = x^2 + 2(6x) + c$$
$$= x^2 + 2(x)(6) + c$$
$$(x + 6)^2 = x^2 + 2(x)(6) + 6^2$$
$$= x^2 + 12x + 36$$

Thus, if $c = 36$, $x^2 + 12x + c$ is a perfect square trinomial.

14. Explain why $(x + 1)(3x - 6)$ is not a complete factorization of $3x^2 - 3x - 6$.

Solution

This factorization is not complete because the second factor, $3x - 6$, has a common factor of 3. The complete factorization of the polynomial is $3(x + 1)(x - 2)$.

15. Solve $(x + 4)(2x - 3) = 0$.

Solution

$$(x + 4)(2x - 3) = 0$$
$$x + 4 = 0 \rightarrow x = -4$$
$$2x - 3 = 0 \rightarrow 2x = 3 \rightarrow x = \frac{3}{2}$$

16. Solve $7x^2 - 14x = 0$.

Solution

$$7x^2 - 14x = 0$$
$$7x(x - 2) = 0$$
$$7x = 0 \rightarrow x = 0$$
$$x - 2 = 0 \rightarrow x = 2$$

17. Solve $3x^2 + 7x - 6 = 0$.

Solution

$$3x^2 + 7x - 6 = 0$$
$$(3x - 2)(x + 3) = 0$$
$$3x - 2 = 0 \rightarrow 3x = 2 \rightarrow x = \frac{2}{3}$$
$$x + 3 = 0 \rightarrow x = -3$$

18. Solve $y(2y - 1) = 6$.

Solution

$$y(2y - 1) = 6$$
$$2y^2 - y - 6 = 0$$
$$2y + 3 = 0 \rightarrow 2y = -3 \rightarrow y = -\frac{3}{2}$$
$$y - 2 = 0 \rightarrow y = 2$$

19. The width of a rectangle is 5 inches less than the length. The area of the rectangle is 84 square inches. Find the dimensions of the rectangle.

Solution

Common formula: $lw = A$

Labels: $l = $ length (inches)
    $w = $ width $= l - 5$ (inches)
    $A = $ area $= 84$ (square inches)

Equation: $l(l - 5) = 84$

$$l^2 - 5l = 84$$

$$l^2 - 5l - 84 = 0$$

$$(l - 12)(l + 7) = 0$$

$$l - 12 = 0 \rightarrow l = 12$$

$$l + 7 = 0 \rightarrow l = -7$$

We discard the negative answer for the length of the rectangle. Thus, $l = 12$ and $l - 5 = 7$. The dimensions of the rectangle are 12 inches by 7 inches.

20. The height of an object dropped from a height of 40 feet is given by

$$\text{Height} = -16t^2 + 40$$

where the height is measured in feet and the time t is measured in seconds. How long will it take the object to fall to a height of 15 feet?

Solution

$$\text{Height} = -16t^2 + 40$$

$$15 = -16t^2 + 40 \qquad \text{(Set height equal to 15.)}$$

$$16t^2 - 25 = 0$$

$$(4t)^2 - (5)^2 = 0$$

$$(4t + 5)(4t - 5) = 0$$

$$4t + 5 = 0 \rightarrow 4t = -5 \rightarrow t = -\tfrac{5}{4}$$

$$4t - 5 = 0 \rightarrow 4t = 5 \rightarrow t = \tfrac{5}{4}$$

We discard the negative answer for time, and therefore, it will take $\tfrac{5}{4}$ of a second for the object to fall to a height of 15 feet.

Cumulative Test for Chapters 4–6

1. Describe how to identify the quadrants in which the points $(-2, y)$ must be located. (y is a real number.)

Solution

$(-2, y)$, y is a real number.

Quadrants II or III

This point is located 2 units to the *left* of the vertical axis. If y is positive, the point would be *above* the horizontal axis and in the second quadrant. If y is negative, the point would be *below* the horizontal axis and in the third quadrant. (If y is zero, the point would be on the horizontal axis *between* the second and third quadrants.)

2. Determine whether the ordered pairs are solution points of the equation $9x - 4y + 36 = 0$.

    (a) $(-1, -1)$ (b) $(8, 27)$ (c) $(-4, 0)$ (d) $(3, -2)$

Solution

    (a) The ordered pair $(-1, -1)$ is *not* a solution because $9(-1) - 4(-1) + 36 = -9 + 4 + 36 \neq 0$.
    (b) The ordered pair $(8, 27)$ *is* a solution because $9(8) - 4(27) + 36 = 72 - 108 + 36 = 0$.
    (c) The ordered pair $(-4, 0)$ *is* a solution because $9(-4) - 4(0) + 36 = -36 - 0 + 36 = 0$.
    (d) The ordered pair $(3, -2)$ is *not* a solution because $9(3) - 4(-2) + 36 = 27 + 8 + 36 \neq 0$.

3. Sketch the graph of $y = 2 - |x|$ and determine any intercepts of the graph.

Solution

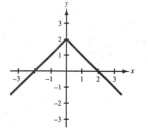

The x-intercepts are $(2, 0)$ and $(-2, 0)$, and the y-intercept is $(0, 2)$.

4. Sketch the graph of $y = \frac{1}{2}x - 2$ and determine any intercepts of the graph.

Solution

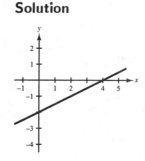

The x-intercepts is $(4, 0)$, and the y-intercept is $(0, -2)$.

5. Use a graphing utility to graph $y = x^2 - 2x - 3$. Use the TRACE features to approximate the x-intercept(s) of the graph. Check your estimates algebraically.

Solution

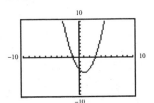

The x-intercepts appear to be $(3, 0)$ and $(-1, 0)$.

$$y = x^2 - 2x - 3$$
$$0 = x^2 - 2x - 3$$
$$0 = (x - 3)(x + 1)$$
$$x - 3 = 0 \rightarrow x = 3$$
$$x + 1 = 0 \rightarrow x = -1$$

This verifies that the x-intercepts are $(3, 0)$ and $(-1, 0)$.

6. Use a graphing utility to graph $y = x^3 - 9x$. Use the TRACE features to approximate the x-intercept(s) of the graph. Check your estimates algebraically.

Solution

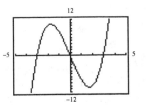

The x-intercepts appear to be $(-3, 0)$, $(0, 0)$, and $(3, 0)$.

$$y = x^3 - 9x$$
$$0 = x^3 - 9x$$
$$0 = x(x^2 - 9)$$
$$0 = x(x - 3)(x + 3)$$
$$x = 0$$
$$x - 3 = 0 \rightarrow x = 3$$
$$x + 3 = 0 \rightarrow x = -3$$

This verifies that the x-intercepts are $(-3, 0)$, $(0, 0)$, and $(3, 0)$.

7. Subtract: $(x^3 - 3x^2) - (x^3 + 2x^2 - 5)$

Solution
$$(x^3 - 3x^2) - (x^3 + 2x^2 - 5)$$
$$= x^3 - 3x^2 - x^3 - 2x^2 + 5$$
$$= (x^3 - x^3) + (-3x^2 - 2x^2) + 5$$
$$= -5x^2 + 5$$

8. Multiply: $(6z)(-7z)(z^2)$

Solution
$$(6z)(-7z)(z^2) = (6)(-7)z \cdot z \cdot z^2$$
$$= -42z^{1+1+2}$$
$$= -42z^4$$

9. Multiply: $(3x + 5)(x - 4)$

Solution
$$\qquad\qquad \text{F} \quad\; \text{O} \quad\; \text{I} \quad\; \text{L}$$
$$(3x + 5)(x - 4) = 3x^2 - 12x + 5x - 20 = 3x^2 - 7x - 20$$

10. Multiply: $(5x - 3)(5x + 3)$

Solution
$$(5x - 3)(5x + 3) = (5x)^2 - 3^2 \qquad \text{Pattern: } (a - b)(a + b) = a^2 - b^2$$
$$= 25x^2 - 9$$

11. Expand: $(5x + 6)^2$

Solution

$$(5x + 6)^2 = (5x)^2 + 2(5x)(6) + 6^2 \qquad \text{Pattern: } (a+b)^2 = a^2 + 2ab + b^2$$

$$= 25x^2 + 60x + 36$$

12. Divide: $(6x^2 + 72x) \div 6x$

Solution

$$(6x^2 + 72x) \div 6x = \frac{6x^2}{6x} + \frac{72x}{6x} = x + 12$$

13. Divide: $\dfrac{x^2 - 3x - 2}{x - 4}$

Solution

$$\begin{array}{r} x + 1 + \dfrac{2}{x-4} \\ x - 4 \overline{\smash{)}\, x^2 - 3x - 2} \\ \underline{x^2 - 4x} \\ x - 2 \\ \underline{x - 4} \\ 2 \end{array}$$

Thus, $\dfrac{x^2 - 3x - 2}{x - 4} = x + 1 + \dfrac{2}{x - 4}$.

14. Evaluate: $(3^2 \cdot 4^{-1})^2$

Solution

$$(3^2 \cdot 4^{-1})^2 = 3^{2 \cdot 2} \cdot 4^{-1 \cdot 2}$$

$$= 3^4 \cdot 4^{-2}$$

$$= 3^4 \cdot \frac{1}{4^2}$$

$$= \frac{3^4}{4^2}$$

$$= \frac{81}{16}$$

15. Factor: $2u^2 - 6u$

Solution

$$2u^2 - 6u = 2u(u - 3)$$

16. Factor and simplify: $(x - 2)^2 - 16$

Solution

$$(x - 2)^2 - 16 = (x - 2)^2 - 4^2$$

$$= [(x - 2) + 4][(x - 2) - 4]$$

$$= (x - 2 + 4)(x - 2 - 4)$$

$$= (x + 2)(x - 6)$$

17. Factor completely: $x^3 + 8x^2 + 16x$

Solution

$$x^3 + 8x^2 + 16x = x(x^2 + 8x + 16)$$

$$= x(x^2 + 2 \cdot x \cdot 4 + 4^2)$$

$$= x(x + 4)^2$$

18. Factor completely: $x^3 + 2x^2 - 4x - 8$

Solution

$$x^3 + 2x^2 - 4x - 8 = (x^3 + 2x^2) - (4x + 8)$$

$$= x^2(x + 2) - 4(x + 2)$$

$$= (x + 2)(x^2 - 4)$$

$$= (x + 2)(x^2 - 2^2)$$

$$= (x + 2)(x + 2)(x - 2)$$

$$= (x + 2)^2(x - 2)$$

19. Solve: $u(u - 12) = 0$

Solution

$$u(u - 12) = 0$$
$$u = 0$$
$$u - 12 = 0 \rightarrow u = 12$$

20. Solve: $5x^2 - 12x - 9 = 0$

Solution

$$5x^2 - 12x - 9 = 0$$
$$(5x + 3)(x - 3) = 0$$
$$5x + 3 = 0 \rightarrow 5x = -3 \rightarrow x = -\tfrac{3}{5}$$
$$x - 3 = 0 \rightarrow \ x = 3$$

The solutions are $-\tfrac{3}{5}$ and 3.

21. Rewrite the following expression using positive exponents.

$$\left(\frac{x}{2}\right)^{-2}$$

Solution

$$\left(\frac{x}{2}\right)^{-2} = \frac{x^{-2}}{2^{-2}} = \frac{2^2}{x^2} = \frac{4}{x^2}$$

22. An inheritance of \$12,000 is divided between two investments earning 7.5% and 9% simple interest. How much is in each investment if the total interest for one year is \$960?

Solution

Verbal model: $\boxed{\text{Interest on first amount}} + \boxed{\text{Interest on second amount}} = \boxed{\text{Total interest}}$

Labels: Both accounts: interest = \$960, principal = \$12,000
First account: interest rate = 0.075, principal = x (dollars)
Second account: interest rate = 0.09, principal = $12,000 - x$ (dollars)

Equation: $0.075(x) + 0.09(12,000 - x) = 960$

$$0.075x + 1080 - 0.09x = 960$$
$$-0.015x + 1080 = 960$$
$$-0.015x = -120$$
$$x = \frac{-120}{-0.015}$$
$$x = 8000$$
$$12,000 - x = 4000$$

Thus, \$8000 is invested at 7.5% and \$4000 is invested at 9%.

23. A sales representative is reimbursed \$125 per day for lodging and meals, plus \$0.35 per mile driven. Write a linear equation giving the daily cost C to the company in terms of x, the number of miles driven. Find the cost for a day when the representative drives 70 miles.

Solution

$$C = 125 + 0.35x$$
$$x = 70 \rightarrow C = 125 + 0.35(70)$$
$$C = 125 + 24.5$$
$$C = \$149.50$$

CHAPTER SEVEN
Functions and Their Graphs

7. Is the relation shown in the textbook a function?

Solution

Yes. No first component has two different second components, so this relation *is* a function.

11. Is the relation shown in the textbook a function?

Solution

No. Since first components, CBS and ABC, are paired with three second components, this relation is *not* a function.

15. Use the Vertical Line Test to determine whether y is a function of x in the graph in the textbook.

Solution

Yes. This graph indicates that y *is* a function of x because *no* vertical line would intersect the graph more than once.

19. Use the Vertical Line Test to determine whether y is a function of x in the graph in the textbook.

Solution

No. This graph indicates that y is *not* a function of x because it *is* possible for a vertical line to intersect the graph more than once.

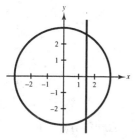

9. Is the relation shown in the textbook a function?

Solution

No. Since one of the first components, -2, is paired with two second components, this relation is *not* a function.

13. Is the relation shown in the textbook a function?

Solution

Yes. No first component has two different second components, so this relation *is* a function.

17. Use the Vertical Line Test to determine whether y is a function of x in the graph in the textbook.

Solution

Yes. This graph indicates that y *is* a function of x because *no* vertical line would intersect the graph more than once.

21. *Graphical Estimation* Use a graphing utility to graph the function $g(x) = 5 - 2x$. Graphically estimate the intercepts of the graph. Explain how to verify your estimates algebraically.

Solution

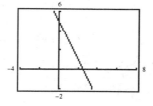

The intercepts are $\left(\frac{5}{2}, 0\right)$ and $(0, 5)$. To verify the x-intercept algebraically, set $g(x) = 0$, or $5 - 2x = 0$, and solve for x. To verify the y-intercept algebraically, set $x = 0$, or $g(x) = 5 - 2(0) = 5$.

23. Use a graphing utility to graph the function $h(x) = x(x - 4)$. Graphically estimate the intercepts of the graph. Explain how to verify your estimates algebraically.

Solution

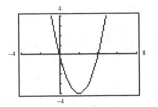

The intercepts are $(0, 0)$ and $(4, 0)$. To verify the x-intercepts algebraically, set $h(x) = 0$, or $x(x - 4) = 0$, and solve for x. To verify the y-intercept algebraically, set $x = 0$, or $h(x) = 0(0 - 4) = 0$.

27. Evaluate $g(u) = |u + 2|$ for (a) $g(2)$, (b) $g(-2)$, (c) $g(10)$, and (d) $g\left(-\frac{5}{2}\right)$.

Solution

(a) $g(2) = |2 + 2| = |4| = 4$

(b) $g(-2) = |-2 + 2| = |0| = 0$

(c) $g(10) = |10 + 2| = |12| = 12$

(d) $g\left(-\frac{5}{2}\right) = \left|-\frac{5}{2} + 2\right| = \left|-\frac{1}{2}\right| = \frac{1}{2}$

31. Determine the range R of $g(x) = 4 - x$ for the domain $D = \{0, 1, 2, 3, 4\}$. Sketch a graphical representation of the function.

Solution

$R = \{4 - 0,\ 4 - 1,\ 4 - 2,\ 4 - 3,\ 4 - 4\}$

$= \{4, 3, 2, 1, 0\}$

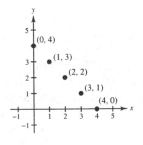

25. Evaluate $f(x) = 2x - 1$ for (a) $f(0)$, (b) $f(3)$, (c) $f(-3)$, and (d) $f\left(-\frac{1}{2}\right)$.

Solution

(a) $f(0) = 2(0) - 1 = -1$

(b) $f(3) = 2(3) - 1 = 5$

(c) $f(-3) = 2(-3) - 1 = -7$

(d) $f\left(-\frac{1}{2}\right) = 2\left(-\frac{1}{2}\right) - 1 = -1 - 1 = -2$

29. Evaluate $f(v) = \frac{1}{2}v^2$ for (a) $f(-4)$, (b) $f(4)$, (c) $f(0)$, and (d) $f(1) + f(2)$.

Solution

(a) $f(-4) = \frac{1}{2}(-4)^2 = \frac{1}{2}(16) = 8$

(b) $f(4) = \frac{1}{2}(4)^2 = \frac{1}{2}(16) = 8$

(c) $f(0) = \frac{1}{2}(0)^2 = \frac{1}{2}(0) = 0$

(d) $f(1) + f(2) = \frac{1}{2}(1)^2 + \frac{1}{2}(2)^2$

$\qquad = \frac{1}{2}(1) + \frac{1}{2}(4) = \frac{1}{2} + 2 = 2\frac{1}{2}$

33. Determine the range R of $g(s) = |s|$ for the domain $D = \{-2, -1, 0, 1, 2\}$. Sketch a graphical representation of the function.

Solution

$R = \{|-2|, |-1|, |0|, |1|, |2|\}$

$= \{2, 1, 0, 1, 2\}$

$= \{2, 1, 0\}$

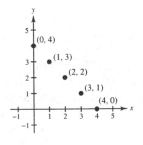

35. *Interpreting a Graph*   Use the information in the graph in the textbook. (*Source:* National Center for Education Statistics). Is the high school enrollment a function of the year?

Solution

Yes, the high school enrollment is a function of the year.

37. *Interpreting a Graph*   Use the information in the graph in the textbook. (*Source:* National Center for Education Statistics). Let $f(t)$ represent the number of high school students in year t. Find $f(1990)$.

Solution

$f(1990) \approx 15,000,000$

39. *Demand Function*   The demand for a product is a function of its price. Consider the demand function $f(p) = 20 - 0.5p$, where p is the price in dollars.

(a) Find $f(10)$ and $f(15)$.

(b) Describe the effect of a price increase on demand.

Solution

(a) $f(10) = 20 - 0.5(10)$ $f(15) = 20 - 0.5(15)$

$\qquad = 20 - 5$ $\qquad = 20 - 7.5$

$\qquad = 15$ $\qquad = 12.5$

(b) As the price increases, the demand decreases.

41. Solve $3x + 9 = 0$. Check your solution by substituting into the original equation.

Solution

$3x + 9 = 0$

$\quad 3x = -9$

$\quad\; x = \dfrac{-9}{3}$

$\quad\; x = -3$

43. Solve $m^2 - 25m = 0$. Check your solution by substituting into the original equation.

Solution

$m^2 - 25m = 0$

$m(m - 25) = 0$

$\qquad m = 0$

$m - 25 = 0 \rightarrow m = 25$

45. *Simple Interest*   An inheritance of $10,000 is divided into two investments earning 6.5% and 8% simple interest. How much is in each investment if the total interest for 1 year is $687.50?

Solution

Verbal model: 0.065 $\boxed{\begin{array}{c}\text{Amount invested}\\ \text{at 6.5\%}\end{array}}$ $+\, 0.08$ $\boxed{\begin{array}{c}\text{Amount invested}\\ \text{at 8\%}\end{array}}$ $=$ $\boxed{\begin{array}{c}\text{Total interest}\\ \text{for one year}\end{array}}$

Labels: Amount invested at 6.5% = x (dollars)
                 Amount invested at 8% = 10,000 − x (dollars)
                 Total interest for one year = 687.50 (dollars)

Equation: $0.065x + 0.08(10,000 - x) = 687.50$

$0.065x + 800 - 0.08x = 687.50$

$-0.015x + 800 = 687.50$

$-0.105x + 800 - 800 = 687.50 - 800$

$-0.015x = -112.50$

$x = \dfrac{-112.50}{-0.015}$

$x = 7500 \text{ and } 10,000 - x = 2500$

Therefore, $7500 is invested at 6.5% and $2500 is invested at 8%.

47. Is the following relation a function?

{(0, 25), (2, 25), (4, 30), (6, 30), (8, 30)}

Solution

Yes. No first component has two different second components, so the relation *is* a function.

49. Is the relation shown in the textbook a function?

Solution

No. One first component, 0, is paired with two second components, so the relation is *not* a function.

51. Is the relation shown in the textbook a function?

Solution

Yes. No first component has two different second components, so the relation *is* a function.

53. Is the relation shown in the textbook a function?

Solution

No. Two first components, 1 and 3, are paired with two second components, so the relation is *not* a function.

55. Use the Vertical Line Test to determine whether y is a function of x when $y = |x|$ (see figure in textbook).

Solution

Yes, y *is* a function of x. No vertical line would intersect the graph at more than one point.

57. Use the Vertical Line Test to determine whether y is a function of x when $x = y^2 + 1$ (see figure in textbook).

Solution

No, y is *not* a function of x. Some vertical lines could intersect the graph at more than one point.

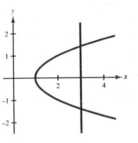

59. Use the Vertical Line Test to determine whether y is a function of x when $y = x^3$ (see figure in textbook).

Solution

Yes, y *is* a function of x. No vertical line would intersect the graph at more than one point.

61. *Graphical Estimation* Use a graphing utility to graph $h(x) = \frac{1}{2}(x + 1)$. Graphically estimate the intercepts of the graph. Explain how to verify your estimates algebraically.

Solution

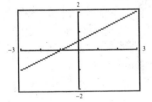

The intercepts are $(-1, 0)$ and $\left(0, \frac{1}{2}\right)$. To verify the x-intercept algebraically, set $h(x) = 0$ or $\frac{1}{2}(x + 1) = 0$, and solve for x. To verify the y-intercept algebraically, set $x = 0$, or $h(x) = \frac{1}{2}(0 + 1) = \frac{1}{2}$.

63. *Graphical Estimation*   Use a graphing utility to graph $g(x) = 2x^2 + 9x - 5$. Graphically estimate the intercepts of the graph. Explain how to verify your estimates algebraically.

Solution

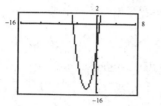

The intercepts are $(-5, 0)$, $\left(\frac{1}{2}, 0\right)$, and $(0, -5)$. To verify the x-intercepts algebraically, set $g(x) = 0$ or $2x^2 + 9x - 5 = 0$, and solve for x. To verify the y-intercept algebraically, set $x = 0$, or $g(x) = 2(0)^2 + 9(0) - 5 = -5$.

65. *Graphical Estimation*   Use a graphing utility to graph $f(x) = x(x - 3)^2$. Graphically estimate the intercepts of the graph. Explain how to verify your estimates algebraically.

Solution

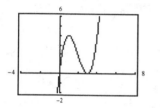

The intercepts are $(0, 0)$ and $(3, 0)$. To verify the x-intercepts algebraically, set $f(x) = 0$ or $x(x - 3)^2 = 0$, and solve for x. To verify the y-intercept algebraically, set $x = 0$, or $f(x) = 0(0 - 3)^2 = 0$.

67. *Graphical Estimation*   Use a graphing utility to graph $h(x) = |x| - 3$. Graphically estimate the intercepts of the graph. Explain how to verify your estimates algebraically.

Solution

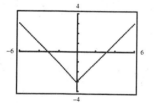

The intercepts are $(-3, 0)$, $(3, 0)$ and $(0, -3)$. To verify the x-intercepts algebraically, set $h(x) = 0$, or $|x| - 3 = 0$, or $|x| = 3$, and solve for x. To verify the y-intercept algebraically, set $x = 0$, or $h(x) = |0| - 3 = -3$.

69. Evaluate the function $f(x) = \frac{3}{4}x + 1$ for (a) $f(1)$, (b) $f(-1)$, (c) $f(-4)$, and (d) $f\left(-\frac{4}{3}\right)$.

Solution

(a) $f(1) = \frac{3}{4}(1) + 1 = 1\frac{3}{4}$

(b) $f(-1) = \frac{3}{4}(-1) + 1 = -\frac{3}{4} + 1 = \frac{1}{4}$

(c) $f(-4) = \frac{3}{4}(-4) + 1 = -3 + 1 = -2$

(d) $f\left(-\frac{4}{3}\right) = \frac{3}{4}\left(-\frac{4}{3}\right) + 1 = -1 + 1 = 0$

71. Evaluate the function $h(t) = 8$ for (a) $h(200)$, (b) $h(-10)$, (c) $h(6)$, and (d) $h\left(-\frac{5}{8}\right)$.

Solution

(a) $h(200) = 8$

(b) $h(-10) = 8$

(c) $h(6) = 8$

(d) $h\left(-\frac{5}{8}\right) = 8$

73. Evaluate the function $g(u) = u(u - 2)$ for (a) $g(0)$, (b) $g(2)$, (c) $g(6)$, and (d) $g(-5)$.

Solution

(a) $g(0) = 0(0 - 2) = 0(-2) = 0$

(b) $g(2) = 2(2 - 2) = 2(0) = 0$

(c) $g(6) = 6(6 - 2) = 6(4) = 24$

(d) $g(-5) = -5(-5 - 2) = -5(-7) = 35$

75. Evaluate the function $h(x) = x^3 - 1$ for (a) $h(0)$, (b) $h(1)$, (c) $h(3)$, and (d) $h\left(\frac{1}{2}\right)$.

Solution

(a) $h(0) = 0^3 - 1 = -1$

(b) $h(1) = 1^3 - 1 = 0$

(c) $h(3) = 3^3 - 1 = 26$

(d) $h\left(\frac{1}{2}\right) = \left(\frac{1}{2}\right)^3 - 1 = \frac{1}{8} - 1 = -\frac{7}{8}$

77. Determine the range R of the function $g(x) = x + 1$ for the domain $D = \{-2, -1, 0, 1, 2\}$.

Solution

$R = \{-2 + 1, -1 + 1, 0 + 1, 1 + 1, 2 + 1\} = \{-1, 0, 1, 2, 3\}$

79. Determine the range R of the function $g(s) = |s + 2| - |s|$ for the domain $D = \{-3, -1, 1, 3, 5\}$.

Solution

$R = \{|-3 + 2| - |-3|, |-1 + 2| - |-1|, |1 + 2| - |1|, |3 + 2| - |3|, |5 + 2| - |5|\}$

$= \{1 - 3, 1 - 1, 3 - 1, 5 - 3, 7 - 5\}$

$= \{-2, 0, 2\}$

81. *Distance*   The function $d(t) = 50t$ gives the distance (in miles) that a car will travel in t hours at an average speed of 50 miles per hour. Find the distance traveled for (a) $t = 2$, (b) $t = 4$, and (c) $t = 10$.

Solution

(a) $d(2) = 50(2) = 100$

When $t = 2$ hours, the car will travel 100 miles.

(b) $d(4) = 50(4) = 200$

When $t = 4$ hours, the car will travel 200 miles.

(c) $d(10) = 50(10) = 500$

When $t = 10$ hours, the car will travel 500 miles.

83. *Geometry*   Write the formula for the perimeter P of a square with side of length s.

Solution

$P = 4s$

85. *Terminology*   Do the statements use the word *function* in a way that is mathematically correct? Explain your reasoning.

(a) The amount of money in your savings account is a function of your salary.

(b) The speed at which a free-falling baseball strikes the ground is a function of the height from which it is dropped.

Solution

(a) No. This statement does not use the word "function" in a mathematically correct way. People earning the same salary can have different amounts of money in their savings accounts.

(b) Yes. This statement uses the word "function" in a mathematically correct way. Each height would determine only *one* speed.

7.2 | Slopes and Graphs of Linear Functions

7. Estimate the slope of the line from its graph.

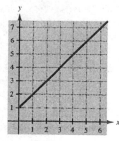

Solution

$m = \frac{1}{1} = 1$

9. Estimate the slope of the line from its graph.

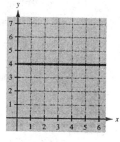

Solution

$m = 0$

11. Estimate the slope of the line from its graph.

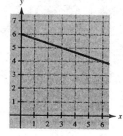

Solution

$m = \frac{-1}{3} = -\frac{1}{3}$

13. Plot the points $(0, 0)$ and $(4, 5)$ and find the slope (if possible) of the line passing through the pair of points. State whether the line rises, falls, is horizontal, or is vertical.

Solution

$m = \frac{5 - 0}{4 - 0} = \frac{5}{4}$

The line rises.

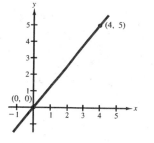

15. Plot the points $(-6, -1)$ and $(-6, 4)$ and find the slope (if possible) of the line passing through the pair of points. State whether the line rises, falls, is horizontal, or is vertical.

Solution

$m = \frac{4 - (-1)}{-6 - (-6)} = \frac{5}{0}$ (undefined)

m is undefined. The line is vertical.

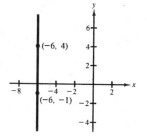

17. Plot the points $(3, -4)$ and $(8, -4)$ and find the slope (if possible) of the line passing through the pair of points. State whether the line rises, falls, is horizontal, or is vertical.

Solution

$$m = \frac{-4 - (-4)}{8 - 3} = \frac{0}{5} = 0$$

The line is horizontal.

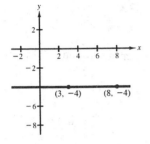

19. *Think About It* Find the value of y so the line through the points $(3, -2)$ and $(0, y)$ will have the slope $m = -8$. Explain how you found the value of y.

Solution

$$\frac{y - (-2)}{0 - 3} = -8$$

$$\frac{y + 2}{-3} = -8$$

$$y + 2 = (-3)(-8)$$

$$y + 2 = 24$$

$$y = 22$$

The value of y can be found by using the formula for slope.

21. Given the point $(2, 1)$ and slope $m = 0$, find two additional points on the line. (There are many correct answers.)

Solution

$(3, 1), (10, 1), (-2, 1)$, etc.

23. Given the point $(0, 1)$ and slope $m = -2$, find two additional points on the line. (There are many correct answers.)

Solution

$(1, -1), (2, -3), (3, -5)$, etc.

25. Given the point $(-4, 0)$ and slope $m = \frac{2}{3}$, find two additional points on the line. (There are many correct answers.)

Solution

$(-1, 2), (2, 4), (5, 6)$, etc.

27. Write $2x - y - 3 = 0$ in slope-intercept form. Use the slope and y-intercept to graph the line.

Solution

$$2x - y - 3 = 0$$

$$-y = -2x + 3$$

$$y = 2x - 3$$

$$m = 2$$

y-intercept: $(0, b) = (0, -3)$

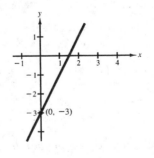

29. Write $3x - 4y + 2 = 0$ in slope-intercept form. Use the slope and y-intercept to graph the line.

Solution

$$3x - 4y + 2 = 0$$

$$-4y = -3x - 2$$

$$y = \tfrac{3}{4}x + \tfrac{1}{2}$$

$$m = \tfrac{3}{4}$$

y-intercept: $(0, b) = \left(0, \tfrac{1}{2}\right)$

31. Write $3x + 5y - 15 = 0$ in slope-intercept form. Use a graphing utility to graph the line. Determine any intercepts of the line.

Solution

$$3x + 5y - 15 = 0$$

$$5y = -3x + 15$$

$$y = \frac{-3x}{5} + \frac{15}{5}$$

$$y = -\frac{3}{5}x + 3$$

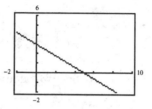

The intercepts are $(5, 0)$ and $(0, 3)$.

33. Write $y - 7 = 0$ in slope-intercept form. Use a graphing utility to graph the line. Determine any intercepts of the line.

Solution

$$y - 7 = 0$$

$$y = 7 \text{ or } y = 0x - 7$$

The y-intercept is $(0, 7)$.
There is no x-intercept.

35. Sketch the graphs of the two lines on the same rectangular coordinate system. Determine whether the lines are parallel, perpendicular, or neither.

Solution

L_1: $y = 2x - 3$
  $m_1 = 2$

L_2: $y = 2x + 1$
  $m_2 = 2$

Since the two slopes are equal, the two lines are *parallel*.

37. Sketch the graphs of the two lines on the same rectangular coordinate system. Determine whether the lines are parallel, perpendicular, or neither.

Solution

L_1: $y = 2x - 3$
$m_1 = 2$

L_2: $y = -\frac{1}{2}x + 1$
$m_2 = -\frac{1}{2}$

Since the two slopes are negative reciprocals of each other, the two lines are *perpendicular.*

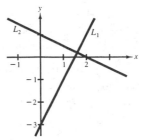

39. *Roof Pitch*   Find the slope (pitch) of the roof of the house in the figure in the textbook.

Solution

The vertical change, or rise, in the roof is $26 - 20$ feet. The horizontal change, or run, in the roof is half of 30 feet. The slope of the roof is the ratio of 6 feet to 15 feet.

$$m = \frac{26 - 20}{(1/2)(30)} = \frac{6}{15} = \frac{2}{5}$$

The slope, or pitch, of the roof is $\frac{2}{5}$.

41. *Graphical Interpretation*   The graph in the textbook gives the revenues (in billions of dollars) for Walt Disney Company from 1989 through 1993. (*Source:* Walt Disney Company)

(a) Find the slope between consecutive points for consecutive years.

(b) Find the slope of the line segment connecting the first and last points of the line graph. Explain the meaning of this slope.

Solution

(a) $m_1 \approx \dfrac{5800 - 4600}{1990 - 1989} = 1200$ million

$m_2 \approx \dfrac{6100 - 5800}{1991 - 1990} = 300$ million

$m_3 \approx \dfrac{7500 - 6100}{1992 - 1991} = 1400$ million

$m_4 \approx \dfrac{8500 - 7500}{1993 - 1992} = 1000$ million

(b) $m \approx \dfrac{8500 - 4600}{1993 - 1989} = \dfrac{3900}{4} = 975$ million

This represents the average increase in revenues of approximately \$975 million per year.

43. Find the product: $-12x(x - 3)$.

Solution

$-12x(x - 3) = -12x^2 + 36x$

45. Find the product: $(2x + 3)(x - 2)$.

Solution

$(2x + 3)(x - 2) = 2x^2 - 4x + 3x - 6$
$= 2x^2 - x - 6$

47. Factor completely: $x^3 + 3x^2 + 2x$.

Solution

$x^3 + 3x^2 + 2x = x(x^2 + 3x + 2) = x(x + 1)(x + 2)$

49. *Construction*   A builder must cut a 10-foot board into three pieces. Two have the same length and the third is three times as long as the two of equal length. Find the lengths of the three pieces.

Solution

Verbal model: $\boxed{\begin{array}{c}\text{Length of}\\\text{first board}\end{array}} + \boxed{\begin{array}{c}\text{Length of}\\\text{second board}\end{array}} + \boxed{\begin{array}{c}\text{Length of}\\\text{third board}\end{array}} = 10$

Labels: Length of first board $= x$ (feet)
               Length of second board $= x$ (feet)
               Length of third board $= 3x$ (feet)

Equation: $x + x + 3x = 10$

$$5x = 10$$

$$x = 2 \text{ and } 3x = 6$$

Therefore, the first two boards are 2 feet long and the third board is 6 feet long.

51. Match each line in the figure in the textbook with its slope.

(a) $m = \frac{3}{2}$ (b) $m = 0$ (c) $m = -2$

Solution

(a) L_2 has slope $m = \frac{3}{2}$.   (b) L_3 has slope $m = 0$.   (c) L_1 has slope $m = -2$.

53. Plot the points $(0, 6)$ and $(8, 0)$ and find the slope (if possible) of the line passing through them. State whether the line through the pair of points rises, falls, is horizontal, or is vertical.

Solution

$$m = \frac{0 - 6}{8 - 0} = \frac{-6}{8} = -\frac{3}{4}$$

The line falls.

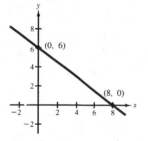

55. Plot the points $(-3, -2)$ and $(1, 6)$ and find the slope (if possible) of the line passing through them. State whether the line through the pair of points rises, falls, is horizontal, or is vertical.

Solution

$$m = \frac{6 - (-2)}{1 - (-3)} = \frac{8}{4} = 2$$

The line rises.

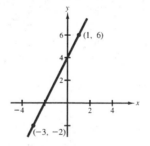

57. Plot the points $\left(\frac{1}{4}, \frac{3}{2}\right)$ and $\left(\frac{9}{2}, -3\right)$ and find the slope (if possible) of the line passing through them. State whether the line through the pair of points rises, falls, is horizontal, or is vertical.

Solution

$$m = \frac{-3 - \frac{3}{2}}{\frac{9}{2} - \frac{1}{4}} = \frac{-\frac{6}{2} - \frac{3}{2}}{\frac{18}{4} - \frac{1}{4}}$$

$$= \frac{-\frac{9}{2}}{\frac{17}{4}} = -\frac{9}{2} \cdot \frac{4}{17}$$

$$= -\frac{9 \cdot 4}{2 \cdot 17} = -\frac{18}{17}$$

The line falls.

59. Plot the points $(3.2, -1)$ and $(-3.2, 4)$ and find the slope (if possible) of the line passing through them. State whether the line through the pair of points rises, falls, is horizontal, or is vertical.

Solution

$$m = \frac{4 - (-1)}{-3.2 - 3.2}$$

$$= \frac{5}{-6.4}$$

$$= -\frac{50}{64}$$

$$= -\frac{25}{32}$$

The line falls.

61. Plot the points $(3.5, -1)$ and $(5.75, 4.25)$ and find the slope (if possible) of the line passing through them. State whether the line through the pair of points rises, falls, is horizontal, or is vertical.

Solution

$$m = \frac{4.25 - (-1)}{5.75 - 3.5}$$

$$= \frac{5.25}{2.25}$$

$$= \frac{525}{225} = \frac{7}{3}$$

The line rises.

63. Plot the points $(a, 3)$ and $(4, 3)$, $a \neq 4$ and find the slope (if possible) of the line passing through them. State whether the line through the pair of points rises, falls, is horizontal, or is vertical.

Solution

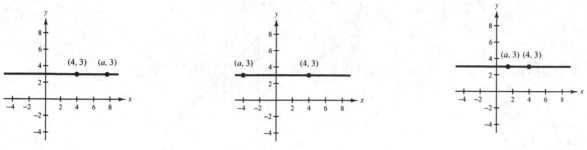

These graphs illustrate some possible locations for the point $(a, 3)$. Regardless of the value of a, the point $(a, 3)$ lies on the horizontal line through $(4, 3)$.

$$m = \frac{3 - 3}{a - 4} = \frac{0}{a - 4} = 0$$

The line is horizontal.

65. *Think About It* Find the value of y so the line through the points $(0, 10)$ and $(6, y)$ will have the slope $m = -\frac{1}{3}$. Explain how you found the value of y.

Solution

$$\frac{y - 10}{6 - 0} = -\frac{1}{3}$$

$$\frac{y - 10}{6} = \frac{-1}{3}$$

$$3(y - 10) = -6 \qquad \text{(Cross-multiply)}$$

$$3y - 30 = -6$$

$$3y = 24$$

$$y = 8$$

The value of y can be found by using the formula for slope.

67. Given the point $(1, -6)$ and slope $m = 2$, find two additional points on the line. (There are many correct answers.)

Solution

$(2, -4), (3, -2), (4, 0)$, etc.

69. Given the point $(3, 5)$ and slope $m = -\frac{1}{2}$, find two additional points on the line. (There are many correct answers.)

Solution

$(5, 4), (7, 3), (9, 2), (11, 1)$, etc.

71. Given the point $(-8, 1)$ and slope m is undefined, find two additional points on the line. (There are many correct answers.)

Solution

$(-8, 2), (-8, 10), (-8, -3)$, etc.

73. Sketch the graph of a line through the point $(0, 2)$ having the slope $m = 0$.

Solution

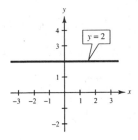

75. Sketch the graph of a line through the point $(0, 2)$ having the slope $m = -\frac{2}{3}$.

Solution

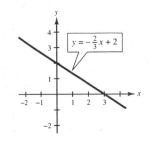

77. Plot the x- and y-intercepts and sketch the graph of $2x - 3y + 6 = 0$.

Solution

x-intercept	y-intercept	
Let $y = 0$.	Let $x = 0$.	
$2x - 3(0) + 6 = 0$	$2(0) - 3y + 6 = 0$	
$2x + 6 = 0$	$-3y + 6 = 0$	
$2x = -6$	$-3y = -6$	
$x = -3$	$y = 2$	
$(-3, 0)$	$(0, 2)$	

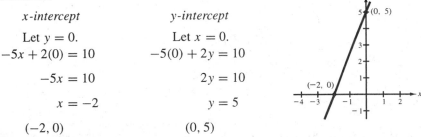

79. Plot the x- and y-intercepts and sketch the graph of $-5x + 2y = 10$.

Solution

x-intercept

Let $y = 0$.

$-5x + 2(0) = 10$

$-5x = 10$

$x = -2$

$(-2, 0)$

y-intercept

Let $x = 0$.

$-5(0) + 2y = 10$

$2y = 10$

$y = 5$

$(0, 5)$

81. Write $x + y = 0$ in slope-intercept form. Then use the slope and y-intercept to graph the line.

Solution

$x + y = 0$

$y = -x$

$m = -1$

y-intercept: $(0, b) = (0, 0)$

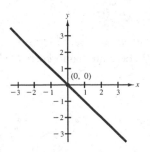

83. Write $x - 3y + 6 = 0$ in slope-intercept form. Then use the slope and y-intercept to graph the line.

Solution

$$x - 3y + 6 = 0$$

$$-3y = -x - 6$$

$$\frac{-3y}{-3} = \frac{-x}{-3} + \frac{-6}{-3}$$

$$y = \frac{1}{3}x + 2$$

$$m = \frac{1}{3}$$

y-intercept: $(0, b) = (0, 2)$

85. Write $3x - 2y - 2 = 0$ in slope-intercept form. Then use the slope and y-intercept to graph the line.

Solution

$$3x - 2y - 2 = 0$$

$$-2y = -3x + 2$$

$$y = \frac{3}{2}x - 1$$

$$m = \frac{3}{2}$$

y-intercept: $(0, b) = (0, -1)$

87. Write $2x + 3y = 0$ in slope-intercept form. Then use the slope and y-intercept to graph the line.

Solution

$$2x + 3y = 0$$

$$3y = -2x$$

$$y = -\frac{2}{3}x$$

$$m = -\frac{2}{3}$$

y-intercept: $(0, b) = (0, 0)$

89. Write $y + 6 = 0$ in slope-intercept form. Then use a graphing utility to graph the line. Determine any intercepts of the line.

Solution

$$y + 6 = 0$$

$$y = -6 \text{ or } y = 0x - 6$$

The y-intercept is $(0, -6)$. (There is no x-intercept.)

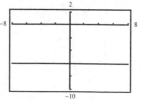

91. Write $1.6x + 0.7y = 5.6$ in slope-intercept form. Then use a graphing utility to graph the line. Determine any intercepts of the line.

Solution

$$1.6x + 0.7y = 5.6$$

$$0.7y = -1.6x + 5.6$$

$$y = \frac{-1.6}{0.7}x + \frac{5.6}{0.7}$$

$$y = -\frac{16}{7}x + 8$$

The intercepts are $\left(\frac{7}{2}, 0\right)$ and $(0, 8)$.

93. Determine whether the lines L_1 and L_2 passing through the given pairs of points are parallel, perpendicular, or neither.

L_1: $(0, -1)$, $(5, 9)$, L_2: $(0, 3)$, $(4, 1)$

Solution

L_1: $(0, -1)$, $(5, 9)$

$$m_1 = \frac{9 - (-1)}{5 - 0} = \frac{10}{5} = 2$$

L_2: $(0, 3)$, $(4, 1)$

$$m_2 = \frac{1 - 3}{4 - 0} = \frac{-2}{4} = -\frac{1}{2}$$

Since the two slopes are negative reciprocals of each other, L_1 and L_2 are *perpendicular*.

95. Determine whether the lines L_1 and L_2 passing through the given pairs of points are parallel, perpendicular, or neither.

L_1: $(3, 6)$, $(-6, 0)$, L_2: $(0, -1)$, $\left(5, \frac{7}{3}\right)$

Solution

L_1: $(3, 6)$, $(-6, 0)$

$$m_1 = \frac{0 - 6}{-6 - 3} = \frac{-6}{-9} = \frac{2}{3}$$

L_2: $(0, -1)$, $\left(5, \frac{7}{3}\right)$

$$m_2 = \frac{\frac{7}{3} - (-1)}{5 - 0} = \frac{\frac{7}{3} + \frac{3}{3}}{5} = \frac{\frac{10}{3}}{5} = \frac{10}{3} \cdot \frac{1}{5} = \frac{10}{15} = \frac{2}{3}$$

Since the two slopes are equal, L_1 and L_2 are *parallel*.

97. *Sketching a Diagram*   A subway track rises 3 feet over a 200-foot horizontal distance.

(a) Sketch a diagram of the track and label the rise and run.

(b) Find the slope of the track.

(c) Would the slope be steeper if the track rose 3 feet over a distance of 100 feet? Explain.

Solution

(a)

(b) $m = \frac{3}{200}$

(c) Yes. If the track rose 3 feet over a distance of 100 feet, the track would rise a total of 6 feet over the 200-foot distance. This is twice as much rise over that distance, and thus it would be a steeper slope.

99. *Comparing Two Models*   Based on different economic assumptions, the marketing department of a company develops two models to predict the annual profit of the company over the next 10 years. The models are $P_1 = 0.2t + 2.4$ and $P_2 = 0.3t + 2.4$ where P_1 and P_2 represent profit in millions of dollars and t is time in years ($0 \le t \le 10$).

(a) Interpret the slopes of the two linear models.

(b) Which model predicts a faster increase in profits?

(c) Use each model to predict profits when $t = 10$.

(d) Use a graphing utility to graph the models on the same viewing rectangle. Use the following range settings.

```
RANGE
Xmin=0
Xmax=10
Xscl=1
Ymin=0
Ymax=7
Yscl=1
```

Solution

(a) The slopes of the linear models represent the estimated yearly increase in profits.

(b) Model P_2 represents a faster increase in profits.

(c) $P_1 = 0.2(10) + 2.4$ $P_2 = 0.3(10) + 2.4$

    $= 2 + 2.4$ $= 3 + 2.4$

    $= 4.4$ million dollars $= 5.4$ million dollars

(d)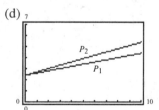

Mid-Chapter Quiz for Chapter 7

1. Does the table in the textbook represent y as a function of x? Explain.

Solution

Yes. No first component has two different second components, so this relation *is* a function.

2. Does the table in the textbook represent y as a function of x? Explain.

Solution

No. Since two of the first components, 0 and 1, are paired with two second components, this relation is *not* a function.

3. Does the graph in the textbook represent y as a function of x? Explain.

   Solution

   No. Any vertical line to the right of the origin would intersect the graph at more than one point, so the graph does *not* represent y as a function of x.

4. Does the graph table in the textbook represent y as a function of x? Explain.

   Solution

   Yes. No vertical line would cross the graph more than once, so the graph *does* represent y as a function of x.

5. Evaluate $f(x) = 3x - 2$ for the values (a) $f(-2)$, (b) $f(0)$, (c) $f(5)$, and (d) $f\left(-\frac{1}{3}\right)$.

   Solution

   (a) $f(-2) = 3(-2) - 2 = -6 - 2 = -8$

   (c) $f(5) = 3(5) - 2 = 15 - 2 = 13$

   (b) $f(0) = 3(0) - 2 = 0 - 2 = -2$

   (d) $f\left(-\frac{1}{3}\right) = 3\left(-\frac{1}{3}\right) - 2 = -1 - 2 = -3$

6. Evaluate $g(t) = 2t^2 - |t|$ for the values (a) $g(-2)$, (b) $g(2)$, (c) $g(0)$, and (d) $g\left(-\frac{1}{2}\right)$.

   Solution

   (a) $g(-2) = 2(-2)^2 - |-2|$
   $= 2(4) - 2$
   $= 8 - 2$
   $= 6$

   (b) $g(2) = 2(2)^2 - |2|$
   $= 2(4) - 2$
   $= 8 - 2$
   $= 6$

   (c) $g(0) = 2(0)^2 - |0|$
   $= 2(0) - 0$
   $= 0 - 0$
   $= 0$

   (d) $g\left(\frac{1}{2}\right) = 2\left(\frac{1}{2}\right)^2 - \left|\frac{1}{2}\right|$
   $= 2\left(\frac{1}{4}\right) - \frac{1}{2}$
   $= \frac{1}{2} - \frac{1}{2}$
   $= 0$

7. Find the range of $f(x) = x^2 - x$ for the domain $D = \{-2, -1, 0, 1, 2\}$.

   Solution

   $R = \{(-2)^2 - (-2),\ (-1)^2 - (-1),\ 0^2 - 0,\ 1^2 - 1,\ 2^2 - 2\}$

   $= \{4 + 2,\ 1 + 1,\ 0 - 0,\ 1 - 1,\ 4 - 2\}$

   $= \{6, 2, 0\}$

8. Use a graphing utility to graph $h(x) = 3x^2 - 4x - 7$. Graphically estimate the intercepts of the graph. Explain how to verify your estimates algebraically.

   Solution

   The intercepts are $(-1, 0)$, $\left(\frac{7}{3}, 0\right)$, and $(0, -7)$. To verify the x-intercepts algebraically, set $h(x) = 0$, or $3x^2 - 4x - 7 = 0$, and solve for x. To verify the y-intercept algebraically, set $x = 0$, or $h(x) = 3(0)^2 - 4(0) - 7 = -7$.

9. Find the slope of the line through $(-3, 7)$ and $(5, 0)$.

Solution

$$m = \frac{0 - 7}{5 - (-3)}$$

$$= \frac{-7}{5 + 3}$$

$$= -\frac{7}{8}$$

10. Find the value of y so that the line through the points $(3, -2)$ and $(6, y)$ has a slope of $m = 2$.

Solution

$$\frac{y - (-2)}{6 - 3} = 2$$

$$\frac{y + 2}{3} = 2$$

$$3\left(\frac{y + 2}{3}\right) = 3(2)$$

$$y + 2 = 6$$

$$y = 4$$

11. A line passes through the point $(1, 5)$ and its slope is $m = -\frac{2}{3}$. Find two additional points on the line. (There are many correct answers.)

Solution

$(4, 3)$, $(7, 1)$, $(10, -1)$, $(13, -3)$, etc.

12. Sketch the line passing through the point $(-2, -1)$ with slope $m = \frac{1}{2}$.

Solution

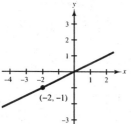

13. Write $2x + y - 4 = 0$ in slope-intercept form. Then use the slope and y-intercept to graph the line.

Solution

$2x + y - 4 = 0$

$y = -2x + 4$

$m = -2$

y-intercept: $(0, 4)$

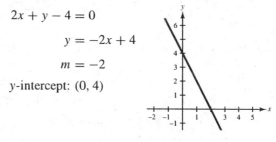

14. Write $3x - 2y + 6 = 0$ in slope-intercept form. Then use the slope and y-intercept to graph the line.

Solution

$3x - 2y + 6 = 0$

$-2y = -3x - 6$

$\dfrac{-2y}{-2} = \dfrac{-3x}{-2} + \dfrac{-6}{-2}$

$y = \dfrac{3}{2}x + 3$

$m = \dfrac{3}{2}$

y-intercept: $(0, 3)$

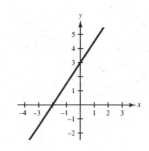

15. Write $3x - 4y = 0$ in slope-intercept form. Then use the slope and y-intercept to graph the line.

Solution

$$3x - 4y = 0$$

$$-4y = -3x$$

$$\frac{-4y}{-4} = \frac{-3x}{-4}$$

$$y = \frac{3}{4}x$$

$$m = \frac{3}{4}$$

y-intercept: $(0, 0)$

16. Write $y - 5 = 0$ in slope-intercept form. Then use the slope and y-intercept to graph the line.

Solution

$$y - 5 = 0$$

$$y = 5 \text{ or } y = 0x + 5$$

$$m = 0$$

y-intercept: $(0, 5)$

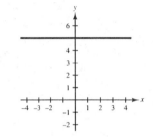

17. Determine whether the lines L_1: $(-2, 0)$, $(4, 8)$ and L_2: $(1, 5)$, $(5, 2)$ passing through the given points are parallel, perpendicular, or neither.

Solution

$$m_1 = \frac{8 - 0}{4 - (-2)} \qquad m_2 = \frac{2 - 5}{5 - 1}$$

$$= \frac{8}{6} \qquad\qquad = \frac{-3}{4}$$

$$= \frac{4}{3} \qquad\qquad = -\frac{3}{4}$$

The lines are perpendicular; their slopes are negative reciprocals of each other.

18. Determine whether the lines L_1: $(-1, 3)$, $(5, 5)$ and L_2: $(4, -1)$, $(7, 0)$ passing through the given points are parallel, perpendicular, or neither.

Solution

$$m_1 = \frac{5 - 3}{5 - (-1)} \qquad m_2 = \frac{0 - (-1)}{7 - 4}$$

$$= \frac{2}{6} \qquad\qquad = \frac{1}{3}$$

$$= \frac{1}{3}$$

The lines are parallel; their slopes are equal.

19. Write the equation of the line $1.3x - 0.8y = 4.2$ in slope-intercept form. Use a graphing utility to graph the line and approximate its intercepts.

Solution

$$1.3x - 0.8y = 4.2$$

$$-0.8y = -1.3x + 4.2$$

$$\frac{-0.8y}{-0.8} = \frac{-1.3x}{-0.8} + \frac{4.2}{-0.8}$$

$$y = \frac{13}{8}x - \frac{42}{8}$$

$$y = \frac{13}{8}x - \frac{21}{4}$$

The intercepts are approximately $(3.23, 0)$ and $(0, -5.25)$.

20. The sales y of a product are modeled by $y = 230x + 5000$, where x is time in years. Interpret the meaning of the slope in this model.

Solution

The slope of 230 represents the increase in sales per year.

7.3 Equations of Lines

7. Match $y = \frac{1}{2}x + 1$ with its line. (See graphs in textbook.)

Solution

Graph (c)

9. Match $y = x + 2$ with its line. (See graphs in textbook.)

Solution

Graph (b)

11. Find the slope of $y - 2 = 5(x + 3)$. If it is not possible, explain why.

Solution

$$y - 2 = 5(x + 3)$$

$$m = 5$$

The equation is in point-slope form, $y - y_1 = m(x - x_1)$.

13. Find the slope of $3x - 2y + 10 = 0$. If it is not possible, explain why.

Solution

$$3x - 2y + 10 = 0$$

$$-2y = -3x - 10$$

$$y = \frac{-3x}{-2} - \frac{10}{-2} = \frac{3}{2}x + 5$$

$$m = \frac{3}{2}$$

The equation was rewritten in slope-intercept form, $y = mx + b$.

15. Write the slope-intercept form of the line that passes through the point $(0, 0)$ and has slope $m = -2$. Sketch the line.

Solution

$$y = mx + b$$

$$y = -2x + 0$$

$$y = -2x$$

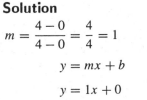

17. Write the slope-intercept form of the line that passes through the point $(0, -2)$ and has slope $m = 3$. Sketch the line.

Solution

$$y = mx + b$$

$$y = 3x - 2$$

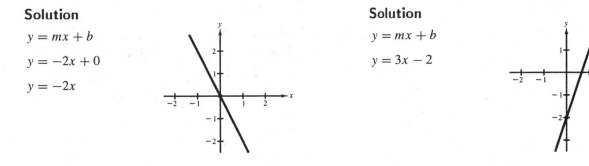

19. Write an equation of the line through the points $(0, 0)$ and $(4, 4)$ and sketch a graph of the line. (Write your answer in general form.)

Solution

$$m = \frac{4 - 0}{4 - 0} = \frac{4}{4} = 1$$

$$y = mx + b$$

$$y = 1x + 0$$

$$y = x$$

$$-x + y = 0 \text{ or } x - y = 0$$

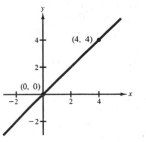

21. Write an equation of the line through the points $(0, 3)$ and $(5, 3)$ and sketch a graph of the line. (Write your answer in general form.)

Solution

$$m = \frac{3 - 3}{5 - 0} = \frac{0}{5} = 0$$

The slope $m = 0$ indicates that the line is a horizontal line with an equation in $y = b$ form. Thus, the equation is $y = 3$.

$$y = 3$$

$$y - 3 = 0 \qquad \text{(General form)}$$

23. Write an equation of the line passing through the point $(-2, 1)$ with slope $m = \frac{2}{3}$. (Write your answer in general form.)

Solution

$$y - y_1 = m(x - x_1)$$

$$y - 1 = \tfrac{2}{3}[x - (-2)]$$

$$y - 1 = \tfrac{2}{3}(x + 2)$$

$$y - 1 = \tfrac{2}{3}x + \tfrac{4}{3}$$

$$3(y - 1) = 3\left(\tfrac{2}{3}x + \tfrac{4}{3}\right)$$

$$3y - 3 = 2x + 4$$

$$-2x + 3y - 7 = 0$$

$$2x - 3y + 7 = 0$$

25. Write an equation of the line passing through the point $(-10, 4)$ with slope $m = 0$. (Write your answer in general form.)

Solution

The slope $m = 0$ indicates a horizontal line with an equation in the form $y = b$. Thus, the equation is $y = 4$.

$$y = 4$$

$$y - 4 = 0 \qquad \text{(General form)}$$

27. Write an equation of the line passing through the points $(5, -1)$ and $(-5, 5)$. (Write your answer in general form.)

Solution

$$m = \frac{5 + 1}{-5 - 5} = \frac{6}{-10} = -\frac{3}{5}$$

$$y - y_1 = m(x - x_1)$$

$$y + 1 = -\frac{3}{5}(x - 5)$$

$$y + 1 = -\frac{3}{5}x + 3$$

$$5(y + 1) = 5\left(-\frac{3}{5}x + 3\right)$$

$$5y + 5 = -3x + 15$$

$$3x + 5y - 10 = 0$$

29. Write an equation of the line passing through the points $(5, 4)$ and $(-2, -4)$. (Write your answer in general form.)

Solution

$$m = \frac{-4 - 4}{-2 - 5} = \frac{-8}{-7} = \frac{8}{7}$$

$$y - y_1 = m(x - x_1)$$

$$y - 4 = \frac{8}{7}(x - 5)$$

$$y - 4 = \frac{8}{7}x - \frac{40}{7}$$

$$7(y - 4) = 7\left(\frac{8}{7}x - \frac{40}{7}\right)$$

$$7y - 28 = 8x - 40$$

$$-8x + 7y + 12 = 0 \text{ or } 8x - 7y - 12 = 0$$

31. Write an equation of the line through the point $(2, 1)$ (a) parallel to and (b) perpendicular to the line $4x - 2y = 3$.

Solution

$$4x - 2y = 3$$

$$-2y = -4x + 3$$

$$y = \frac{-4x}{-2} + \frac{3}{-2}$$

$$y = 2x - \frac{3}{2}$$

$$m = 2$$

(a) Parallel line: $m = 2$, $(2, 1)$

$$y - y_1 = m(x - x_1)$$

$$y - 1 = 2(x - 2)$$

$$y - 1 = 2x - 4$$

$$-2x + y + 3 = 0 \text{ or}$$

$$2x - y - 3 = 0$$

(b) Perpendicular line: $m = -\frac{1}{2}$, $(2, 1)$

$$y - y_1 = m(x - x_1)$$

$$y - 1 = -\frac{1}{2}(x - 2)$$

$$y - 1 = -\frac{1}{2}x + 1$$

$$2(y - 1) = 2\left(-\frac{1}{2}x + 1\right)$$

$$2y - 2 = -x + 2$$

$$x + 2y - 4 = 0$$

33. Write an equation of the line through the point $(-1, 0)$ (a) parallel to and (b) perpendicular to the line $y + 3 = 0$.

Solution

$y + 3 = 0$ (a) Parallel line: $m = 0$, $(-1, 0)$ (b) Perpendicular line: m undefined, $(-1, 0)$

$\quad y = -3$ $y = 0$ (Horizontal line) $x = -1$ (Vertical line)

$\quad$ or $x + 1 = 0$ (General form)

$\quad y = 0x - 3$

$\quad m = 0$ (Horizontal line)

35. Use a graphing utility to graph the lines $y = -0.4x + 3$ and $y = \frac{5}{2}x - 1$. Use a square setting. Are the lines parallel, perpendicular, or neither?

Solution

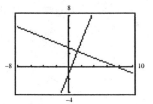

The lines are perpendicular. (Their slopes are negative reciprocals of each other.)

37. Use a graphing utility to graph the lines $y = 0.4x + 1$ and $y = x + 2.5$. Use a square setting. Are the lines parallel, perpendicular, or neither?

Solution

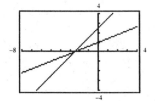

The lines are neither parallel nor perpendicular.

39. *Writing a Linear Model*   A car travels for t hours at an average speed of 50 miles per hour. Write the distance d as a linear function of t. Graph the function for $0 \le t \le 5$.

Solution

$d = 50t$

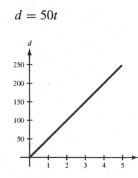

41. *Writing and Using a Linear Model*   A real estate office operates an apartment complex with 50 units. The relationship between the rent p and the demand x is linear. When the rent is $480 per unit, all 50 units are occupied. When the rent is $525 per unit, only 47 units are occupied.

    (a) Represent the given information as two ordered pairs of the form (x, p). Plot these ordered pairs.

    (b) Write an equation of the line that passes through the ordered pairs. Graph the line and describe the relationship between the rent and the demand.

    (c) *Linear Extrapolation*   Predict the number of units occupied when the rent is $555.

    (d) *Linear Interpolation*   Predict the number of units occupied when the rent is $495.

Solution

(a) When $x = 50$, $p = 480$. When $x = 47$, $p = 525$.
The two points are $(50, 480)$ and $(47, 525)$.

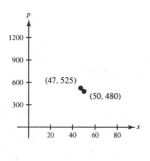

(b)
$$m = \frac{525 - 480}{47 - 50} = \frac{45}{-3} = -15$$

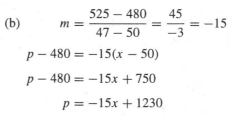

$$p - 480 = -15(x - 50)$$
$$p - 480 = -15x + 750$$
$$p = -15x + 1230$$

As the rent increases, the demand decreases.

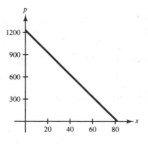

(c) $555 = -15x + 1230$

$\quad -675 = -15x$

$\quad \dfrac{-675}{-15} = x$

$\qquad 45 = x$

Thus, 45 units would be rented if the rent were $555.

(d) $495 = -15x + 1230$

$\quad -735 = -15x$

$\quad \dfrac{-735}{-15} = x$

$\qquad 49 = x$

Thus, 49 units would be rented if the rent were $495.

43. Completely factor $3x^2 + 7x$.

Solution

$3x^2 + 7x = x(3x + 7)$

45. Completely factor $x^2 + 7x - 18$.

Solution

$x^2 + 7x - 18 = (x + 9)(x - 2)$

47. *Exploration* Consider the equation $3x^3 - 4x^2 - 12x + 16 = (3x - 4)(x + 2)(x - 2)$.

   (a) Use a graphing utility to graphically verify the equation by graphing the left and right sides on the same screen.

   (b) Verify the equation algebraically by multiplying the factors on the right side of the equation.

Solution

(a)
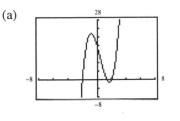

(b) $(3x - 4)(x + 2)(x - 2) = (3x - 4)(x^2 - 4)$

$$= 3x^3 - 12x - 4x^2 + 16$$

$$= 3x^3 - 4x^2 - 12x + 16$$

49. Find the slope of the line $y - 10 = 3(x + 7)$. If it is not possible, explain why.

Solution

$y - 10 = 3(x + 7)$

$m = 3$

The equation is in point-slope form, $y - y_1 = m(x - x_1)$.

51. Find the slope of the line $y + \frac{5}{6} = \frac{2}{3}(x + 4)$. If it is not possible, explain why.

Solution

$y + \frac{5}{6} = \frac{2}{3}(x + 4)$

$m = \frac{2}{3}$

The equation is in point-slope form, $y - y_1 = m(x - x_1)$.

53. Find the slope of the line $y = \frac{3}{8}x - 4$. If it is not possible, explain why.

Solution

$y = \frac{3}{8}x - 4$

$m = \frac{3}{8}$

The equation is in slope-intercept form, $y = mx + b$.

55. Find the slope of the line $2x - y = 0$. If it is not possible, explain why.

Solution

$2x - y = 0$

$-y = -2x$

$y = 2x$

$m = 2$

The equation was rewritten in slope-intercept form, $y = mx + b$.

57. Write an equation of the line that passes through the point $\left(0, \frac{3}{2}\right)$ and has the slope $m = 2$. (Write your answer in slope-intercept form.)

Solution

$y = mx + b$

$m = 2$ and $(0, b) = \left(0, \frac{3}{2}\right)$

$y = 2x + \frac{3}{2}$

59. Write an equation of the line that passes through the point $(0, 0)$ and has the slope $m = -0.8$. (Write your answer in slope-intercept form.)

Solution

$y = mx + b$

$m = -0.8$ and $(0, b) = (0, 0)$

$y = -0.8x + 0$

$y = -0.8x$

61. Write an equation of the line through the points $(0, 0)$ and $(2, -4)$ and sketch a graph of the line. (Write your answer in general form.)

Solution

$$m = \frac{-4 - 0}{2 - 0} = \frac{-4}{2} = -2$$

$$y = mx + b$$

$$y = -2x + 0$$

$$y = -2x$$

$$2x + y = 0$$

63. Write an equation of the line through the points $(2, 3)$ and $(6, 0)$ and sketch a graph of the line. (Write your answer in general form.)

Solution

$$m = \frac{0 - 3}{6 - 2} = \frac{-3}{4} = -\frac{3}{4}$$

$$y - y_1 = m(x - x_1)$$

$$y - 3 = -\frac{3}{4}(x - 2)$$

$$y - 3 = -\frac{3}{4}x + \frac{3}{2}$$

$$4(y - 3) = 4\left(-\frac{3}{4}x + \frac{3}{2}\right)$$

$$4y - 12 = -3x + 6$$

$$3x + 4y - 18 = 0$$

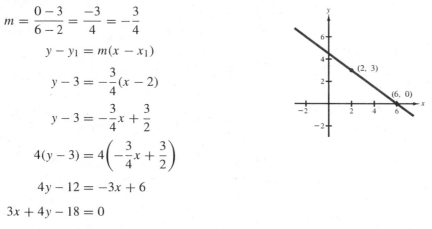

65. Write an equation of the line through the points $(-6, 2)$ and $(3, 5)$ and sketch a graph of the line. (Write your answer in general form.)

Solution

$$m = \frac{5 - 2}{3 + 6} = \frac{3}{9} = \frac{1}{3}$$

$$y - y_1 = m(x - x_1)$$

$$y - 2 = \frac{1}{3}(x + 6)$$

$$y - 2 = \frac{1}{3}x + 2$$

$$3(y - 2) = 3\left(\frac{1}{3}x + 2\right)$$

$$3y - 6 = x + 6$$

$$-x + 3y - 12 = 0 \text{ or } x - 3y + 12 = 0$$

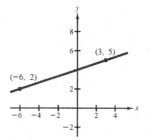

67. Write an equation of the line through the points $(-3, 4)$ and $(1, 4)$ and sketch a graph of the line. (Write your answer in general form.)

Solution

$$m = \frac{4 - 4}{1 - (-3)} = \frac{0}{4} = 0$$

Because $m = 0$, the line is horizontal. A horizontal line has an equation of the form $y = b$. Since the line passes through points $(-3, 4)$ and $(1, 4)$, the equation could be written as $y = 4$, or $y - 4 = 0$.

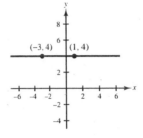

69. Write an equation of the line passing through the point $(0, -4)$ with slope $m = \frac{1}{2}$. (Write your answer in general form.)

Solution

$$y = mx + b$$

$$y = \tfrac{1}{2}x + (-4)$$

$$y = \tfrac{1}{2}x - 4$$

$$2y = 2\left(\tfrac{1}{2}x - 4\right)$$

$$2y = x - 8$$

$$-x + 2y + 8 = 0 \text{ or } x - 2y - 8 = 0$$

71. Write an equation of the line passing through the point $(-3, 6)$ with slope $m = -2$. (Write your answer in general form.)

Solution

$$y - y_1 = m(x - x_1)$$

$$y - 6 = -2(x + 3)$$

$$y - 6 = -2x - 6$$

$$2x + y = 0$$

73. Write an equation of the line passing through the point $(4, 0)$ with slope $m = -\frac{1}{3}$. (Write your answer in general form.)

Solution

$$y - y_1 = m(x - x_1)$$

$$y - 0 = -\tfrac{1}{3}(x - 4)$$

$$y = -\tfrac{1}{3}x + \tfrac{4}{3}$$

$$3y = 3\left(-\tfrac{1}{3}x + \tfrac{4}{3}\right)$$

$$3y = -x + 4$$

$$x + 3y - 4 = 0$$

75. Write an equation of the line passing through the point $(-2, -5)$ with slope $m = 0$. (Write your answer in general form.)

Solution

Because $m = 0$, the line is horizontal. A horizontal line has an equation of the form $y = b$. Since the line passes through the point $(-2, -5)$, the equation could be written as $y = -5$, or $y + 5 = 0$.

77. Write an equation of the line passing through the points $(2, 3)$ and $(3, 2)$. (Write your answer in general form.)

Solution

$$m = \frac{2-3}{3-2} = \frac{-1}{1} = -1$$

$$y - y_1 = m(x - x_1)$$

$$y - 3 = -1(x - 2)$$

$$y - 3 = -x + 2$$

$$x + y - 5 = 0$$

79. Write an equation of the line passing through the points $\left(2, \frac{1}{2}\right)$ and $\left(\frac{1}{2}, \frac{5}{4}\right)$. (Write your answer in general form.)

Solution

$$m = \frac{\frac{5}{4} - \frac{1}{2}}{\frac{1}{2} - 2} = \frac{\frac{5}{4} - \frac{2}{4}}{\frac{1}{2} - \frac{4}{2}} = \frac{\frac{3}{4}}{-\frac{3}{2}}$$

$$= \frac{3}{4} \cdot \left(-\frac{2}{3}\right) = -\frac{6}{12} = -\frac{1}{2}$$

$$y - y_1 = m(x - x_1)$$

$$y - \frac{1}{2} = -\frac{1}{2}(x - 2)$$

$$y - \frac{1}{2} = -\frac{1}{2}x + 1$$

$$2\left(y - \frac{1}{2}\right) = 2\left(-\frac{1}{2}x + 1\right)$$

$$2y - 1 = -x + 2$$

$$x + 2y - 3 = 0$$

81. Write an equation of the line passing through the points $(-8, 1)$ and $(-8, 7)$. (Write your answer in general form.)

Solution

$$m = \frac{7-1}{-8+8} = \frac{6}{0} \quad \text{(Undefined)}$$

The undefined slope indicates a vertical line with an equation in $x = a$ form. Thus, the equation is $x = -8$.

$$x = -8$$

$$x + 8 = 0 \quad \text{(General form)}$$

83. Write an equation of the line passing through the points $(1, 0.6)$ and $(2, -0.6)$. (Write your answer in general form.)

Solution

$$m = \frac{-0.6 - 0.6}{2 - 1} = \frac{-1.2}{1} = -1.2$$

$$y - y_1 = m(x - x_1)$$

$$y - 0.6 = -1.2(x - 1)$$

$$y - 0.6 = -1.2x + 1.2$$

$$1.2y + y - 1.8 = 0$$

Note: The equation could also be written as

$$12x + 10y - 18 = 0 \text{ or}$$

$$6x + 5y - 9 = 0.$$

85. Write an equation of the line through the point $(1, 3)$ (a) parallel to, and (b) perpendicular to the line $2x + y = 0$.

Solution

$$2x + y = 0$$

$$y = -2x$$

$$m = -2$$

(a) Parallel line: $m = -2$, $(1, 3)$

$$y - y_1 = m(x - x_1)$$

$$y - 3 = -2(x - 1)$$

$$y - 3 = -2x + 2$$

$$2x + y - 5 = 0$$

(b) Perpendicular line: $m = \frac{1}{2}$, $(1, 3)$

$$y - y_1 = m(x - x_1)$$

$$y - 3 = \frac{1}{2}(x - 1)$$

$$y - 3 = \frac{1}{2}x - \frac{1}{2}$$

$$2y - 6 = x - 1$$

$$0 = x - 2y + 5$$

87. Write an equation of the line through the point $(-6, 4)$ (a) parallel to, and (b) perpendicular to the line $3x + 4y = 7$.

Solution

$$3x + 4y = 7$$

$$4y = -3x + 7$$

$$y = -\frac{3}{4}x + \frac{7}{4}$$

$$m = -\frac{3}{4}$$

(a) Parallel line: $m = -\frac{3}{4}$, $(-6, 4)$

$$y - y_1 = m(x - x_1)$$

$$y - 4 = -\frac{3}{4}(x + 6)$$

$$y - 4 = -\frac{3}{4}x - \frac{9}{2}$$

$$4(y - 4) = 4\left(-\frac{3}{4}x - \frac{9}{2}\right)$$

$$4y - 16 = -3x - 18$$

$$3x + 4y + 2 = 0$$

(b) Perpendicular line: $m = \frac{4}{3}$, $(-6, 4)$

$$y - y_1 = m(x - x_1)$$

$$y - 4 = \frac{4}{3}(x + 6)$$

$$y - 4 = \frac{4}{3}x + 8$$

$$3(y - 4) = 3\left(\frac{4}{3}x + 8\right)$$

$$3y - 12 = 4x + 24$$

$$-4x + 3y - 36 = 0 \text{ or}$$

$$4x - 3y + 36 = 0$$

89. Use a graphing utility to graph the following equations on the same screen. (Set the ZOOM feature to the *square* setting.) What can you conclude?

(a) $y = \frac{1}{3}x + 2$ (b) $y = 4x + 2$ (c) $y = -3x + 2$ (d) $y = -\frac{1}{4}x + 2$

Solution

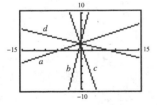

Graphs (a) and (c) are perpendicular. (Note that their slopes are negative reciprocals of each other.) Graphs (b) and (d) are perpendicular. (Note that their slopes are negative reciprocals of each other.)

91. *Research Project*   In a news magazine or newspaper, find an example of data that is *increasing* linearly with time. Write a linear equation that models the data.

Solution

Answers will vary.

93. *Writing a Linear Model*   A sales representative receives a salary of $2000 per month plus a commission of 2% of the total monthly sales. Write a linear equation giving the wages W in terms of sales S.

Solution

$W = 2000 + 0.02S$

95. *Writing and Using a Linear Model*   A store is offering a 20% discount on all items in its inventory.

(a) Write a linear equation giving the sale price S for an item in terms of its list price, L.

(b) Use a graphing utility to graph your model.

(c) Use the graph to estimate the sale price of an item whose list price is $49.98. Confirm your estimate algebraically.

Solution

(a) $S = L - 0.2L = 0.8L$ (b)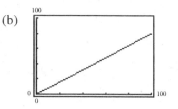

(c) The sale price is approximately $40. You can estimate it more accurately with your graphing calculator, and you can confirm it algebraically.

$S = 0.8(49.98) \approx 39.98$

97. *Writing and Using a Linear Model*   A small liberal arts college had an enrollment of 1200 students in 1990. During the next 10 years the enrollment increased by approximately 50 students per year.

(a) Write a linear equation giving the enrollment N in terms of the year t. (Let $t = 0$ represent 1990.)

(b) *Linear Extrapolation*   Use the model to predict the enrollment in the year 2001.

(c) *Linear Interpolation*   Use the model to estimate the enrollment in 1998.

Solution

(a) $N = 50t + 1200$

(b) The variable $t = 11$ for the year 2001.

$N = 50(11) + 1200$

$= 550 + 1200$

$= 1750$

(c) The variable $t = 8$ for the year 1998.

$N = 50(8) + 1200$

$= 400 + 1200$

$= 1600$

7.4   Graphs of Linear Inequalities

7. Determine which of the following points are solutions of the inequality $x + y > 5$.

(a) $(0, 0)$ (b) $(3, 6)$ (c) $(-6, 20)$ (d) $(3, 2)$

Solution

(a) $(0, 0)$

$$0 + 0 \overset{?}{>} 5$$

$$0 \not> 5$$

The point $(0, 0)$ is *not* a solution.

(c) $(-6, 20)$

$$-6 + 20 \overset{?}{>} 5$$

$$14 > 5$$

The point $(-6, 20)$ *is* a solution.

(b) $(3, 6)$

$$3 + 6 \overset{?}{>} 5$$

$$9 > 5$$

The point $(3, 6)$ *is* a solution.

(d) $(3, 2)$

$$3 + 2 \overset{?}{>} 5$$

$$5 \not> 5$$

The point $(3, 2)$ is *not* a solution.

9. Determine which of the following points are solutions of the inequality $-3x + 5y \leq 12$.

(a) $(1, 2)$ (b) $(2, -3)$ (c) $(1, 3)$ (d) $(2, 8)$

Solution

(a) $(1, 2)$

$$-3(1) + 5(2) \overset{?}{\leq} 12$$

$$-3 + 10 \overset{?}{\leq} 12$$

$$7 \leq 12$$

The point $(1, 2)$ *is* a solution.

(c) $(1, 3)$

$$-3(1) + 5(3) \overset{?}{\leq} 12$$

$$-3 + 15 \overset{?}{\leq} 12$$

$$12 \leq 12$$

The point $(1, 3)$ *is* a solution.

(b) $(2, -3)$

$$-3(2) + 5(-3) \overset{?}{\leq} 12$$

$$-6 - 15 \overset{?}{\leq} 12$$

$$-21 \leq 12$$

The point $(2, -3)$ *is* a solution.

(d) $(2, 8)$

$$-3(2) + 5(8) \overset{?}{\leq} 12$$

$$-6 + 40 \overset{?}{\leq} 12$$

$$34 \not\leq 12$$

The point $(2, 8)$ is *not* a solution.

11. State whether the boundary of the graph of the inequality $2x + 3y < 6$ should be dashed or solid.

Solution

The boundary of the graph of $2x + 3y < 6$ should be *dashed*.

13. State whether the boundary of the graph of the inequality $2x + 3y \geq 6$ should be dashed or solid.

Solution

The boundary of the graph of $2x + 3y \geq 6$ should be *solid*.

15. Match $x + y < 4$ with its graph. [The graphs are labeled (a), (b), (c), and (d) in the textbook.]

Solution

Graph (b)

17. Match $x > 1$ with its graph. [The graphs are labeled (a), (b), (c), and (d) in the textbook.]

Solution

Graph (d)

19. Graph the inequality $y \geq 3$.

Solution

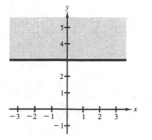

21. Graph the inequality $x - y < 0$.

Solution

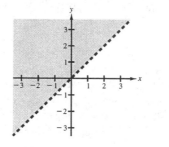

23. Graph the inequality $2x + y - 3 \geq 3$.

Solution

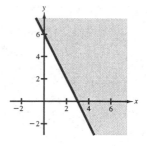

25. Graph the following inequality.

$$\frac{x}{3} + \frac{y}{4} < 1$$

Solution

$$\frac{x}{3} + \frac{y}{4} < 1$$

$$12\left(\frac{x}{3} + \frac{y}{4}\right) < 12(1)$$

$$4x + 3y < 12$$

This could be written in slope-intercept form as $y < -\frac{4}{3}x + 4$.

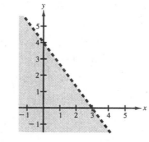

27. Use a graphing utility to graph $y \geq 2x - 1$.

Solution

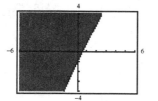

29. Use a graphing utility to graph $6x + 10y - 15 \leq 0$.

Solution

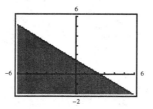

31. *Think About It*   Does $2x < 2y$ have the same graph as $y > x$? Explain your reasoning.

Solution

Yes, the two graphs are the same. If $2x < 2y$, then $x < y$, and this means the same thing as $y > x$.

33. *Graphical Reasoning*   Write an inequality that represents the graph shown in the textbook.

Solution

The inequality can be written in slope-intercept form. The slope is -2 and the y-intercept is 2.

$$y \leq -2x + 2$$

(This could also be written as $2x + y \leq 2$ or as $2x + y - 2 \leq 0$.)

35. *Graphical Reasoning* Write an inequality that represents the graph shown in the textbook.

Solution

The inequality can be written in slope-intercept form. The slope is 2 and the y-intercept is 0.

$y < 2x$

(This could also be written as $2x - y > 0$.)

37. *Creating a Model* Suppose you have two part-time jobs. One is at a grocery store, which pays $7 per hour, and the other is mowing lawns, which pays $5 per hour. Between the two jobs you want to earn at least $140 a week. Write an inequality that shows the different numbers of hours you can work at each job and sketch the graph of the inequality. From the graph, find several ordered pairs with positive integer coordinates that are solutions of the inequality.

Solution

$7x + 5y \geq 140$

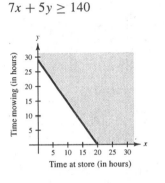

Representative solutions

$(20, 0)$

$(0, 28)$

$(10, 15)$

$(5, 25)$

$(15, 7)$

39. Use the properties of exponents to simplify $(x^2)^3 \cdot x^3$.

Solution

$(x^2)^3 \cdot x^3 = x^6 \cdot x^3 = x^9$

41. Use the properties of exponents to simplify $\dfrac{u^4 v^2}{(uv)^2}$.

Solution

$\dfrac{u^4 v^2}{(uv)^2} = \dfrac{u^4 v^2}{u^2 v^2} = u^2$

43. *Amount of Sales* The sales commission rate is 4.5%. Determine the sales of an employee who earned a commission of $544.50.

Solution

Verbal model: | Commission rate | $\times$ | Sales | $=$ | Commission |

Labels: Commission rate $= 0.045$ (percent in decimal form)
Sales $= x$ (dollars)
Commission $= 544.50$ (dollars)

Equation: $0.045x = 544.50$

$\dfrac{0.045x}{0.045} = \dfrac{544.50}{0.045}$

$x = 12{,}100$

The sales of the employee were $12,100.

45. Which of the following points are solutions of $3x - 2y < 2$?

(a) $(1, 3)$ (b) $(2, 0)$ (c) $(0, 0)$ (d) $(3, -5)$

Solution

(a) $3(1) - 2(3) \overset{?}{<} 2$

$3 - 6 \overset{?}{<} 2$

$-3 < 2$

The point $(1, 3)$ *is* a solution.

(b) $3(2) - 2(0) \overset{?}{<} 2$

$6 - 0 \overset{?}{<} 2$

$6 \not< 2$

The point $(2, 0)$ is *not* a solution.

(c) $3(0) - 2(0) \overset{?}{<} 2$

$0 - 0 \overset{?}{<} 2$

$0 < 2$

The point $(0, 0)$ *is* a solution.

(d) $3(3) - 2(-5) \overset{?}{<} 2$

$9 + 10 \overset{?}{<} 2$

$19 \not< 2$

The point $(3, -5)$ is *not* a solution.

47. Which of the following points are solutions of $5x + 4y \geq 6$?

(a) $(-2, 4)$ (b) $(5, 5)$ (c) $(7, 0)$ (d) $(-2, 5)$

Solution

(a) $5(-2) + 4(4) \overset{?}{\geq} 6$

$-10 + 16 \overset{?}{\geq} 6$

$6 \geq 6$

The point $(-2, 4)$ *is* a solution.

(b) $5(5) + 4(5) \overset{?}{\geq} 6$

$25 + 20 \overset{?}{\geq} 6$

$45 \geq 6$

The point $(5, 5)$ *is* a solution.

(c) $5(7) + 4(0) \overset{?}{\geq} 6$

$35 + 0 \overset{?}{\geq} 6$

$35 \geq 6$

The point $(7, 0)$ *is* a solution.

(d) $5(-2) + 4(5) \overset{?}{\geq} 6$

$-10 + 20 \overset{?}{\geq} 6$

$10 \geq 6$

The point $(-2, 5)$ *is* a solution.

49. Graph $x > \frac{3}{2}$.

Solution

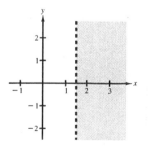

51. Graph $x - 3y < 0$.

Solution

Note: $x - 3y < 0$ could be written as $y > \frac{1}{3}x$.

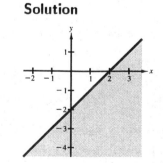

53. Graph $y \leq x - 2$.

Solution

55. Graph $y > x - 2$.

Solution

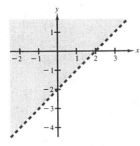

57. Graph $y \geq \frac{2}{3}x + \frac{1}{3}$.

Solution

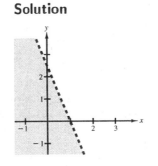

59. Graph $y > -2x + 10$.

Solution

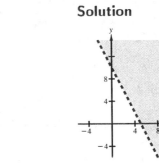

61. Graph $-3x + 2y - 6 < 0$.

Solution

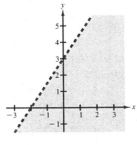

63. Graph $5x + 2y < 5$.

Solution

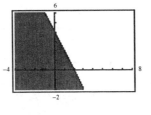

65. Use a graphing utility to graph $y \leq -2x + 4$.

Solution

Keystrokes:

Y= (−) 2 X|T + 4 GRAPH

DRAW cursor to Shade (ENTER

(−) 10 ALPHA , (−) 2 X|T

+ 4) ENTER

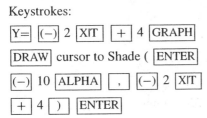

67. Use a graphing utility to graph $y \geq \frac{1}{2}x + 2$.

Solution

Keystrokes:

Y= X|T ÷ 2 + 2 GRAPH

DRAW cursor to Shade (ENTER

X|T ÷ 2 + 2 ALPHA , 10)

ENTER

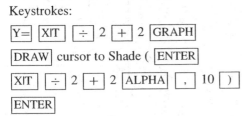

69. *Writing a Model*   Each table produced by a furniture company requires 1 hour in the assembly center, and the matching chair requires $1\frac{1}{2}$ hours in the assembly center. Twelve hours per day are available in the assembly center. Write an inequality that shows the different numbers of hours that can be spent assembling tables and chairs, and sketch a graph of the inequality. From the graph, find several ordered pairs with positive integer coordinates that are solutions of the inequality.

Solution

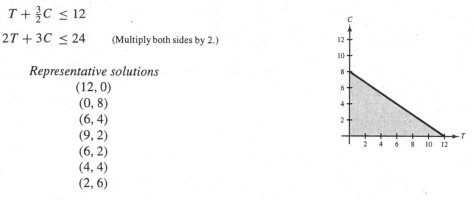

$T + \frac{3}{2}C \le 12$

$2T + 3C \le 24$ (Multiply both sides by 2.)

Representative solutions
(12, 0)
(0, 8)
(6, 4)
(9, 2)
(6, 2)
(4, 4)
(2, 6)

71. *Dietetics*   A dietitian is asked to design a special diet supplement using two foods. Each ounce of food X contains 20 units of calcium and each ounce of food Y contains 10 units of calcium. The minimum daily requirement in the diet is 300 units of calcium. Write an inequality that shows the different numbers of ounces of food X and food Y required. Sketch the graph of the inequality. From the graph, find several ordered pairs with positive integer coordinates that are solutions of the inequality.

Solution

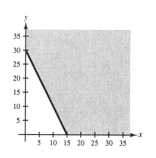

Labels: x = number of ounces of food X
             y = number of ounces of food Y

Inequality: $20x + 10y \ge 300$
Representative solutions
     (10, 20)
     (12, 10)
     (20, 0)
     (0, 30)
     (5, 25)

Review Exercises for Chapter 7

1. Determine whether the relation shown in the textbook is a function.

Solution

Yes. No first component has two different second components, so this relation *is* a function.

3. Determine whether the graph shown in the textbook represents y as a function of x.

Solution

No. A vertical line could cross the graph more than once, so this graph does *not* represent y as a function of x.

5. Use a graphing utility to graph $f(x) = \frac{3}{2}x + 2$. Identify any intercepts of the graph.

Solution

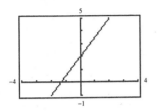

Intercepts: $\left(-\frac{4}{3}, 0\right), (0, 2)$

7. Use a graphing utility to graph $g(x) = x^2(x - 4)$. Identify any intercepts of the graph.

Solution

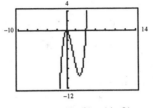

Intercepts: $(0, 0), (4, 0)$

9. Evaluate $f(x) = |2x + 3|$ for (a) $f(0)$ and (b) $f(5)$.

Solution

(a) $f(0) = |2(0) + 3|$

$\quad = |0 + 3| = |3| = 3$

(b) $f(5) = |2(5) + 3|$

$\quad = |10 + 3| = |13| = 13$

11. Evaluate $h(u) = u(u - 3)^2$ for (a) $h(0)$ and (b) $h(3)$.

Solution

(a) $h(0) = 0(0 - 3)^2$

$\quad = 0(-3)^2 = 0(9) = 0$

(b) $h(3) = 3(3 - 3)^2$

$\quad = 3(0)^2 = 3(0) = 0$

13. *Writing a Model*   The cost for sending a package is \$2.25 plus 75¢ per pound. Write the total cost C as a function of its weight x. Use a graphing utility to graph the function for $0 < x \leq 20$.

Solution

$C = 2.25 + 0.75x, \ 0 < x \leq 20$

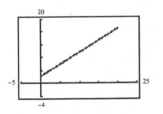

15. *Geometry*   A 24-inch wire is cut into four pieces to form a rectangle (see figure). Express the area A of the rectangle as a function of x.

Solution

The perimeter of the rectangle is 24. Thus, $24 = 2l + 2w$ or:

$24 = 2l + 2x$ $A = lw$

$24 - 2x = 2l$ $A = (12 - x)x$

$\dfrac{24 - 2x}{2} = l$ $A = 12x - x^2$

$12 - x = l$

17. Find the slope of the line through the points (2, 1) and (14, 6).

Solution

$$m = \frac{6 - 1}{14 - 2} = \frac{5}{12}$$

19. Find the slope of the line through the points (−1, 0) and (6, 2).

Solution

$$m = \frac{2 - 0}{6 - (-1)} = \frac{2}{7}$$

21. Find the slope of the line through the points (4, 0) and (4, 6).

Solution

$$m = \frac{6 - 0}{4 - 4} = \frac{6}{0} \text{ (Undefined)}$$

m is undefined.

23. Find the slope of the line through the points (−2, 5) and (1, 1).

Solution

$$m = \frac{1 - 5}{1 - (-2)} = -\frac{4}{3}$$

25. Find the slope of the line through the points (1, −4) and (5, 10).

Solution

$$m = \frac{10 - (-4)}{5 - 1} = \frac{14}{4} = \frac{7}{2}$$

27. Find the slope of the line through the points $\left(0, \frac{5}{2}\right)$ and $\left(\frac{5}{6}, 0\right)$.

Solution

$$m = \frac{0 - \frac{5}{2}}{\frac{5}{6} - 0} = \frac{-\frac{5}{2}}{\frac{5}{6}}$$

$$= -\frac{5}{2} \div \frac{5}{6} = -\frac{5}{2} \cdot \frac{6}{5}$$

$$= -\frac{30}{10} = -3$$

29. *Slope of a Ramp* The floor of a moving van is 4 feet above ground level. The end of the ramp used in loading the truck rests on the ground 6 feet behind the truck. Determine the slope of the ramp.

Solution

$m = \frac{4}{6} = \frac{2}{3}$

31. Given a point on a line (3, −1) and the slope $m = -2$, find two additional points on the line. (There are many correct answers.)

Solution

(4, −3), (5, −5), (6, −7), etc.

33. Given a point on a line (2, 3) and the slope $m = \frac{3}{4}$, find two additional points on the line. (There are many correct answers.)

Solution

(6, 6), (10, 9), (14, 12), etc.

35. Find an equation of the line passing through the point (4, −1) with slope $m = 2$. (Write your answer in general form.)

Solution

$$y + 1 = 2(x - 4)$$

$$y + 1 = 2x - 8$$

$$-2x + y + 9 = 0 \text{ or}$$

$$2x - y - 9 = 0$$

37. Find an equation of the line passing through the point (1, 2) with slope $m = -4$. (Write your answer in general form.)

Solution

$$y - 2 = -4(x - 1)$$

$$y - 2 = -4x + 4$$

$$4x + y - 6 = 0$$

39. Find an equation of the line passing through the point $(-1, -2)$ with slope $m = \frac{4}{5}$. (Write your answer in general form.)

Solution

$$y + 2 = \tfrac{4}{5}(x + 1)$$

$$y + 2 = \tfrac{4}{5}x + \tfrac{4}{5}$$

$$5(y + 2) = 5\left(\tfrac{4}{5}x + \tfrac{4}{5}\right)$$

$$5y + 10 = 4x + 4$$

$$-4x + 5y + 6 = 0 \text{ or}$$

$$4x - 5y - 6 = 0$$

41. Find an equation of the line passing through the point $\left(\frac{7}{2}, 3\right)$ with slope $m = -\frac{8}{3}$. (Write your answer in general form.)

Solution

$$y - 3 = -\tfrac{8}{3}\left(x - \tfrac{7}{2}\right)$$

$$y - 3 = -\tfrac{8}{3}x + \tfrac{28}{3}$$

$$3(y - 3) = 3\left(-\tfrac{8}{3}x + \tfrac{28}{3}\right)$$

$$3y - 9 = -8x + 28$$

$$8x + 3y - 37 = 0$$

43. Find an equation of the line passing through the points $(-4, 0), (0, -2)$. (Write your answer in general form.)

Solution

$$m = \frac{-2 - 0}{0 - (-4)} = \frac{-2}{4} = -\frac{1}{2}$$

$$y - 0 = -\frac{1}{2}(x + 4)$$

$$y = -\frac{1}{2}x - 2$$

$$2y = 2\left(-\frac{1}{2}x - 2\right)$$

$$2y = -x - 4$$

$$x + 2y + 4 = 0$$

Note: We could also obtain the solution using $m = -\frac{1}{2}$ and the y-intercept $(0, -2)$.

$$y = mx + b$$

$$y = -\frac{1}{2}x - 2$$

$$2y = -x - 4$$

$$x + 2y + 4 = 0$$

45. Find an equation of the line passing through the points $(0, 8), (6, 8)$. (Write your answer in general form.)

Solution

$$m = \frac{8 - 8}{6 - 0} = \frac{0}{6} = 0$$

$$y - 8 = 0(x - 0)$$

$$y - 8 = 0$$

Note: The slope of 0 indicates a horizontal line with an equation of the form $y = b$. The horizontal line through $(0, 8)$ has the equation $y = 8$ or $y - 8 = 0$.

47. Find an equation of the line through the point $(-1, 3)$ (a) parallel to and (b) perpendicular to the line $2x + 3y = 1$. (Write your answer in general form.)

Solution

$$2x + 3y = 1$$
$$3y = -2x + 1$$
$$y = -\tfrac{2}{3}x + \tfrac{1}{3}$$
$$m = -\tfrac{2}{3}$$

(a) Parallel line: $m = -\tfrac{2}{3}$

$$y - 3 = -\tfrac{2}{3}(x + 1)$$
$$y - 3 = -\tfrac{2}{3}x - \tfrac{2}{3}$$
$$3(y - 3) = 3\left(-\tfrac{2}{3}x - \tfrac{2}{3}\right)$$
$$3y - 9 = -2x - 2$$
$$2x + 3y - 7 = 0$$

(b) Perpendicular line: $m = \tfrac{3}{2}$

$$y - 3 = \tfrac{3}{2}(x + 1)$$
$$y - 3 = \tfrac{3}{2}x + \tfrac{3}{2}$$
$$2(y - 3) = 2\left(\tfrac{3}{2}x + \tfrac{3}{2}\right)$$
$$2y - 6 = 3x + 3$$
$$-3x + 2y - 9 = 0 \text{ or}$$
$$3x - 2y + 9 = 0$$

49. Find an equation of the line through the point $\left(\tfrac{5}{8}, 4\right)$ (a) parallel to and (b) perpendicular to the line $4x + 3y = 16$. (Write your answer in general form.)

Solution

$$4x + 3y = 16$$
$$3y = -4x + 16$$
$$y = -\tfrac{4}{3}x + \tfrac{16}{3}$$
$$m = -\tfrac{4}{3}$$

(a) Parallel line: $m = -\tfrac{4}{3}$

$$y - 4 = -\tfrac{4}{3}\left(x - \tfrac{5}{8}\right)$$
$$y - 4 = -\tfrac{4}{3}x + \tfrac{5}{6}$$
$$6(y - 4) = 6\left(-\tfrac{4}{3}x + \tfrac{5}{6}\right)$$
$$6y - 24 = -8x + 5$$
$$8x + 6y - 29 = 0$$

(b) Perpendicular line: $m = \tfrac{3}{4}$

$$y - 4 = \tfrac{3}{4}\left(x - \tfrac{5}{8}\right)$$
$$y - 4 = \tfrac{3}{4}x - \tfrac{15}{32}$$
$$32(y - 4) = 32\left(\tfrac{3}{4}x - \tfrac{15}{32}\right)$$
$$32y - 128 = 24x - 15$$
$$-24x + 32y - 113 = 0 \text{ or}$$
$$24x - 32y + 113 = 0$$

51. *Graphical Interpretation*   The velocity (in feet per second) of a ball thrown upward from ground level is modeled by $v = -32t + 48$, where t is time in seconds.

(a) Graph the expression for the velocity.

(b) Interpret the slope of the line in the context of this real-life setting.

(c) Find the velocity when $t = 0$ and $t = 1$.

(d) Find the time when the ball reaches its maximum height. (*Hint:* Find the time when $v = 0$.)

Solution

(a)

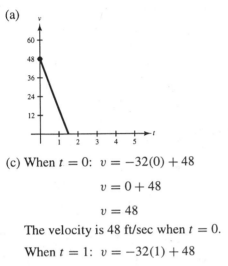

(b) Each second the velocity decreases by 32 feet per second.

(c) When $t = 0$: $v = -32(0) + 48$

$v = 0 + 48$

$v = 48$

The velocity is 48 ft/sec when $t = 0$.

When $t = 1$: $v = -32(1) + 48$

$v = -32 + 48$

$v = 16$

The velocity is 16 ft/sec when $t = 1$.

(d) When $v = 0$: $0 = -32t + 48$

$-48 = -32t$

$\dfrac{-48}{-32} = t$

$\dfrac{3}{2} = t$

The time $t = \frac{3}{2}$ seconds when $v = 0$.

53. Graph $x - 2 \geq 0$.

Solution

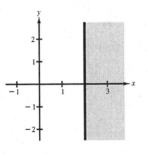

55. Graph $2x + y < 1$.

Solution

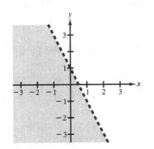

57. Graph $x \leq 4y - 2$.

Solution

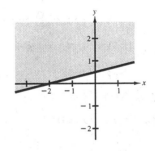

59. *Writing a Model*   Each week a company produces x VCRs and y camcorders. The assembly times for the two types of units are 2 and 3 hours, respectively. The time available in a week is 120 hours. Write an inequality that shows the different numbers of VCRS and camcorders that can be produced. Sketch a graph of the inequality. From the graph, find several ordered pairs that are solutions of the inequality.

Solution

$2x + 3y \leq 120$

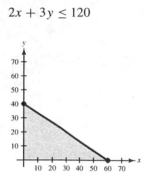

Representative solutions

$(20, 25)$
$(60, 0)$
$(10, 30)$
$(40, 10)$
$(0, 40)$

Test for Chapter 7

1. Does the table in the textbook represent y as a function of x? Explain your reasoning.

Solution

No. Some first components, 0 and 1, have two different second components, so the table does not represent y as a function of x.

2. Does the graph in the textbook represent y as a function of x? Explain your reasoning.

Solution

Yes. No vertical line would cross the graph more than once.

3. Evaluate $f(x) = x^3 - 2x^2$ at the indicated values.

(a) $f(0)$ (b) $f(2)$ (c) $f(-2)$ (d) $f\left(\frac{1}{2}\right)$

Solution

(a) $f(0) = 0^3 - 2(0)^2$

$\quad = 0 - 2(0)$

$\quad = 0$

(b) $f(2) = 2^3 - 2(2)^2$

$\quad = 8 - 2(4)$

$\quad = 8 - 8$

$\quad = 0$

(c) $f(-2) = (-2)^3 - 2(-2)^2$

$\quad = -8 - 2(4)$

$\quad = -8 - 8$

$\quad = -16$

(d) $f\left(\frac{1}{2}\right) = \left(\frac{1}{2}\right)^3 - 2\left(\frac{1}{2}\right)^2$

$\quad = \frac{1}{8} - 2\left(\frac{1}{4}\right)$

$\quad = \frac{1}{8} - \frac{1}{2}$

$\quad = \frac{1}{8} - \frac{4}{8}$

$\quad = -\frac{3}{8}$

4. Find the slope of the line passing through the points $(-5, 0)$ and $\left(2, \frac{3}{2}\right)$.

Solution

$$m = \frac{\frac{3}{2} - 0}{2 - (-5)} = \frac{\frac{3}{2}}{7} = \frac{3}{2} \cdot \frac{1}{7} = \frac{3}{14}$$

5. A line with slope $m = \frac{3}{4}$ passes through the points $(1, -2)$ and $(9, y)$. Find the y-coordinate of the second point.

Solution

$$\frac{3}{4} = \frac{y + 2}{9 - 1}$$

$$\frac{3}{4} = \frac{y + 2}{8}$$

$$4(y + 2) = 3 \cdot 8 \qquad \text{(Cross-multiply)}$$

$$4y + 8 = 24$$

$$4y = 16$$

$$y = 4$$

6. A line with slope $m = -2$ passes through the point $(-3, 4)$. Find two additional points on the line. (The problem has many correct answers.)

Solution

$(-2, 2)$, $(-1, 0)$, $(0, -2)$, etc.

7. Sketch the graph of $y = \frac{1}{4}x - 1$.

Solution

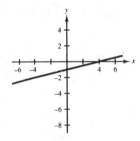

8. Sketch the graph of $y - 4 = 0$.

Solution

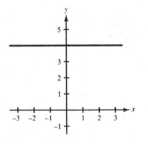

9. Sketch the graph of $4x + 3y - 12 = 0$.

Solution

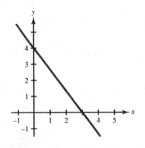

10. Sketch the graph of $y = |x - 2| - 2$.

Solution

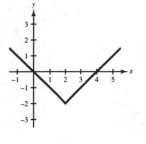

11. Sketch the graph of a line whose slope is undefined. Then write an equation for the graph.

Solution

Answers can vary. A representative graph is shown below. The graph would be a vertical line.

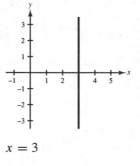

$x = 3$

12. Find an equation of the line that passes through the point $(0, 6)$ with slope $m = -\frac{3}{8}$.

Solution

$$y = mx + b$$

$$m = -\frac{3}{8} \text{ and } (0, b) = (0, 6)$$

$$y = -\frac{3}{8}x + 6 \text{ or}$$

$$3x + 8y - 48 = 0$$

13. Find an equation of the line through the points $(2, 1)$ and $(6, 6)$. Write your answer in general form.

Solution

$$m = \frac{6 - 1}{6 - 2} = \frac{5}{4}$$

$$y - 1 = \frac{5}{4}(x - 2)$$

$$y - 1 = \frac{5}{4}x - \frac{5}{2}$$

$$4(y - 1) = 4\left(\frac{5}{4}x - \frac{5}{2}\right)$$

$$4y - 4 = 5x - 10$$

$$-5x + 4y + 6 = 0 \text{ or}$$

$$5x - 4y - 6 = 0$$

14. Find the slope of the line *perpendicular* to the line $3x - 5y + 2 = 0$.

Solution

$$3x - 5y + 2 = 0$$

$$-5y = -3x - 2$$

$$y = \frac{-3}{-5}x - \frac{2}{-5}$$

$$y = \frac{3}{5}x + \frac{2}{5}$$

$$m = \frac{3}{5}$$

A perpendicular line would have slope $m = -\frac{5}{3}$ (the negative reciprocal of $\frac{3}{5}$).

15. Which points are solutions of the inequality $3x + 5y \leq 16$?

(a) $(2, 2)$ (b) $(6, -1)$ (c) $(-2, 4)$ (d) $(7, -1)$

Solution

(a) $(2, 2)$

$$3 \cdot 2 + 5 \cdot 2 \overset{?}{\leq} 16$$

$$6 + 10 \overset{?}{\leq} 16$$

$$16 \leq 16$$

The point $(2, 2)$ *is* a solution.

(b) $(6, -1)$

$$3 \cdot 6 + 5(-1) \overset{?}{\leq} 16$$

$$18 + (-5) \overset{?}{\leq} 16$$

$$18 - 5 \overset{?}{\leq} 16$$

$$13 \leq 16$$

The point $(6, -1)$ *is* a solution.

(c) $(-2, 4)$

$$3(-2) + 5 \cdot 4 \overset{?}{\leq} 16$$

$$-6 + 20 \overset{?}{\leq} 16$$

$$14 \leq 16$$

The point $(-2, 4)$ *is* a solution.

(d) $(7, -1)$

$$3 \cdot 7 + 5(-1) \overset{?}{\leq} 16$$

$$21 - 5 \overset{?}{\leq} 16$$

$$16 \leq 16$$

The point $(7, -1)$ *is* a solution.

16. Graph the inequality $y \geq -2$.

Solution

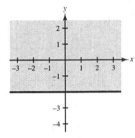

17. Graph the inequality $y < 5 - 2x$.

Solution

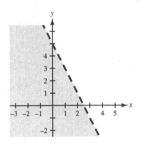

18. Graph the inequality $x \geq 2$.

Solution

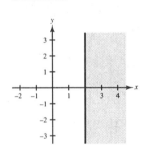

19. Graph the inequality $y \leq 5$.

Solution

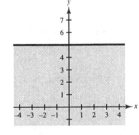

20. The value of a new plain paper copier is $4200. It is estimated that during the first 5 years the copier will depreciate at the rate of $800 per year. Write the value V of the copier as a function of time t in years for $0 < t \leq 5$. Find the value of the copier after 3 years.

Solution

$$V = 4200 - 800t, \quad 0 < t \leq 5$$

When $t = 3$: $V = 4200 - 800(3)$

$$V = 4200 - 2400$$

$$V = 1800$$

After 3 years, the value of the copier will be $1800.

CHAPTER EIGHT
Systems of Linear Equations

8.1   Solving Systems of Equations by Graphing

9. Which ordered pair (a) (2, 3) or (b) (5, 4) is a solution of the following system of equations?

$$x + 3y = 11$$
$$-x + 3y = 7$$

Solution

(a) (2, 3)

$$2 + 3(3) \overset{?}{=} 11$$
$$2 + 9 = 11$$
$$-2 + 3(3) \overset{?}{=} 7$$
$$-2 + 9 = 7$$

(2, 3) *is* a solution.

(b) (5, 4)

$$5 + 3(4) \overset{?}{=} 11$$
$$5 + 12 \neq 11$$
$$-5 + 3(4) \overset{?}{=} 7$$
$$-5 + 12 = 7$$

(5, 4) is *not* a solution.

11. Which ordered pair (a) (5, −3) or (b) (−1, 2) is a solution of the following system of equations?

$$2x - 3y = -8$$
$$x + y = 1$$

Solution

(a) (5, −3)

$$2(5) - 3(-3) \overset{?}{=} -8$$
$$10 + 9 \neq -8$$
$$5 + (-3) \overset{?}{=} 1$$
$$5 - 3 \neq 1$$

(5, −3) is *not* a solution.

(b) (−1, 2)

$$2(-1) - 3(2) \overset{?}{=} -8$$
$$-2 - 6 = -8$$
$$-1 + 2 \overset{?}{=} 1$$
$$-1 + 2 = 1$$

(−1, 2) *is* a solution.

13. *Graphing and Checking*   Use the graph shown in the textbook to determine the solution of the system of linear equations. Check your solution.

$$2x + y = 4$$
$$x - y = 2$$

Solution

The graph indicates that the solution is (2, 0).

15. *Graphing and Checking*   Use the graph shown in the textbook to determine the solution of the system of linear equations. Check your solution.

$$x - y = 0$$
$$3x - 2y = -1$$

Solution

The graph indicates that the solution is (−1, −1).

17. Solve the following system of linear equations by graphing. Check your solution.

$$y = -x + 3$$
$$y = x + 1$$

Solution

The solution is $(1, 2)$.

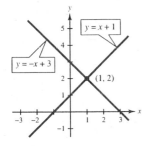

21. Solve the following system of linear equations by graphing. Check your solution.

$$3x - 4y = 5$$
$$x = 3$$

Solution

The solution is $(3, 1)$.

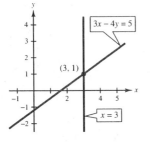

19. Solve the following system of linear equations by graphing. Check your solution

$$x - y = 2$$
$$x + y = 2$$

Solution

The solution is $(2, 0)$.

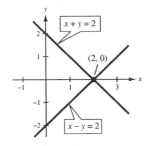

23. Solve the following system of linear equations by graphing. Check your solution.

$$4x + 5y = 20$$
$$\tfrac{4}{5}x + y = 4$$

Solution

There are *infinitely many* solutions. The solution set consists of all ordered pairs (x, y) such that $4x + 5y = 20$.

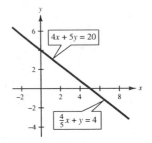

25. Solve the following system of linear equations by graphing. Check your solution.

$$x + 2y = -4$$

$$5x - y = 13$$

Solution

The solution is $(2, -3)$.

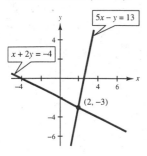

29. Use a graphing utility to solve the following system. Check your solution.

$$y = 2x - 1$$

$$y = -3x + 9$$

Solution

$$y_1 = 2x - 1$$

$$y_2 = -3x + 9$$

The solution is $(2, 3)$.

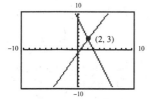

33. Find the slopes of the following equations. What can you conclude about the number of solutions of the system?

$$2x - 3y = -12$$

$$-8x + 12y = -12$$

Solution

$2x - 3y = -12$	$-8x + 12y = -12$
$-3y = -2x - 12$	$12y = 8x - 12$
$y = \frac{2}{3}x + 4$	$y = \frac{2}{3}x - 1$
$\left(m = \frac{2}{3}\right)$	$\left(m = \frac{2}{3}\right)$

These two lines are parallel; they have the *same* slope and *different* y-intercepts. Thus, the system has *no* solution. The system is inconsistent.

27. Solve the following system of linear equations by graphing. Check your solution.

$$2x + y = 14$$

$$3x - y = 11$$

Solution

The solution is $(5, 4)$.

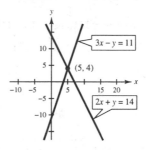

31. Use a graphing utility to solve the following system. Check your solution.

$$2x + 3y = 8$$

$$4x - 2y = 0$$

Solution

$$y_1 = -\frac{2}{3}x + \frac{8}{3}$$

$$y_2 = 2x$$

The solution is $(1, 2)$.

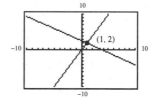

35. Find the slopes of the following equations. What can you conclude about the number of solutions of the system?

$$-x + 4y = 7$$

$$3x - 12y = -21$$

Solution

$-x + 4y = 7$	$3x - 12y = -21$
$4y = x + 7$	$-12y = -3x - 21$
$y = \frac{1}{4}x + \frac{7}{4}$	$y = \frac{1}{4}x + \frac{7}{4}$
$\left(m = \frac{1}{4}\right)$	$\left(m = \frac{1}{4}\right)$

These two lines coincide; they have the *same* slope and the *same* y-intercept. Thus, the system has *infinitely many* solutions.

37. *Writing a Model*   The sum of two numbers is 20 and the difference of the two numbers is 2. Write a system of equations that models this problem and solve the system graphically.

Solution

$x + y = 20$

$x - y = 2$

The solution is $(11, 9)$. The two numbers are 11 and 9.

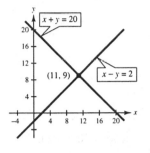

41. Solve $\frac{1}{2}x - \frac{1}{5}x = 15$ and check your answer.

Solution

$$\frac{1}{2}x - \frac{1}{5}x = 15$$

$$10\left(\frac{1}{2}x - \frac{1}{5}x\right) = 10(15)$$

$$5x - 2x = 150$$

$$3x = 150$$

$$x = \frac{150}{3}$$

$$x = 50$$

45. *Graphical Reasoning*   Use the graph shown in the textbook to determine how many solutions the following system has.

$$x - 2y = -4$$

$$-0.5x + y = 2$$

Solution

The system has infinitely many solutions.

39. Solve $y - 3(4y - 2) = 1$ and check your answer.

Solution

$$y - 3(4y - 2) = 1$$

$$y - 12y + 6 = 1$$

$$-11y + 6 = 1$$

$$-11y = -5$$

$$y = \frac{-5}{-11}$$

$$y = \frac{5}{11}$$

43. Translate the following phrase into an algebraic expression.

Time to travel 250 miles at r miles per hour.

Solution

$$\text{Distance } = (\text{Rate})(\text{Time})$$

$$\frac{\text{Distance}}{\text{Rate}} = \text{Time}$$

The time can be expressed as $\dfrac{250}{r}$.

47. *Graphical Reasoning*   Use the graph shown in the textbook to determine how many solutions the following system has.

$$2x - 3y = 6$$

$$4x + 3y = 12$$

Solution

The system has one solution.

49. Solve the following system of linear equations by graphing. Check your solution.

$$y = 2x - 1$$
$$y = x + 1$$

Solution

The solution is $(2, 3)$.

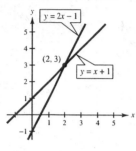

51. Solve the following system of linear equations by graphing. Check your solution.

$$x + 2y = 4$$
$$2x - y = 3$$

Solution

The solution is $(2, 1)$.

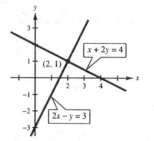

53. Solve the following system of linear equations by graphing. Check your solution.

$$4x - 5y = 0$$
$$6x - 5y = 10$$

Solution

The solution is $(5, 4)$.

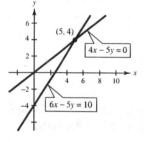

55. Solve the following system of linear equations by graphing. Check your solution.

$$4x - 3y = -1$$
$$4x - 3y = 0$$

Solution

The system has no solution. (The lines are parallel.)

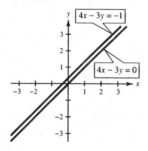

57. Solve the following system of linear equations by graphing. Check your solution.

$$x + 7y = -5$$
$$3x - 2y = 8$$

Solution

The solution is $(2, -1)$.

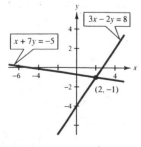

59. Solve the following system of linear equations by graphing. Check your solution.

$$x + 2y = 3$$
$$x - 3y = 13$$

Solution

The solution is $(7, -2)$.

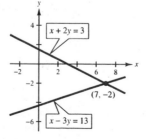

61. Solve the following system of linear equations by graphing. Check your solution.

$$-x + 10y = 30$$
$$x + 10y = 10$$

Solution

The solution is $(-10,\ 2)$.

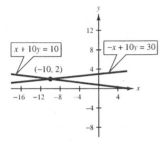

65. Solve the following system of linear equations by graphing. Check your solution.

$$x + 2y = 4$$
$$2x - 2y = -1$$

Solution

The solution is $\left(1, \frac{3}{2}\right)$.

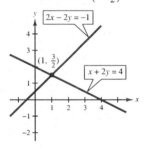

69. Use a graphing utility to solve the following system. Check your solution.

$$y = x - 1$$
$$y = -2x + 8$$

Solution

$y_1 = x - 1$

$y_2 = -2x + 8$

The solution is $(3, 2)$.

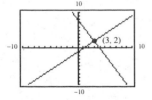

63. Solve the following system of linear equations by graphing. Check your solution.

$$2x + 3y = 10$$
$$y = 3$$

Solution

The solution is $\left(\frac{1}{2}, 3\right)$.

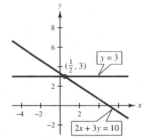

67. Solve the following system of linear equations by graphing. Check your solution.

$$-3x + 10y = 15$$
$$3x - 10y = 15$$

Solution

There is *no* solution.
The system is inconsistent.

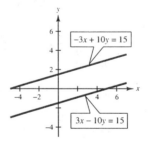

71. Use a graphing utility to solve the following system. Check your solution.

$$3x + 2y = 6$$
$$x - 3y = 13$$

Solution

$y_1 = -\frac{3}{2}x + 3$

$y_2 = \frac{1}{3}x - \frac{13}{3}$

The solution is $(4, -3)$.

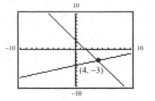

73. Find the slopes of the following equations. What can you conclude?

$$-2x + 3y = 4$$

$$2x + 3y = 8$$

Solution

$-2x + 3y = 4$ $\qquad$ $2x + 3y = 8$

$\qquad 3y = 2x + 4$ $\qquad\qquad 3y = -2x + 8$

$\qquad\quad y = \frac{2}{3}x + \frac{4}{3}$ $\qquad\qquad\quad y = -\frac{2}{3}x + \frac{8}{3}$

$\qquad\left(m = \frac{2}{3}\right)$ $\qquad\qquad\quad\left(m = -\frac{2}{3}\right)$

These two lines do *not* have the same slope, so the two lines intersect. The system has *one* solution.

75. Find the slopes of the following equations. What can you conclude?

$$-6x + 8y = \quad 9$$

$$3x - 4y = -4.5$$

Solution

$-6x + 8y = 9$ $\qquad\qquad 3x - 4y = -4.5$

$\qquad 8y = 6x + 9$ $\qquad\quad -4y = -3x - 4.5$

$\qquad\quad y = \frac{3}{4}x + \frac{9}{8}$ $\qquad\qquad y = \frac{3}{4}x + \frac{4.5}{4}$ or $y = \frac{3}{4}x + \frac{9}{8}$

$\qquad\left(m = \frac{3}{4}\right)$ $\qquad\qquad\quad\left(m = \frac{3}{4}\right)$

These two lines coincide; they have the *same* slope and the *same* y-intercept. Thus, the system has infinitely many solutions.

77. *Think About It* The graphs of the two equations appear parallel. Are the two lines actually parallel? Does the system have a solution? If so, find the solution.

$$x - 200y = -200$$

$$x - 199y = \quad 198$$

Solution

$x - 200y = -200$ $\qquad\qquad x - 199y = 198$

$\quad -200y = -x - 200$ $\qquad\quad -199y = -x + 198$

$\qquad\quad y = \frac{1}{200}x + 1$ $\qquad\qquad\quad y = \frac{1}{199}x - \frac{198}{199}$

$\qquad\left(m = \frac{1}{200}\right)$ $\qquad\qquad\quad\left(m = \frac{1}{199}\right)$

The graphical method is difficult to apply because the slopes of the two lines are nearly equal, and therefore, the lines *appear* to be parallel on the portion of the graph pictured. The point where the lines intersect is far from this portion of the graph. (Using techniques found in the next sections of this chapter, we can determine that the solution of this system is (79,400, 398); these are the coordinates of the point of intersection of the two lines.)

79. *Chemical Mixture*   Nine gallons of a 30% acid solution is obtained by mixing a 20% solution with a 50% solution. Let x represent the number of gallons of the 20% solution and y represent the number of gallons of the 50% solution. A mathematical model for the problem is

$$x + \quad y = 9$$
$$0.20x + 0.50y = 0.30(9).$$

Solve this system graphically.

Solution

The solution is (6, 3).

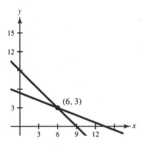

81. *Writing a Model*   The sum of two numbers is 35 and the difference of the two numbers is 11. Write a system of equations that models this problem and solve the system graphically.

Solution

$$x + y = 35$$
$$x - y = 11$$

The solution of the system is (23, 12).

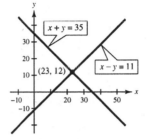

8.2   Solving Systems of Equations by Substitution

7. Use substitution to solve the following system. Use the graph shown in the textbook to check the solution.

$$2x + y = 4$$
$$-x + y = 1$$

Solution

$$2x + y = 4 \quad \rightarrow \quad y = 4 - 2x$$
$$-x + y = 1$$

$$-x + (4 - 2x) = 1 \qquad \text{Replace } y \text{ by } 4 - 2x \text{ in second equation.}$$

$$-x + 4 - 2x = 1$$

$$-3x + 4 = 1$$

$$-3x = -3$$

$$x = 1 \qquad y = 4 - 2(1) \qquad \text{Replace } x \text{ by } 1 \text{ in revised first equation.}$$

$$y = 4 - 2$$

$$(1, 2) \qquad y = 2$$

9. Use substitution to solve the following system. Use the graph shown in the textbook to check the solution.

$$2x - y = 2$$
$$4x + 3y = 9$$

Solution

$$2x - y = 2 \quad \rightarrow \quad -y = 2 - 2x \quad \rightarrow \quad y = -2 + 2x$$
$$4x + 3y = 9$$

$$4x + 3(-2 + 2x) = 9 \qquad\qquad \text{Replace } y \text{ by } -2 + 2x \text{ in second equation.}$$
$$4x - 6 + 6x = 9$$
$$10x - 6 = 9$$
$$10x = 15$$
$$x = \tfrac{3}{2} \qquad y = -2 + 2\left(\tfrac{3}{2}\right) \qquad \text{Replace } x \text{ by } \tfrac{3}{2} \text{ in revised first equation.}$$
$$y = -2 + 3$$
$$\left(\tfrac{3}{2}, 1\right) \qquad\qquad y = 1$$

11. Use substitution to solve the following system.

$$x - y = 2$$
$$2x + y = 1$$

Solution

$$x - y = 2 \quad \rightarrow \quad x = y + 2$$
$$2x + y = 1$$

$$2(y + 2) + y = 1 \qquad\qquad \text{Replace } x \text{ by } y + 2 \text{ in second equation.}$$
$$2y + 4 + y = 1$$
$$3y + 4 = 1$$
$$3y = -3$$
$$y = -1 \qquad x = 2 + (-1) \qquad \text{Replace } y \text{ by } -1 \text{ in revised first equation.}$$
$$(1, -1) \qquad\qquad x = 1$$

13. Use substitution to solve the following system.

$$x - y = 0$$
$$5x - 3y = 10$$

Solution

$$x - y = 0 \quad \rightarrow \quad x = y$$
$$5x - 3y = 10$$
$$5(y) - 3y = 10 \qquad\qquad \text{Replace } x \text{ by } y \text{ in second equation.}$$
$$2y = 10$$
$$y = 5 \qquad x = 5 \qquad \text{Replace } y \text{ by 5 in revised first equation.}$$
$$(5, 5)$$

15. Use substitution to solve the following system.

$$-5x + 4y = 14$$
$$5x - 4y = 4$$

Solution

$$-5x + 4y = 14 \quad \rightarrow \quad 4y = 5x + 14 \quad \rightarrow \quad y = \tfrac{5}{4}x + \tfrac{7}{2}$$
$$5x - 4y = 4$$
$$5x - 4\left(\tfrac{5}{4}x + \tfrac{7}{2}\right) = 4 \qquad\qquad \text{Replace } y \text{ by } \tfrac{5}{4}x + \tfrac{7}{2} \text{ in second equation.}$$
$$5x - 5x - 14 = 4$$
$$-14 = 4 \qquad\qquad \text{False}$$

The system has *no* solution. The system is inconsistent.

17. Use substitution to solve the following system.

$$5x + 3y = 11$$
$$x - 5y = 5$$

Solution

$$5x + 3y = 11$$
$$x - 5y = 5 \quad \rightarrow \quad x = 5y + 5$$
$$5(5y + 5) + 3y = 11 \qquad\qquad \text{Replace } x \text{ by } 5y + 5 \text{ in first equation.}$$
$$25y + 25 + 3y = 11$$
$$28y + 25 = 11$$
$$28y = -14$$
$$y = -\tfrac{1}{2} \qquad x = 5\left(-\tfrac{1}{2}\right) + 5 \qquad \text{Replace } y \text{ by } -\tfrac{1}{2} \text{ in revised second equation.}$$
$$x = -\tfrac{5}{2} + \tfrac{10}{2}$$
$$\left(\tfrac{5}{2}, -\tfrac{1}{2}\right) \qquad\qquad x = \tfrac{5}{2}$$

19. Use substitution to solve the following system. Use a graphing utility to check the solution graphically.

$$3x + 2y = 12$$
$$x - y = 3$$

Solution

$$3x + 2y = 12$$
$$x - y = 3 \quad \rightarrow \quad x = y + 3$$
$$3(y + 3) + 2y = 12 \qquad \text{Replace } x \text{ by } y + 3 \text{ in first equation.}$$
$$3y + 9 + 2y = 12$$
$$5y + 9 = 12$$
$$5y = 3$$
$$y = \tfrac{3}{5} \qquad x = \tfrac{3}{5} + 3 \qquad \text{Replace } y \text{ by } \tfrac{3}{5} \text{ in revised second equation.}$$
$$x = \tfrac{3}{5} + \tfrac{15}{5}$$
$$\left(\tfrac{18}{5}, \tfrac{3}{5}\right) \qquad x = \tfrac{18}{5}$$

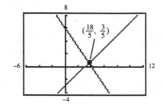

21. Use substitution to solve the following system. Use a graphing utility to check the solution graphically.

$$3x + 2y = 0$$
$$x + 2y = 4$$

Solution

$$3x + 2y = 0$$
$$x + 2y = 4 \quad \rightarrow \quad x = 4 - 2y$$
$$3(4 - 2y) + 2y = 0 \qquad \text{Replace } x \text{ by } 4 - 2y \text{ in first equation.}$$
$$12 - 6y + 2y = 0$$
$$12 - 4y = 0$$
$$-4y = -12$$
$$y = 3 \qquad x = 4 - 2(3) \qquad \text{Replace } y \text{ by } 3 \text{ in revised second equation.}$$
$$x = 4 - 6$$
$$(-2, 3) \qquad x = -2$$

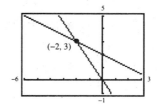

23. Find a system of linear equations that has the solution (2, 1). (*Note:* Each problem has many correct answers.)

Solution

Here are *some* systems that have the solution (2, 1). There are many other correct answers.

$$x + y = 3 \qquad \text{or} \qquad 3x + 2y = 8 \qquad \text{or} \qquad x - 3y = -1$$
$$x - y = 1 \qquad\qquad\qquad 5x - 4y = 6 \qquad\qquad\qquad 8x + 7y = 23$$

25. Find a system of linear equations that has the solution $\left(\frac{7}{2}, -3\right)$. (*Note:* Each problem has many correct answers.)

Solution

Here are *some* systems that have the solution $\left(\frac{7}{2}, -3\right)$. There are many other correct answers.

$$2x + y = 4 \qquad \text{or} \qquad -2x + y = -10 \qquad \text{or} \qquad 6x + 5y = 6$$
$$2x - y = 10 \qquad\qquad\qquad 4x + 3y = 5 \qquad\qquad\qquad 2x + 3y = -2$$

27. *Problem Solving*   Six people ate a buffet dinner for $61.70. The price for adults was $11.95 and the price for children was $6.95. How many adults were there?

Solution

Verbal model: $\boxed{\text{Number of adults}} + \boxed{\text{Number of children}} = \boxed{\text{Total number of people}}$

$\boxed{\$11.95} \cdot \boxed{\text{Number of adults}} + \boxed{\$6.95} \cdot \boxed{\text{Number of children}} = \boxed{\text{Total price}}$

Labels: Number of adults $= x$
Number of children $= y$
Total number of people $= 6$
Total price $= \$61.70$

System of equations: $\quad x + \quad y = \ 6$
$11.95x + 6.95y = 61.70$

Solving by substitution:

$$x + \quad y = 6 \quad \rightarrow \quad x = -y + 6$$

$$11.95x + 6.95y = 61.70 \quad \rightarrow \quad 1195x + 695y = 6170 \qquad \text{Multiply by 100.}$$

$$1195(-y + 6) + 695y = 6170 \qquad\qquad \text{Replace } x \text{ by } -y + 6 \\ \text{in revised second equation.}$$

$$-1195y + 7170 + 695y = 6170$$

$$-500y + 7170 = 6170$$

$$-500y = -1000$$

$$y = 2 \qquad x = -2 + 6 \qquad \text{Replace } y \text{ by 2} \\ \text{in revised first equation.}$$

$$(4, \ 2) \qquad\qquad x = 4$$

Thus, four people were charged the adult price. (Two people were charged the children's price.)

29. Find an equation of the line passing through $(0, 4)$ and $(10, 0)$.

Solution

$$m = \frac{y_2 - y_1}{x_2 - x_1}$$

$$m = \frac{0 - 4}{10 - 0}$$

$$m = \frac{-4}{10} = -\frac{2}{5}$$

$$y - y_1 = m(x - x_1)$$

$$y - 4 = -\frac{2}{5}(x - 0)$$

$$y - 4 = -\frac{2}{5}x$$

$$5(y - 4) = 5\left(-\frac{2}{5}x\right)$$

$$5y - 20 = -2x$$

$$2x + 5y - 20 = 0$$

31. Find an equation of the line passing through $(-1, 3)$ and $(4, 8)$.

Solution

$$m = \frac{y_2 - y_1}{x_2 - x_1}$$

$$m = \frac{8 - 3}{4 - (-1)}$$

$$m = \frac{5}{5}$$

$$m = 1$$

$$y - y_1 = m(x - x_1)$$

$$y - 3 = 1(x + 1)$$

$$y - 3 = x + 1$$

$$x - y + 4 = 0$$

33. *Geometry* The length of each edge of a cube is x inches. Write the surface area A as a function of x.

Solution

Each of the six faces of the cube has an area of $(x)(x) = x^2$. Therefore, the surface area of the entire cube is $6x^2$.

35. Use substitution to solve the following system. Use the graph shown in the textbook to check the solution.

$$x - y = 0$$
$$x + y = 2$$

Solution

$x - y = 0 \quad \rightarrow \quad x = y$

$x + y = 2$

$y + y = 2$ Replace x by y in second equation.

$2y = 2$

$y = 1 \quad\quad x = 1$ Replace y by 1 in revised first equation.

$(1, 1)$

37. Use substitution to solve the following system. Use the graph shown in the textbook to check the solution.

$$-x + y = 1$$
$$x - y = 1$$

Solution

$$-x + y = 1 \quad \rightarrow \quad y = 1 + x$$
$$x - y = 1$$
$$x - (1 + x) = 1 \qquad\qquad \text{Replace } y \text{ by } 1 + x \text{ in second equation.}$$
$$x - 1 - x = 1$$
$$-1 = 1 \qquad\qquad \text{False}$$

The system has *no* solution. The system is inconsistent.

39. Use substitution to solve the following system.

$$x - y = 0$$
$$2x + y = 0$$

Solution

$$x - y = 0 \quad \rightarrow \quad x = y$$
$$2x + y = 0$$
$$2(y) + y = 0 \qquad\qquad \text{Replace } x \text{ by } y \text{ in second equation.}$$
$$2y + y = 0$$
$$3y = 0$$
$$y = 0 \qquad x = 0 \qquad \text{Replace } y \text{ by } 0 \text{ in revised first equation.}$$
$$(0, \ 0)$$

41. Use substitution to solve the following system.

$$2x - y = -2$$
$$4x + y = 5$$

Solution

$$2x - y = -2$$
$$4x + y = 5 \quad \rightarrow \quad y = -4x + 5$$
$$2x - (-4x + 5) = -2 \qquad\qquad \text{Replace } y \text{ by } -4x + 5 \text{ in first equation.}$$
$$2x + 4x - 5 = -2$$
$$6x - 5 = -2$$
$$6x = 3$$
$$x = \tfrac{3}{6}$$
$$x = \tfrac{1}{2} \qquad y = -4\left(\tfrac{1}{2}\right) + 5 \qquad \text{Replace } x \text{ by } \tfrac{1}{2} \text{ in revised second equation.}$$
$$y = -2 + 5$$
$$\left(\tfrac{1}{2}, \ 3\right) \qquad\qquad y = 3$$

43. Use substitution to solve the following system.

$$\tfrac{1}{5}x + \tfrac{1}{2}y = 8$$
$$x + y = 20$$

Solution

$$\tfrac{1}{5}x + \tfrac{1}{2}y = 8 \quad \rightarrow \quad 2x + 5y = 80 \qquad \text{Multiply by 10.}$$
$$x + y = 20 \quad \rightarrow \quad x = -y + 20$$
$$2(-y + 20) + 5y = 80 \qquad\qquad\qquad \text{Replace } x \text{ by } -y + 20 \text{ in revised first equation.}$$
$$-2y + 40 + 5y = 80$$
$$3y + 40 = 80$$
$$3y = 40$$
$$y = \tfrac{40}{3} \qquad x = -\tfrac{40}{3} + 20 \qquad \text{Replace } y \text{ by } \tfrac{40}{3} \text{ in revised second equation.}$$
$$x = -\tfrac{40}{3} + \tfrac{60}{3}$$
$$\left(\tfrac{20}{3},\ \tfrac{40}{3}\right) \qquad\qquad x = \tfrac{20}{3}$$

45. Use substitution to solve the following system.

$$4x - y = 2$$
$$2x - \tfrac{1}{2}y = 1$$

Solution

$$4x - y = 2 \quad \rightarrow \quad -y = -4x + 2 \quad \rightarrow \quad y = 4x - 2$$
$$2x - \tfrac{1}{2}y = 1$$
$$2x - \tfrac{1}{2}(4x - 2) = 1 \qquad\qquad \text{Replace } y \text{ by } 4x - 2 \text{ in second equation.}$$
$$2x - 2x + 1 = 1$$
$$1 = 1$$

The system has *infinitely many* solutions. The solution set consists of all ordered pairs (x, y) such that $4x - y = 2$.

47. Use substitution to solve the following system.

$$2x + y = 8$$
$$5x + 2.5y = 10$$

Solution

$$2x + y = 8 \quad \rightarrow \quad y = 8 - 2x$$
$$5x + 2.5y = 10$$
$$5x + 2.5(8 - 2x) = 10 \qquad \text{Replace } y \text{ by } 8 - 2x \text{ in second equation.}$$
$$5x + 20 - 5x = 10$$
$$20 = 10 \qquad \text{False}$$

The system has no solution.

49. Use substitution to solve the following system.

$$0.5x + 0.5y = 4$$
$$x + y = -1$$

Solution

$$0.5x + 0.5y = 4$$
$$x + y = -1 \quad \rightarrow \quad x = -y - 1$$
$$0.5x(-y - 1) + 0.5y = 4 \qquad \text{Replace } x \text{ by } -y-1 \text{ in first equation.}$$
$$-0.5y - 0.5 + 0.5y = 4$$
$$-0.5 = 4 \qquad \text{False}$$

The system has no solution.

51. Use substitution to solve the following system.

$$2x = 5$$
$$x + y = 1$$

Solution

$$2x = 5 \quad \rightarrow \quad x = \tfrac{5}{2}$$
$$x + y = 1$$
$$\tfrac{5}{2} + y = 1 \qquad\qquad\qquad \text{Replace } x \text{ by } \tfrac{5}{2} \text{ in second equation.}$$
$$y = 1 - \tfrac{5}{2}$$
$$y = \tfrac{2}{2} - \tfrac{5}{2}$$
$$y = -\tfrac{3}{2}$$
$$\left(\tfrac{5}{2}, -\tfrac{3}{2}\right)$$

53. Use substitution to solve the following system.

$$8x + 5y = 100$$
$$9x - 10y = 50$$

Solution

$$8x + 5y = 100 \quad \rightarrow \quad 5y = -8x + 100 \quad \rightarrow \quad y = -\tfrac{8}{5}x + 20$$
$$9x - 10y = 50$$
$$9x - 10\left(-\tfrac{8}{5}x + 20\right) = 50 \qquad\qquad \text{Replace } y \text{ by } -\tfrac{8}{5}x + 20$$
$$\qquad\qquad\qquad\qquad\qquad\qquad\qquad\qquad \text{in second equation.}$$
$$9x + 16x - 200 = 50$$
$$25x - 200 = 50$$
$$25x = 250$$
$$x = 10 \qquad\qquad y = -\tfrac{8}{5}(10) + 20 \qquad \text{Replace } x \text{ by } 10 \text{ in}$$
$$\qquad\qquad\qquad\qquad\qquad\qquad\qquad \text{revised first equation.}$$
$$y = -16 + 20$$
$$y = 4$$
$$(10, 4)$$

55. Use substitution to solve the system. Use a graphing utility to check the solution graphically.

$$y = -2x + 10$$
$$y = \quad x + 4$$

Solution

$$y = -2x + 10$$
$$y = x + 4$$
$$x + 4 = -2x + 10 \qquad \text{Replace } y \text{ by } x + 4 \text{ in first equation.}$$
$$3x + 4 = 10$$
$$3x = 6$$
$$x = 2 \quad \rightarrow \quad y = 2 + 4 \qquad \text{Replace } x \text{ by } 2 \text{ in the second equation.}$$
$$\qquad\qquad\qquad y = 6$$

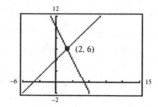

$(2, 6)$

57. Use substitution to solve the system. Use a graphing utility to check the solution graphically.

$$5x + 3y = 15$$
$$2x - 3y = \quad 6$$

Solution

$$5x + 3y = 15$$
$$2x - 3y = 6 \quad \rightarrow \quad -3y = -2x + 6$$
$$y = \frac{-2x}{-3} + \frac{6}{-3}$$
$$y = \frac{2}{3}x - 2$$
$$5x + 3\left(\frac{2}{3}x - 2\right) = 15 \qquad \text{Replace } y \text{ by } \tfrac{2}{3}x - 2 \text{ in first equation.}$$
$$5x + 2x - 6 = 15$$
$$7x - 6 = 15$$
$$7x = 21$$
$$x = 3 \quad \rightarrow \quad y = \frac{2}{3}(3) - 2 \qquad \text{Replace } x \text{ by } 3 \text{ in revised second equation.}$$
$$y = 2 - 2$$
$$y = 0$$

$(3, 0)$

59. *Think About It* Find the values of a or b such that the following system is inconsistent.

$$x + by = 1$$
$$x + 2y = 2$$

Solution

$$x + by = 1 \quad \rightarrow \quad x = -by + 1$$
$$x + 2y = 2$$
$$-by + 1 + 2y = 2 \qquad \text{Replace } x \text{ by } -by + 1 \text{ in second equation.}$$
$$-by + 2y + 1 = 2$$
$$y(-b + 2) + 1 = 2$$

The variable y will "drop out" of this equation when $-b + 2 = 0$.

$$-b + 2 = 0 \quad \rightarrow \quad -b = -2 \quad \rightarrow \quad b = 2$$

Therefore, when $b = 2$, the equation becomes $1 = 2$, a *false* statement. The system is inconsistent when $b = 2$.

61. *Think About It* Find the values of a or b such that the following system is inconsistent.

$$-6x + y = 4$$
$$2x + by = 3$$

Solution

$$-6x + y = 4$$
$$2x + by = 3 \quad \rightarrow \quad 2x = -by + 3 \quad \rightarrow \quad x = -\frac{b}{2}y + \frac{3}{2}$$
$$-6\left(-\frac{b}{2}y + \frac{3}{2}\right) + y = 4 \qquad \text{Replace } x \text{ by } -\frac{b}{2}y + \frac{3}{2} \text{ in first equation.}$$
$$3by - 9 + y = 4$$
$$3by + y - 9 = 4$$
$$y(3b + 1) - 9 = 4$$

The variable y will "drop out" of this equation when $3b + 1 = 0$.

$$3b + 1 = 0 \quad \rightarrow \quad 3b = -1 \quad \rightarrow \quad b = -\frac{1}{3}$$

Therefore, when $b = -\frac{1}{3}$, the equation becomes $-9 = 4$, which is a *false* statement. The system is inconsistent when $b = -\frac{1}{3}$.

63. *Problem Solving*   You are selling football tickets. Student tickets cost \$2 and general admission tickets cost \$3. You sell 1957 tickets and collect \$5035. How many of each type of ticket did you sell?

Solution

$$s + g = 1957 \quad \rightarrow \quad s = 1957 - g$$
$$2s + 3g = 5035$$
$$2(1957 - g) + 3g = 5035 \qquad \qquad \text{Replace } s \text{ by } 1957 - g \text{ in second equation.}$$
$$3914 - 2g + 3g = 5035$$
$$3914 + g = 5035$$
$$g = 5035 - 3914$$
$$g = 1211 \quad \rightarrow \quad s = 1957 - 1121$$
$$s = 836$$

(836, 1121)

You sold 836 student tickets and 1121 general admission tickets.

65. *Analyzing Data*   United States exports (in billions of dollars) for the years 1989 through 1992 are given in the table in the textbook.

(a) Plot the points (x, y) on a rectangular coordinate system where x represents the year, with $x = 0$ corresponding to 1990, and y represents exports.

(b) The line $y = mx + b$ that best fits the given data is given by the following system of linear equations.

$$4b + 2m = 1626.2$$
$$2b + 6m = 953.1$$

Use the method of substitution to solve the system and find the equation of the required line. Graph the line on the coordinate system of part (a).

(c) Interpret the meaning of the slope of the line in the context of this problem.

Solution

(a)

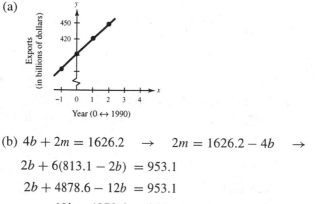

(b) $4b + 2m = 1626.2 \quad \rightarrow \quad 2m = 1626.2 - 4b \quad \rightarrow \quad m = 813.1 - 2b$

$$2b + 6(813.1 - 2b) = 953.1 \qquad \qquad \text{Replace } m \text{ by } 813.1 - 2b \text{ in second equation.}$$
$$2b + 4878.6 - 12b = 953.1$$
$$-10b + 4878.6 = 953.1$$
$$-10b = -3925.5$$
$$b = 392.55 \quad \rightarrow \quad m = 813.1 - 2(392.55)$$
$$m = 813.1 - 785.1$$
$$m = 28$$

The equation of the line is $y = 28x + 392.55$. The graph of the line is shown on the graph in part (a).

(c) The slope represents the annual increase in exports in billions of dollars.

Mid-Chapter Quiz for Chapter 8

1. Is $(4, -2)$ a solution of $3x + 4y = 4$ *and* $5x - 3y = 14$? Explain your reasoning.

Solution

$$3x + 4y = 4 \qquad\qquad 5x - 3y = 14$$

$$3(4) + 4(-2) \overset{?}{=} 4 \qquad 5(4) - 3(-2) \overset{?}{=} 14$$

$$12 - 8 = 4 \qquad\qquad 20 + 6 \ne 14$$

$(4, -2)$ is not a solution.

2. Use the graph shown in the textbook to solve the following system.

$$x + \ \ y = \ \ 5$$
$$x - 3y = -3$$

Solution

The graph indicates that the solution of the system is $(3, 2)$.

3. Use the graph shown in the textbook to solve the following system.

$$x + 2y = 6$$
$$3x - 4y = 8$$

Solution

The graph indicates that the solution of the system is $(4, 1)$.

4. Use the graph shown in the textbook to solve the following system.

$$y = \tfrac{3}{2}$$
$$x - 2y = 0$$

Solution

The graph indicates that the solution of the system is $\left(3, \tfrac{3}{2}\right)$.

5. Solve the following system graphically. Check the solution.

$$x \qquad\ \ = 6$$
$$x + y = 8$$

Solution

The solution is $(6, 2)$.

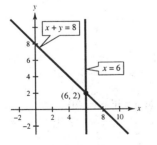

6. Solve the following system graphically. Check the solution.

$$y = \tfrac{3}{2}x - 1$$
$$y = -x + 4$$

Solution

The solution is $(2, 2)$.

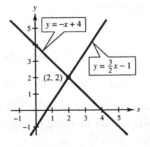

7. Solve the following system graphically. Check the solution.

$$4x + y = 0$$

$$-x + y = 5$$

Solution

The solution is $(-1, 4)$.

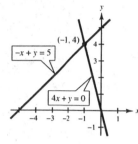

8. Solve the following system graphically. Check the solution.

$$x + y = -2$$

$$x - y = \quad 4$$

Solution

The solution is $(1, -3)$.

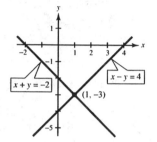

9. Use substitution to solve the following system.

$$x - y = 4$$

$$y = 2$$

Solution

$$x - y = 4$$

$$y = 2$$

$$x - 2 = 4 \qquad \text{Replace } y \text{ by 2 in first equation.}$$

$$x = 6$$

The solution is $(6, 2)$.

10. Use substitution to solve the following system.

$$y = -\tfrac{2}{3}x + 5$$

$$y = \quad 2x - 3$$

Solution

$$y = -\tfrac{2}{3}x + 5$$

$$y = 2x - 3$$

$$2x - 3 = -\tfrac{2}{3}x + 5 \qquad \text{Replace } y \text{ by } 2x - 3 \text{ in first equation.}$$

$$3(2x - 3) = 3\left(-\tfrac{2}{3}x + 5\right)$$

$$6x - 9 = -2x + 15$$

$$8x - 9 = 15$$

$$8x = 24$$

$$x = 3 \quad \rightarrow \quad y = 2(3) - 3$$

$$(3, 3) \qquad\qquad y = 3$$

The solution is $(3, 3)$.

11. Use substitution to solve the following system.

$$2x - y = -7$$
$$4x + 3y = 16$$

Solution

$$2x - y = -7 \quad \rightarrow \quad -y = -2x - 7 \quad \rightarrow \quad y = 2x + 7$$

$$4x + 3y = 16$$

$$4x + 3(2x + 7) = 16 \qquad\qquad\qquad \text{Replace } y \text{ by } 2x + 7 \text{ in second equation.}$$

$$4x + 6x + 21 = 16$$

$$10x + 21 = 16$$

$$10x = -5$$

$$x = -\tfrac{1}{2} \quad \rightarrow \quad y = 2\left(-\tfrac{1}{2}\right) + 7$$

$$y = -1 + 7$$

$$y = 6$$

$$\left(-\tfrac{1}{2}, 6\right)$$

The solution is $\left(-\tfrac{1}{2}, 6\right)$.

12. Use substitution to solve the following system.

$$-x + 3y = 10$$
$$9x - 4y = 5$$

Solution

$$-x + 3y = 10 \quad \rightarrow \quad -x = -3y + 10 \quad \rightarrow \quad x = 3y - 10$$

$$9x - 4y = 5$$

$$9(3y - 10) - 4y = 5 \qquad\qquad\qquad \text{Replace } x \text{ by } 3y - 10 \text{ in second equation.}$$

$$27y - 90 - 4y = 5$$

$$23y - 90 = 5$$

$$23y = 95$$

$$y = \tfrac{95}{23} \quad \rightarrow \quad x = 2\left(\tfrac{95}{23}\right) - 10$$

$$x = \tfrac{285}{23} - \tfrac{230}{23}$$

$$x = \tfrac{55}{23}$$

$$\left(\tfrac{55}{23}, \tfrac{95}{23}\right)$$

The solution is $\left(\tfrac{55}{23}, \tfrac{95}{23}\right)$.

13. Find a system of linear equations that has $(0, 0)$ as its only solution. (There are many correct answers.)

Solution

There are many correct answers. Here are some examples.

$$3x - y = 0 \qquad\qquad x + y = 0$$
$$4x + y = 0 \qquad -2x + 5y = 0$$

14. Find a system of linear equations that has $(6, -8)$ as its only solution. (There are many correct answers.)

Solution

There are many correct answers. Here are some examples.

$$4x + 3y = 0 \qquad x - y = 14$$
$$2x + y = 4 \qquad 3x + 2y = 2$$

15. Find a system of linear equations that has $\left(2, \frac{5}{2}\right)$ as its only solution. (There are many correct answers.)

Solution

There are many correct answers. Here are some examples.

$$x + 2y = 7 \qquad 5x - 4y = 0$$
$$3x - 2y = 1 \qquad -2x + 6y = 11$$

16. Find a system of linear equations that has $(0.8, 3.4)$ as its only solution. (There are many correct answers.)

Solution

There are many correct answers. Here are some examples.

$$-x + 2y = 6 \qquad 2x + y = 5$$
$$3x - y = -1 \qquad x + 3y = 11$$

17. Find the value of k so that the following system is inconsistent.

$$5x + ky = 3$$
$$10x - 4y = 1$$

Solution

$$5x + ky = 3 \quad \rightarrow \quad 5x = -ky + 3 \quad \rightarrow \quad x = \frac{-ky}{5} + \frac{3}{5}$$

$$10x - 4y = 1$$

$$10\left(\frac{-ky}{5} + \frac{3}{5}\right) - 4y = 1$$

$$-2ky + 6 - 4y = 1$$

$$-(2k + 4)y + 6 = 1$$

The variable y will "drop out" of this equation when $2k + 4 = 0$.

$$2k + 4 = 0 \quad \rightarrow \quad 2k = -4 \quad \rightarrow \quad k = -2$$

Therefore, when $k = -2$, the equation becomes $6 = 1$, which is a *false* statement. The system is inconsistent when $k = -2$.

18. Find the value of k so that the following system is inconsistent.

$$8x - 5y = 16$$

$$kx - 0.5y = 3$$

Solution

$$8x - 5y = 16$$

$$kx - 0.5y = 3 \quad \rightarrow \quad -0.5y = -kx + 3 \quad \rightarrow \quad y = 2kx - 6$$

$$8x - 5(2kx - 6) = 16 \qquad \qquad \text{Replace } y \text{ by } 2kx - 6 \text{ in first equation.}$$

$$8x - 10kx + 30 = 16$$

$$(8 - 10k)x + 30 = 16$$

The variable x will "drop out" of this equation when $8 - 10k = 0$.

$$8 - 10k = 0 \quad \rightarrow \quad 8 = 10k \quad \rightarrow \quad 0.8 = k$$

Therefore, when $k = 0.8$, the equation becomes $30 = 16$, which is a *false* statement. The system is inconsistent when $k = 0.8$.

19. The sum of two numbers is 50 and their difference is 22. Write a system of equations that models this problem and solve the system.

Solution

$$x + y = 50 \quad \rightarrow \quad y = -x + 50$$

$$x - y = 22$$

$$x - (-x + 50) = 22 \qquad \text{Replace } y \text{ by } -x + 50 \text{ in the second equation.}$$

$$x + x - 50 = 22$$

$$2x - 50 = 22$$

$$2x = 72$$

$$x = 36 \quad \rightarrow \quad y = -36 + 50$$

$$y = 14$$

(36, 14)

The two numbers are 36 and 14.

20. A student spent a total of $32 for a book and a calendar. The price of the book was $2 more than four times the price of the calendar. Write a system of equations that models this problem, and solve the system.

Solution

$$b + c = 32$$

$$b = 4c + 2$$

$$(4c + 2) + c = 32 \qquad \text{Replace } b \text{ by } 4c + 2 \text{ in first equation.}$$

$$5c + 2 = 32$$

$$5c = 30$$

$$c = 6 \text{ and } 4c + 2 = 4(6) + 2 = 26$$

(26, 6)

The price of the book was $26, and the price of the calendar was $6.

8.3 | Solving Systems of Equations by Elimination

7. Use elimination to solve the following system. Use the graph in the textbook to check your solution.

$$2x + y = 4$$
$$x - y = 2$$

Solution

$$\begin{array}{l} 2x + y = 4 \\ \underline{x - y = 2} \\ 3x \quad\; = 6 \\ x \quad\;\; = 2 \end{array}$$

$$2 - y = 2 \qquad \text{Replace } x \text{ by 2 in second equation.}$$

$$-y = 0$$

$$y = 0$$

$$(2, \; 0)$$

9. Use elimination to solve the following system. Use the graph in the textbook to check your solution.

$$9x - 3y = -1$$
$$3x + 6y = -5$$

Solution

$$\begin{array}{llll} 9x - 3y = -1 & \rightarrow & 18x - 6y = -2 & \text{Multiply equation by 2.} \\ 3x + 6y = -5 & \rightarrow & \underline{3x + 6y = -5} \\ & & 21x \quad\;\;\; = -7 \\ & & x \quad\quad\;\; = -\frac{1}{3} \end{array}$$

$$9\left(-\tfrac{1}{3}\right) - 3y = -1 \qquad \text{Replace } x \text{ by } -\tfrac{1}{3} \text{ in first equation.}$$

$$-3 - 3y = -1$$

$$-3y = 2$$

$$y = -\tfrac{2}{3}$$

$$\left(-\tfrac{1}{3}, \; -\tfrac{2}{3}\right)$$

11. Use elimination to solve the following system. Check your result algebraically.

$$x - y = 4$$
$$x + y = 12$$

Solution

$$\begin{array}{r} x - y = 4 \\ \underline{x + y = 12} \\ 2x \quad = 16 \\ x \quad = 8 \end{array}$$

$$8 + y = 12 \qquad \text{Replace } x \text{ by 8 in second equation.}$$

$$y = 4$$

$$(8, 4)$$

13. Use elimination to solve the following system. Check your result algebraically.

$$x + 7y = 12$$
$$3x - 5y = 10$$

Solution

$$\begin{array}{llll} x + 7y = 12 & \rightarrow & -3x - 21y = -36 & \text{Multiply equation by } -3. \\ 3x - 5y = 10 & \rightarrow & \underline{3x - 5y = 10} \\ & & \quad -26y = -26 \\ & & \qquad y = 1 \end{array}$$

$$x + 7(1) = 12 \qquad \text{Replace } y \text{ by 1 in first equation.}$$

$$x + 7 = 12$$

$$x = 5$$

$$(5, 1)$$

15. Use elimination to solve the following system. Check your result algebraically.

$$6r - 5s = 3$$
$$-12r + 10s = 5$$

Solution

$$\begin{array}{llll} 6r - 5s = 3 & \rightarrow & 12r - 10s = 6 & \text{Multiply equation by 2.} \\ -12r + 10s = 5 & \rightarrow & \underline{-12r + 10s = 5} \\ & & \quad\quad 0 = 11 & \text{False} \end{array}$$

The system has *no* solution; it is *inconsistent*.

17. Use elimination to solve the following system. Check your result algebraically.

$$\frac{1}{2}s - t = \frac{3}{2}$$
$$4s + 2t = 27$$

Solution

$\frac{1}{2}s - t = \frac{3}{2}$ → $s - 2t = 3$ Multiply equation by 2.

$4s + 2t = 27$ → $\underline{4s + 2t = 27}$

$ 5s = 30$

$ s = 6$

$4(6) + 2t = 27$ Replace s by 6 in second equation.

$24 + 2t = 27$

$2t = 3$

$t = \frac{3}{2}$

$\left(6, \frac{3}{2}\right)$

19. Use elimination to solve the following system. Use a graphing utility to check your solution.

$$7x + 8y = 6$$
$$3x - 4y = 10$$

Solution

$7x + 8y = 6$ → $7x + 8y = 6$

$3x - 4y = 10$ → $\underline{6x - 8y = 20}$ Multiply equation by 2.

$ 13x = 26$

$ x = 2$

$7(2) + 8y = 6$ Replace x by 2 in first equation.

$14 + 8y = 6$

$8y = -8$

$y = -1$

The solution is $(2, -1)$.

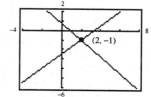

21. Use elimination to solve the following system. Use a graphing utility to check your solution.

$$5x + 2y = 7$$
$$3x - 6y = -3$$

Solution

$$5x + 2y = 7 \quad \rightarrow \quad 15x + 6y = 21 \quad \text{Multiply equation by 3.}$$
$$3x - 6y = -3 \quad \rightarrow \quad \underline{3x - 6y = -3}$$
$$18x = 18$$
$$x = 1$$

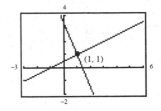

$$5(1) + 2y = 7 \quad \text{Replace } x \text{ by 1 in first equation.}$$
$$5 + 2y = 7$$
$$2y = 2$$
$$y = 1$$

The solution is $(1, 1)$.

23. Use the most convenient method (graphing, substitution, or elimination) to solve the following system.

$$x - y = 2$$
$$y = 3$$

Solution

$$x - y = 2$$
$$y = 3$$
$$x - 3 = 2 \quad \text{Replace } y \text{ by 3 in first equation.}$$
$$x = 5$$

$(5, 3)$

25. Use the most convenient method (graphing, substitution, or elimination) to solve the following system.

$$6x + 21y = 132$$
$$6x - 4y = 32$$

Solution

$$6x + 21y = 132 \quad \rightarrow \quad 6x + 21y = 132$$
$$6x - 4y = 32 \quad \rightarrow \quad \underline{-6x + 4y = -32} \quad \text{Multiply equation by } -1.$$
$$25y = 100$$
$$y = 4$$

$$6x + 21(4) = 132 \quad \text{Replace } y \text{ by 4 in first equation.}$$
$$6x + 84 = 132$$
$$6x = 48$$
$$x = 8$$

$(8, 4)$

27. Use the most convenient method (graphing, substitution, or elimination) to solve the following system.

$$y = 2x - 1$$
$$y = x + 1$$

Solution

$$y = 2x - 1$$
$$y = x + 1$$
$$2x - 1 = x + 1 \qquad \text{Replace } y \text{ by } 2x - 1 \text{ in second equation.}$$
$$x - 1 = 1$$
$$x = 2 \qquad y = 2 + 1 \qquad \text{Replace } x \text{ by 2 in second equation.}$$

$$(2, 3) \qquad\qquad y = 3$$

29. *Think About It*   Find a system of linear equations that has the solution $\left(6, \frac{4}{3}\right)$. (*Note:* The problem has many correct answers.)

Solution

Here are *some* systems that have the solution $\left(6, \frac{4}{3}\right)$. There are many other correct answers.

$$\begin{array}{ccccc} x + 3y = 10 & \text{or} & 2x - 3y = 8 & \text{or} & x - 3y = 2 \\ x - 6y = -2 & & x + 9y = 18 & & -x + 12y = 10 \end{array}$$

31. *Ticket Sales*   Ticket sales for a play were \$3799 on the first night and \$4905 on the second night. The first night 213 student tickets and 632 general admission tickets were sold. The second night 275 student tickets and 816 general admission tickets were sold. Find the price of each type of ticket.

Solution

Verbal model: $\boxed{213} \cdot \boxed{\begin{array}{c}\text{Price of} \\ \text{student ticket}\end{array}} + \boxed{632} \cdot \boxed{\begin{array}{c}\text{Price of general} \\ \text{admission ticket}\end{array}} = \boxed{\begin{array}{c}\text{Income on} \\ \text{first night}\end{array}}$

$\boxed{275} \cdot \boxed{\begin{array}{c}\text{Price of} \\ \text{student ticket}\end{array}} + \boxed{816} \cdot \boxed{\begin{array}{c}\text{Price of general} \\ \text{admission ticket}\end{array}} = \boxed{\begin{array}{c}\text{Income on} \\ \text{second night}\end{array}}$

Labels: Price of student ticket $= x$ (dollars)
Price of general admission ticket $= y$ (dollars)
Income on first night $= \$3799$
Income on second night $= \$4905$

System of equations: $213x + 632y = 3799$

$$275x + 816y = 4905$$

Solving by elimination:

$$58{,}575x + 173{,}800y = 1{,}044{,}725 \qquad \text{Multiply equation by 275.}$$
$$-58{,}575x - 173{,}808y = -1{,}044{,}765 \qquad \text{Multiply equation by } -213.$$

$$\rule{6cm}{0.4pt}$$

$$-8y = -40$$
$$y = 5$$

$$213x + 632(5) = 3799 \qquad \text{Replace } y \text{ by 5 in first equation.}$$
$$213x + 3160 = 3799$$
$$213x = 639$$
$$x = 3$$

Thus, student tickets cost \$3 and general admission tickets cost \$5.

33. Plot $(-6, 4)$ and $(-3, -4)$ in a rectangular coordinate system. Find the slope of the line passing through the points.

Solution

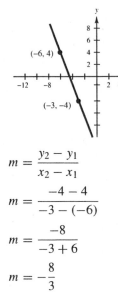

$$m = \frac{y_2 - y_1}{x_2 - x_1}$$

$$m = \frac{-4 - 4}{-3 - (-6)}$$

$$m = \frac{-8}{-3 + 6}$$

$$m = -\frac{8}{3}$$

35. Plot $\left(\frac{7}{2}, \frac{9}{2}\right)$ and $\left(\frac{4}{3}, -3\right)$ in a rectangular coordinate system. Find the slope of the line passing through the points.

Solution

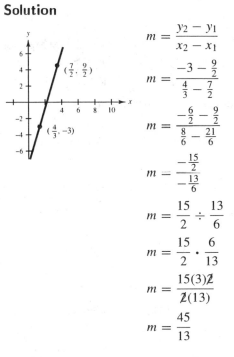

$$m = \frac{y_2 - y_1}{x_2 - x_1}$$

$$m = \frac{-3 - \frac{9}{2}}{\frac{4}{3} - \frac{7}{2}}$$

$$m = \frac{-\frac{6}{2} - \frac{9}{2}}{\frac{8}{6} - \frac{21}{6}}$$

$$m = \frac{-\frac{15}{2}}{-\frac{13}{6}}$$

$$m = \frac{15}{2} \div \frac{13}{6}$$

$$m = \frac{15}{2} \cdot \frac{6}{13}$$

$$m = \frac{15(3)\cancel{2}}{\cancel{2}(13)}$$

$$m = \frac{45}{13}$$

37. *Defective Units* A quality control engineer finds three defective units in a sample of 100. How many would you expect to find in a sample of 5000 units?

Solution

$$\frac{3}{100} = \frac{x}{5000}$$

$$100x = 3(5000) = 15,000$$

$$x = \frac{15,000}{100} = 150$$

39. Use elimination to solve the following system. Use the graph shown in the textbook to check your solution.

$$x - y = 0$$
$$3x - 2y = -1$$

Solution

$$\begin{array}{ll} x - y = 0 & \rightarrow \quad -2x + 2y = 0 \quad \text{Multiply equation by } -2. \\ 3x - 2y = -1 & \rightarrow \quad \underline{\quad 3x - 2y = -1} \\ & \qquad\qquad\quad x \qquad = -1 \end{array}$$

$$-1 - y = 0 \qquad \text{Replace } x \text{ by } -1 \text{ in first equation.}$$

$$-y = 1$$

$$y = -1$$

$$(-1, -1)$$

41. Use elimination to solve the following system. Use the graph shown in the textbook to check your solution.

$$x - y = 1$$
$$-2x + 2y = 5$$

Solution

$$
\begin{array}{lcl}
x - y = 1 & \rightarrow & 2x - 2y = 2 \quad \text{Multiply equation by 2.}\\
-2x + 2y = 5 & \rightarrow & \underline{-2x + 2y = 5}\\
& & \qquad\quad 0 = 7 \quad \text{False}
\end{array}
$$

The system has *no* solution; it is *inconsistent*.

43. Use elimination to solve the following system. Check your result algebraically.

$$3x - 5y = 1$$
$$2x + 5y = 9$$

Solution

$$
\begin{array}{ll}
3x - 5y = \ 1 &\\
\underline{2x + 5y = \ 9} &\\
5x \qquad\ = 10 &\\
x \qquad\ = \ 2 &
\end{array}
$$

$$3(2) - 5y = 1 \qquad \text{Replace } x \text{ by 2 in first equation.}$$
$$6 - 5y = 1$$
$$-5y = -5$$
$$y = 1$$

$$(2, \ 1)$$

45. Use elimination to solve the following system. Check your result algebraically.

$$5x + 2y = \ 7$$
$$3x - y = 13$$

Solution

$$
\begin{array}{lcl}
5x + 2y = \ 7 & \rightarrow & 5x + 2y = \ 7\\
3x - y = 13 & \rightarrow & \underline{6x - 2y = 26} \quad \text{Multiply equation by 2.}\\
& & 11x \qquad = 33\\
& & \ x \qquad\ \ = \ 3
\end{array}
$$

$$5(3) + 2y = 7 \qquad \text{Replace } x \text{ by 3 in first equation.}$$
$$15 + 2y = 7$$
$$2y = -8$$
$$y = -4$$

$$(3, \ -4)$$

47. Use elimination to solve the following system. Check your result algebraically.

$$3x + 2y = 10$$
$$2x + 5y = 3$$

Solution

$$3x + 2y = 10 \quad \rightarrow \quad 6x + 4y = 20 \qquad \text{Multiply equation by 2.}$$
$$2x + 5y = 3 \quad \rightarrow \quad \underline{-6x - 15y = -9} \qquad \text{Multiply equation by } -3.$$
$$-11y = 11$$
$$y = -1$$

$$3x + 2(-1) = 10 \qquad \text{Replace } y \text{ by } -1 \text{ in first equation.}$$
$$3x - 2 = 10$$
$$3x = 12$$
$$x = 4$$

$$(4, \ -1)$$

49. Use elimination to solve the following system. Check your result algebraically.

$$3x - 2y = 6$$
$$-6x + 4y = -12$$

Solution

$$3x - 2y = 6 \quad \rightarrow \quad 6x - 4y = 12$$
$$-6x + 4y = -12 \quad \rightarrow \quad \underline{-6x + 4y = -12}$$
$$0 = 0$$

The system is dependent and has infinitely many solutions. The solution consists of all ordered pairs (x, y) such that $3x - 2y = 6$.

51. Use elimination to solve the following system. Check your solution algebraically.

$$2u + v = 120$$
$$u + 2v = 120$$

Solution

$$2u + v = 120 \quad \rightarrow \quad -4u - 2v = -240 \qquad \text{Multiply equation by } -2.$$
$$u + 2v = 120 \quad \rightarrow \quad \underline{u + 2v = 120}$$
$$-3u = -120$$
$$u = 40$$

$$2(40) + v = 120 \qquad \text{Replace } u \text{ by 40 in first equation.}$$
$$80 + v = 120$$
$$v = 40$$

$$(40, \ 40)$$

53. Use elimination to solve the following system. Check your solution algebraically.

$$3a + 3b = 7$$
$$3a + 5b = 3$$

Solution

$$3a + 3b = 7 \quad \rightarrow \quad -3a - 3b = -7 \qquad \text{Multiply equation by } -1.$$
$$3a + 5b = 3 \quad \rightarrow \quad \underline{3a + 5b = 3}$$
$$2b = -4$$
$$b = -2$$

$$3a + 3(-2) = 7 \qquad \text{Replace } b \text{ by } -2 \text{ in first equation.}$$
$$3a - 6 = 7$$
$$3a = 13$$
$$a = \tfrac{13}{3}$$
$$\left(\tfrac{13}{3},\ -2\right)$$

55. Use elimination to solve the following system. Check your solution algebraically.

$$10x - 8y = 18$$
$$5x - 4y = 9$$

Solution

$$10x - 8y = 18 \quad \rightarrow \quad 10x - 8y = 18$$
$$5x - 4y = 9 \quad \rightarrow \quad \underline{-10x + 8y = -18} \qquad \text{Multiply equation by } -2.$$
$$0 = 0$$

The system is dependent and has infinitely many solutions. The solution consists of all ordered pairs, (x, y) such that $5x - 4y = 9$.

57. Use elimination to solve the following system. Check your solution algebraically.

$$0.02x - 0.05y = -0.19$$
$$0.03x + 0.04y = 0.52$$

Solution

Multiply both equations by 100.

$$2x - 5y = -19 \quad \rightarrow \quad 8x - 20y = -76 \qquad \text{Multiply equation by 4.}$$
$$3x + 4y = 52 \quad \rightarrow \quad \underline{15x + 20y = 260} \qquad \text{Multiply equation by 5.}$$
$$23x = 184$$
$$x = 8$$

$$2(8) - 5y = -19 \qquad \text{Replace } x \text{ by 8 in first revised equation.}$$
$$16 - 5y = -19$$
$$-5y = -35$$
$$y = 7$$
$$(8,\ 7)$$

59. Use elimination to solve the following system. Use a graphing utility to check your solution.

$$x + 2y = 3$$
$$-x - y = -1$$

Solution

$$x + 2y = 3$$
$$-x - y = -1$$
$$y = 2$$
$$x + 2(2) = 3$$
$$x + 4 = 3$$
$$x = -1$$

$(-1, 2)$

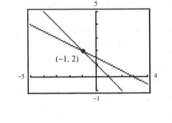

$(-1, 2)$

61. Use elimination to solve the following system. Use a graphing utility to check your solution.

$$8x - 4y = 7$$
$$5x + 2y = 1$$

Solution

$$8x - 4y = 7 \quad \rightarrow \quad 8x - 4y = 7$$
$$5x + 2y = 1 \quad \rightarrow \quad \underline{10x + 4y = 2} \quad \text{Multiply equation by 2.}$$
$$18x \quad\quad\ = 9$$
$$x \quad\quad = \tfrac{1}{2}$$

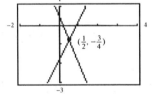

$$8\left(\tfrac{1}{2}\right) - 4y = 7$$
$$4 - 4y = 7$$
$$-4y = 3$$
$$y = -\tfrac{3}{4}$$

$\left(\tfrac{1}{2}, -\tfrac{3}{4}\right)$

63. Use the most convenient method (graphing, substitution or elimination) to solve the following system.

$$2x - y = 4$$
$$y = x$$

Solution

$$2x - x = 4 \qquad\qquad \text{Replace } y \text{ by } x \text{ in first equation.}$$
$$x = 4 \qquad y = 4 \qquad \text{Replace } x \text{ by 4 in second equation.}$$

$(4, \ 4)$

65. Use the most convenient method (graphing, substitution or elimination) to solve the following system.

$$-4x + 3y = 11$$
$$3x - 10y = 15$$

Solution

$-4x + 3y = 11$	$\rightarrow$	$-12x + 9y = 33$	Multiply equation by 3.
$3x - 10y = 15$	$\rightarrow$	$12x - 40y = 60$	Multiply equation by 4.

$$-31y = 93$$
$$y = -3$$

$-4x + 3(-3) = 11$ Replace y by -3 in first equation.

$$-4x - 9 = 11$$
$$-4x = 20$$
$$x = -5$$

$$(-5, -3)$$

67. Use the most convenient method (graphing, substitution, or elimination) to solve the following system.

$$y = \tfrac{1}{2}x - 2$$
$$x = 2(4 - y)$$

Solution

$$y = \tfrac{1}{2}x - 2$$
$$x = 2(4 - y)$$
$$x = 2\left[4 - \left(\tfrac{1}{2}x - 2\right)\right] \qquad \text{Replace } y \text{ by } \tfrac{1}{2}x - 2 \text{ in second equation.}$$
$$x = 2\left(4 - \tfrac{1}{2}x + 2\right)$$
$$x = 2\left(6 - \tfrac{1}{2}x\right)$$
$$x = 12 - x$$
$$2x = 12$$
$$x = 6 \quad \rightarrow \quad y = \tfrac{1}{2}(6) - 2$$
$$y = 3 - 2$$
$$y = 1$$

$$(6, 1)$$

69. Use the most convenient method (graphing, substitution, or elimination) to solve the following system.

$$x + 0.3y = 5$$
$$0.5x + y = 11$$

Solution

$$x + 0.3y = 5 \quad \rightarrow \quad x + 0.3y = 5$$
$$0.5x + y = 11 \quad \rightarrow \quad \underline{-x - 2y = -22} \quad \text{Multiply equation by } -2.$$
$$-1.7y = -17$$
$$y = 10$$

$$0.5x + 10 = 11 \qquad \text{Replace } y \text{ by 10 in second equation.}$$
$$0.5x = 1$$
$$x = 2$$

$(2, 10)$

71. *Focal Length of a Camera* When parallel rays of light pass through a convex lens, they are bent inward and meet at a *focus* (see figure shown in textbook). The distance from the center of the lens to the focus is called the *focal length*. The equations of the lines containing the two bent rays in the camera are

$$x + 3y = 1$$
$$-x + 3y = -1$$

where x and y are measured in inches. Which of these equations is the upper ray? What is the focal length?

Solution

The upper ray has a negative slope. The first equation can be written in slope-intercept form as $y = -\frac{1}{3}x + \frac{1}{3}$, and this indicates a negative slope of $-\frac{1}{3}$. Therefore, the first equation is the upper ray.

$$x + 3y = 1$$
$$\underline{-x + 3y = -1}$$
$$6y = 0$$
$$y = 0$$
$$x + 3(0) = 1 \qquad \text{Replace } y \text{ by 0 in first equation.}$$
$$x = 1$$

$(1, 0)$

The rays meet at $(1, 0)$. This focus is 1 inch from the center of the lens at $(0, 0)$. Therefore, the focal length is 1.

73. *Problem Solving* The sum of two numbers is 154, and the difference of the numbers is 38. Find the numbers.

Solution

Let x and y represent the two numbers.

$$x + y = 154$$
$$\underline{x - y = 38}$$
$$2x = 192$$
$$x = 96$$
$$96 + y = 154 \qquad \text{Replace } x \text{ by 96 in first equation.}$$
$$y = 58$$

The two numbers are 96 and 58.

8.4 Applications of Systems of Linear Equations

5. *Modeling* The total cost of 15 gallons of regular gasoline and 10 gallons of premium gasoline is $35.50. Premium costs $0.20 more per gallon than regular. What is the cost per gallon of each type of gasoline?

(a) Write a verbal model for this problem.
(b) Assign labels to the verbal model.
(c) Use the labels to write a linear system.
(d) Solve the system and answer the question.

Solution

(a) *Verbal model:* $15 \boxed{\text{Price of regular gasoline}} + 10 \boxed{\text{Price of premium gasoline}} = \boxed{\text{Total cost}}$

$\boxed{\text{Price of premium gasoline}} = \boxed{\text{Price of regular gasoline}} + 0.20$

(b) *Labels:* Price of regular gasoline $= x$ (dollars per gallon)
Price of premium gasoline $= y$ (dollars per gallon)
Total cost $= 35.50$ (dollars)

(c) *System:* $15x + 10y = 35.50$
$$y = x + 0.20$$

(d) $15x + 10y = 35.50$
$$y = x + 0.20$$
$$15x + 10(x + 0.20) = 35.50 \qquad \text{Replace } y \text{ by } x + 0.20 \text{ in first equation.}$$
$$15x + 10x + 2 = 35.50$$
$$25x + 2 = 35.50$$
$$25x = 33.50$$
$$x = \frac{33.50}{25}$$
$$x = 1.34 \text{ and } y = 1.34 + 0.20$$
$$y = 1.54$$

$(1.34, 1.54)$

Regular gasoline costs $1.34 per gallon, and premium gasoline costs $1.54 per gallon.

7. The sum of the larger number and twice the smaller is 100, and their difference is 10. Find the numbers.

Solution

Verbal model: | Larger number | + | 2 | · | Smaller number | = | 100 |

| Larger number | − | Smaller number | = | 10 |

Labels: Larger number $= x$
Smaller number $= y$

System of equations: $x + 2y = 100$

$x - y = 10$

Solving by elimination:

$$
\begin{aligned}
x + 2y &= 100 &\rightarrow&& x + 2y &= 100 \\
x - y &= 10 &\rightarrow&& -x + y &= -10 &\quad&\text{Multiply by } -1. \\
\cline{5-5}
&&&& 3y &= 90 \\
&&&& y &= 30 \\
&&&& x - 30 &= 10 &\quad&\text{Replace } y \text{ by 30 in first equation.} \\
&&&& x &= 40
\end{aligned}
$$

The two numbers are 40 and 30.

9. *Geometry* A rectangle is 4 feet longer than it is wide. Its perimeter is 40 feet. Find the dimensions of the rectangle.

Solution

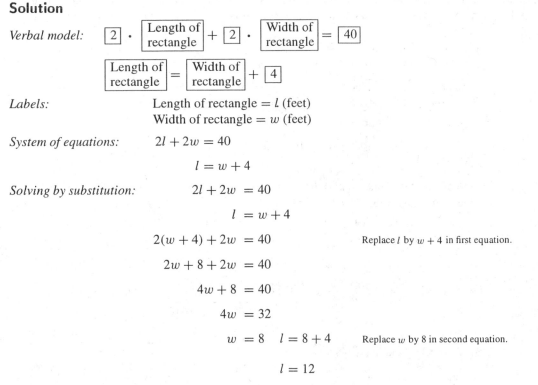

Verbal model: | 2 | · | Length of rectangle | + | 2 | · | Width of rectangle | = | 40 |

| Length of rectangle | = | Width of rectangle | + | 4 |

Labels: Length of rectangle $= l$ (feet)
Width of rectangle $= w$ (feet)

System of equations: $2l + 2w = 40$

$l = w + 4$

Solving by substitution:

$$
\begin{aligned}
2l + 2w &= 40 \\
l &= w + 4 \\
2(w + 4) + 2w &= 40 &\quad&\text{Replace } l \text{ by } w + 4 \text{ in first equation.} \\
2w + 8 + 2w &= 40 \\
4w + 8 &= 40 \\
4w &= 32 \\
w &= 8 \quad l = 8 + 4 &\quad&\text{Replace } w \text{ by 8 in second equation.} \\
l &= 12
\end{aligned}
$$

The length of the rectangle is 12 feet and the width of the rectangle is 8 feet.

11. *Wholesale Cost*   A watch sells for $108.75. The markup rate is 45% of the wholesale cost. Find the wholesale cost.

Solution

Verbal model: $\boxed{\text{Wholesale cost}} + \boxed{\text{Markup}} = \boxed{\text{Selling price}}$

$\boxed{\text{Markup}} = \boxed{0.45} \cdot \boxed{\text{Wholesale cost}}$

Labels: Wholesale cost $= C$ (dollars)
Markup $= M$ (dollars)
Selling price $= \$108.75$

System of equations: $C + M = 108.75$

$M = 0.45C$

Solving by substitution: $C + M\ = 108.75$

$M\ = 0.45C$

$C + 0.45C\ = 108.75$ Replace M by $0.45C$ in first equation.

$1.45C\ = 108.75$

$C\ = 75$

The wholesale cost of the watch is $75.00.

13. *Nut Mixture*   Ten pounds of mixed nuts sell for $5.86 per pound. The mixture contains two kinds of nuts: one costs $4.25 per pound and the other costs $6.55 per pound. How many pounds of each kind of nut were used in the mixture?

Solution

Verbal model: $\boxed{\begin{array}{c}\text{Pounds of}\\\$4.25/\text{lb nuts}\end{array}} + \boxed{\begin{array}{c}\text{Pounds of}\\\$6.55/\text{lb nuts}\end{array}} = \boxed{\begin{array}{c}\text{Pounds of}\\\text{mixture}\end{array}}$

$\boxed{\begin{array}{c}\text{Value of}\\\$4.25/\text{lb nuts}\end{array}} + \boxed{\begin{array}{c}\text{Value of}\\\$6.55/\text{lb nuts}\end{array}} = \boxed{\begin{array}{c}\text{Value of}\\\text{mixture}\end{array}}$

Labels: Pounds of $4.25/lb nuts $= x$
Pounds of $6.55/lb nuts $= y$
Pounds of mixture $= 10$
Value of $4.25/lb $= 4.25x$ (dollars)
Value of $6.55/lb nuts $= 6.55y$ (dollars)
Value of mixture $= \$5.86(10) = \$58.60$

System of equations: $x + \quad y = \quad 10$

$4.25x + 6.55y = 58.60$

Solving by elimination: $x + \quad y = \quad 10 \quad \rightarrow \quad -425x - 425y = -4250$ Multiply by -425.

$4.25x + 6.55y = 58.60 \quad \rightarrow \quad \underline{425x + 655y = \quad 5860}$ Multiply by 100.

$230y = \quad 1610$

$y = \quad 7$

$x + 7 = 10$ Replace y by 7 in first equation.

$x = 3$

Thus, 3 pounds of the $4.25 nuts and 7 pounds of the $6.55 nuts were used in the mixture.

15. Evaluate $2(4 - 3^2)$.

Solution

$$2(4 - 3^2) = 2(4 - 9)$$
$$= 2(-5)$$
$$= -10$$

17. Evaluate $\dfrac{3 - (5 - 20)}{6}$.

Solution

$$\frac{3 - (5 - 20)}{6} = \frac{3 - (-15)}{6}$$
$$= \frac{18}{6}$$
$$= 3$$

19. *Travel Reimbursement* You are reimbursed $130 per day for lodging and meals plus $0.32 per mile driven. Write a linear model giving the daily cost C in terms of x, the number of miles driven.

Solution

$$C = 130 + 0.32x$$

21. *Problem Solving* The sum of two numbers is 67, and their difference is 17. Find the numbers.

Solution

Verbal model: $\boxed{\text{Larger number}} + \boxed{\text{Smaller number}} = \boxed{67}$

$\boxed{\text{Larger number}} - \boxed{\text{Smaller number}} = \boxed{17}$

Labels: Larger number $= x$
Smaller number $= y$

System of equations: $x + y = 67$
$x - y = 17$

Solving by elimination: $x + y = 67$
$\underline{x - y = 17}$
$2x = 84$
$x = 42$

$42 + y = 67$ Replace x by 42 in first equation.
$y = 25$

The two numbers are 42 and 25.

23. *Problem Solving* The sum of two numbers is 132, and the larger number is 6 more than twice the smaller. Find the numbers.

Solution

Verbal model:

| Larger number | + | Smaller number | = | 132 |

| Larger number | = | 2 | · | Smaller number | + | 6 |

Labels: Larger number $= x$
Smaller number $= y$

System of equations: $x + y = 132$

$x = 2y + 6$

Solving by substitution: $x + y = 132$

$x = 2y + 6$

$(2y + 6) + y = 132$ Replace x by $2y + 6$ in first equation.

$3y + 6 = 132$

$3y = 126$

$y = 42$ $x = 2(42) + 6$ Replace y by 42 in second equation.

$x = 90$

The two numbers are 90 and 42.

25. *Problem Solving* A cash register has 35 coins: nickels and quarters. The value of the coins is \$5.75. How many coins of each type are in the register?

Solution

Verbal model:

| Number of nickels | + | Number of quarters | = | Number of coins |

| Value of nickels | + | Value of quarters | = | Value of coins |

Labels: Number of nickels $= x$
Number of quarters $= y$
Number of coins $= 35$
Value of nickels $= 0.05x$ (dollars)
Value of quarters $= 0.25y$ (dollars)
Total value of coins $= \$5.75$

System of equations: $x + y = 35$

$0.05x + 0.25y = 5.75$

Solving by elimination: $x + y = 35 \rightarrow -5x - 5y = -175$ Multiply equation by -5.

$0.05x + 0.25y = 5.75 \rightarrow \underline{5x + 25y = 575}$ Multiply equation by 100.

$20y = 400$

$y = 20$

$x + 20 = 35$

$x = 15$

$(15, 20)$

There are 15 nickels and 20 quarters in the cash register.

27. *Problem Solving*   A cash register has 44 coins: nickels and dimes. The value of the coins is $3.00. How many coins of each type are in the register?

Solution

Verbal model:

$$\boxed{\text{Number of nickels}} + \boxed{\text{Number of dimes}} = \boxed{\text{Number of coins}}$$

$$\boxed{\text{Value of nickels}} + \boxed{\text{Value of dimes}} = \boxed{\text{Value of coins}}$$

Labels: Number of nickels $= x$
Number of dimes $= y$
Number of coins $= 44$
Value of nickels $= 0.05x$ (dollars)
Value of dimes $= 0.10y$ (dollars)
Total value of coins $= \$3.00$

System of equations: $x + y = 44$

$0.05x + 0.10y = 3.00$

Solving by elimination: $x + y = 44 \quad \rightarrow \quad -5x - 5y = -220$ Multiply equation by -5.

$0.05x + 0.10 = 3.00 \quad \rightarrow \quad \underline{5x + 10y = 300}$ Multiply equation by 100.

$5y = 80$

$y = 16 \quad \rightarrow \quad x + 16 = 44$

$x = 28$

$(28, 16)$

There are 28 nickels and 16 dimes in the cash register.

29. *Geometry*   The width of a rectangle if $\frac{6}{10}$ of its length. The perimeter is 16 yards. Find the dimensions of the rectangle.

Solution

Verbal model:

$$\boxed{2} \cdot \boxed{\text{Length of rectangle}} + \boxed{2} \cdot \boxed{\text{Width of rectangle}} = \boxed{16}$$

$$\boxed{\text{Width of rectangle}} = \boxed{0.6} \cdot \boxed{\text{Length of rectangle}}$$

Labels: Length of rectangle $= l$ (yards)
Width of rectangle $= w$ (yards)

System of equations: $2l + 2w = 16$

$w = 0.6l$

Solving by substitution: $2l + 2w = 16$

$w = 0.6l$

$2l + 2(0.6l) = 16$ Replace w by $0.6l$ in first equation.

$2l + 1.2l = 16$

$3.2l = 16$

$l = 5 \quad w = 0.6(5)$ Replace l by 5 in second equation.

$w = 3$

The length of the rectangle is 5 yards and the width of the rectangle is 3 yards.

31. *Problem Solving*   The sum of the digits of a given two-digit number is 12. If the digits are reversed, the number is increased by 36. Find the number.

Solution

Verbal model:

$$\boxed{\text{Ten's digit of number}} + \boxed{\text{One's digit of number}} = \boxed{12}$$

$$\boxed{\text{Number with digits reversed}} = \boxed{\text{Original number}} + \boxed{36}$$

Labels:

Ten's digit of original number $= x$
One's digit of original number $= y$
Original number $= 10x + y$
Number with digits reversed $= 10y + x$

System of equations:

$$x + y = 12$$
$$10y + x = 10x + y + 36$$

Solving by elimination:

$$x + y = 12$$
$$10y + x = 10x + y + 36$$

$$10y + x = 10x + y + 36 \qquad \text{Equation can be simplified.}$$

$$-9x + 9y = 36 \qquad \text{Combine like terms.}$$

$$-x + y = 4 \qquad \text{Divide by 9.}$$

$$x + y = 12$$
$$\underline{-x + y = 4}$$
$$2y = 16$$

$$y = 8 \qquad x + y = 12 \qquad \text{Replace } y \text{ by 8 in first equation.}$$

$$x + 8 = 12$$

$$x = 4$$

The ten's digit is 4 and the one's digit is 8, so the original number is 48.

33. *Mixture Problem*   Ten gallons of 30% acid solution is obtained by mixing a 20% solution with a 50% solution. How many gallons of each solution must be used to obtain the desired mixture?

Solution

Verbal model:

$$\boxed{\begin{array}{c}\text{Gallons of}\\ \text{20\% solution}\end{array}} + \boxed{\begin{array}{c}\text{Gallons of}\\ \text{50\% solution}\end{array}} = \boxed{\begin{array}{c}\text{Gallons of}\\ \text{30\% solution}\end{array}}$$

$$\boxed{\begin{array}{c}\text{Acid in}\\ \text{20\% solution}\end{array}} = \boxed{\begin{array}{c}\text{Acid in}\\ \text{50\% solution}\end{array}} + \boxed{\begin{array}{c}\text{Acid in}\\ \text{30\% solution}\end{array}}$$

Labels:

Number of gallons of 20% solution $= x$
Number of gallons of 50% solution $= y$
Number of gallons of 30% solution $= 10$
Acid in 20% solution $= 0.20x$ (gallons)
Acid in 50% solution $= 0.50y$ (gallons)
Acid in 30% solution $= 0.30(10) = 3$ (gallons)

33. —CONTINUED—

System of equations:

$$x + y = 10$$
$$0.20x + 0.50y = 0.30(10)$$

Solving by elimination:

$$x + y = 12 \quad \rightarrow \quad -20x - 20y = -200 \qquad \text{Multiply by } -20.$$
$$0.20x + 0.50y = 3 \quad \rightarrow \quad \underline{20x + 50y = 300} \qquad \text{Multiply by } 100.$$
$$30y = 100$$
$$y = \tfrac{10}{3} \text{ or } 3\tfrac{1}{3}$$

$$x + \tfrac{10}{3} = 10 \qquad \text{Replace } y \text{ by } \tfrac{10}{3} \text{ in first equation.}$$
$$x = \tfrac{30}{3} - \tfrac{10}{3}$$
$$x = \tfrac{20}{3} \text{ or } 6\tfrac{2}{3}$$

Thus, $6\tfrac{2}{3}$ gallons of the 20% solution and $3\tfrac{1}{3}$ gallons of the 50% solution must be used.

35. *Airplane Speed*   An airplane flying into a headwind travels 1800 miles in 3 hours and 36 minutes. On the return flight, the same distance is traveled in 3 hours. Find the speed of the plane in still air and the speed of the wind, assuming that both remain constant throughout the round trip.

Solution

Verbal model:

| Distance of first flight | = | Rate of first flight | · | Time of first flight |

| Distance of return flight | = | Rate of return flight | · | Time of return flight |

Labels:

Speed of plane in still air $= x$ (miles per hour)
Speed of wind $= y$ (miles per hour)
Distance of first flight $= 1800$ (miles)
Rate of first flight (into head wind) $= x - y$ (miles per hour)
Time of first flight $= 3\tfrac{36}{60} = 3.6$ (hours)
Distance of return flight $= 1800$ (miles)
Rate of return flight (with tail wind) $= x + y$ (miles per hour)
Time of return flight $= 3$ (hours)

System of equations:

$$1800 = (x - y)3.6$$
$$1800 = (x + y)3$$

Solving by elimination:

$$1800 = 3.6x - 3.6y \quad \rightarrow \quad 18{,}000 = 36x - 36y \qquad \text{Multiply by } 10.$$
$$1800 = 3x + 3y \quad \rightarrow \quad \underline{21{,}600 = 36x + 36y} \qquad \text{Multiply by } 12.$$
$$39{,}600 = 72x$$
$$550 = x$$

$$1800 = 3(550) + 3y \qquad \text{Replace } x \text{ by } 550 \text{ in second equation.}$$
$$1800 = 1650 + 3y$$
$$150 = 3y$$
$$50 = y$$

Thus, the speed of the plane in still air is 550 miles per hour and the speed of the wind is 50 miles per hour.

37. *Wholesale Cost*   The selling price of a cordless phone is $119.91. The markup rate is 40% of the wholesale cost. Find the wholesale cost.

Solution

Verbal model: $\boxed{\text{Wholesale cost}} + \boxed{\text{Markup}} = \boxed{\text{Selling price}}$

$\boxed{\text{Markup}} = \boxed{0.4} \cdot \boxed{\text{Wholesale cost}}$

Labels: Wholesale cost $= C$ (dollars)
Markup $= M$ (dollars)
Selling price $= \$119.91$

System of equations: $C + M = 119.91$

$M = 0.4C$

Solving by substitution: $C + M = 119.91$

$M = 0.4C$

$C + 0.4C = 119.91$ Replace M by $0.4C$ in first equation.

$1.4C = 119.91$

$C = 85.65$

The wholesale cost of the cordless phone is $85.65.

39. *List Price*   The sale price of a stereo system is $320. The discount is 20% of the list price. Find the list price.

Solution

Verbal model: $\boxed{\text{List price}} - \boxed{\text{Discount}} = \boxed{\text{Sale price}}$

$\boxed{\text{Discount}} = 0.20 \boxed{\text{List price}}$

Labels: List price $= x$ (dollars)
Discount $= d$ (dollars)
Sale price $= 320$ (dollars)

System of equations: $x - d = 320$

$d = 0.20x$

Solving by substitution: $x - d = 320$

$d = 0.20x$

$x - 0.20x = 320$ Substitute $0.20x$ for d in first equation.

$0.80x = 320$

$x = \dfrac{320}{0.80}$

$x = 400$ and $d = 80$

Therefore, the list price of the stereo system is $400.

41. *Investment*   A combined total of $8000 is invested in two bonds that pay 7% and 8.5% simple interest. The annual interest $635. How much is invested in each bond?

Solution

Verbal model:

$$\boxed{\begin{array}{l}\text{Amount in}\\7\%\text{ bond}\end{array}} + \boxed{\begin{array}{l}\text{Amount in}\\8.5\%\text{ bond}\end{array}} = \boxed{\begin{array}{l}\text{Total}\\\text{amount}\end{array}}$$

$$\boxed{\begin{array}{l}\text{Interest from}\\7\%\text{ bond}\end{array}} + \boxed{\begin{array}{l}\text{Interest from}\\8.5\%\text{ bond}\end{array}} = \boxed{\begin{array}{l}\text{Total}\\\text{interest}\end{array}}$$

Labels:

Amount in 7% bond $= x$ (dollars)
Amount in 8.5% bond $= y$ (dollars)
Total amount $= \$8000$
Interest from 7% bond $= 0.07x$ (dollars)
Interest from 8.5% bond $= 0.085y$ (dollars)
Total interest $= \$635$

System of equations:

$$x + y = 8000$$
$$0.07x + 0.085y = 635$$

Solving by elimination:

$$x + y = 8000 \quad \rightarrow \quad -70x - 70y = -560{,}000 \qquad \text{Multiply by } -70$$
$$0.07x + 0.085y = 635 \quad \rightarrow \quad \underline{70x + 85y = 635{,}000} \qquad \text{Multiply by 1000.}$$
$$15y = 75{,}000$$
$$y = 5000$$

$$x + 5000 = 8000 \qquad \text{Replace } y \text{ by 5000 in first equation.}$$
$$x = 3000$$

Thus, $3000 is invested in the 7% bond and $5000 is invested in the 8.5% bond.

43. *Gasoline Mixture*   The total cost of 8 gallons of regular unleaded gasoline and 12 gallons of premium unleaded gasoline is $27.84. Premium unleaded gasoline costs $0.17 more per gallon than regular unleaded. Find the price per gallon for each grade of gasoline.

Solution

Verbal model:

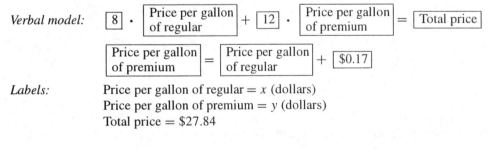

Labels:

Price per gallon of regular $= x$ (dollars)
Price per gallon of premium $= y$ (dollars)
Total price $= \$27.84$

43. **—CONTINUED—**

System of equations: $8x + 12y = 27.84$

$$y = x + 0.17$$

Solving by substitution: $8x + 12y = 27.84$

$$y = x + 0.17$$

$$8x + 12(x + 0.17) = 27.84 \qquad \text{Replace } y \text{ by } x + 0.17 \text{ in first equation.}$$

$$8x + 12x + 2.04 = 27.84$$

$$20x + 2.04 = 27.84$$

$$20x = 25.80$$

$$x = 1.29$$

$$y = 1.29 + 0.17 \qquad \text{Replace } x \text{ by } 1.29 \text{ in second equation.}$$

$$y = 1.46$$

Thus, the regular gasoline costs \$1.29 per gallon and the premium gasoline costs \$1.46 per gallon.

45. *Food Costs* You and your friend go to a Mexican restaurant. You order 2 tacos and 3 enchiladas, and your friend orders 3 tacos and 5 enchiladas. Your bill is \$7.80 plus tax, and the other bill is \$12.70 plus tax. How much is each taco and each enchilada?

Solution

Verbal model: 2 | Cost per taco | + 3 | Cost per enchilada | = | Your bill |

3 | Cost per taco | + 4 | Cost per enchilada | = | Your friend's bill |

Labels: Cost per taco = t (dollars)
Cost per enchilada = e (dollars)
Your bill = 7.80 (dollars)
Your friend's bill = 12.70 (dollars)

System of equations: $2t + 3e = 7.80$

$$3t + 5e = 12.70$$

Solving by elimination: $2t + 3e = 7.80 \quad \rightarrow \quad 6t + 9e = 23.40 \qquad \text{Multiply equation by 3.}$

$$3t + 5e = 12.70 \quad \rightarrow \quad \underline{-6t - 10e = -25.40} \qquad \text{Multiply equation by } -2.$$

$$-e = -2.00$$

$$e = 2.00$$

$$2t + 3(2.00) = 7.80 \qquad \text{Replace } e \text{ by } 2.00 \text{ in first equation.}$$

$$2t + 6.00 = 7.80$$

$$2t = 1.80$$

$$t = 0.90$$

(0.90, 2.00)
Each taco was \$0.90, and each enchilada was \$2.00.

47. *Airplane Speed* Two planes start from the same airport and fly in opposite directions. The second plane starts $\frac{1}{2}$ hour after the first plane, but its speed is 50 miles per hour faster. Two hours after the first plane starts, the planes are 2000 miles apart. Find the ground speed of each plane.

Solution

Verbal model:

$$\boxed{\text{Speed of second plane}} = \boxed{\text{Speed of first plane}} + \boxed{50}$$

$$\boxed{\text{Distance of first plane}} + \boxed{\text{Distance of second plane}} = \boxed{2000}$$

Labels:

Speed of first plane $= x$ (miles per hour)
Speed of second plane $= y$ (miles per hour)
Time of first plane's flight $= 2$ (hours)
Time of second plane's flight $= 1.5$ (hours)
Distance of first plane $= 2x$ (miles)
Distance of second plane $= 1.5y$ (miles)

System of equations:

$$y = x + 50$$
$$2x + 1.5y = 2000$$

Solving by substitution:

$$y = x + 50$$
$$2x + 1.5y = 2000$$
$$2x + 1.5(x + 50) = 2000 \qquad \text{Replace } y \text{ by } x + 50 \text{ in second equation.}$$
$$2x + 1.5x + 75 = 2000$$
$$3.5x + 75 = 2000$$
$$3.5x = 1925$$
$$x = 550 \quad y = 550 + 50 \qquad \text{Replace } x \text{ by } 550 \text{ in first equation.}$$
$$y = 600$$

The speed of the first plane is 550 mph and the speed of the second plane is 600 mph.

49. *Best-Fitting Line* The line $y = mx + b$ that best fits the three noncollinear points $(0, 0)$, $(1, 2)$, and $(2, 2)$ is given by the following system.

$$3b + 3m = 4$$
$$3b + 5m = 6$$

(a) Solve the system and find the equation of the best-fitting line.
(b) Plot the three points and sketch the graph of the best-fitting line.

Solution

(a) $3b + 3m = 4 \rightarrow -3b - 3m = -4$ Multiply by -1.

$3b + 5m = 6 \rightarrow$ $\underline{3b + 5m = 6}$

$$2m = 2$$
$$m = 1$$

$3b + 3(1) = 4$ Replace m by 1 in first equation.

$$3b + 3 = 4$$
$$3b = 1$$
$$b = \tfrac{1}{3}$$

(b)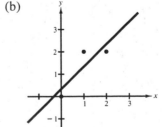

The equation of the line $y = mx + b$ is $y = 1x + \frac{1}{3}$ or $y = x + \frac{1}{3}$.

51. *Graphical Interpretation*   A homeowner monitored the average daily temperature x and the amount y (in gallons) of heating oil to heat the house. After three consecutive days the ordered pairs generated were $(0, 2)$, $(10, 1.2)$, and $(6, 1.4)$.

(a) Plot the points.

(b) The line $y = mx + b$ that best fits the data is given by the following system.

$$3b + 16m = 4.6$$

$$4b + 34m = 5.1$$

Solve the system and find the equation of the required line. Graph the line on the coordinate system of part (a).

(c) Interpret the meaning of the slope of the line in the context of this problem.

Solution

(a)

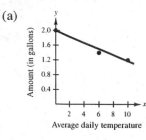

(b) $3b + 16m = 4.6$ $\rightarrow$ $12b + 64m = 18.4$ Multiply equation by 4.

$\,4b + 34m = 5.1$ $\rightarrow$ $\underline{-12b - 102m = -15.3}$ Multiply equation by 3.

$$-38m = 3.1$$

$$m = \frac{3.1}{-38}$$

$$m \approx -0.0816$$

$3b + 16(-0.0816) \approx 4.6$ Replace m by -0.0816 in first equation.

$3b - 1.3056 \approx 4.6$

$3b \approx 5.9056$

$b \approx \dfrac{5.9056}{3}$

$b \approx 1.968$

The equation of the line is $y = -0.08x + 1.97$.

(c) The slope indicates how much of a change in gallons of heating oil resulted from each degree of temperature change. It shows that the amount of heating oil decreased by 0.08 gallons for each 1 degree increase in temperature.

Review Exercises for Chapter 8

1. Which ordered pair (a) $(2, -1)$ or (b) $(3, -2)$ is a solution of the following system of equations?

$$3x - 5y = 11$$
$$-x + 2y = -4$$

Solution

(a) $(2, -1)$

$$3(2) - 5(-1) \stackrel{?}{=} 11$$
$$6 + 5 = 11$$
$$-2 + 2(-1) = -4$$
$$-2 - 2 = -4$$

$(2, -1)$ *is* a solution.

(b) $(3, -2)$

$$3(3) - 5(-2) \stackrel{?}{=} 11$$
$$9 + 10 \neq 11$$

$(3, -2)$ is *not* a solution.

3. Which ordered pair (a) $(0.5, -0.7)$ or (b) $(15, 5)$ is a solution of the following system of equations?

$$0.2x + 0.4y = 5$$
$$x + 3y = 30$$

Solution

(a) $(0.5, -0.7)$

$$0.2(0.5) + 0.4(-0.7) \stackrel{?}{=} 5$$
$$0.01 - 0.28 \neq 5$$

$(0.5, -0.7)$ is *not* a solution.

(b) $(15, 5)$

$$0.2(15) + 0.4(5) \stackrel{?}{=} 5$$
$$3 + 2 = 5$$
$$15 + 3(5) \stackrel{?}{=} 30$$
$$15 + 15 = 30$$

$(15, 5)$ *is* a solution.

5. Match the following system with its graph shown in the textbook. [The graphs are labeled (a), (b), (c), and (d).]

$$x + 2y = 6$$
$$x + 2y = 2$$

Solution

Graph (d)

7. Match the following system with its graph shown in the textbook. [The graphs are labeled (a), (b), (c), and (d).]

$$2x + y = 4$$
$$-4x - 2y = -8$$

Solution

Graph (a)

9. Solve the following system graphically.

$$x + y = 2$$
$$x - y = 0$$

Solution

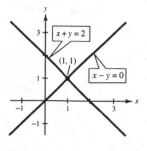

The solution is (1, 1).

11. Solve the following system graphically.

$$x - y = 9$$
$$-x + y = 1$$

Solution

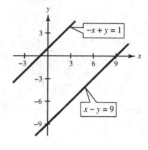

There is *no* solution. The system is inconsistent.

13. Solve the following system by substitution.

$$y = 2x$$
$$y = x + 4$$

Solution

$$y = 2x$$
$$y = x + 4$$

$2x = x + 4$ Replace y by $2x$ in second equation.

$x = 4$ $y = 2(4)$ Replace x by 4 in first equation.

$y = 8$

$(4, \ 8)$

15. Solve the following system by substitution.

$$2x - y = 2$$
$$6x + 8y = 39$$

Solution

$2x - y = 2 \quad \rightarrow \quad -y = -2x + 2 \quad \rightarrow \quad y = 2x - 2$

$6x + 8y = 39$

$6x + 8(2x - 2) = 39$ Replace y by $2x - 2$ in second equation.

$6x + 16x - 16 = 39$

$22x - 16 = 39$

$22x = 55$

$x = \frac{55}{22}$

$x = \frac{5}{2}$ $y = 2\left(\frac{5}{2}\right) - 2$ Replace x by $\frac{5}{2}$ in revised first equation.

$y = 5 - 2$

$y = 3$

$\left(\frac{5}{2}, \ 3\right)$

17. Solve the following system by elimination.

$$5x + 4y = 2$$
$$-x + y = -22$$

Solution

$$5x + 4y = 2 \quad \rightarrow \quad 5x + 4y = 2$$
$$-x + y = -22 \quad \rightarrow \quad \underline{-5x + 5y = -110} \quad \text{Multiply equation by 5.}$$
$$9y = -108$$
$$y = -12$$

$$-x + (-12) = -22 \quad \text{Replace } y \text{ by } -12 \text{ in second equation.}$$
$$-x = -10$$
$$x = 10$$

$$(10, -12)$$

19. Solve the following system by elimination.

$$2x + y = 0.3$$
$$3x - y = -1.3$$

Solution

$$2x + y = 0.3$$
$$\underline{3x - y = -1.3}$$
$$5x = -1$$
$$x = -\tfrac{1}{5}$$
$$2\left(-\tfrac{1}{5}\right) + y = 0.3 \quad \text{Replace } x \text{ by } -\tfrac{1}{5} \text{ in first equation.}$$
$$-\tfrac{2}{5} + y = 0.3$$
$$y = 0.3 + \tfrac{2}{5}$$
$$y = \tfrac{3}{10} + \tfrac{4}{10}$$
$$y = \tfrac{7}{10}$$

$$\left(-\tfrac{1}{5}, \tfrac{7}{10}\right) \text{ or } (-0.2, 0.7)$$

21. Use the method of your choice to solve the following system.

$$6x - 5y = 0$$
$$y = 6$$

Solution

$$6x - 5y = 0$$
$$y = 6$$

$$6x - 5(6) = 0 \quad \text{Replace } y \text{ by 6 in first equation.}$$
$$6x - 30 = 0$$
$$6x = 30$$
$$x = 5$$

$$(5, 6)$$

23. Use the method of your choice to solve the following system.

$$x - y = 0$$
$$x - 6y = 5$$

Solution

$$x - y = 0 \quad \rightarrow \quad x = y$$
$$x - 6y = 5$$

$$y - 6y = 5 \qquad \text{Replace } x \text{ by } y \text{ in second equation.}$$

$$-5y = 5$$

$$y = -1 \qquad x = -1 \qquad \text{Replace } y \text{ by } -1 \text{ in revised first equation.}$$

$$(-1, -1)$$

25. Use the method of your choice to solve the following system.

$$6x - 3y = 27$$
$$-2x + y = -9$$

Solution

$$6x - 3y = 27 \quad \rightarrow \quad 6x - 3y = 27$$
$$-2x + y = -9 \quad \rightarrow \quad \underline{-6x + 3y = -27} \qquad \text{Multiply equation by 3.}$$
$$0 = 0$$

The system has infinitely many solutions. The solution set consists of all ordered pairs (x, y) such that $-2x + y = -9$.

27. Use the method of your choice to solve the following system.

$$5x + 8y = 8$$
$$x - 8y = 16$$

Solution

$$5x + 8y = 8$$
$$\underline{x - 8y = 16}$$
$$6x = 24$$

$$x = 4$$

$$5(4) + 8y = 8 \qquad \text{Replace } x \text{ by 4 in first equation.}$$

$$20 + 8y = 8$$

$$8y = -12$$

$$y = -\frac{3}{2}$$

$$\left(4, -\frac{3}{2}\right)$$

29. Use the method of your choice to solve the following system.

$$2x + 5y = 20$$
$$4x + 5y = 10$$

Solution

$$2x + 5y = 20 \quad \rightarrow \quad -2x - 5y = -20 \qquad \text{Multiply equation by } -1.$$
$$4x + 5y = 10 \quad \rightarrow \quad \underline{4x + 5y = 10}$$
$$2x = -10$$

$$x = -5$$

$$2(-5) + 5y = 20 \qquad \text{Replace } x \text{ by } -5 \text{ in first equation.}$$

$$-10 + 5y = 20$$

$$5y = 30$$

$$y = 6$$

$$(-5, 6)$$

31. Use the method of your choice to solve the following system.

$$-x + 4y = 4$$
$$x + y = 6$$

Solution

$$-x + 4y = 4$$
$$\underline{x + y = 6}$$
$$5y = 10$$

$$y = 2$$

$$x + 2 = 6 \qquad \text{Replace } y \text{ by 2 in second equation.}$$

$$x = 4$$

$$(4, 2)$$

33. Use the method of your choice to solve the following system.

$$x + y = 0$$
$$2x + y = 0$$

Solution

$$x + y = 0 \quad \rightarrow \quad -x - y = 0 \qquad \text{Multiply equation by } -1.$$
$$2x + y = 0 \quad \rightarrow \quad \underline{2x + y = 0}$$
$$x = 0$$

$$0 + y = 0 \qquad \text{Replace } x \text{ by 0 in first equation.}$$

$$y = 0$$

$$(0, 0)$$

35. Use the method of your choice to solve the following system.

$$\tfrac{1}{3}x + \tfrac{4}{7}y = 3$$

$$2x + 3y = 15$$

Solution

$$\tfrac{1}{3}x + \tfrac{4}{7}y = 3 \quad \rightarrow \quad 7x + 12y = 63 \qquad \text{Multiply equation by 21.}$$

$$\underline{2x + 3y = 15 \quad \rightarrow \quad -8x - 12y = -60} \qquad \text{Multiply equation by } -4.$$

$$-x \quad\quad = 3$$

$$x \quad\quad = -3$$

$$2(-3) + 3y = 15 \qquad \text{Replace } x \text{ by } -3 \text{ in second equation.}$$

$$-6 + 3y = 15$$

$$3y = 21$$

$$y = 7$$

$$(-3,\ 7)$$

37. Use the method of your choice to solve the following system.

$$12s + 42t = -17$$

$$30s - 18t = 19$$

Solution

$$12s + 42t = -17 \quad \rightarrow \quad -60s - 210t = 85 \qquad \text{Multiply equation by } -5.$$

$$\underline{30s - 18t = 19 \quad \rightarrow \quad 60s - 36t = 38} \qquad \text{Multiply equation by 2.}$$

$$-246t = 123$$

$$t = -\tfrac{123}{246}$$

$$t = -\tfrac{1}{2}$$

$$30s - 18\left(-\tfrac{1}{2}\right) = 19 \qquad \text{Replace } t \text{ by } -\tfrac{1}{2} \text{ in second equation.}$$

$$30s + 9 = 19$$

$$30s = 10$$

$$s = \tfrac{1}{3}$$

$$\left(\tfrac{1}{3}, -\tfrac{1}{2}\right)$$

39. Use the method of your choice to solve the following system.

$$-x + 2y = 1.5$$

$$2x - 4y = 3$$

Solution

$$-x + 2y = 1.5 \quad \rightarrow \quad -2x + 4y = 3 \qquad \text{Multiply equation by 2.}$$

$$\underline{2x - 4y = 3 \quad \rightarrow \quad 2x - 4y = 3}$$

$$0 = 6 \qquad \text{False}$$

There is *no* solution. The system is inconsistent.

41. Use the method of your choice to solve the following system.

$$x + 2y = 7$$
$$2x + y = 8$$

Solution

$$x + 2y = 7 \quad \rightarrow \quad -2x - 4y = -14 \qquad \text{Multiply equation by } -2.$$
$$2x + y = 8 \quad \rightarrow \quad \underline{2x + y = 8}$$
$$-3y = -6$$
$$y = 2$$

$$x + 2(2) = 7 \qquad \text{Replace } y \text{ by 2 in first equation.}$$

$$x + 4 = 7$$
$$x = 3$$

$$(3, 2)$$

43. Use a graphing utility to solve the following system. Check your solution algebraically.

$$y = -5x + 4$$
$$y = 3(x + 4)$$

Solution

$$y_1 = -5x + 4$$
$$y_2 = 3(x + 4)$$

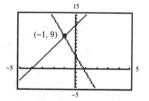

45. Use a graphing utility to solve the following system. Check your solution algebraically.

$$x - y = 1$$
$$-2x + y = 2$$

Solution

$$y_1 = x - 1$$
$$y_2 = 2x + 2$$

47. Find a system of linear equations that has the solution (5, 10). (There are many correct answers.)

Solution

Here are *some* systems of linear equations that have a solution of (5, 10). (*Note:* There are many other correct answers.)

$$\begin{array}{lll} 3x + y = 25 & x - y = -5 & -2x + y = 0 \\ 2x - y = 0 & 4x - y = 10 & x + y = 15 \end{array}$$

49. Find a system of linear equations that has the solution $\left(3, \frac{8}{3}\right)$. (There are many correct answers.)

Solution

Here are *some* systems that have the solution $\left(3, \frac{8}{3}\right)$. There are many other correct answers.

$$\begin{array}{lll} x + 3y = 11 & -x + 6y = 13 & 2x - 3y = -2 \\ x - 3y = -5 & 5x - 6y = -1 & -7x + 9y = 3 \end{array}$$

51. Find a linear system for the graphical model shown in the textbook. If only one line is shown, find two different equations for the line.

Solution

Here are *some* systems of linear equations for the graphical model. (*Note:* There are many other correct answers.)

$$y = 3x - 2 \qquad\qquad 3x - y = 2$$
$$6x - 2y = 4 \qquad\qquad -9x + 3y = -6$$

53. Find a linear system for the graphical model shown in the textbook. If only one line is shown, find two different equations for the line.

Solution

Here are *some* systems of linear equations for the graphical model. (*Note:* There are many other correct answers.)

$$y = -\tfrac{1}{2}x + 4 \qquad\qquad x + 2y = 8$$
$$y = -\tfrac{1}{2}x - 4 \qquad\qquad x + 2y = -8$$

55. *Geometry* A rectangular sign (see the figure shown in the textbook) has a perimeter of 120 inches. The height of the sign is two-thirds of its width. Find the dimensions of the sign.

Solution

Verbal model: $\boxed{2} \cdot \boxed{\text{Width of rectangle}} + \boxed{2} \cdot \boxed{\text{Height of rectangle}} = \boxed{\text{Perimeter of rectangle}}$

$\boxed{\text{Height of rectangle}} = \boxed{\tfrac{2}{3}} \cdot \boxed{\text{Width of rectangle}}$

Labels: Height of rectangle = h (inches)
Width of rectangle = w (inches)
Perimeter of rectangle = 120 (inches)

System of equations: $2w + 2h = 120$

$h = \tfrac{2}{3}w$

Solving by substitution: $2w + 2h = 120$

$h = \tfrac{2}{3}w$

$2w + 2\left(\tfrac{2}{3}w\right) = 120$ Replace h by $\tfrac{2}{3}w$ in first equation.

$2w + \tfrac{4}{3}w = 120$

$3\left(2w + \tfrac{4}{3}w\right) = 3(120)$

$6w + 4w = 360$

$10w = 360$

$w = 36 \qquad h = \tfrac{2}{3}(36)$ Replace w by 36 in second equation.

$h = 24$

$(36, 24)$

The width of the rectangle is 36 inches, and the height of the rectangle is 24 inches.

57. *Price per Gallon* You buy 2 gallons of gasoline for your lawn mower and 5 gallons of diesel fuel for your garden tractor. The total bill is $9.75. Gasoline costs $0.08 more per gallon than diesel fuel. Find the price per gallon for each type of fuel.

Solution

Verbal model:

$$\boxed{2} \cdot \boxed{\begin{array}{c}\text{Price per gallon}\\\text{of gasoline}\end{array}} + \boxed{5} \cdot \boxed{\begin{array}{c}\text{Price per gallon}\\\text{of diesel fuel}\end{array}} = \boxed{\$9.75}$$

$$\boxed{\begin{array}{c}\text{Price per gallon}\\\text{of gasoline}\end{array}} = \boxed{\begin{array}{c}\text{Price per gallon}\\\text{of diesel fuel}\end{array}} + \boxed{\$0.08}$$

Labels: Price per gallon of gasoline $= x$ (dollars)
Price per gallon of diesel fuel $= y$ (dollars)

System of equations: $2x + 5y = 9.75$

$$x = y + 0.08$$

Solving by substitution: $2x + 5y = 9.75$

$$x = y + 0.08$$

$$2(y + 0.08) + 5y = 9.75 \qquad \text{Replace } x \text{ by } y + 0.08 \text{ in first equation.}$$

$$2y + 0.16 + 5y = 9.75$$

$$7y + 0.16 = 9.75$$

$$7y = 9.59$$

$$y = 1.37 \qquad x = 1.37 + 0.08 \qquad \text{Replace } y \text{ by } 1.37 \text{ in second equation.}$$

$$x = 1.45$$

The price per gallon of gasoline is $1.45 and the price per gallon of diesel fuel is $1.37.

59. *Wholesale Cost* The selling price of a VCR is $434. The markup rate is 40% of the wholesale cost. Find the wholesale cost.

Solution

Verbal model:

$$\boxed{\text{Wholesale cost}} + \boxed{\text{Markup}} = \boxed{\text{Selling price}}$$

$$\boxed{\text{Markup}} = \boxed{0.40} \cdot \boxed{\text{Wholesale cost}}$$

Labels: Wholesale cost $= C$ (dollars)
Markup $= M$ (dollars)
Selling price $= \$434$

System of equations: $C + M = 434$

$$M = 0.40C$$

Solving by substitution: $C + M = 434$

$$M = 0.40C$$

$$C + 0.40C = 434 \qquad \text{Replace } M \text{ by } 0.40C \text{ in first equation.}$$

$$1.4C = 434$$

$$C = \frac{434}{1.4}$$

$$C = 310$$

The wholesale cost of the VCR is $310.

61. Consider the following system of linear equations.

$$2x + 3y = 8$$

$$6x + ky = 12$$

(a) Find the value of k for which the system is inconsistent.

(b) Find a value of k for which the system has a unique solution.

(c) Is there a value of k for which the system has an infinite number of solutions? Why or why not?

Solution

$$2x + 3y = 8 \quad \rightarrow \quad -6x - 9y = -24 \qquad \text{Multiply equation by } -3.$$

$$6x + ky = 12 \quad \rightarrow \quad \underline{6x + ky = 12}$$

$$ky - 9y = -12$$

$$y(k - 9) = -12$$

(a) The system is inconsistent if the variables "drop out" in the elimination step and the remaining equation is a false statement. The variable y drops out of the last equation when $k - 9 = 0$.

$$k - 9 = 0 \quad \rightarrow \quad k = 9$$

Therefore, when $k = 9$, the equation becomes $0 = -12$, a false statement. The system is inconsistent when $k = 9$.

(b) The system has a unique solution when $k - 9 \neq 0$. Therefore, the system has a unique solution when k is *any number except* 9. For example, when $k = 3$,

$$y(3 - 9) = -12 \quad \rightarrow \quad -6y = -12 \quad \rightarrow \quad y = 2$$

and

$$2x + 3(2) = 8 \quad \rightarrow \quad 2x + 6 = 8 \quad \rightarrow \quad 2x = 2 \quad \rightarrow \quad x = 1.$$

When $k = 3$, the unique solution is $(1, 2)$.

(c) No, the system will never have infinitely many solutions. When $k = 9$, the system is inconsistent, and when k has any other value, the system has a unique solution.

Note: This question could also be answered by rewriting each equation in slope-intercept form and considering the graphs of the two equations.

(a) The system is inconsistent when the graphs of the two equations are parallel lines. Parallel lines have the same slope.

$$2x + 3y = 8 \qquad\qquad 6x + ky = 12$$

$$3y = -2x + 8 \qquad\qquad ky = -6x + 12$$

$$y = -\frac{2}{3}x + \frac{8}{3} \qquad\qquad y = \frac{-6}{k} + \frac{12}{k}$$

$$m = -\frac{2}{3} \qquad\qquad m = -\frac{6}{k}$$

The lines are parallel if $-2/3 = -6/k$.

$$\frac{-2}{3} = \frac{-6}{k}$$

$$-2k = -18 \qquad \text{(Cross-multiply.)}$$

$$k = 9$$

If $k = 9$, the *first* line has a slope of $-2/3$ and y-intercept of $8/3$, and the *second* line has a slope of $-6/k$ or $-6/9 = -2/3$ and y-intercept of $12/k$ or $12/9$ or $4/3$. Thus, when $k = 9$, the lines are parallel and the system is inconsistent.

61. —CONTINUED—

(b) The system has a unique solution when k is *any* real number *except* 9. When $k \neq 9$, the slopes of the lines are not equal and the lines must intersect. The coordinates of the point of intersection provide the unique solution to the system.

(c) No, there is *no* value of k for which the system has an infinite number of solutions. The system would have infinitely many solutions if the graphs of the two equations were the same line. In other words, the two lines would have to have the same slope *and* the same y-intercept. In part (a), we saw that the two lines have the same slope only when $k = 9$. However, when $k = 9$, the y-intercepts are $\frac{8}{3}$ and $\frac{4}{3}$. Thus, the two lines cannot coincide, and the system *cannot* have infinitely many solutions.

Test for Chapter 8

1. Which is the solution of the system $x - 6y = -19$ and $4x - 5y = 0$: $(3, -2)$ or $(5, 4)$? Explain your reasoning.

Solution

(a) $3 - 6(-2) \overset{?}{=} -19 \qquad 4(3) - 5(-2) \overset{?}{=} 0$

$\qquad 3 + 12 \overset{?}{=} -19 \qquad 12 + 10 \overset{?}{=} 0$

$\qquad 15 \neq -19 \qquad\qquad 22 \neq 0$

$(3, -2)$ is *not* a solution.

(b) $5 - 6(4) \overset{?}{=} -19 \qquad 4(5) - 5(4) \overset{?}{=} 0$

$\qquad 5 - 24 \overset{?}{=} -19 \qquad 20 - 20 \overset{?}{=} 0$

$\qquad -19 = -19 \qquad\qquad 0 = 0$

$(5, 4)$ *is* a solution.

2. Determine the number of solutions of the following system.

$$3x + 4y = 16$$
$$3x - 4y = 8$$

Solution

The system of equations has *one* solution, $(4, 1)$.

3. Determine the number of solutions of the following system.

$$x - 2y = -4$$
$$x - 2y = 2$$

Solution

The system of equations has *no* solution. (The system is inconsistent.)

4. Determine the number of solutions for the following system.

$$x + 2y = 4$$
$$x + 2y = -2$$

Solution

The system of equations has *no* solution. (The system is inconsistent.)

5. Solve the following system of equations graphically.

$$x - 2y = -2$$
$$x + y = 4$$

Solution

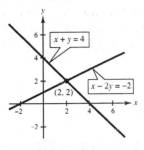

The solution is $(2, 2)$.

6. Solve the following system of equations graphically.

$$2x + y = 4$$
$$x - 2y = -3$$

Solution

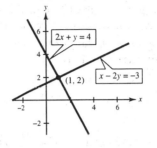

The solution is $(1, 2)$.

7. Solve the following system of equations graphically.

$$x - 3y = -2$$
$$2x + y = 10$$

Solution

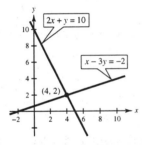

The solution is $(4, 2)$.

8. Solve the following system of equations graphically.

$$2x = 3$$
$$2x + 3y = 9$$

Solution

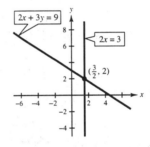

The solution is $\left(\frac{3}{2}, 2\right)$.

9. Solve the following system of equations by substitution.

$$x + 5y = 10$$
$$4x - 5y = 15$$

Solution

$$x + 5y = 10 \quad \rightarrow \quad x = -5y + 10$$
$$4x - 5y = 15$$
$$4(-5y + 10) - 5y = 15 \qquad \text{Replace } x \text{ by } -5y + 10 \text{ in second equation.}$$
$$-20y + 40 - 5y = 15$$
$$-25y + 40 = 15$$
$$-25y = -25$$
$$y = 1 \qquad x = -5(1) + 10 \qquad \text{Replace } y \text{ by 1 in revised first equation.}$$
$$x = -5 + 10$$
$$x = 5$$
$$(5, \ 1)$$

10. Solve the following system of equations by substitution.

$$x + 3y = 15$$
$$-2x + 5y = 14$$

Solution

$$x + 3y = 15 \quad \rightarrow \quad x = -3y + 15$$
$$-2x + 5y = 14$$
$$-2(-3y + 15) + 5y = 14 \qquad \text{Replace } x \text{ by } -3y + 15 \text{ in second equation.}$$
$$6y - 30 + 5y = 14$$
$$11y = 44$$
$$y = 4 \quad \rightarrow \quad x = -3(4) + 15$$
$$x = -12 + 15$$
$$x = 3$$

$$(3, 4)$$

11. Solve the following system of equations by substitution.

$$0.5x + 0.3y = 3$$
$$7x - y = 16$$

Solution

$$0.5x + 0.3y = 3$$
$$7x - y = 16 \quad \rightarrow \quad -y = -7x + 16$$
$$y = 7x - 16$$
$$0.5x + 0.3(7x - 16) = 3 \qquad \text{Replace } y \text{ by } 7x - 16 \text{ in first equation.}$$
$$0.5x + 2.1x - 4.8 = 3$$
$$2.6x - 4.8 = 3$$
$$2.6x = 7.8$$
$$x = 3 \quad \rightarrow \quad y = 7(3) - 16$$
$$y = 21 - 16$$
$$y = 5$$

$$(3, 5)$$

17. Find the value of a such that the following system is inconsistent.

$$ax - 8y = 9$$
$$3x + 4y = 0$$

Describe the method you used to find a.

Solution

$$ax - 8y = 9 \quad \rightarrow \quad ax - 8y = 9$$
$$3x + 4y = 0 \quad \rightarrow \quad \underline{6x + 8y = 0} \qquad \text{Multiply equation by 2.}$$
$$ax + 6x = 9$$
$$x(a + 6) = 9$$

The variable x will "drop out" of this equation when $a + 6 = 0$.

$$a + 6 = 0 \quad \rightarrow \quad a = -6$$

Therefore, when $a = -6$, the equation becomes $0 = -9$, a *false* statement. The system is inconsistent when $a = -6$.

18. Find a system of linear equations that has the solution $(-3, 4)$. (There are many correct solutions.)

Solution

Here are *some* systems that have the solution $(-3, 4)$. There are many other correct answers.

$$\begin{array}{lll} x + y = 1 & x + 2y = 5 & 2x + 5y = 14 \\ x - y = -7 & -x + y = 7 & 3x + 2y = -1 \end{array}$$

19. A rectangle has a perimeter of 40 meters. The length of the rectangle is three times its width. Find the dimensions of the rectangle.

Solution

Verbal model: $\boxed{2} \cdot \boxed{\text{Length of rectangle}} + \boxed{2} \cdot \boxed{\text{Width of rectangle}} = \boxed{\text{Perimeter of rectangle}}$

$\boxed{\text{Length of rectangle}} = \boxed{3} \cdot \boxed{\text{Width of rectangle}}$

Labels: Length of rectangle = x (meters)
Width of rectangle = y (meters)
Perimeter of rectangle = 40 (meters)

System of equations: $2x + 2y = 40$
$x = 3y$

Solving by substitution: $2x + 2y = 40$
$x = 3y$

$2(3y) + 2y = 40 \qquad \text{Replace } x \text{ by } 3y \text{ in first equation.}$
$6y + 2y = 40$
$8y = 40$
$y = 5 \qquad x = 3(5) \qquad \text{Replace } y \text{ by 5 in second equation.}$
$x = 15$

The length of the rectangle is 15 meters and the width of the rectangle is 5 meters.

20. Twenty liters of 20% acid solution is obtained by mixing a 30% solution and a 5% solution. How many liters of each solution are needed to obtain the specified mixture?

Solution

Verbal model: | Liters of 30% solution | + | Liters of 5% solution | = | Liters of 20% solution |

| Acid in 30% solution | + | Acid in 5% solution | = | Acid in 20% solution |

Labels: Liters of 30% solution $= x$
Liters of 5% solution $= y$
Liters of 20% solution $= 20$
Acid in 30% solution $= 0.30x$ (liters)
Acid in 5% solution $= 0.05y$ (liters)
Acid in 20% solution $= 0.20(20) = 4$ (liters)

System of equations:
$$x + y = 20$$
$$0.30x + 0.05y = 4$$

Solving by elimination:

$$x + y = 20 \quad \rightarrow \quad 5x + 5y = 100 \qquad \text{Multiply by 5.}$$
$$0.30x + 0.05y = 4 \quad \rightarrow \quad \underline{-30x - 5y = -400} \qquad \text{Multiply by } -100.$$
$$-25x \qquad\quad = -300$$
$$x \qquad\quad = 12$$

$$x + y = 20 \qquad \text{Replace } x \text{ by 12 in first equation.}$$
$$12 + y = 20$$
$$y = 8$$
$$(12, 8)$$

The mixture contains 12 liters of the 30% solution and 8 liters of the 5% solution.

CHAPTER NINE
Rational Expressions and Equations

<table>
<tr><td>9.1</td><td></td></tr>
</table>

9.1   Simplifying Rational Expressions

7. Find the domain of the following expression.

$$\frac{5}{x-4}$$

Solution

The denominator is zero when $x - 4 = 0$ or $x = 4$. Thus, the domain is all real values of x such that $x \neq 4$.

9. Find the domain of the following expression.

$$\frac{4t}{t^2 - 25}$$

Solution

The denominator $t^2 - 25 = (t + 5)(t - 5)$ is zero when $t = -5$ or 5. Thus, the domain is all real values of t such that $t \neq -5$ and $t \neq 5$.

11. Evaluate the following expression for the values (a) $x = 0$, (b) $x = 3$, (c) $x = 10$, and (d) $x = -3$. (If not possible, state the reason.)

$$\frac{x}{x-3}$$

Solution

(a) $\dfrac{x}{x-3} = \dfrac{0}{0-3}$

$= \dfrac{0}{-3}$

$= 0$

(b) $\dfrac{x}{x-3} = \dfrac{3}{3-3}$

$= \dfrac{3}{0}$

Undefined
Division by 0
is undefined.

(c) $\dfrac{x}{x-3} = \dfrac{10}{10-3}$

$= \dfrac{10}{7}$

(d) $\dfrac{x}{x-3} = \dfrac{-3}{-3-3}$

$= \dfrac{-3}{-6}$

$= \dfrac{1}{2}$

13. Evaluate the following expression for the values (a) $x = 2$, (b) $x = 1$, (c) $x = -5$, and (d) $x = -2$. (If not possible, state the reason.)

$$\frac{x+1}{x^2-4}$$

Solution

(a) $\dfrac{x+1}{x^2-4} = \dfrac{2+1}{2^2-4}$

$= \dfrac{3}{0}$

Undefined
Division by 0
is undefined.

(b) $\dfrac{x+1}{x^2-4} = \dfrac{1+1}{1^2-4}$

$= \dfrac{2}{-3}$

$= -\dfrac{2}{3}$

(c) $\dfrac{x+1}{x^2-4} = \dfrac{-5+1}{(-5)^2-4}$

$= \dfrac{-4}{21}$

$= -\dfrac{4}{21}$

(d) $\dfrac{x+1}{x^2-4} = \dfrac{-2+1}{(-2)^2-4}$

$= \dfrac{-1}{0}$

Undefined
Division by 0
is undefined.

15. *Think About It*   Write two equivalent versions of the following expression by changing signs of the numerator, the denominator, or the fraction.

$$\frac{x}{12}$$

Solution

$$\frac{-x}{-12}, \quad -\frac{-x}{12}, \quad -\frac{x}{-12}$$

19. Find the missing factor.

$$\frac{5}{2x} = \frac{5\boxed{}}{6x^2}$$

Solution

$$\frac{5}{2x} = \frac{5(3x)}{2x(3x)} = \frac{5(3x)}{6x^2}$$

The missing factor is $3x$.

23. Simplify the following expression.

$$\frac{4x}{12}$$

Solution

$$\frac{4x}{12} = \frac{\cancel{4}(x)}{\cancel{4}(3)} = \frac{x}{3}$$

27. Simplify the following expression.

$$\frac{x-5}{2x-10}$$

Solution

$$\frac{x-5}{2x-10} = \frac{x-5}{2(x-5)}$$

$$= \frac{1\cancel{(x-5)}}{2\cancel{(x-5)}}$$

$$= \frac{1}{2}, \quad x \neq 5$$

31. Simplify the following expression.

$$\frac{a+2}{a^2+4a+4}$$

Solution

$$\frac{a+2}{a^2+4a+4} = \frac{a+2}{(a+2)(a+2)}$$

$$= \frac{1\cancel{(a+2)}}{\cancel{(a+2)}(a+2)}$$

$$= \frac{1}{a+2}$$

17. *Think About It*   Write two equivalent versions of the following expression by changing signs of the numerator, the denominator, or the fraction.

$$-\frac{t+2}{t^2-1}$$

Solution

$$\frac{-t-2}{t^2-1}, \quad -\frac{-t-2}{-t^2+1}, \quad \frac{t+2}{-t^2+1}, \quad \frac{t+2}{1-t^2}$$

21. Find the missing factor.

$$\frac{x+1}{x} = \frac{(x+1)\boxed{}}{x(x-2)}$$

Solution

$$\frac{x+1}{x} = \frac{(x+1)(x-2)}{x(x-2)}$$

The missing factor is $(x-2)$.

25. Simplify the following expression.

$$\frac{15x^2}{10x}$$

Solution

$$\frac{15x^2}{10x} = \frac{\cancel{(5)}(3)\cancel{(x)}(x)}{\cancel{(5)}(2)\cancel{(x)}} = \frac{3x}{2}, \quad x \neq 0$$

29. Simplify the following expression.

$$\frac{5-x}{2x-10}$$

Solution

$$\frac{5-x}{2x-10} = \frac{5-x}{2(x-5)}$$

$$= \frac{-1\cancel{(x-5)}}{2\cancel{(x-5)}}$$

$$= -\frac{1}{2}, \quad x \neq 5$$

33. Simplify the following expression.

$$\frac{y^2-4}{y^2+3y-10}$$

Solution

$$\frac{y^2-4}{y^2+3y-10} = \frac{(y+2)(y-2)}{(y+5)(y-2)}$$

$$= \frac{(y+2)\cancel{(y-2)}}{(y+5)\cancel{(y-2)}}$$

$$= \frac{y+2}{y+5}, \quad y \neq 2$$

35. Simplify the following expression.

$$\frac{a^3 - 8}{a^2 - 4}$$

Solution

$$\frac{a^3 - 8}{a^2 - 4} = \frac{(a - 2)(a^2 + 2a + 4)}{(a + 2)(a - 2)}$$

$$= \frac{a^2 + 2a + 4}{a + 2}, \quad a \neq 2$$

37. Simplify the following expression.

$$\frac{x^3 + 2x^2 + x + 2}{x^2 + 1}$$

Solution

$$\frac{x^3 + 2x^2 + x + 2}{x^2 + 1} = \frac{x^2(x + 2) + 1(x + 2)}{x^2 + 1}$$

$$= \frac{(x + 2)(x^2 + 1)}{x^2 + 1}$$

$$= x + 2$$

39. *Creating a Table* Use a calculator to complete the table shown in the textbook. Explain why the values of the expressions agree for all values of x except one.

Solution

x	2	2.5	3	3.5	4
$\dfrac{x^3 - 3x^2}{x - 3}$	4	6.25	Undefined	12.25	16
x^2	4	6.25	9	12.25	16

$$\frac{x^3 - 3x^2}{x - 3} = \frac{x^2(x - 3)}{x - 3} = x^2, \quad x \neq 3$$

The two expressions are equal for all values of x except 3. When $x = 3$, the first expression is undefined.

41. Simplify the following expression. Then use a graphing utility to check your result.

$$\frac{2x^2 + 4x}{2x}$$

Solution

$$\frac{2x^2 + 4x}{2x} = \frac{2x(x + 2)}{2x} = x + 2, \quad x \neq 0$$

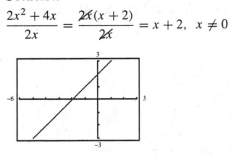

43. *Using a Model*   As air pressure increases, the temperature at which water boils also increases. (This is the purpose of pressure canners.) A model that relates air pressure to boiling temperature is

$$B = \frac{156.89x + 7.34x^2}{x + 0.017x^2}, \quad 10 \leq x \leq 100$$

where B is measured in degrees Fahrenheit and x is measured in pounds per square inch.

(a) Simplify the rational expression.

(b) Use the model to estimate the boiling temperature of water when $x = 14.7$ (approximate air pressure at sea level).

Solution

(a) $B = \dfrac{156.89x + 7.34x^2}{x + 0.017x^2}$

$= \dfrac{\cancel{(x)}(156.89 + 7.34x)}{\cancel{(x)}(1 + 0.017x)}$

$= \dfrac{156.89 + 7.34x}{1 + 0.017x}, \quad 10 \leq x \leq 100$

(b) $B = \dfrac{156.89 + 7.34(14.7)}{1 + 0.017(14.7)}$

$= \dfrac{264.788}{1.2499}$

$= 211.847°F$

45. Solve the following equation and check your answer.

$$\frac{x}{3} + 5 = 8$$

Solution

$\dfrac{x}{3} + 5 = 8$

$3\left(\dfrac{x}{3} + 5\right) = 3(8)$

$x + 15 = 24$

$x = 9$

47. Solve $14 - 2x = x + 2$ and check your answer.

Solution

$14 - 2x = x + 2$

$14 - 14 - 2x = x + 2 - 14$

$-2x = x - 12$

$-2x - x = x - 12 - x$

$-3x = -12$

$x = \dfrac{-12}{-3}$

$x = 4$

49. *Insurance*   The annual insurance premium for a policyholder is normally $645. However, after having an automobile accident, the policyholder was charged an additional 25%. What is the new premium?

Solution

Verbal model: $\boxed{\text{Original premium}} + 0.25 \boxed{\text{Original premium}} = \boxed{\text{New premium}}$

Labels: Original premium = 645 (dollars)
New premium = x (dollars)

Equation: $645 + 0.25(645) = x$

$645 + 161.25 = x$

$806.25 = x$

The new premium is $806.25.

51. Determine whether the following expression is a rational expression. If not, explain why.

$$\frac{x^2 + 1}{5x - 2}$$

Solution

Yes, the expression is a rational expression. It is one polynomial divided by another.

53. Determine whether the following expression is a rational expression. If not, explain why.

$$\frac{x^{1/2} - 2x}{x + 1}$$

Solution

No, the expression is not a rational expression because the numerator is not a polynomial. (A polynomial has only positive integral exponents.)

55. Find the domain of the following expression.

$$\frac{x}{x + 2}$$

Solution

The denominator is zero when $x + 2 = 0$ or $x = -2$. Thus, the domain is all real values of x such that $x \neq -2$.

57. Find the domain of the following expression.

$$\frac{-5(y + 2)}{y - 3}$$

Solution

The denominator is zero when $y - 3 = 0$ or $y = 3$. Thus, the domain is all real values of y such that $y \neq 3$.

59. Find the domain of the following expression.

$$\frac{3}{x^2 + 4}$$

Solution

The denominator is zero when $x^2 + 4 = 0$, but $x^2 + 4$ is never zero for any real values of x. Thus, the domain is the set of *all* real numbers.

61. Find the missing factor.

$$\frac{-7}{3x} = \frac{7\boxed{}}{3x^3}$$

Solution

$$\frac{-7}{3x} = \frac{-7(x^2)}{3x(x^2)} = \frac{7(-x^2)}{3x^3}$$

The missing factor is $(-x^2)$.

63. Find the missing factor.

$$\frac{3}{4} = \frac{3\boxed{}}{4(x + 1)}$$

Solution

$$\frac{3}{4} = \frac{3(x + 1)}{4(x + 1)}$$

The missing factor is $(x + 1)$.

65. Find the missing factor.

$$\frac{x}{2} = \frac{x(x + 2)}{2\boxed{}}$$

Solution

$$\frac{x}{2} = \frac{x(x + 2)}{2(x + 2)}$$

The missing factor is $(x + 2)$.

67. Find the missing factor.

$$\frac{3x}{x - 3} = \frac{3x\boxed{}}{x^2 - x - 6}$$

Solution

$$\frac{3x}{x - 3} = \frac{3x(x + 2)}{(x - 3)(x + 2)}$$

$$= \frac{3x(x + 2)}{x^2 - x - 6}$$

The missing factor is $(x + 2)$.

69. Simplify the following rational expression.

$$\frac{75x}{15}$$

Solution

$$\frac{75x}{15} = \frac{\cancel{15}(5x)}{\cancel{15}(1)} = \frac{5x}{1} = 5x$$

71. Simplify the following rational expression.

$$\frac{2y^2}{y}$$

Solution

$$\frac{2y^2}{y} = \frac{\cancel{y}(2y)}{\cancel{y}(1)} = \frac{2y}{1} = 2y, \quad y \neq 0$$

73. Simplify the following rational expression.

$$\frac{x^2(x+1)}{x(x+1)}$$

Solution

$$\frac{x^2(x+1)}{x(x+1)} = \frac{x\cancel{(x)}\cancel{(x+1)}}{1\cancel{(x)}\cancel{(x+1)}}$$

$$= \frac{x}{1} = x, \quad x \neq 0, \ x \neq -1$$

75. Simplify the following rational expression.

$$\frac{3xy}{xy + x}$$

Solution

$$\frac{3xy}{xy + x} = \frac{3xy}{x(y+1)}$$

$$= \frac{\cancel{(x)}(3y)}{\cancel{(x)}(y+1)}$$

$$= \frac{3y}{y+1}, \quad x \neq 0$$

77. Simplify the following rational expression.

$$\frac{y^2 - 16}{3y + 12}$$

Solution

$$\frac{y^2 - 16}{3y + 12} = \frac{(y+4)(y-4)}{3(y+4)}$$

$$= \frac{\cancel{(y+4)}(y-4)}{3\cancel{(y+4)}}$$

$$= \frac{y-4}{3}, \quad y \neq -4$$

79. Simplify the following rational expression.

$$\frac{x^2 - 5x}{x^2 - 10x + 25}$$

Solution

$$\frac{x^2 - 5x}{x^2 - 10x + 25} = \frac{x(x-5)}{(x-5)(x-5)}$$

$$= \frac{x\cancel{(x-5)}}{\cancel{(x-5)}(x-5)}$$

$$= \frac{x}{x-5}$$

81. Simplify the following rational expression.

$$\frac{3-x}{x^2 - 5x + 6}$$

Solution

$$\frac{3-x}{x^2 - 5x + 6} = \frac{-1(x-3)}{(x-3)(x-2)}$$

$$= \frac{-1\cancel{(x-3)}}{\cancel{(x-3)}(x-2)}$$

$$= -\frac{1}{x-2}, \quad x \neq 3$$

83. Simplify the following rational expression.

$$\frac{x^2 + 8x - 20}{x^2 + 11x + 10}$$

Solution

$$\frac{x^2 + 8x - 20}{x^2 + 11x + 10} = \frac{(x+10)(x-2)}{(x+10)(x+1)}$$

$$= \frac{\cancel{(x+10)}(x-2)}{\cancel{(x+10)}(x+1)}$$

$$= \frac{x-2}{x+1}, \quad x \neq -10$$

85. Simplify the following rational expression.

$$\frac{x^3 + 5x^2 + 6x}{x^2 - 4}$$

Solution

$$\frac{x^3 + 5x^2 + 6x}{x^2 - 4} = \frac{x(x^2 + 5x + 6)}{(x+2)(x-2)}$$

$$= \frac{x(x+3)(x+2)}{(x+2)(x-2)}$$

$$= \frac{x(x+3)\cancel{(x+2)}}{\cancel{(x+2)}(x-2)}$$

$$= \frac{x(x+3)}{x-2}, \quad x \neq -2$$

87. Simplify the following rational expression.

$$\frac{x^3 - 2x^2 + x - 2}{x - 2}$$

Solution

$$\frac{x^3 - 2x^2 + x - 2}{x - 2} = \frac{(x^3 - 2x^2) + (x - 2)}{x - 2}$$

$$= \frac{x^2(x - 2) + 1(x - 2)}{x - 2}$$

$$= \frac{(x - 2)(x^2 + 1)}{x - 2}$$

$$= \frac{\cancel{(x - 2)}(x^2 + 1)}{1\cancel{(x - 2)}}$$

$$= x^2 + 1, \quad x \neq 2$$

89. Simplify the following expression. Then use a graphing utility to check your result.

$$\frac{x^2 - 3x}{2x}$$

Solution

$$\frac{x^2 - 3x}{2x} = \frac{\cancel{x}(x - 3)}{2\cancel{(x)}}$$

$$= \frac{x - 3}{2}, \quad x \neq 0$$

91. Simplify the following expression. Then use a graphing utility to check your result.

$$\frac{x^2 + x - 2}{x^3 + 8}$$

Solution

$$\frac{x^2 + x - 2}{x^3 + 8} = \frac{\cancel{(x + 2)}(x - 1)}{\cancel{(x + 2)}(x^2 - 2x + 4)}$$

$$= \frac{x - 1}{x^2 - 2x + 4}, \quad x \neq -2$$

93. *Creating a Model*   A machine shop has a setup cost of $3000 for the production of a new product. The cost for producing each unit is $7.50.

(a) Write a rational expression that gives the average cost per unit when x units are produced.

(b) Find the domain of the expression of part (a).

(c) Find the average cost per unit when $x = 100$ units are produced.

Solution

(a) Average cost per unit $= \dfrac{3000 + 7.50x}{x}$

(b) The variable x represents the number of units produced. If portions of units can be produced, the domain is $x > 0$; if complete units must be produced, the domain is $\{1, 2, 3, 4, \ldots\}$.

(c) $\dfrac{3000 + 7.50(100)}{100} = \dfrac{3000 + 750}{100} = \dfrac{3750}{100} = \$37.50$ per unit

95. *Comparing Distances*   You start a trip and drive at an average speed of 50 miles per hour. Two hours later, a friend starts a trip on the same road and drives at an average speed of 60 miles per hour.

(a) Find polynomial expressions that represent the distance each of you has driven when your friend has been driving for *t* hours.

(b) Use the result of part (a) to determine the ratio of the distance your friend has driven to the distance you have driven.

(c) Evaluate the ratio described in part (b) when $t = 5$ and $t = 10$.

Solution

Verbal models: $\boxed{\text{Distance friend drives}} = \boxed{\text{Friend's speed}} \cdot \boxed{\text{Friend's driving time}}$

$\boxed{\text{Distance you drive}} = \boxed{\text{Your speed}} \cdot \boxed{\text{Your driving time}}$

Labels: Distance friend drives $= D$ (in miles)
Distance you drive $= d$ (in miles)
Friend's rate $= 60$ mph
Your rate $= 50$ mph
Friend's time $= t$ (in hours)
Your time $= t + 2$ (in hours)

(a) $D = 60t$

$d = 50(t + 2)$ or $50t + 100$

(b) $\dfrac{60t}{50(t+2)} = \dfrac{\cancel{10}(6)(t)}{\cancel{10}(5)(t+2)}$

$= \dfrac{6t}{5(t+2)}$

(c) $\dfrac{6(5)}{5(5+2)} = \dfrac{30}{5(7)}$

$= \dfrac{30}{35} = \dfrac{\cancel{5}(6)}{\cancel{5}(7)} = \dfrac{6}{7}$

$\dfrac{6(10)}{5(10+2)} = \dfrac{60}{5(12)} = \dfrac{60}{60} = 1$

9.2   Multiplying and Dividing Rational Expressions

7. Evaluate the following expression for the values (a) $x = 5$, (b) $x = 0$, (c) $x = -5$, and (d) $x = 6$. If it is not possible, state the reason.

$$\frac{x - 5}{3x}$$

Solution

(a) $\dfrac{5 - 5}{3(5)} = \dfrac{0}{15}$

$= 0$

(b) $\dfrac{0 - 5}{3(0)} = \dfrac{-5}{0}$

Undefined
Division by 0
is undefined.

(c) $\dfrac{-5 - 5}{3(-5)} = \dfrac{-10}{-15}$

$= \dfrac{2}{3}$

(d) $\dfrac{6 - 5}{3(6)} = \dfrac{6 - 5}{18}$

$= \dfrac{1}{18}$

9. Find the missing factor:

$$\frac{3}{7x} = \frac{15}{7x\boxed{}}$$

Solution

$\dfrac{3}{7x} = \dfrac{3(5)}{7x(5)} = \dfrac{15}{7x(5)}$

The missing factor is 5.

11. Find the missing factor:

$$\frac{x}{x+1} = \frac{x\boxed{}}{(x+1)^2}$$

Solution

$\dfrac{x}{x+1} = \dfrac{x(x+1)}{(x+1)(x+1)} = \dfrac{x(x+1)}{(x+1)^2}$

The missing factor is $x + 1$.

13. Multiply and simplify:

$$\frac{6x}{5} \cdot \frac{1}{x}$$

Solution

$$\frac{6x}{5} \cdot \frac{1}{x} = \frac{6x}{5x}$$

$$= \frac{6\cancel{(x)}}{5\cancel{(x)}}$$

$$= \frac{6}{5}, \quad x \neq 0$$

15. Multiply and simplify:

$$\frac{x+1}{2} \cdot \frac{4x}{x+1}$$

Solution

$$\frac{x+1}{2} \cdot \frac{4x}{x+1} = \frac{\cancel{(x+1)}\cancel{(2)}(2)(x)}{\cancel{2}\cancel{(x+1)}}$$

$$= 2x, \quad x \neq -1$$

17. Multiply and simplify:

$$\frac{1-r}{3} \cdot \frac{3}{r-1}$$

Solution

$$\frac{1-r}{3} \cdot \frac{3}{r-1} = \frac{3(1-r)}{3(r-1)}$$

$$= \frac{\cancel{3}(-1)\cancel{(r-1)}}{\cancel{3}\cancel{(r-1)}}$$

$$= -1, \quad r \neq 1$$

19. Multiply and simplify:

$$\frac{r}{r-1} \cdot \frac{r^2-1}{r^2}$$

Solution

$$\frac{r}{r-1} \cdot \frac{r^2-1}{r^2} = \frac{r(r+1)(r-1)}{r^2(r-1)}$$

$$= \frac{\cancel{r}(r+1)\cancel{(r-1)}}{r\cancel{(r)}\cancel{(r-1)}}$$

$$= \frac{r+1}{r}, \quad r \neq 1$$

21. Multiply and simplify:

$$(x^2 - 4) \cdot \frac{x}{(x-2)^2}$$

Solution

$$(x^2 - 4) \cdot \frac{x}{(x-2)^2} = \frac{(x+2)(x-2)}{1} \cdot \frac{x}{(x-2)^2}$$

$$= \frac{(x+2)\cancel{(x-2)}(x)}{1\cancel{(x-2)}(x-2)}$$

$$= \frac{x(x+2)}{x-2}$$

23. Multiply and simplify:

$$\frac{a+1}{a-1} \cdot \frac{a^2-2a+1}{a} \cdot (3a^2 + 3a)$$

Solution

$$\frac{a+1}{a-1} \cdot \frac{a^2-2a+1}{a} \cdot (3a^2 + 3a) = \frac{a+1}{a-1} \cdot \frac{a^2-2a+1}{a} \cdot \frac{3a^2+3a}{1}$$

$$= \frac{(a+1)(a-1)(a-1)(3a)(a+1)}{(a-1)(a)}$$

$$= \frac{(a+1)\cancel{(a-1)}(a-1)(3)\cancel{(a)}(a+1)}{\cancel{(a-1)}\cancel{(a)}}$$

$$= 3(a+1)^2(a-1), \quad a \neq 0, \ a \neq 1$$

25. Divide and simplify:

$$\frac{2x}{3} \div \frac{4x^2}{15}$$

Solution

$$\frac{2x}{3} \div \frac{4x^2}{15} = \frac{2x}{3} \cdot \frac{15}{4x^2}$$

$$= \frac{(2x)(3)(5)}{3(2x)(2x)}$$

$$= \frac{5}{2x}$$

27. Divide and simplify:

$$\frac{a}{a+1} \div \frac{6}{(a+1)^2}$$

Solution

$$\frac{a}{a+1} \div \frac{6}{(a+1)^2} = \frac{a}{a+1} \cdot \frac{(a+1)^2}{6}$$

$$= \frac{a(a+1)(a+1)}{(a+1)(6)}$$

$$= \frac{a(a+1)}{6}, \quad a \neq -1$$

29. Divide and simplify:

$$\frac{y^2 - 4}{y^2} \div \frac{y - 2}{3y}$$

Solution

$$\frac{y^2 - 4}{y^2} \div \frac{y - 2}{3y} = \frac{y^2 - 4}{y^2} \cdot \frac{3y}{y - 2}$$

$$= \frac{(y - 2)(y + 2)(3)(y)}{y(y)(y - 2)}$$

$$= \frac{3(y + 2)}{y}, \quad y \neq 2$$

31. Divide and simplify:

$$\frac{\left(\dfrac{5x^2 + 30x + 40}{x + 2}\right)}{15 - x}$$

Solution

$$\frac{\left(\dfrac{5x^2 + 30x + 40}{x + 2}\right)}{15 - x} = \frac{5x^2 + 30x + 40}{x + 2} \div \frac{15 - x}{1}$$

$$= \frac{5(x^2 + 6x + 8)}{x + 2} \cdot \frac{1}{15 - x}$$

$$= \frac{5(x + 4)(x + 2)(1)}{(x + 2)(15 - x)}$$

$$= \frac{5(x + 4)}{15 - x}, \quad x \neq -2$$

33. Divide and simplify:

$$\frac{x + 3}{\left(\dfrac{x^2 + 6x + 9}{x^2 + 1}\right)}$$

Solution

$$\frac{x + 3}{\left(\dfrac{x^2 + 6x + 9}{x^2 + 1}\right)} = \frac{x + 3}{1} \div \frac{x^2 + 6x + 9}{x^2 + 1}$$

$$= \frac{x + 3}{1} \cdot \frac{x^2 + 1}{x^2 + 6x + 9}$$

$$= \frac{(x + 3)(x^2 + 1)}{1(x + 3)(x + 3)}$$

$$= \frac{x^2 + 1}{x + 3}$$

35. Divide and simplify:

$$\frac{\left(\dfrac{2x - 10}{x + 1}\right)}{\left(\dfrac{(x - 5)^2}{x + 1}\right)}$$

Solution

$$\frac{\left(\dfrac{2x - 10}{x + 1}\right)}{\left(\dfrac{(x - 5)^2}{x + 1}\right)} = \frac{2x - 10}{x + 1} \div \frac{(x - 5)^2}{x + 1}$$

$$= \frac{2x - 10}{x + 1} \cdot \frac{x + 1}{(x - 5)^2}$$

$$= \frac{2(x - 5)(x + 1)}{(x + 1)(x - 5)(x - 5)}$$

$$= \frac{2}{x - 5}, \quad x \neq -1$$

37. Perform the indicated operations and simplify your answer.

$$\left(\frac{5x}{3}\right)^2 \div \left(\frac{5x}{2}\right)^3$$

Solution

$$\left(\frac{5x}{3}\right)^2 \div \left(\frac{5x}{2}\right)^3 = \frac{(5x)^2}{3^2} \div \frac{(5x)^3}{2^3}$$

$$= \frac{25x^2}{9} \cdot \frac{8}{125x^3}$$

$$= \frac{8(25)x^2}{9(25)(5)(x^2)(x)}$$

$$= \frac{8}{45x}$$

39. Perform the indicated operations and simplify your answer.

$$\left(\frac{x^2}{5} \cdot \frac{x+a}{2}\right) \div \frac{x}{30}$$

Solution

$$\left(\frac{x^2}{5} \cdot \frac{x+a}{2}\right) \div \frac{x}{30} = \left(\frac{x^2}{5} \cdot \frac{x+a}{2}\right) \cdot \frac{30}{x}$$

$$= \frac{x(x)(x+a)(5)(2)(3)}{(5)(2)(x)}$$

$$= 3x(x+a), \quad x \neq 0$$

41. Find the ratio of the area of the blue region to the total area of the figure shown in the textbook.

Solution

Area of blue region $= (x+1)(x+2)$

Total area of figure $= (3x+6)(2x+2)$

$$\frac{(x+1)(x+2)}{(3x+6)(2x+2)} = \frac{(x+1)(x+2)}{3(x+2)(2)(x+1)} = \frac{1}{6}$$

The ratio of the blue region to the total area of the figure is $\frac{1}{6}$.

43. Determine the slope of the line passing through the points $(-3, 2)$ and $(5, 0)$.

Solution

$$m = \frac{y_2 - y_1}{x_2 - x_1}$$

$$m = \frac{0 - 2}{5 - (-3)}$$

$$= \frac{-2}{8} = -\frac{1}{4}$$

45. Determine the slope of the line passing through the points $(-4, -4)$ and $(-4, 6)$.

Solution

$$m = \frac{y_2 - y_1}{x_2 - x_1}$$

$$m = \frac{6 - (-4)}{-4 - (-4)} = \frac{10}{0}$$

Undefined

47. Sketch the graph of the line through the point $(2, 3)$ with slope (a) $m = 0$ and (b) $m = 1$.

Solution

(a) $m = 0$. This is a horizontal line.

(b) $m = 1$ or $m = \frac{1}{1}$

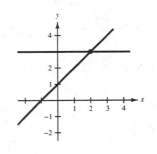

49. Find the missing factor:

$$\frac{3ab}{7} = \frac{3ab\boxed{}}{7ab}$$

Solution

$$\frac{3ab}{7} = \frac{3ab(ab)}{7(ab)}$$

$$= \frac{3ab(ab)}{7ab}$$

The missing factor is ab.

53. Find the missing factor:

$$\frac{3t + 6}{t} = \frac{(3t + 6)\boxed{}}{5t^2}$$

Solution

$$\frac{3t + 6}{t} = \frac{(3t + 6)(5t)}{t(5t)}$$

$$= \frac{(3t + 6)(5t)}{5t^2}$$

The missing factor is $5t$.

57. Multiply and simplify:

$$\frac{12x^2}{6y} \cdot \frac{12y}{8x^2}$$

Solution

$$\frac{12x^2}{6y} \cdot \frac{12y}{8x^2} = \frac{12(12)x^2 y}{6(8)x^2 y}$$

$$= \frac{\cancel{6}(2)(\cancel{4})(3)\cancel{(x^2)}\cancel{(y)}}{\cancel{6}(\cancel{4})(2)\cancel{(x^2)}\cancel{(y)}}$$

$$= 3, \quad x \neq 0, \ y \neq 0$$

61. Multiply and simplify:

$$3(a + 2) \cdot \frac{1}{3a + 6}$$

Solution

$$3(a + 2) \cdot \frac{1}{3a + 6} = \frac{3(a + 2)}{1} \cdot \frac{1}{3a + 6}$$

$$= \frac{3(a + 2)}{3(a + 2)}$$

$$= \frac{\cancel{3(a + 2)}}{\cancel{3(a + 2)}}$$

$$= 1, \quad a \neq -2$$

51. Find the missing factor:

$$\frac{2x}{x + 2} = \frac{2x\boxed{}}{4 - x^2}$$

Solution

$$\frac{2x}{x + 2} = \frac{2x}{2 + x}$$

$$= \frac{2x(2 - x)}{(2 + x)(2 - x)} = \frac{2x(2 - x)}{4 - x^2}$$

The missing factor is $(2 - x)$.

55. Multiply and simplify:

$$\frac{8x^2}{3} \cdot \frac{9}{16x}$$

Solution

$$\frac{8x^2}{3} \cdot \frac{9}{16x} = \frac{8(9)x^2}{3(16)x}$$

$$= \frac{\cancel{8}\cancel{(3)}(3)\cancel{(x)}(x)}{\cancel{3}\cancel{(8)}(2)\cancel{(x)}}$$

$$= \frac{3x}{2}, \quad x \neq 0$$

59. Multiply and simplify:

$$\frac{8}{2 + 3x} \cdot (8 + 12x)$$

Solution

$$\frac{8}{2 + 3x} \cdot (8 + 12x) = \frac{8}{2 + 3x} \cdot \frac{8 + 12x}{1}$$

$$= \frac{(8)(4)(2 + 3x)}{(2 + 3x)(1)}$$

$$= \frac{(8)(4)\cancel{(2 + 3x)}}{\cancel{(2 + 3x)}(1)}$$

$$= 32, \quad x \neq -\frac{2}{3}$$

63. Multiply and simplify:

$$\frac{5}{x - 1} \cdot \frac{x - 1}{25(x - 2)}$$

Solution

$$\frac{5}{x - 1} \cdot \frac{x - 1}{25(x - 2)} = \frac{5(x - 1)}{25(x - 1)(x - 2)}$$

$$= \frac{\cancel{5}\cancel{(x - 1)}}{5\cancel{(5)}\cancel{(x - 1)}(x - 2)}$$

$$= \frac{1}{5(x - 2)}, \quad x \neq 1$$

65. Multiply and simplify:

$$(u-2)^2 \cdot \frac{u+2}{u-2}$$

Solution

$$(u-2)^2 \cdot \frac{u+2}{u-2} = \frac{(u-2)^2}{1} \cdot \frac{u+2}{u-2}$$

$$= \frac{(u-2)(u-2)(u+2)}{1(u-2)}$$

$$= (u-2)(u+2), \quad u \neq 2$$

67. Multiply and simplify:

$$\frac{(x+5)(x-3)}{x+2} \cdot \frac{1}{(x+5)(x+2)}$$

Solution

$$\frac{(x+5)(x-3)}{x+2} \cdot \frac{1}{(x+5)(x+2)}$$

$$= \frac{(x+5)(x-3)}{(x+2)(x+5)(x+2)}$$

$$= \frac{(x+5)(x-3)}{(x+2)(x+5)(x+2)}$$

$$= \frac{x-3}{(x+2)^2}, \quad x \neq -5$$

69. Multiply and simplify:

$$\frac{y^2-16}{2y^3} \cdot \frac{4y}{y^2-6y+8}$$

Solution

$$\frac{y^2-16}{2y^3} \cdot \frac{4y}{y^2-6y+8} = \frac{4y(y+4)(y-4)}{2y^3(y-4)(y-2)} = \frac{2(2)(y)(y+4)(y-4)}{2(y)(y^2)(y-4)(y-2)} = \frac{2(y+4)}{y^2(y-2)}, \quad y \neq 4$$

71. Multiply and simplify:

$$\frac{2}{z+3} \cdot \frac{z^2+6z+9}{z-3} \cdot \frac{4}{z^2-9}$$

Solution

$$\frac{2}{z+3} \cdot \frac{z^2+6z+9}{z-3} \cdot \frac{4}{z^2-9} = \frac{2(z+3)(z+3)(4)}{(z+3)(z-3)(z+3)(z-3)} = \frac{8}{(z-3)^2}, \quad z \neq -3$$

73. Perform the indicated operations and simplify your answer.

$$\frac{7x^2}{10} \div \frac{14x^3}{15}$$

Solution

$$\frac{7x^2}{10} \div \frac{14x^3}{15} = \frac{7x^2}{10} \cdot \frac{15}{14x^3}$$

$$= \frac{7(15)(x^2)}{10(14)(x^3)}$$

$$= \frac{7(5)(3)(x^2)}{5(2)(7)(2)(x^2)(x)}$$

$$= \frac{3}{4x}$$

75. Perform the indicated operations and simplify your answer.

$$\frac{3(x+4)}{4} \div \frac{x+4}{2}$$

Solution

$$\frac{3(x+4)}{4} \div \frac{x+4}{2} = \frac{3(x+4)}{4} \cdot \frac{2}{x+4}$$

$$= \frac{3(2)(x+4)}{4(x+4)}$$

$$= \frac{3(2)(x+4)}{2(2)(x+4)}$$

$$= \frac{3}{2}, \quad x \neq -4$$

77. Perform the indicated operations and simplify your answer.

$$\frac{x^3}{6} \div \frac{x^2}{3}$$

Solution

$$\frac{x^3}{6} \div \frac{x^2}{3} = \frac{x^3}{6} \cdot \frac{3}{x^2} = \frac{x(x^2)(3)}{3(2)(x^2)} = \frac{x}{2}, \quad x \neq 0$$

79. Perform the indicated operations and simplify your answer.

$$\frac{(x+y)^2}{x^2+y^2} \div \frac{x+y}{x^3y+xy^3}$$

Solution

$$\frac{(x+y)^2}{x^2+y^2} \div \frac{x+y}{x^3y+xy^3}$$

$$= \frac{(x+y)^2}{x^2+y^2} \cdot \frac{x^3y+xy^3}{x+y}$$

$$= \frac{(x+y)^2(xy)(x^2+y^2)}{(x^2+y^2)(x+y)}$$

$$= \frac{xy(x+y)(x+y)(x^2+y^2)}{(x^2+y^2)(x+y)}$$

$$= xy(x+y), \quad x \neq -y, \ x \neq 0, \ y \neq 0$$

81. Perform the indicated operations and simplify your answer.

$$\frac{x^2-y^2}{xy} \div \frac{(x-y)^2}{xy}$$

Solution

$$\frac{x^2-y^2}{xy} \div \frac{(x-y)^2}{xy} = \frac{x^2-y^2}{xy} \cdot \frac{xy}{(x-y)^2}$$

$$= \frac{(x+y)(x-y)(xy)}{xy(x-y)^2}$$

$$= \frac{(x+y)(x-y)(xy)}{(xy)(x-y)(x-y)}$$

$$= \frac{x+y}{x-y}, \quad x \neq 0, \ y \neq 0$$

83. Perform the indicated operations and simplify your answer.

$$\frac{\left(\dfrac{x^3}{4}\right)}{\left(\dfrac{x}{8}\right)}$$

Solution

$$\frac{\left(\dfrac{x^3}{4}\right)}{\left(\dfrac{x}{8}\right)} = \frac{x^3}{4} \div \frac{x}{8}$$

$$= \frac{x^3}{4} \cdot \frac{8}{x}$$

$$= \frac{8x^3}{4x}$$

$$= \frac{4(2)(x)(x^2)}{4(x)}$$

$$= 2x^2, \quad x \neq 0$$

85. Perform the indicated operations and simplify your answer.

$$\left[\left(\frac{x+2}{3}\right)^2 \cdot \left(\frac{x+1}{2}\right)^2\right] \div \frac{(x+1)(x+2)}{36}$$

Solution

$$\left[\left(\frac{x+2}{3}\right)^2 \cdot \left(\frac{x+1}{2}\right)^2\right] \div \frac{(x+1)(x+2)}{36} = \left[\frac{(x+2)^2}{3^2} \cdot \frac{(x+1)^2}{2^2}\right] \div \frac{(x+1)(x+2)}{36}$$

$$= \frac{(x+2)^2(x+1)^2}{9(4)} \cdot \frac{36}{(x+1)(x+2)}$$

$$= \frac{36(x+2)^2(x+1)^2}{36(x+1)(x+2)}$$

$$= \frac{36(x+2)(x+2)(x+1)(x+1)}{36(x+1)(x+2)}$$

$$= (x+2)(x+1), \quad x \neq -1, \ x \neq -2$$

87. *Photocopy Rate* A photocopier produces copies at a rate of 12 pages per minute. Determine the time required to copy (a) 1 page, (b) x pages, and (c) 32 pages.

Solution

(a) $\dfrac{1}{12}$ min

(b) $x\left(\dfrac{1}{12}\right) = \dfrac{x}{1} \cdot \dfrac{1}{12} = \dfrac{x}{12}$ min

(c) $32\left(\dfrac{1}{12}\right) = \dfrac{32}{12} = \dfrac{8}{3}$ min or $2\frac{2}{3}$ min

89. *Analyzing Data* The number P (in thousands) and total sales S (in millions of dollars) of franchised printing and copying businesses in the United States over the period from 1988 to 1991 can be modeled by

$$P = \frac{5176 - 307t}{10 - t} \text{ and } S = \frac{100,000}{7(20 - t)}$$

where $t = 0$ represents 1990. (*Source:* U.S. Bureau of the Census)

(a) Find a model for the average sales per business.

(b) Use the model of part (a) to complete the table shown in the textbook for average sales (in millions of dollars).

Solution

(a) Sales per business (in millions) $= \dfrac{S}{P} = \dfrac{\left(\dfrac{100,000}{7(20 - t)}\right)}{\left(\dfrac{5176 - 307t}{10 - t}\right)}$

$$= \frac{100,000}{7(20 - t)} \div \frac{5176 - 307t}{10 - t}$$

$$= \frac{100,000}{7(20 - t)} \cdot \frac{10 - t}{5176 - 307t}$$

$$= \frac{100,000(10 - t)}{7(20 - t)(5176 - 307t)}$$

(b)

Year, t	-2	-1	0	1
Average sales (in millions)	1.35	1.36	1.38	1.39

Mid-Chapter Quiz for Chapter 9

1. In your own words, explain the meaning of *domain*. Find the domain of

(a) $\dfrac{x^2}{x^2 + 4}$ and (b) $\dfrac{x^2}{x^2 - 4}$.

Solution

(a) For this rational expression, the denominator is not zero for any real values of x, so the domain is the set of *all* real numbers.

(b) For this rational expression, the denominator, $x^2 - 4 = (x + 2)(x - 2)$, is zero when $x = -2$ or 2. Thus, the domain is all real values of x such that $x \neq -2$ and $x \neq 2$.

2. Evaluate $\dfrac{y - 3}{y - 2}$ for (a) $y = 10$, (b) $y = 3$, and (c) $y = -2$.

Solution

(a) $\dfrac{10 - 3}{10 + 2} = \dfrac{7}{12}$

(b) $\dfrac{3 - 3}{3 + 2} = \dfrac{0}{5} = 0$

(c) $\dfrac{-2 - 3}{-2 + 2} = \dfrac{-5}{0}$

Undefined

3. Simplify $\dfrac{14z^2}{35z}$.

Solution

$\dfrac{14z^4}{35z} = \dfrac{7(2)(z)(z^3)}{5(7)(z)}$

$= \dfrac{2z^3}{5}, \quad z \neq 0$

4. Simplify $\dfrac{15u(u - 3)^2}{25u^2(u - 3)}$.

Solution

$\dfrac{15u(u - 3)^2}{25u^2(u - 3)} = \dfrac{3(5)(u)(u - 3)(u - 3)}{5(5)(u)(u)(u - 3)}$

$= \dfrac{3(u - 3)}{5u}, \quad u \neq 3$

5. Simplify $\dfrac{24(9 - x)}{15(x - 9)}$.

Solution

$\dfrac{24(9 - x)}{15(x - 9)} = \dfrac{8(3)(-1)(x - 9)}{5(3)(x - 9)}$

$= -\dfrac{8}{5}, \quad x \neq 9$

6. Simplify $\dfrac{y^2 - 4}{8 - 4y}$.

Solution

$\dfrac{y^2 - 4}{8 - 4y} = \dfrac{(y + 2)(y - 2)}{4(2 - y)}$

$= \dfrac{(y + 2)(y - 2)}{4(-1)(y - 2)}$

$= -\dfrac{y + 2}{4}, \quad y \neq 2$

7. Simplify $\dfrac{b^2 + 3b}{b^3 + 2b^2 - 3b}$.

Solution

$\dfrac{b^2 + 3b}{b^3 + 2b^2 - 3b} = \dfrac{b(b + 3)}{b(b^2 + 2b - 3)}$

$= \dfrac{(b)(b + 3)}{b(b + 3)(b - 1)}$

$= \dfrac{1}{b - 1}, \quad b \neq 0, b \neq -3$

8. Simplify $\dfrac{4x^2 - 12x + 9}{2x^2 - x - 3}$.

Solution

$\dfrac{4x^2 - 12x + 9}{2x^2 - x - 3} = \dfrac{(2x - 3)(2x - 3)}{(2x - 3)(x + 1)}$

$= \dfrac{2x - 3}{x + 1}, \quad x \neq \dfrac{3}{2}$

9. Simplify $\dfrac{s^3 - 8}{s^3 + 2s^2 + 4s}$.

Solution

$$\frac{s^3 - 8}{s^3 + 2s^2 + 4s} = \frac{(s-2)\cancel{(s^2 + 2s + 4)}}{s\cancel{(s^2 + 2s + 4)}}$$

$$= \frac{s-2}{s}$$

10. Simplify $\dfrac{x^3 + 2x^2 - 3x - 6}{x(x-1) - 6}$.

Solution

$$\frac{x^3 + 2x^2 - 3x - 6}{x(x-1) - 6} = \frac{x^2(x+2) - 3(x+2)}{x^2 - x - 6}$$

$$= \frac{\cancel{(x+2)}(x^2 - 3)}{(x-3)\cancel{(x+2)}}$$

$$= \frac{x^2 - 3}{x - 3}, \quad x \neq -2$$

11. Perform the specified operation and simplify:

$$\frac{3y^3}{5} \cdot \frac{25}{9y}$$

Solution

$$\frac{3y^3}{5} \cdot \frac{25}{9y} = \frac{\cancel{(3y)}(y^2)\cancel{(5)}(5)}{\cancel{5}(3)\cancel{(3y)}}$$

$$= \frac{5y^2}{3}, \quad y \neq 0$$

12. Perform the specified operation and simplify:

$$\frac{s-5}{15} \cdot \frac{12s}{25 - s^2}$$

Solution

$$\frac{s-5}{15} \cdot \frac{12s}{25 - s^2} = \frac{(s-5)(3)(4s)}{5(3)(5+s)(5-s)}$$

$$= \frac{\cancel{(s-5)}\cancel{(3)}(4s)}{5\cancel{(3)}(5+s)(-1)\cancel{(s-5)}}$$

$$= -\frac{4s}{5(s+5)}, \quad s \neq 5$$

13. Perform the specified operation and simplify:

$$(x^3 + 4x^2) \cdot \frac{5x}{x^2 + 2x - 8}$$

Solution

$$(x^3 + 4x^2) \cdot \frac{5x}{x^2 + 2x - 8} = \frac{x^3 + 4x^2}{1} \cdot \frac{5x}{x^2 + 2x - 8}$$

$$= \frac{x^2\cancel{(x+4)}(5x)}{\cancel{(x+4)}(x-2)}$$

$$= \frac{5x^3}{x-2}, \quad x \neq -4$$

14. Perform the specified operation and simplify:

$$\left(\frac{2\pi}{\theta}\right)\left(\frac{u+v}{2uv}\right)\left(\frac{uv\theta}{4u + 4v}\right)$$

Solution

$$\left(\frac{2\pi}{\theta}\right)\left(\frac{u+v}{2uv}\right)\left(\frac{uv\theta}{4u + 4v}\right)$$

$$= \frac{\cancel{(2)}(\pi)\cancel{(u+v)}\cancel{(uv)}\cancel{(\theta)}}{\cancel{(\theta)}\cancel{(2)}\cancel{(uv)}(4)\cancel{(u+v)}}$$

$$= \frac{\pi}{4}, \quad u \neq 0, \ v \neq 0, \ \theta \neq 0, \ u \neq -v$$

15. Perform the specified operation and simplify:

$$\frac{x}{25} \div \frac{x^2 + 2x}{10}$$

Solution

$$\frac{x}{25} \div \frac{x^2 + 2x}{10} = \frac{x}{25} \cdot \frac{10}{x^2 + 2x}$$

$$= \frac{\cancel{x}\cancel{(5)}(2)}{5\cancel{(5)}\cancel{(x)}(x+2)}$$

$$= \frac{2}{5(x+2)}, \quad x \neq 0$$

16. Perform the specified operation and simplify:

$$\frac{r^2 - 16}{r} \div (r+4)^2$$

Solution

$$\frac{r^2 - 16}{r} \div (r+4)^2 = \frac{r^2 - 16}{r} \div \frac{(r+4)^2}{1}$$

$$= \frac{r^2 - 16}{r} \cdot \frac{1}{(r+4)^2}$$

$$= \frac{\cancel{(r+4)}(r-4)}{r\cancel{(r+4)}(r+4)}$$

$$= \frac{r-4}{r(r+4)}$$

17. Perform the specified operation and simplify:

$$\frac{10x^2}{3y} \div \left(\frac{y}{x} \cdot \frac{x^3 y}{6}\right)$$

Solution

$$\frac{10x^2}{3y} \div \left(\frac{y}{x} \cdot \frac{x^3 y}{6}\right) = \frac{10x^2}{3y} \div \left(\frac{x^3 y^2}{6x}\right)$$

$$= \frac{10x^2}{3y} \cdot \frac{6x}{x^3 y^2}$$

$$= \frac{10(x^2)(3)(2)(x)}{3(y)(x^2)(x)(y^2)}$$

$$= \frac{20}{y^3}, \quad x \neq 0$$

18. Perform the specified operation and simplify:

$$\frac{\left[\dfrac{3x - 12}{(x + 1)^2}\right]}{\left(\dfrac{x^2 - 8x + 16}{x^2 + x}\right)}$$

Solution

$$\frac{\left[\dfrac{3x - 12}{(x + 1)^2}\right]}{\left(\dfrac{x^2 - 8x + 16}{x^2 + x}\right)} = \frac{3x - 12}{(x + 1)^2} \div \frac{x^2 - 8x + 16}{x^2 + x}$$

$$= \frac{3x - 12}{(x + 1)^2} \cdot \frac{x^2 + x}{x^2 - 8x + 16}$$

$$= \frac{3(x - 4)(x)(x + 1)}{(x + 1)(x + 1)(x - 4)(x - 4)}$$

$$= \frac{3x}{(x + 1)(x - 4)}, \quad x \neq 0$$

19. A small business has a setup cost of $10,000 for the production of a new product. The cost of labor and materials for producing each unit is $25.

(a) Write an expression for the average cost per unit $\overline{C}$, when x number of units are produced.

(b) Complete the table shown in the textbook and describe any trends you find.

Solution

(a) Cost $C = 10,000 + 25x$

  Average cost per unit:

$$\overline{C} = \frac{C}{x}$$

$$= \frac{10,000 + 25x}{x}$$

(b)

x	2000	3000	4000	5000
$\overline{C}$	$30.00	$28.33	$27.50	$27.00

As the number of units increases, the cost per unit decreases.

20. Find the ratio of the area of the shaded region to the total area of the figure shown in the textbook.

Solution

Area of shaded region $= (x + 3)(x)$
Total area of figure $= (x + 10)(x + 3)$

$$\frac{x(x + 3)}{(x + 10)(x + 3)} = \frac{x}{x + 10}$$

9.3 | Adding and Subtracting Rational Expressions

7. Combine and simplify:

$$\frac{3}{32} + \frac{5}{32}$$

Solution

$$\frac{3}{32} + \frac{5}{32} = \frac{3+5}{32} = \frac{8}{32} = \frac{1}{4}$$

11. Combine and simplify:

$$\frac{4z}{3} - \frac{4z-3}{3}$$

Solution

$$\frac{4z}{3} - \frac{4z-3}{3} = \frac{4z - (4z-3)}{3}$$

$$= \frac{4z - 4z + 3}{3}$$

$$= \frac{3}{3} = 1$$

15. Find the least common multiple of $2x$ and x^3.

Solution

$$2x = 2 \cdot x$$

$$x^3 = x \cdot x \cdot x$$

The different factors are 2 and x. Using the highest powers of these factors, we conclude that the least common multiple is $2x^3$.

9. Combine and simplify:

$$\frac{9}{x} - \frac{4}{x}$$

Solution

$$\frac{9}{x} - \frac{4}{x} = \frac{9-4}{x} = \frac{5}{x}$$

13. Find the least common multiple of 45 and 75.

Solution

$$45 = 3 \cdot 3 \cdot 5 = 3^2 \cdot 5$$

$$75 = 3 \cdot 5 \cdot 5 = 3 \cdot 5^2$$

The different factors are 3 and 5. Using the highest powers of these factors, we conclude that the least common multiple is $3^2 \cdot 5^2 = 225$.

17. Find the least common multiple of $x + 2$, $x^2 - 4$, and x.

Solution

$$x + 2$$

$$x^2 - 4 = (x+2)(x-2)$$

$$x$$

The least common multiple is $x(x+2)(x-2)$.

19. Rewrite one or both of the following fractions so that they have the same denominator.

$$\frac{x+5}{3x-6}, \quad \frac{10}{x-2}$$

Solution

The denominators are $3x - 6 = 3(x-2)$ and $(x-2)$. The least common multiple of these denominators is $3(x-2)$.

$$\frac{x+5}{3x-6} = \frac{x+5}{3(x-2)}$$

$$\frac{10}{x-2} = \frac{10(3)}{(x-2)(3)} = \frac{30}{3(x-2)}$$

21. Rewrite one or both of the following fractions so that they have the same denominator.

$$\frac{2}{x^2}, \quad \frac{5}{x(x+3)}$$

Solution

The denominators are x^2 and $x(x+3)$. The least common multiple of the denominators is $x^2(x+3)$.

$$\frac{2}{x^2} = \frac{2(x+3)}{x^2(x+3)} \text{ or } \frac{2x+6}{x^2(x+3)}$$

$$\frac{5}{x(x+3)} = \frac{5(x)}{x(x+3)(x)} = \frac{5x}{x^2(x+3)}$$

23. Rewrite one or both of the following fractions so that they have the same denominator.

$$\frac{x-8}{x^2-16}, \quad \frac{9x}{x^2-8x+16}$$

Solution

The denominators are $x^2 - 16 = (x-4)(x+4)$ and $x^2 - 8x + 16 = (x-4)^2$. The least common multiple of the denominators is $(x-4)^2(x+4)$.

$$\frac{x-8}{x^2-16} = \frac{x-8}{(x-4)(x+4)} = \frac{(x-8)(x-4)}{(x-4)(x+4)(x-4)} = \frac{(x-8)(x-4)}{(x-4)^2(x+4)}$$

$$\text{or } \frac{x^2-12x+32}{(x-4)^2(x+4)}$$

$$\frac{9x}{x^2-8x+16} = \frac{9x}{(x-4)^2} = \frac{9x(x+4)}{(x-4)^2(x+4)} \text{ or } \frac{9x^2+36x}{(x-4)^2(x+4)}$$

25. Combine and simplify:

$$\frac{1}{5x} - \frac{3}{5}$$

Solution

The denominators are $5x = 5 \cdot x$ and 5. The least common denominator is $5x$.

$$\frac{1}{5x} - \frac{3}{5} = \frac{1}{5x} - \frac{3(x)}{5(x)}$$

$$= \frac{1}{5x} - \frac{3x}{5x}$$

$$= \frac{1-3x}{5x} \text{ or } \frac{-3x+1}{5x}$$

27. Combine and simplify:

$$\frac{4}{x-3} + \frac{4}{3-x}$$

Solution

The denominators are $(x-3)$ and $(3-x) = -1(x-3)$. The least common denominator is $x-3$.

$$\frac{4}{x-3} + \frac{4}{3-x} = \frac{4}{x-3} + \frac{(-1)4}{(-1)(3-x)}$$

$$= \frac{4}{x-3} + \frac{-4}{x-3}$$

$$= \frac{4-4}{x-3}$$

$$= \frac{0}{x-3} = 0, \quad x \neq 3$$

29. Combine and simplify:

$$\frac{1}{x-1} - \frac{1}{x+2}$$

Solution

The denominators are $(x-1)$ and $(x+2)$. The least common denominator is $(x-1)(x+2)$.

$$\frac{1}{x-1} - \frac{1}{x+2} = \frac{1(x+2)}{(x-1)(x+2)} - \frac{1(x-1)}{(x+2)(x-1)}$$

$$= \frac{1(x+2) - 1(x-1)}{(x-1)(x+2)}$$

$$= \frac{x+2-x+1}{(x-1)(x+2)}$$

$$= \frac{3}{(x-1)(x+2)}$$

31. Combine and simplify:

$$7 + \frac{2}{x-3}$$

Solution

The denominators are 1 and $x-3$. The least common denominator is $x-3$.

$$7 + \frac{2}{x-3} = \frac{7}{1} + \frac{2}{x-3}$$

$$= \frac{7(x-3)}{x-3} + \frac{2}{x-3}$$

$$= \frac{7x-21+2}{x-3}$$

$$= \frac{7x-19}{x-3}$$

33. Combine and simplify: $\dfrac{3}{x} - \dfrac{1}{x^2} + \dfrac{1}{x+1}$

Solution

The denominators are x, x^2, and $x+1$. The least common denominator is $x^2(x+1)$.

$$\frac{3}{x} - \frac{1}{x^2} + \frac{1}{x+1} = \frac{3(x)(x+1)}{x(x)(x+1)} - \frac{1(x+1)}{x^2(x+1)} + \frac{1(x^2)}{(x+1)(x^2)}$$

$$= \frac{3x(x+1) - (x+1) + x^2}{x^2(x+1)}$$

$$= \frac{3x^2 + 3x - x - 1 + x^2}{x^2(x+1)}$$

$$= \frac{4x^2 + 2x - 1}{x^2(x+1)}$$

35. Combine and simplify: $\dfrac{2}{x^2-4} - \dfrac{1}{x^2-3x+2}$

Solution

The denominators are $x^2 - 4 = (x+2)(x-2)$ and $x^2 - 3x + 2 = (x-2)(x-1)$. The least common denominator is $(x+2)(x-2)(x-1)$.

$$\frac{2}{x^2-4} - \frac{1}{x^2-3x+2} = \frac{2}{(x+2)(x-2)} - \frac{1}{(x-2)(x-1)}$$

$$= \frac{2(x-1)}{(x+2)(x-2)(x-1)} - \frac{1(x+2)}{(x+2)(x-2)(x-1)}$$

$$= \frac{2(x-1) - 1(x+2)}{(x+2)(x-2)(x-1)}$$

$$= \frac{2x - 2 - x - 2}{(x+2)(x-2)(x-1)}$$

$$= \frac{x-4}{(x+2)(x-2)(x-1)}$$

37. Simplify the following complex fraction.

$$\frac{\left(\dfrac{3}{x}\right)}{\left(\dfrac{6}{x^2}\right)}$$

Solution

$$\frac{\left(\dfrac{3}{x}\right)}{\left(\dfrac{6}{x^2}\right)} = \frac{3}{x} \div \frac{6}{x^2}$$

$$= \frac{3}{x} \cdot \frac{x^2}{6}$$

$$= \frac{3x^2}{6x}$$

$$= \frac{\cancel{3}(x)\cancel{(x)}}{\cancel{3}(2)\cancel{(x)}}$$

$$= \frac{x}{2}, \quad x \neq 0$$

Alternate Method:

$$\frac{\left(\dfrac{3}{x}\right)}{\left(\dfrac{6}{x^2}\right)} = \frac{\left(\dfrac{3}{x}\right) \cdot x^2}{\left(\dfrac{6}{x^2}\right) \cdot x^2}$$

$$= \frac{3x}{6} = \frac{\cancel{3}x}{\cancel{3}(2)} = \frac{x}{2}, \quad x \neq 0$$

39. Simplify the following complex fraction.

$$\frac{\left(\dfrac{3}{2}\right)}{\left(2 + \dfrac{3}{x}\right)}$$

Solution

$$\frac{\left(\dfrac{3}{2}\right)}{\left(2 + \dfrac{3}{x}\right)} = \frac{\left(\dfrac{3}{2}\right)}{\left(\dfrac{2x}{x} + \dfrac{3}{x}\right)}$$

$$= \frac{3}{2} \div \frac{2x + 3}{x}$$

$$= \frac{3}{2} \cdot \frac{x}{2x + 3}$$

$$= \frac{3x}{2(2x + 3)}$$

$$= \frac{3x}{4x + 6}, \quad x \neq 0$$

Alternate Method:

$$\frac{\left(\dfrac{3}{2}\right)}{\left(2 + \dfrac{3}{x}\right)} = \frac{\left(\dfrac{3}{2}\right) 2x}{\left(2 + \dfrac{3}{x}\right) 2x}$$

$$= \frac{3x}{4x + 6}, \quad x \neq 0$$

41. Simplify the following complex fraction.

$$\frac{\left(\dfrac{x}{3}-4\right)}{\left(5+\dfrac{1}{x}\right)}$$

Solution

$$\frac{\left(\dfrac{x}{3}-4\right)}{\left(5+\dfrac{1}{x}\right)} = \frac{\left(\dfrac{x}{3}-\dfrac{12}{3}\right)}{\left(\dfrac{5x}{x}+\dfrac{1}{x}\right)} = \frac{\left(\dfrac{x-12}{3}\right)}{\left(\dfrac{5x+1}{x}\right)}$$

$$= \frac{x-12}{3} \div \frac{5x+1}{x}$$

$$= \frac{x-12}{3} \cdot \frac{x}{5x+1}$$

$$= \frac{x(x-12)}{3(5x+1)}, \quad x \neq 0$$

Alternate Method:

$$\frac{\left(\dfrac{x}{3}-4\right)}{\left(5+\dfrac{1}{x}\right)} = \frac{\left(\dfrac{x}{3}-4\right)3x}{\left(5+\dfrac{1}{x}\right)3x}$$

$$= \frac{x^2-12x}{15x+3}$$

$$= \frac{x(x-12)}{3(5x+1)}, \quad x \neq 0$$

43. Simplify the following complex fraction.

$$\frac{\left(\dfrac{10}{x+1}\right)}{\left(\dfrac{1}{2}+\dfrac{3}{x+1}\right)}$$

Solution

$$\frac{\left(\dfrac{10}{x+1}\right)}{\left(\dfrac{1}{2}+\dfrac{3}{x+1}\right)} = \frac{\left(\dfrac{10}{x+1}\right)}{\left(\dfrac{1(x+1)}{2(x+1)}+\dfrac{3(2)}{2(x+1)}\right)}$$

$$= \frac{10}{(x+1)} \div \frac{(x+1)+6}{2(x+1)}$$

$$= \frac{10}{x+1} \cdot \frac{2(x+1)}{x+7}$$

$$= \frac{20(x+1)}{(x+1)(x+7)}$$

$$= \frac{20}{x+7}, \quad x \neq -1$$

Alternate Method:

$$\frac{\left(\dfrac{10}{x+1}\right)}{\left(\dfrac{1}{2}+\dfrac{3}{x+1}\right)} = \frac{\left(\dfrac{10}{x+1}\right)(2)(x+1)}{\left(\dfrac{1}{2}+\dfrac{3}{x+1}\right)(2)(x+1)}$$

$$= \frac{20}{(x+1)+6}$$

$$= \frac{20}{x+7}, \quad x \neq -1$$

45. *Work Rate*   After two people work together for t hours on a common task, the fractional parts of the job done by each of the workers are $t/8$ and $t/7$. What fractional part of the task has been completed?

Solution

$$\frac{t}{8}+\frac{t}{7} = \frac{t(7)}{8(7)}+\frac{t(8)}{7(8)} = \frac{7t+8t}{56} = \frac{15t}{56}$$

47. Factor $9x^2 - 4y^2$ completely.

Solution

$$9x^2 - 4y^2 = (3x)^2 - (2y)^2$$
$$= (3x + 2y)(3x - 2y)$$

49. Factor $15x^2 - 11x - 14$ completely.

Solution

$$15x^2 - 11x - 14 = (3x + 2)(5x - 7)$$

51. *Markup* A coat that sells for $234 was marked up 30% of the wholesale cost. Find the wholesale cost.

Solution

Verbal model: $\boxed{\text{Wholesale cost}} + \boxed{\text{Markup}} = \boxed{\text{Selling price}}$

Labels: Wholesale cost $= x$ (dollars)
Markup $= 0.30x$ (dollars)
Selling price $= 234$ (dollars)

Equation: $x + 0.30x = 234$

$$1.30x = 234$$
$$x = \frac{234}{1.30}$$
$$x = 180$$

The wholesale cost of the coat is $180.

53. *Geometry* Find expressions for the perimeter and area of the region shown in the figure in the textbook. Then simplify the expressions.

Solution

Perimeter $= 2(\text{Length}) + 2(\text{Width})$

$$= 2(x + 3) + 2(2x - 1)$$
$$= 2x + 6 + 4x - 2$$
$$= 6x + 4$$

Area $= (\text{Length})(\text{Width})$

$$= (x + 3)(2x - 1)$$
$$= 2x^2 - x + 6x - 3$$
$$= 2x^2 + 5x - 3$$

55. Combine and simplify:

$$\frac{7}{9} + \frac{2}{9}$$

Solution

$$\frac{7}{9} + \frac{2}{9} = \frac{9}{9} = 1$$

57. Combine and simplify:

$$\frac{y}{4} + \frac{3y}{4}$$

Solution

$$\frac{y}{4} + \frac{3y}{4} = \frac{y + 3y}{4}$$
$$= \frac{4y}{4} = \frac{\cancel{4}y}{\cancel{4}} = y$$

59. Combine and simplify:

$$\frac{5}{3a} + \frac{9}{3a}$$

Solution

$$\frac{5}{3a} + \frac{9}{3a} = \frac{5 + 9}{3a} = \frac{14}{3a}$$

61. Combine and simplify:

$$\frac{x}{3} + \frac{1-x}{3}$$

Solution

$$\frac{x}{3} + \frac{1-x}{3} = \frac{x + (1-x)}{3}$$

$$= \frac{x + 1 - x}{3} = \frac{1}{3}$$

65. Combine and simplify:

$$\frac{5y+2}{y-1} - \frac{4y+1}{y-1}$$

Solution

$$\frac{5y+2}{y-1} - \frac{4y+1}{y-1} = \frac{(5y+2) - (4y+1)}{y-1}$$

$$= \frac{5y + 2 - 4y - 1}{y-1}$$

$$= \frac{y+1}{y-1}$$

69. Find the least common multiple of x and $3(x+5)$.

Solution

The different factors are x, 3, and $(x+5)$. The least common multiple is $3x(x+5)$.

73. Combine and simplify:

$$\frac{22}{5} - \frac{8}{35}$$

Solution

$$\frac{22}{5} - \frac{8}{35} = \frac{22(7)}{5(7)} - \frac{8}{35}$$

$$= \frac{154}{35} - \frac{8}{35}$$

$$= \frac{146}{35}$$

63. Combine and simplify:

$$\frac{5}{x-1} + \frac{x}{x-1}$$

Solution

$$\frac{5}{x-1} + \frac{x}{x-1} = \frac{5+x}{x-1} \text{ or } \frac{x+5}{x-1}$$

67. Find the least common multiple of $9y^2$ and $12y$.

Solution

$$9y^2 = 3 \cdot 3 \cdot y \cdot y = 3^2 \cdot y^2$$

$$12y = 2 \cdot 2 \cdot 3 \cdot y = 2^2 \cdot 3 \cdot y$$

The different factors are 2, 3, and y. Using the highest powers of these factors, we conclude that the least common multiple is $2^2 \cdot 3^2 \cdot y^2$, or $36y^2$.

71. Find the least common multiple of $x^2 - 4$ and $x(x+2)$.

Solution

$$x^2 - 4 = (x+2)(x-2)$$

$$x(x+2)$$

The different factors are $(x+2)$, $(x-2)$, and x. Using the highest powers of these factors, we conclude that the least common multiple is $x(x+2)(x-2)$.

75. Combine and simplify:

$$\frac{5}{z} + \frac{6}{z^2}$$

Solution

The denominators are z and z^2. The least common denominator is z^2.

$$\frac{5}{z} + \frac{6}{z^2} = \frac{5(z)}{z(z)} + \frac{6}{z^2}$$

$$= \frac{5z}{z^2} + \frac{6}{z^2}$$

$$= \frac{5z+6}{z^2}$$

77. Combine and simplify:

$$\frac{2x}{x-5} - \frac{5}{5-x}$$

Solution

The denominators are $x - 5$ and $5 - x = -1(x - 5)$. The least common denominator is $x - 5$.

$$\frac{2x}{x-5} - \frac{5}{5-x} = \frac{2x}{x-5} - \frac{(-1)5}{(-1)(5-x)}$$

$$= \frac{2x}{x-5} - \frac{-5}{x-5}$$

$$= \frac{2x-(-5)}{x-5}$$

$$= \frac{2x+5}{x-5}$$

81. Combine and simplify:

$$\frac{3}{2(x-4)} - \frac{1}{2x}$$

Solution

The denominators are $2(x - 4)$ and $2x$. The least common denominator is $2x(x - 4)$.

$$\frac{3}{2(x-4)} - \frac{1}{2x} = \frac{3(x)}{2(x-4)(x)} - \frac{1(x-4)}{2x(x-4)}$$

$$= \frac{3x-(x-4)}{2x(x-4)}$$

$$= \frac{3x-x+4}{2x(x-4)}$$

$$= \frac{2x+4}{2x(x-4)}$$

$$= \frac{\cancel{2}(x+2)}{\cancel{2}(x)(x-4)}$$

$$= \frac{x+2}{x(x-4)}$$

79. Combine and simplify:

$$6 - \frac{5}{x+3}$$

Solution

The least common denominator is $x + 3$.
(*Note:* $6 = \frac{6}{1}$.)

$$6 - \frac{5}{x+3} = \frac{6}{1} - \frac{5}{x+3}$$

$$= \frac{6(x+3)}{1(x+3)} - \frac{5}{x+3}$$

$$= \frac{6(x+3)-5}{x+3}$$

$$= \frac{6x+18-5}{x+3}$$

$$= \frac{6x+13}{x+3}$$

83. Combine and simplify:

$$\frac{x}{x^2-9} + \frac{3}{x+3}$$

Solution

The denominators are $x^2 - 9 = (x + 3)(x - 3)$ and $(x + 3)$. The least common denominator is $(x + 3)(x - 3)$.

$$\frac{x}{x^2-9} + \frac{3}{x+3} = \frac{x}{(x+3)(x-3)} + \frac{3(x-3)}{(x+3)(x-3)}$$

$$= \frac{x+3(x-3)}{(x+3)(x-3)}$$

$$= \frac{x+3x-9}{(x+3)(x-3)}$$

$$= \frac{4x-9}{(x+3)(x-3)}$$

85. Combine and simplify: $\dfrac{5v}{v(v+4)} + \dfrac{2v}{v^2}$

Solution

The denominators are $v(v+4)$ and v^2. The least common denominator is $v^2(v+4)$.

$$\dfrac{5v}{v(v+4)} + \dfrac{2v}{v^2} = \dfrac{5v(v)}{v(v+4)(v)} + \dfrac{2v(v+4)}{v^2(v+4)}$$

$$= \dfrac{5v^2 + 2v(v+4)}{v^2(v+4)}$$

$$= \dfrac{5v^2 + 2v^2 + 8v}{v^2(v+4)}$$

$$= \dfrac{7v^2 + 8v}{v^2(v+4)}$$

$$= \dfrac{\cancel{v}(7v+8)}{\cancel{v}(v)(v+4)}$$

$$= \dfrac{7v+8}{v(v+4)}$$

Note: The fractions in the original problem could have been simplified.

$$\dfrac{5v}{v(v+4)} + \dfrac{2v}{v^2} = \dfrac{5\cancel{v}}{\cancel{v}(v+4)} + \dfrac{2\cancel{v}}{\cancel{v}(v)}$$

$$= \dfrac{5}{v+4} + \dfrac{2}{v} \qquad \text{(The common denominator is now } v(v+4).)$$

$$= \dfrac{5(v)}{(v+4)(v)} + \dfrac{2(v+4)}{v(v+4)}$$

$$= \dfrac{5v + 2(v+4)}{v(v+4)}$$

$$= \dfrac{5v + 2v + 8}{v(v+4)}$$

$$= \dfrac{7v+8}{v(v+4)}$$

87. Combine and simplify: $-\dfrac{1}{x} + \dfrac{2}{x^2+1}$

Solution

The denominators are x and x^2+1. The least common denominator is $x(x^2+1)$.

$$-\dfrac{1}{x} + \dfrac{2}{x^2+1} = -\dfrac{1(x^2+1)}{x(x^2+1)} + \dfrac{2(x)}{(x^2+1)(x)}$$

$$= \dfrac{-1(x^2+1) + 2x}{x(x^2+1)}$$

$$= \dfrac{-x^2 - 1 + 2x}{x(x^2+1)}$$

$$= \dfrac{-x^2 + 2x - 1}{x(x^2+1)} \quad \text{or} \quad -\dfrac{x^2 - 2x + 1}{x(x^2+1)} = -\dfrac{(x-1)^2}{x(x^2+1)}$$

89. Combine and simplify:

$$\frac{2}{x-1} - \frac{1}{(x-1)^2}$$

Solution

The denominators are $(x-1)$ and $(x-1)^2$. The least common denominator is $(x-1)^2$.

$$\frac{2}{x-1} - \frac{1}{(x-1)^2} = \frac{2(x-1)}{(x-1)(x-1)} - \frac{1}{(x-1)^2}$$

$$= \frac{2(x-1)-1}{(x-1)^2}$$

$$= \frac{2x-2-1}{(x-1)^2}$$

$$= \frac{2x-3}{(x-1)^2}$$

91. Combine and simplify:

$$\frac{4}{t^2+2} - \frac{4}{t^2}$$

Solution

The denominators are (t^2+2) and t^2. The least common denominator is $t^2(t^2+2)$.

$$\frac{4}{t^2+2} - \frac{4}{t^2} = \frac{4(t^2)}{(t^2+2)(t^2)} - \frac{4(t^2+2)}{t^2(t^2+2)}$$

$$= \frac{4t^2 - 4(t^2+2)}{t^2(t^2+2)}$$

$$= \frac{4t^2 - 4t^2 - 8}{t^2(t^2+2)}$$

$$= -\frac{8}{t^2(t^2+2)}$$

93. Combine and simplify:

$$\frac{2}{x+1} + \frac{1-x}{x^2-2x+3}$$

Solution

$$\frac{2}{x+1} + \frac{1-x}{x^2-2x+3} = \frac{2(x^2-2x+3)}{(x+1)(x^2-2x+3)} + \frac{(1-x)(x+1)}{(x+1)(x^2-2x+3)}$$

$$= \frac{2x^2 - 4x + 6 + x + 1 - x^2 - x}{(x+1)(x^2-2x+3)}$$

$$= \frac{x^2 - 4x + 7}{(x+1)(x^2-2x+3)}$$

95. Combine and simplify:

$$\frac{3u}{u^2-4u+4} + \frac{2}{u-2}$$

Solution

The denominators are $u^2 - 4u + 4$ and $u - 2$. The least common denominator is $(u-2)^2$.

$$\frac{3u}{u^2-4u+4} + \frac{2}{u-2} = \frac{3u}{(u-2)(u-2)} + \frac{2(u-2)}{(u-2)(u-2)}$$

$$= \frac{3u + 2(u-2)}{(u-2)^2}$$

$$= \frac{3u + 2u - 4}{(u-2)^2}$$

$$= \frac{5u - 4}{(u-2)^2}$$

97. Simplify the complex fraction: $\dfrac{\left(1+\dfrac{3}{y}\right)}{y}$

Solution

$$\frac{\left(1+\dfrac{3}{y}\right)}{y} = \frac{\left(1+\dfrac{3}{y}\right)\cdot y}{(y)\cdot y}$$

$$= \frac{1(y)+\dfrac{3}{y}(y)}{y\cdot y}$$

$$= \frac{y+3}{y^2}$$

Alternate Method:

$$\frac{\left(1+\dfrac{3}{y}\right)}{y} = \frac{\left(\dfrac{y}{y}+\dfrac{3}{y}\right)}{y} = \frac{\left(\dfrac{y+3}{y}\right)}{\left(\dfrac{y}{1}\right)}$$

$$= \frac{y+3}{y}\div\frac{y}{1} = \frac{y+3}{y}\cdot\frac{1}{y}$$

$$= \frac{(y+3)(1)}{(y)(y)} = \frac{y+3}{y^2}$$

99. Simplify the complex fraction: $\dfrac{\left(\dfrac{x}{4}-1\right)}{(x-4)}$

Solution

$$\frac{\left(\dfrac{x}{4}-1\right)}{(x-4)} = \frac{\left(\dfrac{x}{4}-\dfrac{4}{4}\right)}{(x-4)}$$

$$= \frac{x-4}{4}\div\frac{x-4}{1}$$

$$= \frac{x-4}{4}\cdot\frac{1}{x-4}$$

$$= \frac{(x-4)(1)}{4(x-4)} = \frac{1}{4},\ x\neq 4$$

Alternate Method:

$$\frac{\left(\dfrac{x}{4}-1\right)}{(x-4)} = \frac{\left(\dfrac{x}{4}-1\right)4}{(x-4)4}$$

$$= \frac{x-4}{4(x-4)}$$

$$= \frac{1(x-4)}{4(x-4)} = \frac{1}{4},\ x\neq 4$$

101. Simplify the complex fraction: $\dfrac{\left(z-\dfrac{4}{z}\right)}{\left(\dfrac{1}{z}-4\right)}$

Solution

$$\frac{\left(z-\dfrac{4}{z}\right)}{\left(\dfrac{1}{z}-4\right)} = \frac{\left(\dfrac{z^2}{z}-\dfrac{4}{z}\right)}{\left(\dfrac{1}{z}-\dfrac{4z}{z}\right)} = \frac{\left(\dfrac{z^2-4}{z}\right)}{\left(\dfrac{1-4z}{z}\right)}$$

$$= \frac{z^2-4}{z}\div\frac{1-4z}{z}$$

$$= \frac{(z+2)(z-2)}{z}\cdot\frac{z}{1-4z}$$

$$= \frac{(z+2)(z-2)(z)}{z(1-4z)}$$

$$= \frac{(z+2)(z-2)}{1-4z},\ z\neq 0$$

Alternate Method:

$$\frac{\left(z-\dfrac{4}{z}\right)}{\left(\dfrac{1}{z}-4\right)} = \frac{\left(z-\dfrac{4}{z}\right)\cdot z}{\left(\dfrac{1}{z}-4\right)\cdot z} = \frac{(z)z-\left(\dfrac{4}{z}\right)z}{\left(\dfrac{1}{z}\right)z-(4)z}$$

$$= \frac{z^2-4}{1-4z}\ \text{or}\ \frac{(z+2)(z-2)}{1-4z},\ z\neq 0$$

103. Simplify the following complex fraction.

$$\frac{\left(\dfrac{1}{x} - \dfrac{1}{x+1}\right)}{\left(\dfrac{1}{x+1}\right)}$$

Solution

$$\frac{\left(\dfrac{1}{x} - \dfrac{1}{x+1}\right)}{\left(\dfrac{1}{x+1}\right)} = \frac{\left(\dfrac{1}{x} - \dfrac{1}{x+1}\right) \cdot (x)(x+1)}{\left(\dfrac{1}{x+1}\right) \cdot (x)(x+1)}$$

$$= \frac{\left(\dfrac{1}{x}\right)(x)(x+1) - \left(\dfrac{1}{x+1}\right)(x)(x+1)}{\left(\dfrac{1}{x+1}\right)(x)(x+1)}$$

$$= \frac{(x+1) - x}{x} = \frac{x+1-x}{x}$$

$$= \frac{1}{x}, \quad x \neq -1$$

Alternate Method:

$$\frac{\left(\dfrac{1}{x} - \dfrac{1}{x+1}\right)}{\left(\dfrac{1}{x+1}\right)} = \frac{\left(\dfrac{1(x+1)}{x(x+1)} - \dfrac{1(x)}{(x+1)(x)}\right)}{\left(\dfrac{1}{x+1}\right)}$$

$$= \frac{\left(\dfrac{(x+1) - x}{x(x+1)}\right)}{\left(\dfrac{1}{x+1}\right)}$$

$$= \frac{1}{x(x+1)} \div \frac{1}{x+1}$$

$$= \frac{1}{x(x+1)} \cdot \frac{x+1}{1}$$

$$= \frac{1\cancel{(x+1)}}{x\cancel{(x+1)}} = \frac{1}{x}, \quad x \neq -1$$

105. *Rewriting a Fraction* Consider the following.

$$\frac{x+5}{x^2+x-2} = \frac{x+5}{(x-1)(x+2)} = \frac{A}{x-1} + \frac{B}{x+2}$$

The numbers A and B are solutions of the system

$$A + B = 1$$
$$2A - B = 5.$$

Solve the system and verify that the sum of the two resulting fractions is the original fraction.

Solution

$$\begin{array}{ll} A + B = 1 \\ \underline{2A - B = 5} \\ 3A = 6 \\ A = 2 \qquad 2 + B = 1 \\ B = -1 \\ (2, -1) \end{array}$$

Thus, $\dfrac{x+5}{x^2+x-2} = \dfrac{2}{x-1} + \dfrac{-1}{x+2}$ or $\dfrac{x+5}{x^2+x-2} = \dfrac{2}{x-1} - \dfrac{1}{x+2}$.

Verification: $\dfrac{2}{x-1} - \dfrac{1}{x+2} = \dfrac{2(x+2)}{(x-1)(x+2)} - \dfrac{1(x-1)}{(x-1)(x+2)}$

$$= \frac{2(x+2) - 1(x-1)}{(x-1)(x+2)} = \frac{2x+4-x+1}{(x-1)(x+2)} = \frac{x+5}{x^2+x-2}$$

107. *Average of Two Numbers*   Determine the average of the two real numbers given by $x/5$ and $x/6$.

Solution

$$\dfrac{\left(\dfrac{x}{5}+\dfrac{x}{6}\right)}{2} = \dfrac{\left(\dfrac{6x}{30}+\dfrac{5x}{30}\right)}{2} = \dfrac{\left(\dfrac{6x+5x}{30}\right)}{2} = \dfrac{\left(\dfrac{11x}{30}\right)}{\left(\dfrac{2}{1}\right)}$$

$$= \dfrac{11x}{30} \div \dfrac{2}{1}$$

$$= \dfrac{11x}{30} \cdot \dfrac{1}{2}$$

$$= \dfrac{11x}{60}$$

Alternate Method:

$$\dfrac{\left(\dfrac{x}{5}+\dfrac{x}{6}\right)}{2} = \dfrac{\left(\dfrac{x}{5}+\dfrac{x}{6}\right) \cdot 30}{2 \cdot 30}$$

$$= \dfrac{\left(\dfrac{x}{5}\right)(30)+\left(\dfrac{x}{6}\right)(30)}{60}$$

$$= \dfrac{6x+5x}{60}$$

$$= \dfrac{11x}{60}$$

109. *Equal Lengths*   Find three real numbers that divide the real number line between $x/9$ and $x/6$ into four parts of equal lengths (see figure).

Solution

We begin by finding the difference between $x/6$ and $x/9$, and we divide this difference by 4.

$$\dfrac{\dfrac{x}{6}-\dfrac{x}{9}}{4} = \dfrac{\dfrac{3x}{6(3)}-\dfrac{2x}{9(2)}}{4}$$

$$= \dfrac{\dfrac{3x}{18}-\dfrac{2x}{18}}{4} = \dfrac{\dfrac{x}{18}}{4}$$

$$= \dfrac{x}{18} \div 4 = \dfrac{x}{18} \cdot \dfrac{1}{4} = \dfrac{x}{72}$$

To find x_1: $\dfrac{x}{9}+\dfrac{x}{72} = \dfrac{8x}{9(8)}+\dfrac{x}{72}$

$$= \dfrac{8x+x}{72}$$

$$= \dfrac{9x}{72}$$

$$= \dfrac{9x}{9(8)}$$

$$= \dfrac{x}{8}$$

To find x_2: $\dfrac{x}{8}+\dfrac{x}{72} = \dfrac{9x}{9(8)}+\dfrac{x}{72}$

$$= \dfrac{9x+x}{72}$$

$$= \dfrac{10x}{72}$$

$$= \dfrac{2(5x)}{2(36)}$$

$$= \dfrac{5x}{36}$$

To find x_3: $\dfrac{5x}{36}+\dfrac{x}{72} = \dfrac{(5x)(2)}{36(2)}+\dfrac{x}{72}$

$$= \dfrac{10x+x}{72}$$

$$= \dfrac{11x}{72}$$

The three numbers are $\dfrac{x}{8}$, $\dfrac{5x}{36}$, and $\dfrac{11x}{72}$.

111. *Electrical Resistance*   When two resistors are connected in parallel, the total resistance is given by

$$\frac{1}{\dfrac{1}{R_1} + \dfrac{1}{R_2}}.$$

Simplify this complex fraction.

Solution

$$\frac{1}{\dfrac{1}{R_1} + \dfrac{1}{R_2}} = \frac{1 \cdot R_1 R_2}{\left(\dfrac{1}{R_1} + \dfrac{1}{R_2}\right) \cdot R_1 R_2}$$

$$= \frac{R_1 R_2}{\dfrac{1}{R_1} \cdot R_1 R_2 + \dfrac{1}{R_2} \cdot R_1 R_2}$$

$$= \frac{R_1 R_2}{R_2 + R_1}$$

Alternate Method:

$$\frac{1}{\dfrac{1}{R_1} + \dfrac{1}{R_2}} = \frac{1}{\dfrac{1 R_2}{R_1 R_2} + \dfrac{1 R_1}{R_2 R_1}} = \frac{1}{\dfrac{R_2 + R_1}{R_1 R_2}}$$

$$= 1 \div \frac{R_2 + R_1}{R_1 R_2}$$

$$= 1 \cdot \frac{R_1 R_2}{R_2 + R_1}$$

$$= \frac{R_1 R_2}{R_2 + R_1}$$

9.4   Solving Equations

7. Determine whether the given value of x is a solution to the following equation.

$$\frac{x}{5} - \frac{3}{x} = \frac{1}{10}$$

(a) $x = 0$ (b) $x = -1$ (c) $x = \dfrac{1}{6}$ (d) $x = 6$

Solution

(a)
$$\frac{0}{5} - \frac{3}{0} \overset{?}{=} \frac{1}{10}$$
$$0 - \frac{3}{0} \neq \frac{1}{10}$$
$\frac{3}{0}$ is undefined.
0 is *not* a solution.

(b)
$$\frac{-1}{5} - \frac{3}{-1} \overset{?}{=} \frac{1}{10}$$
$$-\frac{1}{5} + 3 \overset{?}{=} \frac{1}{10}$$
$$-\frac{1}{5} + \frac{15}{5} \overset{?}{=} \frac{1}{10}$$
$$\frac{14}{5} \neq \frac{1}{10}$$
-1 is *not* a solution.

(c)
$$\frac{(1/6)}{5} - \frac{3}{(1/6)} \overset{?}{=} \frac{1}{10}$$
$$\frac{1}{6}\left(\frac{1}{5}\right) - 3\left(\frac{6}{1}\right) \overset{?}{=} \frac{1}{10}$$
$$\frac{1}{30} - 18 \neq \frac{1}{10}$$
$\frac{1}{6}$ is *not* a solution.

(d)
$$\frac{6}{5} - \frac{3}{6} \overset{?}{=} \frac{1}{10}$$
$$\frac{6}{5} - \frac{1}{2} \overset{?}{=} \frac{1}{10}$$
$$\frac{12}{10} + \frac{5}{10} \overset{?}{=} \frac{1}{10}$$
$$\frac{17}{10} \neq \frac{1}{10}$$
6 is *not* a solution.

9. Determine whether the given value of x is a solution to the following equation.

$$\frac{5}{2x} - \frac{4}{x} = 3$$

(a) $x = -\frac{1}{2}$ (b) $x = 4$ (c) $x = 0$ (d) $x = \frac{1}{4}$

Solution

(a) $\dfrac{5}{2(-1/2)} - \dfrac{4}{(-1/2)} \stackrel{?}{=} 3$

$\dfrac{5}{-1} - \dfrac{4}{1} \cdot \dfrac{-2}{1} \stackrel{?}{=} 3$

$-5 + 8 \stackrel{?}{=} 3$

$3 = 3$

$-\frac{1}{2}$ *is* a solution.

(b) $\dfrac{5}{2(4)} - \dfrac{4}{4} \stackrel{?}{=} 3$

$\dfrac{5}{8} - 1 \stackrel{?}{=} 3$

$\dfrac{5}{8} - \dfrac{8}{8} \stackrel{?}{=} 3$

$-\dfrac{3}{8} \neq 3$

4 is *not* a solution.

(c) $\dfrac{5}{2(0)} - \dfrac{4}{0} \stackrel{?}{=} 3$

$\dfrac{5}{0} - \dfrac{4}{0} \stackrel{?}{=} 3$

$\frac{5}{0}$ and $\frac{4}{0}$ are undefined.
0 is *not* a solution.

(d) $\dfrac{5}{2(1/4)} - \dfrac{4}{(1/4)} \stackrel{?}{=} 3$

$\dfrac{5}{(1/2)} - \dfrac{4}{(1/4)} \stackrel{?}{=} 3$

$\dfrac{5}{1}\left(\dfrac{2}{1}\right) - \dfrac{4}{1}\left(\dfrac{4}{1}\right) \stackrel{?}{=} 3$

$10 - 16 \stackrel{?}{=} 3$

$-6 \neq 3$

$\frac{1}{4}$ is *not* a solution.

11. Solve the following equation.

$$\frac{x}{3} = \frac{2}{3}$$

Solution

$\dfrac{x}{3} = \dfrac{2}{3}$

$3\left(\dfrac{x}{3}\right) = 3\left(\dfrac{2}{3}\right)$

$x = 2$

13. Solve the following equation.

$$\frac{t}{3} = 25 - \frac{t}{6}$$

Solution

$\dfrac{t}{3} = 25 - \dfrac{t}{6}$ The least common denominator is 6.

$6\left(\dfrac{t}{3}\right) = 6\left(25 - \dfrac{t}{6}\right)$ Multiply both sides by 6.

$2t = 6(25) - 6\left(\dfrac{t}{6}\right)$

$2t = 150 - t$

$3t = 150$

$t = 50$

15. Solve the equation: $\dfrac{2}{x} = 8$

Solution

$\dfrac{2}{x} = 8$ The least common denominator is x.

$x\left(\dfrac{2}{x}\right) = 8x$ Multiply both sides by x.

$2 = 8x, \;\; x \neq 0$

$\dfrac{2}{8} = x$

$\dfrac{1}{4} = x$

17. Solve the equation: $\dfrac{4}{x+2} - \dfrac{1}{x} = \dfrac{1}{x}$

Solution

$$\dfrac{4}{x+2} - \dfrac{1}{x} = \dfrac{1}{x}$$ The least common denominator is $x(x+2)$.

$$x(x+2)\left(\dfrac{4}{x+2} - \dfrac{1}{x}\right) = x(x+2)\left(\dfrac{1}{x}\right)$$ Multiply both sides by $x(x+2)$.

$$x(x+2)\left(\dfrac{4}{x+2}\right) - x(x+2)\left(\dfrac{1}{x}\right) = x(x+2)\left(\dfrac{1}{x}\right)$$

$$4x - (x+2) = (x+2), \;\; x \neq 0, \; x \neq -2$$

$$4x - x - 2 = x + 2$$

$$3x - 2 = x + 2$$

$$2x - 2 = 2$$

$$2x = 4$$

$$x = 2$$

19. Solve the equation: $2 = \dfrac{18}{x^2}$

Solution

$$2 = \dfrac{18}{x^2}$$

$$\dfrac{2}{1} = \dfrac{18}{x^2}$$

$$2x^2 = 18, \;\; x \neq 0 \quad \text{(Cross-multiply.)}$$

$$x^2 = 9$$

$$x^2 - 9 = 0$$

$$(x+3)(x-3) = 0$$

$$x + 3 = 0 \rightarrow x = -3$$

$$x - 3 = 0 \rightarrow x = 3$$

21. Solve the following equation.

$$2y = \frac{y+6}{y+1}$$

Solution

$$2y = \frac{y+6}{y+1}$$

$$(y+1)2y = (y+1)\left(\frac{y+6}{y+1}\right) \qquad \text{Multiply both sides by } (y+1).$$

$$2y^2 + 2y = y + 6, \quad y \neq -1$$

$$2y^2 + y - 6 = 0$$

$$(2y-3)(y+2) = 0$$

$$2y - 3 = 0 \rightarrow 2y = 3 \rightarrow y = \frac{3}{2}$$

$$y + 2 = 0 \rightarrow y = -2$$

23. Solve the following equation.

$$\frac{x}{2} = \frac{1 + \dfrac{3}{x}}{1 + \dfrac{1}{x}}$$

Solution

$$\frac{x}{2} = \frac{1 + \dfrac{3}{x}}{1 + \dfrac{1}{x}}$$

$$x\left(1 + \frac{1}{x}\right) = 2\left(1 + \frac{3}{x}\right) \qquad \text{(Cross-multiply.)}$$

$$x + 1 = 2 + \frac{6}{x}$$

$$x(x+1) = x\left(2 + \frac{6}{x}\right)$$

$$x^2 + x = 2x + 6, \quad x \neq 0$$

$$x^2 - x - 6 = 0$$

$$(x-3)(x+2) = 0$$

$$x - 3 = 0 \rightarrow x = 3$$

$$x + 2 = 0 \rightarrow x = -2$$

25. Solve the following equation.

$$\frac{x}{x+4} + \frac{4}{x+4} + 2 = 0$$

Solution

$$\frac{x}{x+4} + \frac{4}{x+4} + 2 = 0$$

$$(x+4)\left(\frac{x}{x+4} + \frac{4}{x+4} + 2\right) = (x+4)(0)$$

$$x + 4 + 2(x+4) = 0, \quad x \neq -4$$

$$x + 4 + 2x + 8 = 0$$

$$3x + 12 = 0$$

$$3x = -12$$

$$x = -4 \qquad \text{(Extraneous)}$$

The equation has no solution.

27. (a) Use a graphing utility to graph the following equation and identify the x-intercepts. (b) Solve the fractional equation when $y = 0$. Compare your answer with the result of part (a).

$$y = \frac{5}{x+1} - 1$$

Solution

(a) The x-intercept appears to be $(4, 0)$.

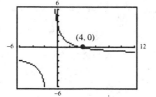

(b)
$$0 = \frac{5}{x+1} - 1$$

$$(x+1)(0) = (x+1)\left(\frac{5}{x+1} - 1\right)$$

$$0 = 5 - 1(x+1), \quad x \neq -1$$

$$0 = 5 - x - 1$$

$$0 = 4 - x$$

$$x = 4$$

This verifies the x-intercept of $(4, 0)$.

29. (a) Use a graphing utility to graph the following equation and identify the x-intercepts. (b) Solve the fractional equation when $y = 0$. Compare your answer with the result of part (a).

$$y = \frac{6}{x} - \frac{3x}{2}$$

Solution

(a) The x-intercepts appear to be $(-2, 0)$ and $(2, 0)$.

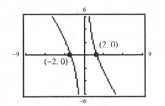

(b)

$$0 = \frac{6}{x} - \frac{3x}{2}$$

$$2x(0) = 2x\left(\frac{6}{x} - \frac{3x}{2}\right)$$

$$0 = 2(6) - x(3x), \quad x \neq 0$$

$$0 = 12 - 3x^2$$

$$3x^2 - 12 = 0$$

$$3(x^2 - 4) = 0$$

$$3(x + 2)(x - 2) = 0$$

$$3 \neq 0$$

$$x + 2 = 0 \to x = -2$$

$$x - 2 = 0 \to x = 2$$

This verifies that the x-intercepts are $(-2, 0)$ and $(2, 0)$.

31. *Problem Solving* Find a number such that the sum of 3 times the number and 25 times the reciprocal of the number is $\frac{65}{2}$.

Solution

Verbal model: $3 \boxed{\text{A number}} + 25 \boxed{\text{Reciprocal of the number}} = \dfrac{65}{2}$

Labels: A number $= x$

The reciprocal $= \dfrac{1}{x}$

Equation:

$$3x + 25\left(\frac{1}{x}\right) = \frac{65}{2}$$

$$2x\left(3x + \frac{25}{x}\right) = 2x\left(\frac{65}{2}\right)$$

$$6x^2 + 50 = 65x, \quad x \neq 0$$

$$6x^2 - 65x + 50 = 0$$

$$(6x - 5)(x - 10) = 0$$

$$6x - 5 = 0 \to 6x = 5 \to x = \frac{5}{6}$$

$$x - 10 = 0 \to x = 10$$

The number is either 10 or $\frac{5}{6}$.

33. *Modeling a Distance Problem*   A car leaves a town 20 minutes after a truck. The speed of the truck is about 10 miles per hour slower than the car. After traveling 100 miles, the car overtakes the truck. Use the following model to find the speed of each vehicle.

$$\boxed{\text{Distance traveled by car}} = \boxed{\text{Distance traveled by truck}}$$

Solution

Verbal model: $\boxed{\text{Distance traveled by car}} = \boxed{\text{Distance traveled by truck}}$

Labels: Speed of car $= x$ (miles per hour)
Speed of truck $= x - 10$ (miles per hour)

Time car travels $= \dfrac{100}{x}$ (hours)

Time truck travels $= \dfrac{100}{x} + \dfrac{1}{3}$ (hours)   *Note:* 20 minutes $= \dfrac{1}{3}$ hour

Distance traveled by car $= 100$ (miles)

Distance traveled by truck $=$ (speed of truck)(time truck travels) $= (x - 10)\left(\dfrac{100}{x} + \dfrac{1}{3}\right)$

Equation:
$$100 = (x - 10)\left(\frac{100}{x} + \frac{1}{3}\right)$$

$$100 = \frac{100(x - 10)}{x} + \frac{x - 10}{3}$$

$$3x(100) = 3x\left(\frac{100(x - 10)}{x} + \frac{x - 10}{3}\right)$$

$$300x = 300(x - 10) + x(x - 10), \quad x \neq 0$$

$$300x = 300x - 3000 + x^2 - 10x$$

$$x^2 - 10x - 3000 = 0$$

$$(x - 60)(x + 50) = 0$$

$$x - 60 = 0 \rightarrow x = 60 \text{ and } x - 10 = 50$$

$$x + 50 = 0 \rightarrow x = -50 \qquad \text{Eliminate negative answer.}$$

The speed of the car is 60 miles per hour, and the speed of the truck is 50 miles per hour.

35. Solve $16x - 3 = 29$.

Solution

$$16x - 3 = 29$$
$$16x = 32$$
$$x = 2$$

37. Solve $x(8 - x) = 16$.

Solution

$$x(8 - x) = 16$$
$$8x - x^2 = 16$$
$$-x^2 + 8x - 16 = 0$$
$$x^2 - 8x + 16 = 0$$
$$(x - 4)(x - 4) = 0$$
$$x - 4 = 0 \rightarrow x = 4$$

39. *Depreciation* A business purchases equipment for $12,000. After 4 years, its depreciated value will be $1000. Write a linear equation giving the value V of the equipment in terms of time t. Estimate the value of the equipment after 1 year.

Solution

When $t = 0$, $V = 12,000$; when $t = 4$, $V = 1000$.   (0, 12,000) and (4, 1000)

$$m = \frac{1000 - 12,000}{4 - 0} = -\frac{11,000}{4} = -2750$$

$$V - V_1 = m(t - t_1)$$

$$V - 12,000 = -2750(t - 0) = -2750t$$

$$V = -2750t + 12,000$$

$$t = 1 \rightarrow V = -2750(1) + 12,000$$

$$V = 9250$$

After 4 years, the value V is $9250.

41. Solve the equation: $\dfrac{x}{2} = \dfrac{1}{5}$

Solution

$\dfrac{x}{2} = \dfrac{1}{5}$

$5x = 2$ (Cross-multiply.)

$x = \dfrac{2}{5}$

43. Solve the equation: $\dfrac{z - 4}{3} - \dfrac{z}{8} = 0$

Solution

$\dfrac{z - 4}{3} - \dfrac{z}{8} = 0$ The least common denominator is 24.

$24\left(\dfrac{z - 4}{3} - \dfrac{z}{8}\right) = 24(0)$ Multiply both sides by 24.

$24\left(\dfrac{z - 4}{3}\right) - 24\left(\dfrac{z}{8}\right) = 0$

$8(z - 4) - 3z = 0$

$8z - 32 - 3z = 0$

$5z - 32 = 0$

$5z = 32$

$z = \dfrac{32}{5}$

45. Solve the equation: $\dfrac{x}{10} + \dfrac{x}{5} = 20$

Solution

$\dfrac{x}{10} + \dfrac{x}{5} = 20$ The least common denominator is 10.

$10\left(\dfrac{x}{10} + \dfrac{x}{5}\right) = 10(20)$ Multiply both sides by 10.

$10\left(\dfrac{x}{10}\right) + 10\left(\dfrac{x}{5}\right) = 200$

$x + 2x = 200$

$3x = 200$

$x = \dfrac{200}{3}$

47. Solve the equation: $\dfrac{3}{y} = \dfrac{12}{5}$

Solution

$\dfrac{3}{y} = \dfrac{12}{5}$

$15 = 12y, \ y \neq 0$ (Cross-multiply.)

$\dfrac{15}{12} = y$

$\dfrac{5}{4} = y$

49. Solve the equation: $\dfrac{1}{3-y} = -\dfrac{1}{10}$

Solution

$$\dfrac{1}{3-y} = -\dfrac{1}{10}$$

$$1(10) = -1(3-y), \quad y \neq 3 \qquad \text{(Cross-multiply.)}$$

$$10 = -3 + y$$

$$13 = y$$

51. Solve the equation: $3 - \dfrac{16}{a} = \dfrac{5}{3}$

Solution

$$3 - \dfrac{16}{a} = \dfrac{5}{3} \qquad \text{The least common denominator is } 3a.$$

$$3a\left(3 - \dfrac{16}{a}\right) = 3a\left(\dfrac{5}{3}\right) \qquad \text{Multiply both sides by } 3a.$$

$$3a(3) - 3a\left(\dfrac{16}{a}\right) = 3a\left(\dfrac{5}{3}\right), \quad a \neq 0$$

$$9a - 48 = 5a$$

$$-48 = -4a$$

$$12 = a$$

53. Solve the equation: $\dfrac{10}{y+3} + \dfrac{10}{3} = 6$

Solution

$$\dfrac{10}{y+3} + \dfrac{10}{3} = 6 \qquad \text{The least common denominator is } 3(y+3).$$

$$3(y+3)\left(\dfrac{10}{y+3} + \dfrac{10}{3}\right) = 3(y+3)(6) \qquad \text{Multiply both sides by } 3(y+3).$$

$$3(y+3)\left(\dfrac{10}{y+3}\right) + 3(y+3)\left(\dfrac{10}{3}\right) = 18(y+3)$$

$$30 + 10(y+3) = 18(y+3), \quad y \neq -3$$

$$30 + 10y + 30 = 18y + 54$$

$$10y + 60 = 18y + 54$$

$$-8y + 60 = 54$$

$$-8y = -6$$

$$y = \dfrac{-6}{-8}$$

$$y = \dfrac{3}{4}$$

55. Solve the equation: $\dfrac{3}{x} = \dfrac{9}{2(x+2)}$

Solution

$$\dfrac{3}{x} = \dfrac{9}{2(x+2)}$$

$3 \cdot 2(x+2) = 9x, \quad x \neq 0, \ x \neq -2$ (Cross-multiply.)

$$6(x+2) = 9x$$

$$6x + 12 = 9x$$

$$12 = 3x$$

$$4 = x$$

57. Solve the equation: $\dfrac{3}{x+5} = \dfrac{2}{x+1}$

Solution

$$\dfrac{3}{x+5} = \dfrac{2}{x+1}$$

$3(x+1) = 2(x+5), \quad x \neq -1, \ x \neq -5$ (Cross-multiply.)

$$3x + 3 = 2x + 10$$

$$x + 3 = 10$$

$$x = 7$$

59. Solve the equation: $\dfrac{50}{t} = 2t$

Solution

$$\dfrac{50}{t} = 2t$$

$t\left(\dfrac{50}{t}\right) = t(2t)$ Multiply both sides by t.

$$50 = 2t^2, \quad t \neq 0$$

$$25 = t^2$$

$$0 = t^2 - 25$$

$$0 = (t-5)(t+5)$$

$$0 = t - 5 \rightarrow t = 5$$

$$0 = t + 5 \rightarrow t = -5$$

61. Solve the equation: $x + 4 = \dfrac{-4}{x}$

Solution

$$x + 4 = \dfrac{-4}{x}$$

$x(x+4) = x\left(\dfrac{-4}{x}\right)$ Multiply both sides by x.

$$x^2 + 4x = -4, \quad x \neq 0$$

$$x^2 + 4x + 4 = 0$$

$$(x+2)^2 = 0$$

$$x + 2 = 0 \rightarrow x = -2$$

63. Solve the equation: $\dfrac{20 - x}{x} = x$

Solution

$$\dfrac{20 - x}{x} = x$$

$$x\left(\dfrac{20 - x}{x}\right) = x(x) \qquad \text{Multiply both sides by } x.$$

$$20 - x = x^2, \quad x \neq 0$$

$$0 = x^2 + x - 20$$

$$0 = (x + 5)(x - 4)$$

$$0 = x + 5 \rightarrow x = -5$$

$$0 = x - 4 \rightarrow x = 4$$

65. Solve the equation: $x + \dfrac{1}{x} = \dfrac{5}{2}$

Solution

$$x + \dfrac{1}{x} = \dfrac{5}{2} \qquad \text{The least common denominator is } 2x.$$

$$2x\left(x + \dfrac{1}{x}\right) = 2x\left(\dfrac{5}{x}\right) \qquad \text{Multiply both sides by } 2x.$$

$$2x(x) + 2x\left(\dfrac{1}{x}\right) = 2x\left(\dfrac{5}{2}\right)$$

$$2x^2 + 2 = 5x, \quad x \neq 0$$

$$2x^2 - 5x + 2 = 0$$

$$(2x - 1)(x - 2) = 0$$

$$2x - 1 = 0 \rightarrow 2x = 1 \rightarrow x = \dfrac{1}{2}$$

$$x - 2 = 0 \rightarrow x = 2$$

67. Solve the equation: $10 - \dfrac{13}{x} = 4 + \dfrac{5}{x}$

Solution

$$10 - \dfrac{13}{x} = 4 + \dfrac{5}{x} \qquad \text{The least common denominator is } x.$$

$$x\left(10 - \dfrac{13}{x}\right) = x\left(4 + \dfrac{5}{x}\right) \qquad \text{Multiply both sides by } x.$$

$$10x - 13 = 4x + 5, \quad x \neq 0$$

$$6x - 13 = 5$$

$$6x = 18$$

$$x = \dfrac{18}{6}$$

$$x = 3$$

69. Solve the equation: $\dfrac{3}{x(x-3)} + \dfrac{4}{x} = \dfrac{1}{x-3}$

Solution

$$\dfrac{3}{x(x-3)} + \dfrac{4}{x} = \dfrac{1}{x-3}$$ The least common denominator is $x(x-3)$.

$$x(x-3)\left(\dfrac{3}{x(x-3)} + \dfrac{4}{x}\right) = x(x-3)\left(\dfrac{1}{x-3}\right)$$ Multiply both sides by $x(x-3)$.

$$x(x-3)\left(\dfrac{3}{x(x-3)}\right) + x(x-3)\left(\dfrac{4}{x}\right) = x(x-3)\left(\dfrac{1}{x-3}\right)$$

$$3 + (x-3)4 = x, \quad x \neq 0, \ x \neq 3$$

$$3 + 4x - 12 = x$$

$$4x - 9 = x$$

$$-9 = -3x$$

$$\dfrac{-9}{-3} = x$$

$$3 = x$$

However, this solution is extraneous because substituting 3 for x in the original equation results in division by 0. Therefore, the equation has *no solution.*

71. Solve the equation: $\dfrac{1}{x-3} + \dfrac{1}{x+3} = \dfrac{10}{x^2-9}$

Solution

$$\dfrac{1}{x-3} + \dfrac{1}{x+3} = \dfrac{10}{x^2-9}$$ The common denominator is $(x-3)(x+3)$.

$$(x-3)(x+3)\left(\dfrac{1}{x-3} + \dfrac{1}{x+3}\right) = (x-3)(x+3)\left(\dfrac{10}{x^2-9}\right)$$ Multiply both sides by $(x-3)(x+3)$.

$$(x-3)(x+3)\left(\dfrac{1}{x-3}\right) + (x-3)(x+3)\left(\dfrac{1}{x+3}\right) = (x-3)(x+3)\left(\dfrac{10}{(x-3)(x+3)}\right)$$

$$x+3 + x-3 = 10, \quad x \neq 3, \ x \neq -3$$

$$2x = 10$$

$$x = \dfrac{10}{2} = 5$$

73. (a) Use a graphing utility to graph the following equation and identify the x-intercepts. (b) Solve the fractional equation when $y = 0$. Compare your answer with the result of part (a).

$$y = 2\left(\frac{3}{x - 4} - 1\right)$$

Solution

(a) The x-intercept appears to be $(7, 0)$.

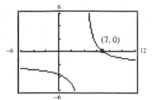

(b)
$$0 = 2\left(\frac{3}{x - 4} - 1\right)$$

$$0 = \frac{6}{x - 4} - 2$$

$$(x - 4)(0) = (x - 4)\left(\frac{6}{x - 4} - 2\right)$$

$$0 = 6 - 2(x - 4), \quad x \neq 4$$

$$0 = 6 - 2x + 8$$

$$0 = 14 - 2x$$

$$2x = 14$$

$$x = 7$$

This verifies that the x-intercept is $(7, 0)$.

75. (a) Use a graphing utility to graph the following equation and identify the x-intercepts. (b) Solve the fractional equation when $y = 0$. Compare your answer with the result of part (a).

$$y = x + \frac{x - 6}{x}$$

Solution

(a) The x-intercepts appear to be $(2, 0)$ and $(-3, 0)$.

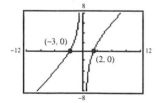

(b)
$$0 = x + \frac{x - 6}{x}$$

$$x(0) = x\left(x + \frac{x - 6}{x}\right)$$

$$0 = x^2 + x - 6, \quad x \neq 0$$

$$(x + 3)(x - 2) = 0$$

$$x + 3 = 0 \rightarrow x = -3$$

$$x - 2 = 0 \rightarrow x = 2$$

This verifies that the x-intercepts are $(2, 0)$ and $(-3, 0)$.

77. Find a number such that the sum of the number and its reciprocal is $\frac{10}{3}$.

Solution

Verbal model: $\boxed{\text{A number}} + \boxed{\begin{array}{c}\text{The reciprocal}\\\text{of the number}\end{array}} = \boxed{\dfrac{10}{3}}$

Labels: Number $= x$

Reciprocal $= \dfrac{1}{x}$

Equation: $x + \dfrac{1}{x} = \dfrac{10}{3}$

$$3x\left(x + \dfrac{1}{x}\right) = 3x\left(\dfrac{10}{3}\right)$$

$$3x(x) + 3x\left(\dfrac{1}{x}\right) = 3x\left(\dfrac{10}{3}\right)$$

$$3x^2 + 3 = 10x, \quad x \neq 0$$

$$3x^2 - 10x + 3 = 0$$

$$(3x - 1)(x - 3) = 0$$

$$3x - 1 = 0 \rightarrow 3x = 1 \rightarrow x = \dfrac{1}{3}$$

$$x - 3 = 0 \rightarrow x = 3$$

The number is $\frac{1}{3}$ or 3.

79. *Wind Speed*   You flew to a meeting in a city 1500 miles away (see figure in textbook). After traveling the same amount of time on the return flight, the pilot mentions that you still have 300 miles to go. If the plane has a speed of 600 miles per hour in still air, how fast is the wind blowing? (Assume that the wind direction is parallel to the flight path and is constant all day.)

Solution

Verbal model: $\boxed{\begin{array}{c}\text{Time for trip}\\\text{to meeting}\end{array}} = \boxed{\begin{array}{c}\text{Time for first}\\\text{portion of return trip}\end{array}}$

Formula: Distance $=$ (Rate)(Time) $\rightarrow$ Time $= \dfrac{\text{Distance}}{\text{Rate}}$

Labels: Wind speed $= w$ (miles per hour)

Trip to meeting: Distance $= 1500$ miles,
   Rate $= 600 + w$ (miles per hour)

Return trip: Distance $= 1500 - 300 = 1200$ miles,
   Rate $= 600 - w$ (miles per hour)

Equation: $\dfrac{1500}{600 + w} = \dfrac{1200}{600 - w}$

$$1500(600 - w) = 1200(600 + w), \quad w \neq 600, \; w \neq -600$$

$$900,000 - 1500w = 720,000 + 1200w$$

$$900,000 - 2700w = 720,000$$

$$-2700w = -180,000$$

$$w = \dfrac{-180,000}{-2700}$$

$$w = \dfrac{200}{3} \text{ or } w = 66\tfrac{2}{3}$$

Thus, the speed of the wind is $66\frac{2}{3}$ mph.

81. *Partnership* A group of people buy a piece of property for $48,000 by sharing the cost equally. To ease the financial burden, they look for two additional partners in order to reduce the per-person cost by $800. How many people are currently in the partnership?

Solution

Verbal model: | Current cost per person | $- \; 800 \; =$ | Decreased cost per person |

Labels: Total cost $= \$48,000$

Current number of persons $= n$

Current cost per person $= \dfrac{48,000}{n}$

Decreased cost per person $= \dfrac{48,000}{n+2}$

Equation:
$$\frac{48,000}{n} - 800 = \frac{48,000}{n+2}$$

$$n(n+2)\left(\frac{48,000}{n} - 800\right) = n(n+2)\left(\frac{48,000}{n+2}\right)$$

$$48,000(n+2) - 800n(n+2) = 48,000n, \quad n \neq 0, \; n \neq -2$$

$$48,000n + 96,000 - 800n^2 - 1600n = 48,000n$$

$$-800n^2 - 1600n + 96,000 = 0$$

$$-800(n^2 + 2n - 120) = 0$$

$$-800(n+12)(n-10) = 0$$

$$-800 \neq 0$$

$$n + 12 = 0 \rightarrow n = -12 \qquad \text{Discard negative solution.}$$

$$n - 10 = 0 \rightarrow n = 10$$

There are currently 10 people in the partnership.

83. *Air Pollution* A utility company burns oil to generate electricity. The cost in dollars of removing p percent of the air pollution in the stack emission of the utility company is given by

$$\text{Cost} = \frac{80,000p}{1-p}.$$

Determine the percent of the stack emission that can be removed for $240,000. (Assume the percent p is in decimal form.)

Solution

$$\text{Cost} = \frac{80,000p}{1-p}$$

$$240,000 = \frac{80,000p}{1-p}$$

$$(1-p)(240,000) = (1-p)\left(\frac{80,000p}{1-p}\right)$$

$$240,000 - 240,000p = 80,000p, \quad p \neq 1$$

$$240,000 = 320,000p$$

$$\frac{240,000}{320,000} = p$$

$$0.75 = p$$

Thus, 75% of the stack emission can be removed for $240,000.

85. The first two columns of the table in the textbook give the times required by each of two people working *alone* to complete a task. Complete the table by finding the time for each pair of individuals working *together* to complete the task. (Assume that when they work together their individual rates do not change.)

Solution

	Person #1	Person #2	Together
(a)	4 days	4 days	2 days
(b)	4 hours	6 hours	$2\frac{2}{5}$ hours
(c)	4 hours	$2\frac{1}{2}$ hours	$1\frac{7}{13}$ hours

Verbal model: $\boxed{\text{Rate for Person \#1}} + \boxed{\text{Rate for Person \#2}} = \boxed{\text{Rate together}}$

(a) *Labels:* Person #1: Time = 4 days, Rate = 1/4 task per day
Person #2: Time = 4 days, Rate = 1/4 task per day
Together: Time = x days, Rate = $1/x$ task per day

Equation:
$$\frac{1}{4} + \frac{1}{4} = \frac{1}{x}$$

$$4x\left(\frac{1}{4} + \frac{1}{4}\right) = 4x\left(\frac{1}{x}\right)$$

$$4x\left(\frac{1}{4}\right) + 4x\left(\frac{1}{4}\right) = 4x\left(\frac{1}{x}\right)$$

$$x + x = 4, \quad x \neq 0$$

$$2x = 4$$

$$x = 2$$

Thus, 2 days are required to complete the task when the two people work together.

(b) *Labels:* Person #1: Time = 4 hours, Rate = 1/4 task per hour
Person #2: Time = 6 hours, Rate = 1/6 task per hour
Together: Time = x hours, Rate = $1/x$ task per hour

Equation:
$$\frac{1}{4} + \frac{1}{6} = \frac{1}{x}$$

$$12x\left(\frac{1}{4} + \frac{1}{6}\right) = 12x\left(\frac{1}{x}\right)$$

$$12x\left(\frac{1}{4}\right) + 12x\left(\frac{1}{6}\right) = 12x\left(\frac{1}{x}\right)$$

$$3x + 2x = 12, \quad x \neq 0$$

$$5x = 12$$

$$x = \frac{12}{5}$$

Thus, $\frac{12}{5}$ or $2\frac{2}{5}$ hours are required to complete the task when the two people work together.

85. —CONTINUED—

(c) *Labels:* Person #1: Time = 4 hours, Rate = 1/4 task per hour

Person #2: Time = $2\frac{1}{2}$ hours, Rate = $1/(2\frac{1}{2})$ task per hour or 2/5 task per hour

Note: $\dfrac{1}{2\frac{1}{2}} = \dfrac{1}{5/2} = 1 \div \dfrac{5}{2} = 1 \cdot \dfrac{2}{5} = \dfrac{2}{5}$

Together: Time = x hours, Rate = 1/x task per hour

Equation:
$$\frac{1}{4} + \frac{2}{5} = \frac{1}{x}$$

$$20x\left(\frac{1}{4} + \frac{2}{5}\right) = 20x\left(\frac{1}{x}\right)$$

$$5x + 8x = 20, \quad x \neq 0$$

$$13x = 20$$

$$x = \frac{20}{13}$$

Thus, $\frac{20}{13}$ or $1\frac{7}{13}$ hours are required to complete the task when the two people work together.

87. *Work Rate* One landscaper works twice as fast as a second. Find their individual times to complete a task if it takes them 8 hours working together.

Solution

Verbal model: $\boxed{\begin{array}{c}\text{Rate of}\\\text{faster worker}\end{array}} + \boxed{\begin{array}{c}\text{Rate of}\\\text{slower worker}\end{array}} = \boxed{\begin{array}{c}\text{Rate}\\\text{together}\end{array}}$

Labels: Time for faster worker = t (hours)
Time for slower worker = 2t (hours)
Time together = 8 (hours)

Rate for faster worker = $\dfrac{1}{t}$ (jobs per hour)

Rate for slower worker = $\dfrac{1}{2t}$ (jobs per hour)

Rate together = $\dfrac{1}{8}$ (jobs per hour)

Equation:
$$\frac{1}{t} + \frac{1}{2t} = \frac{1}{8}$$

$$8t\left(\frac{1}{t} + \frac{1}{2t}\right) = 8t\left(\frac{1}{8}\right)$$

$$8 + 4 = t, \quad t \neq 0$$

$$12 = t \quad \text{and} \quad 2t = 24$$

The faster landscaper could complete the job in 12 hours; the slower landscaper could complete the job in 24 hours.

89. *Average Speed* One car makes a trip of 400 miles. Another car takes the same amount of time to make a trip of 480 miles. The average speed of the second car is 10 miles per hour faster than the average speed of the first car. What is the average speed of each car?

Solution

Verbal model: $\boxed{\text{Time for trip of first car}} = \boxed{\text{Time for trip of second car}}$

Labels: First car: Distance = 400 miles, Rate = x mph → Time = $400/x$ hours
Second car: Distance = 480 miles, Rate $x + 10$ mph → Time = $480/(x + 10)$ hours

Equation:
$$\frac{400}{x} = \frac{480}{x + 10}$$

$$400(x + 10) = 480x, \quad x \neq 0, \ x \neq -10 \qquad \text{(Cross-multiply.)}$$

$$400x + 4000 = 480x$$

$$4000 = 80x$$

$$50 = x$$

$$60 = x + 10$$

Thus, the average speed of the first car is 50 mph, and the average speed of the second car is 60 mph.

Review Exercises for Chapter 9

1. Find the domain of the following expression.

$$\frac{8x}{x - 5}$$

Solution

The denominator is zero when $x - 5 = 0$ or $x = 5$. Thus, the domain is all real values of x such that $x \neq 5$.

3. Find the domain of the following expression.

$$\frac{t}{t^2 - 3t + 2}$$

Solution

The denominator $t^2 - 3t + 2 = (t - 2)(t - 1)$ is zero when $t = 2$ or 1. Thus, the domain is all real values of t such that $t \neq 2$ and $t \neq 1$.

5. Simplify the following expression.

$$\frac{7x^2y}{21xy^2}$$

Solution

$$\frac{7x^2y}{21xy^2} = \frac{(7)(x)(x)(y)(x)}{(7)(3)(x)(y)(y)}$$

$$= \frac{x}{3y}, \quad x \neq 0$$

7. Simplify the following expression.

$$\frac{3b - 6}{4b - 8}$$

Solution

$$\frac{3b - 6}{4b - 8} = \frac{3(b - 2)}{4(b - 2)}$$

$$= \frac{3(b - 2)}{4(b - 2)}$$

$$= \frac{3}{4}, \quad b \neq 2$$

9. Simplify the following expression.

$$\frac{4x - 4y}{y - x}$$

Solution

$$\frac{4x - 4y}{y - x} = \frac{4(x - y)}{y - x}$$

$$= \frac{4(x - y)}{-1(x - y)}$$

$$= -4, \quad x \neq y$$

13. Simplify the following expression.

$$\frac{1 - x^3}{x^2 - 1}$$

Solution

$$\frac{1 - x^3}{x^2 - 1} = \frac{(-1)(x^3 - 1)}{x^2 - 1}$$

$$= \frac{(-1)(x - 1)(x^2 + x + 1)}{(x + 1)(x - 1)}$$

$$= -\frac{x^2 + x + 1}{x + 1}, \quad x \neq 1$$

11. Simplify the following expression.

$$\frac{x^2 - 9}{x^2 - x - 6}$$

Solution

$$\frac{x^2 - 9}{x^2 - x - 6} = \frac{(x + 3)(x - 3)}{(x - 3)(x + 2)}$$

$$= \frac{x + 3}{x + 2}, \quad x \neq 3$$

15. Simplify the following expression.

$$\frac{x(x - 4) + 7(x - 4)}{x^2 + 7x}$$

Solution

$$\frac{x(x - 4) + 7(x - 4)}{x^2 + 7x} = \frac{(x - 4)(x + 7)}{x(x + 7)}$$

$$= \frac{x - 4}{x}, \quad x \neq -7$$

17. Use a calculator to complete the table shown in the textbook. Explain why the values of the expressions agree for all values of x except one.

Solution

x	1	1.5	2	2.5	3
$\dfrac{x - 2}{x^2 - 4}$	$\dfrac{1}{3}$	$\dfrac{2}{7}$	Undefined	$\dfrac{2}{9}$	$\dfrac{1}{5}$
$\dfrac{1}{x + 2}$	$\dfrac{1}{3}$	$\dfrac{2}{7}$	$\dfrac{1}{4}$	$\dfrac{2}{9}$	$\dfrac{1}{5}$

$$\frac{x - 2}{x^2 - 4} = \frac{x - 2}{(x + 2)(x - 2)} = \frac{1}{x + 2}, \quad x \neq 2$$

The expressions are equal for all values of x except 2. When $x = 2$, the first expression is undefined.

19. Perform the indicated operation and simplify.

$$y \cdot \frac{2}{y + 3}$$

Solution

$$y \cdot \frac{2}{y + 3} = \frac{y}{1} \cdot \frac{2}{y + 3}$$

$$= \frac{2y}{y + 3}$$

21. Perform the indicated operation and simplify.

$$\frac{5x^2 y}{4} \cdot \frac{6x}{10y^3}$$

Solution

$$\frac{5x^2 y}{4} \cdot \frac{6x}{10y^3} = \frac{5(6)x^3 y}{4(10)y^3}$$

$$= \frac{(5)(2)(3)(x^3)y}{(4)(5)(2)(y^2)y}$$

$$= \frac{3x^3}{4y^2}$$

23. Perform the indicated operation and simplify.

$$\frac{z}{z+1} \cdot \frac{z^2-1}{5}$$

Solution

$$\frac{z}{z+1} \cdot \frac{z^2-1}{5} = \frac{z(z^2-1)}{5(z+1)}$$

$$= \frac{z(z+1)(z-1)}{5(z+1)}$$

$$= \frac{z\cancel{(z+1)}(z-1)}{5\cancel{(z+1)}}$$

$$= \frac{z(z-1)}{5}, \quad z \neq -1$$

25. Perform the indicated operation and simplify.

Solution

$$\frac{\left(\dfrac{4}{x}\right)}{\left(\dfrac{1}{x^2}\right)} = \frac{4}{x} \div \frac{1}{x^2} = \frac{4}{x} \cdot \frac{x^2}{1}$$

$$= \frac{4x^2}{x} = \frac{4x\cancel{(x)}}{\cancel{x}} = 4x, \quad x \neq 0$$

27. Perform the indicated operation and simplify.

Solution

$$\frac{5x}{\left(\dfrac{x}{y}\right)} = \frac{5x}{1} \div \frac{x}{y}$$

$$= \frac{5x}{1} \cdot \frac{y}{x}$$

$$= \frac{5xy}{x}$$

$$= \frac{\cancel{x}(5y)}{\cancel{x}} = 5y, \quad x \neq 0, \ y \neq 0$$

29. Perform the indicated operation and simplify.

$$10y^2 \div \frac{y}{5}$$

Solution

$$10y^2 \div \frac{y}{5} = \frac{10y^2}{1} \cdot \frac{5}{y}$$

$$= \frac{50y^2}{y}$$

$$= \frac{50y\cancel{(y)}}{\cancel{y}}$$

$$= 50y, \quad y \neq 0$$

31. Perform the indicated operation and simplify.

$$\frac{u^2}{u^2-9} \div \frac{u}{u+3}$$

Solution

$$\frac{u^2}{u^2-9} \div \frac{u}{u+3} = \frac{u^2}{u^2-9} \cdot \frac{u+3}{u}$$

$$= \frac{u^2(u+3)}{u(u^2-9)} = \frac{u(u)(u+3)}{u(u+3)(u-3)} = \frac{\cancel{u}(u)\cancel{(u+3)}}{\cancel{u}\cancel{(u+3)}(u-3)} = \frac{u}{u-3}, \quad u \neq 0, \ u \neq -3$$

33. Perform the indicated operation and simplify.

$$\frac{x^2-36}{6} \cdot \frac{3}{x^2-12x+36}$$

Solution

$$\frac{x^2-36}{6} \cdot \frac{3}{x^2-12x+36} = \frac{(x+6)(x-6)}{6} \cdot \frac{3}{(x-6)(x-6)} = \frac{\cancel{3}(x+6)\cancel{(x-6)}}{\cancel{3}(2)\cancel{(x-6)}(x-6)} = \frac{x+6}{2(x-6)}$$

35. Perform the indicated operation and simplify.

$$\frac{x^2 - 8x}{x - 1} \div \frac{x^2 - 16x + 64}{x^2 - 1}$$

Solution

$$\frac{x^2 - 8x}{x - 1} \div \frac{x^2 - 16x + 64}{x^2 - 1}$$

$$= \frac{x^2 - 8x}{x - 1} \cdot \frac{x^2 - 1}{x^2 - 16x + 64}$$

$$= \frac{x(x - 8)(x + 1)(x - 1)}{(x - 1)(x - 8)(x - 8)}$$

$$= \frac{x(x - 8)(x + 1)(x - 1)}{(x - 1)(x - 8)(x - 8)}$$

$$= \frac{x(x + 1)}{x - 8}, \quad x \neq 1, \ x \neq -1$$

37. Perform the indicated operation and simplify.

$$\left(\frac{x}{y} \cdot \frac{x + 1}{y + 1}\right) \div \frac{x}{y^2 - y}$$

Solution

$$\left(\frac{x}{y} \cdot \frac{x + 1}{y + 1}\right) \div \frac{x}{y^2 - y}$$

$$= \frac{x(x + 1)}{y(y + 1)} \div \frac{x}{y^2 - y}$$

$$= \frac{x(x + 1)}{y(y + 1)} \cdot \frac{y^2 - y}{x}$$

$$= \frac{x(x + 1)(y)(y - 1)}{y(y + 1)(x)}$$

$$= \frac{x(x + 1)(y)(y - 1)}{y(y + 1)(x)}$$

$$= \frac{(x + 1)(y - 1)}{y + 1}, \quad x \neq 0, \ y \neq 0, \ y \neq 1$$

39. Perform the indicated operation and simplify.

$$\frac{5}{8} - \frac{3}{8}$$

Solution

$$\frac{5}{8} - \frac{3}{8} = \frac{5 - 3}{8}$$

$$= \frac{2}{8} = \frac{2(1)}{2(4)} = \frac{1}{4}$$

41. Perform the indicated operation and simplify.

$$\frac{4}{x + 2} + \frac{x}{x + 2}$$

Solution

$$\frac{4}{x + 2} + \frac{x}{x + 2} = \frac{x + 4}{x + 2}$$

43. Perform the indicated operation and simplify.

$$\frac{5}{16} - \frac{5}{24}$$

Solution

$$\frac{5}{16} - \frac{5}{24} = \frac{5(3)}{16(3)} - \frac{5(2)}{24(2)}$$

$$= \frac{15}{48} - \frac{10}{48}$$

$$= \frac{15 - 10}{48} = \frac{5}{48}$$

45. Perform the indicated operation and simplify.

$$\frac{1}{x + 2} - \frac{1}{x + 1}$$

Solution

$$\frac{1}{x + 2} - \frac{1}{x + 1} = \frac{1(x + 1)}{(x + 2)(x + 1)} - \frac{1(x + 2)}{(x + 1)(x + 2)}$$

$$= \frac{1(x + 1) - 1(x + 2)}{(x + 2)(x + 1)}$$

$$= \frac{x + 1 - x - 2}{(x + 2)(x + 1)}$$

$$= \frac{-1}{(x + 2)(x + 1)}$$

$$\text{or} \quad -\frac{1}{(x + 2)(x + 1)}$$

47. Perform the indicated operation and simplify.

$$x - 1 + \frac{1}{x+2} + \frac{1}{x-1}$$

Solution

$$x - 1 + \frac{1}{x+2} + \frac{1}{x-1} = \frac{x-1}{1} + \frac{1}{x+2} + \frac{1}{x-1}$$

$$= \frac{(x-1)(x+2)(x-1)}{1(x+2)(x-1)} + \frac{1(x-1)}{(x+2)(x-1)} + \frac{1(x+2)}{(x-1)(x+2)}$$

$$= \frac{(x-1)^2(x+2)}{(x+2)(x-1)} + \frac{(x-1)}{(x+2)(x-1)} + \frac{(x+2)}{(x+2)(x-1)}$$

$$= \frac{(x-1)^2(x+2) + (x-1) + (x+2)}{(x+2)(x-1)}$$

$$= \frac{(x^2 - 2x + 1)(x+2) + (x-1) + (x+2)}{(x+2)(x-1)}$$

$$= \frac{x^3 + 2x^2 - 2x^2 - 4x + x + 2 + x - 1 + x + 2}{(x+2)(x-1)}$$

$$= \frac{x^3 - x + 3}{(x+2)(x-1)}$$

49. Perform the indicated operation and simplify.

$$\frac{1}{x} - \frac{x-1}{x^2+1}$$

Solution

$$\frac{1}{x} - \frac{x-1}{x^2+1} = \frac{1(x^2+1)}{x(x^2+1)} - \frac{(x-1)(x)}{(x^2+1)(x)}$$

$$= \frac{x^2 + 1 - (x-1)(x)}{x(x^2+1)} = \frac{x^2 + 1 - x^2 + x}{x(x^2+1)} = \frac{x+1}{x(x^2+1)}$$

51. Perform the indicated operation and simplify.

$$\frac{1}{x-2} + \frac{1}{(x-2)^2} + \frac{1}{x+2}$$

Solution

$$\frac{1}{x-2} + \frac{1}{(x-2)^2} + \frac{1}{x+2} = \frac{1(x-2)(x+2)}{(x-2)^2(x+2)} + \frac{1(x+2)}{(x-2)^2(x+2)} + \frac{1(x-2)^2}{(x+2)(x-2)^2}$$

$$= \frac{x^2 - 4 + x + 2 + x^2 - 4x + 4}{(x-2)^2(x+2)}$$

$$= \frac{2x^2 - 3x + 2}{(x-2)^2(x+2)}$$

53. Perform the indicated operation and simplify.

$$\frac{x}{\left(1 - \dfrac{1}{x}\right)}$$

Solution

$$\frac{x}{\left(1 - \dfrac{1}{x}\right)} = \frac{x(x)}{\left(1 - \dfrac{1}{x}\right)x}$$

$$= \frac{x^2}{1(x) - \dfrac{1}{x}(x)}$$

$$= \frac{x^2}{x - 1}, \quad x \neq 0$$

55. Perform the indicated operation and simplify.

$$\frac{\left(\dfrac{1}{x} - \dfrac{1}{y}\right)}{x^2 - y^2}$$

Solution

$$\frac{\left(\dfrac{1}{x} - \dfrac{1}{y}\right)}{x^2 - y^2} = \frac{\left(\dfrac{1}{x} - \dfrac{1}{y}\right) \cdot xy}{(x^2 - y^2) \cdot xy}$$

$$= \frac{\dfrac{1}{x}(xy) - \dfrac{1}{y}(xy)}{(x^2 - y^2)(xy)}$$

$$= \frac{y - x}{(x + y)(x - y)xy}$$

$$= \frac{-1(x - y)}{(x + y)(x - y)xy}$$

$$= -\frac{1}{xy(x + y)}, \quad x \neq y$$

57. Perform the indicated operation and simplify.

$$\frac{\left(\dfrac{1}{x + 1} - \dfrac{1}{4}\right)}{x - 3}$$

Solution

$$\frac{\left(\dfrac{1}{x + 1} - \dfrac{1}{4}\right)}{x - 3} = \frac{\left(\dfrac{4(1)}{4(x + 1)} - \dfrac{1(x + 1)}{4(x + 1)}\right)}{x - 3}$$

$$= \frac{\left(\dfrac{4 - (x + 1)}{4(x + 1)}\right)}{x - 3} = \frac{\left(\dfrac{4 - x - 1}{4(x + 1)}\right)}{x - 3}$$

$$= \frac{\left(\dfrac{3 - x}{4(x + 1)}\right)}{x - 3}$$

$$= \frac{3 - x}{4(x + 1)} \div (x - 3)$$

$$= \frac{-1(x - 3)}{4(x + 1)} \cdot \frac{1}{x - 3}$$

$$= -\frac{(x - 3)(1)}{4(x + 1)(x - 3)}$$

$$= -\frac{1}{4(x + 1)}, \quad x \neq 3$$

59. Perform the indicated operation and simplify.

$$\frac{\left(\dfrac{x^3}{x^2 + x - 2}\right)}{\left(\dfrac{x^2}{x^2 - 1}\right)}$$

Solution

$$\frac{\left(\dfrac{x^3}{x^2 + x - 2}\right)}{\left(\dfrac{x^2}{x^2 - 1}\right)} = \frac{x^3}{x^2 + x - 2} \div \frac{x^2}{x^2 - 1}$$

$$= \frac{x^3}{(x + 2)(x - 1)} \cdot \frac{(x + 1)(x - 1)}{x^2}$$

$$= \frac{x^3(x + 1)(x - 1)}{x^2(x + 2)(x - 1)}$$

$$= \frac{x(x^2)(x + 1)(x - 1)}{x^2(x + 2)(x - 1)}$$

$$= \frac{x(x + 1)}{x + 2}, \quad x \neq 0, \ x \neq 1, \ x \neq -1$$

61. Solve the following equation. Be sure to check for extraneous solutions.

$$\frac{x}{4} = -2$$

Solution

$$\frac{x}{4} = -2$$

$$4\left(\frac{x}{4}\right) = 4(-2)$$

$$x = -8$$

63. Solve the equation. Be sure to check for extraneous solutions.

$$\frac{5x - 4}{5x + 4} = \frac{2}{3}$$

Solution

$$\frac{5x - 4}{5x + 4} = \frac{2}{3}$$

$$3(5x - 4) = 2(5x + 4), \quad x \neq -\frac{4}{5} \qquad \text{(Cross-multiply.)}$$

$$15x - 12 = 10x + 8$$

$$5x - 12 = 8$$

$$5x = 20$$

$$x = 4$$

65. Solve the following equation. Be sure to check for extraneous solutions.

$$3\left(1 - \frac{5}{t}\right) = 0$$

Solution

$$3\left(1 - \frac{5}{t}\right) = 0$$

$$3 - \frac{15}{t} = 0$$

$$t\left(3 - \frac{15}{t}\right) = t \cdot 0$$

$$3t - 15 = 0, \quad t \neq 0$$

$$3t = 15$$

$$t = \frac{15}{3}$$

$$t = 5$$

67. Solve the following equation. Be sure to check for extraneous solutions.

$$\frac{7}{x} - 2 = \frac{3}{x} + 6$$

Solution

$$\frac{7}{x} - 2 = \frac{3}{x} + 6$$

$$x\left(\frac{7}{x} - 2\right) = x\left(\frac{3}{x} + 6\right)$$

$$7 - 2x = 3 + 6x, \quad x \neq 0$$

$$7 - 8x = 3$$

$$-8x = -4$$

$$x = \frac{-4}{-8}$$

$$x = \frac{1}{2}$$

69. Solve the following equation. Be sure to check for extraneous solutions.

$$\frac{t}{t-4}+\frac{3}{t-2}=0$$

Solution

$$\frac{t}{t-4}+\frac{3}{t-2}=0$$

$$(t-4)(t-2)\left(\frac{t}{t-4}+\frac{3}{t-2}\right)=(t-4)(t-2)(0)$$

$$(t-4)(t-2)\left(\frac{t}{t-4}\right)+(t-4)(t-2)\left(\frac{3}{t-2}\right)=0$$

$$t(t-2)+3(t-4)=0,\ \ t\neq 4,\ t\neq 2$$

$$t^2-2t+3t-12=0$$

$$t^2+t-12=0$$

$$(t+4)(t-3)=0$$

$$t+4=0\rightarrow t=-4$$

$$t-3=0\rightarrow t=3$$

71. Solve the following equation. Be sure to check for extraneous solutions.

$$\frac{2}{x}-\frac{x}{6}=\frac{2}{3}$$

Solution

$$\frac{2}{x}-\frac{x}{6}=\frac{2}{3}$$

$$6x\left(\frac{2}{x}-\frac{x}{6}\right)=6x\left(\frac{2}{3}\right)$$

$$6x\left(\frac{2}{x}\right)-6x\left(\frac{x}{6}\right)=6x\left(\frac{2}{3}\right)$$

$$6(2)-x(x)=2x(2),\ \ x\neq 0$$

$$12-x^2=4x$$

$$0=x^2+4x-12$$

$$0=(x+6)(x-2)$$

$$0=x+6\rightarrow x=-6$$

$$0=x-2\rightarrow x=2$$

73. Find the missing factor.

$$\frac{7}{4x}=\frac{7\boxed{}}{12x^3}$$

Solution

$$\frac{7}{4x}=\frac{7(3x^2)}{4x(3x^2)}=\frac{7(3x^2)}{12x^3}$$

The missing factor is $3x^2$.

75. Find the missing factor.

$$\frac{x-3}{x-1}=\frac{(x-3)\boxed{}}{x^2-1}$$

Solution

$$\frac{x-3}{x-1}=\frac{(x-3)(x+1)}{(x-1)(x+1)}=\frac{(x-3)(x+1)}{x^2-1}$$

The missing factor is $x+1$.

77. Find the least common multiple of 20, 24, and 30.

Solution

$20 = 2 \cdot 2 \cdot 5 = 2^2 \cdot 5$

$24 = 2 \cdot 2 \cdot 2 \cdot 3 = 2^3 \cdot 3$

$30 = 2 \cdot 3 \cdot 5$

The different factors are 2, 3, and 5. Using the highest powers of these factors, we conclude that the least common multiple is $2^3 \cdot 3 \cdot 5$, or 120.

79. Find the least common multiple of $x - 5$, $2x^2$, and $x(x + 5)$.

Solution

$x - 5$

$2x^2$

$x(x + 5)$

The different factors are 2, x, $x - 5$, and $x + 5$. Using the highest powers of these factors, we conclude that the least common multiple is $2x^2(x - 5)(x + 5)$ or $2x^2(x^2 - 25)$.

81. *Think About It*   You drive 72 miles one way on a service call for your company. The return trip takes 10 minutes less because you drive an average of 6 miles per hour faster. Is this enough information to determine your average speed on the return trip? If so, what was your average speed on the return trip?

Solution

Verbal model: $\boxed{\begin{array}{c}\text{Time for} \\ \text{return trip}\end{array}} = \boxed{\begin{array}{c}\text{Time for trip} \\ \text{to service call}\end{array}} - \boxed{10 \text{ minutes}}$

Common formula: $\text{Distance} = (\text{Rate})(\text{Time}) \rightarrow \text{Time} = \dfrac{\text{Distance}}{\text{Rate}}$

Labels: Trip to service call: Distance = 72 (miles)

Rate = x (miles per hour)

$\text{Time} = \dfrac{72}{x}$ (hours)

Return trip: Distance = 72 (miles)

Rate = $x + 6$ (miles per hour)

$\text{Time} = \dfrac{72}{x + 6}$ (hours)

Equation: $\dfrac{72}{x + 6} = \dfrac{72}{x} - \dfrac{1}{6}$ (Note: 10 minutes = 1/6 hour)

$$6x(x + 6)\left(\dfrac{72}{x + 6}\right) = 6x(x + 6)\left(\dfrac{72}{x} - \dfrac{1}{6}\right)$$

$$6x(72) = 6(x + 6)(72) - x(x + 6)(1), \quad x \neq 0, \ x \neq -6$$

$$432x = 432(x + 6) - x(x + 6)$$

$$432x = 432x + 2592 - x^2 - 6x$$

$$x^2 + 6x - 2592 = 0$$

$$(x + 54)(x - 48) = 0$$

$$x + 54 = 0 \rightarrow x = -54 \qquad \text{(Discard.)}$$

$$x - 48 = 0 \rightarrow x = 48 \text{ and } x + 6 = 54$$

We discard the negative answer for rate, so $x = 48$. Thus, the rate for the trip to the service call was 48 miles per hour, and the rate for the return trip was 54 miles per hour.

83. *Forming a Partnership*   A group of people agree to share equally in the cost of a $48,000 piece of machinery. If they could find two more partners to join the group, each person's share of the cost would decrease by $4000. How many people are currently in the group?

Solution

Verbal model: $\boxed{\begin{array}{l}\text{Each person's share of}\\\text{cost in larger group}\end{array}} = \boxed{\begin{array}{l}\text{Each person's current}\\\text{share of cost}\end{array}} - \boxed{4000}$

Labels: Total cost $= 48{,}000$ (dollars)

Number currently in group $= x$

Each person's current share of cost $= \dfrac{48{,}000}{x}$ (dollars)

Number of persons in larger group $= x + 2$

Each person's share of costs in larger group $= \dfrac{48{,}000}{x+2}$ (dollars)

Equation:

$$\frac{48{,}000}{x+2} = \frac{48{,}000}{x} - 4000$$

$$x(x+2)\left(\frac{48{,}000}{x+2}\right) = x(x+2)\left(\frac{48{,}000}{x} - 4000\right)$$

$$\frac{x(x+2)}{1} \cdot \frac{48{,}000}{x+2} = \frac{x(x+2)}{1} \cdot \frac{48{,}000}{x} - x(x+2)(4000)$$

$$48{,}000x = 48{,}000(x+2) - 4000x(x+2), \quad x \neq 0, \ x \neq -2$$

$$48{,}000x = 48{,}000x + 96{,}000 - 4000x^2 - 8000x$$

$$48{,}000x = 40{,}000x + 96{,}000 - 4000x^2$$

$$4000x^2 + 8000x - 96{,}000 = 0$$

$$4000(x^2 + 2x - 24) = 0$$

$$4000(x+6)(x-4) = 0$$

$$4000 \neq 0$$

$$x + 6 = 0 \rightarrow x = -6$$

$$x - 4 = 0 \rightarrow x = 4$$

We discard the negative answer for the number of people currently in the group. Thus, there are currently four people in the group.

85. *Work Rate*   Your supervisor takes 10 minutes to complete a task that takes you 8 minutes. Determine the time required to complete the task if you work together.

Solution

Verbal model: $\boxed{\text{Your rate}} + \boxed{\text{Supervisor's rate}} = \boxed{\text{Rate together}}$

Labels: Your rate $= \dfrac{1}{8}$ task per hour

Supervisor's rate $= \dfrac{1}{10}$ task per hour

Time to do the task together $= x$ (hours)

Rate together $= \dfrac{1}{x}$ task per hour

Equation:

$$\frac{1}{8} + \frac{1}{10} = \frac{1}{x}$$

$$40x\left(\frac{1}{8} + \frac{1}{10}\right) = 40x\left(\frac{1}{x}\right)$$

$$\frac{40x}{1} \cdot \frac{1}{8} + \frac{40x}{1} \cdot \frac{1}{10} = \frac{40x}{1} \cdot \frac{1}{x}$$

$$5x + 4x = 40, \quad x \neq 0$$

$$9x = 40$$

$$x = \frac{40}{9}$$

Thus, it will take $4\frac{4}{9}$ minutes to complete the task if you and your supervisor work together.

87. *Batting Average*   After 40 times at bat, a baseball player has a batting average of 0.300. How many additional consecutive times must the player hit safely to obtain a batting average of 0.440?

Solution

Verbal model: $\boxed{\text{Batting average}} = \boxed{\text{Total hits}} \div \boxed{\text{Total times at bat}}$

Labels: Current times at bat $= 40$
Current hits $= 40(0.300) = 12$
Additional consecutive hits $= x$

Equation:

$$0.440 = \frac{x + 12}{x + 40}$$

$$0.440(x + 40) = \left(\frac{x + 12}{x + 40}\right)(x + 40)$$

$$0.44x + 17.6 = x + 12$$

$$17.6 = 0.56x + 12$$

$$5.6 = 0.56x$$

$$10 = x$$

Thus, the player must successfully hit the ball for the next 10 consecutive times at bat.

89. *Ice Cream Production*   Ice cream production y (in millions of gallons) in the United States for the years 1987 through 1991 is approximated by

$$y = \frac{822.45 - 286.83t}{1 - 0.35t - 0.03t^2}$$

where t is time in years with $t = 0$ corresponding to 1990. Use the model to complete the table shown in the textbook. (*Source:* U.S. Department of Agriculture)

Solution

Year	1987	1988	1989	1990	1991
t	-3	-2	-1	0	1
y	945.47	883.61	840.36	822.45	863.90

Test for Chapter 9

1. Find the domain of the rational expression $\dfrac{x}{x - 10}$.

Solution

The denominator is zero when $x - 10 = 0$ or $x = 10$. Thus, the domain is the set of all real values of x such that $x \neq 10$.

2. Complete the statement: $\dfrac{2x^2}{x + 1} = \dfrac{2x^2\left(\boxed{}\right)}{x(x + 1)^2}$

Solution

$$\frac{2x^2}{x + 1} = \frac{2x^2[(x)(x + 1)]}{(x + 1)[(x)(x + 1)]}$$

$$= \frac{2x^2(x^2 + x)}{x(x + 1)^2}$$

The missing factor is $x(x + 1)$ or $x^2 + x$.

3. Simplify: $\dfrac{8x^2(x + 1)}{x(x + 1)^2}$

Solution

$$\frac{8x^2(x + 1)}{x(x + 1)^2} = \frac{8(x)(x)(x + 1)}{x(x + 1)(x + 1)}$$

$$= \frac{8x}{x + 1}, \quad x \neq 0$$

4. Simplify: $\dfrac{x^2 - 64}{x^2 - 3x - 40}$

Solution

$$\frac{x^2 - 64}{x^2 - 3x - 40} = \frac{(x + 8)(x - 8)}{(x - 8)(x + 5)}$$

$$= \frac{(x + 8)(x - 8)}{(x - 8)(x + 5)}$$

$$= \frac{x + 8}{x + 5}, \quad x \neq 8$$

5. Multiply and simplify: $\dfrac{18x}{5} \cdot \dfrac{15}{3x^3}$

Solution

$$\dfrac{18x}{5} \cdot \dfrac{15}{3x^3} = \dfrac{18x \cdot 15}{5 \cdot 3x^3}$$

$$= \dfrac{18(x)(5)(3)}{(5)(3)(x)(x^2)}$$

$$= \dfrac{18}{x^2}$$

6. Multiply and simplify: $(x+2)^2 \cdot \dfrac{x-2}{x^3 + 2x^2}$

Solution

$$(x+2)^2 \cdot \dfrac{x-2}{x^3 + 2x^2} = \dfrac{(x+2)^2}{1} \cdot \dfrac{x-2}{x^2(x+2)}$$

$$= \dfrac{(x+2)(x+2)(x-2)}{x^2(x+2)}$$

$$= \dfrac{(x+2)(x-2)}{x^2}, \quad x \neq -2$$

7. Divide and simplify: $\dfrac{3x^2}{4} \div \dfrac{9x^3}{10}$

Solution

$$\dfrac{3x^2}{4} \div \dfrac{9x^3}{10} = \dfrac{3x^2}{4} \cdot \dfrac{10}{9x^3}$$

$$= \dfrac{3x^2 \cdot 10}{4 \cdot 9x^3}$$

$$= \dfrac{(3)(x^2)(2)(5)}{(2)(2)(3)(3)(x^2)(x)}$$

$$= \dfrac{5}{6x}$$

8. Divide and simplify: $\dfrac{\left(\dfrac{t}{t-5}\right)}{\left(\dfrac{t^2}{5-t}\right)}$

Solution

$$\dfrac{\dfrac{t}{t-5}}{\dfrac{t^2}{5-t}} = \dfrac{t}{t-5} \div \dfrac{t^2}{5-t}$$

$$= \dfrac{t}{t-5} \cdot \dfrac{5-t}{t^2}$$

$$= \dfrac{t \cdot (5-t)}{(t-5)(t^2)}$$

$$= \dfrac{(t)(-1)(t-5)}{(t-5)(t)(t)}$$

$$= -\dfrac{1}{t}, \quad t \neq 5$$

9. Simplify: $\left[\left(\dfrac{x}{x-3}\right)^2 \cdot \dfrac{x^2}{x^2 - 3x}\right] \div (x-3)^5$

Solution

$$\left[\left(\dfrac{x}{x-3}\right)^2 \cdot \dfrac{x^2}{x^2 - 3x}\right] \div (x-3)^5 = \left[\dfrac{x^2}{(x-3)^2} \cdot \dfrac{x^2}{x(x-3)}\right] \div \dfrac{(x-3)^5}{1}$$

$$= \dfrac{x^4}{(x-3)^3(x)} \cdot \dfrac{1}{(x-3)^5}$$

$$= \dfrac{x^4}{(x)(x-3)^8}$$

$$= \dfrac{(x)(x^3)}{(x)(x-3)^8}$$

$$= \dfrac{x^3}{(x-3)^8}, \quad x \neq 0$$

10. Find the least common multiple of $6x(x+3)^2$, $9x^3$, and $12(x+3)$.

Solution

$6x(x+3)^2 = 2 \cdot 3 \cdot x \cdot (x+3)^2$

$9x^3 = 3^2 x^3$

$12(x+3) = 2^2 \cdot 3 \cdot (x+3)$

The different factors are 2, 3, x, and $(x+3)$. Using the highest powers of these factors, we conclude that the least common multiple is $2^2 \cdot 3^2 \cdot x^3(x+3)^2 = 36x^3(x+3)^2$.

11. Add and simplify: $\dfrac{8}{3u^2} + \dfrac{3}{u}$

Solution

$$\dfrac{8}{3u^2} + \dfrac{3}{u} = \dfrac{8}{3u^2} + \dfrac{3(3u)}{u(3u)}$$

$$= \dfrac{8}{3u^2} + \dfrac{9u}{3u^2}$$

$$= \dfrac{8+9u}{3u^2}$$

$$\text{or } \dfrac{9u+8}{3u^2}$$

12. Subtract and simplify: $\dfrac{3}{x+2} - 6$

Solution

$$\dfrac{3}{x+2} - 6 = \dfrac{3}{x+2} - \dfrac{6}{1}$$

$$= \dfrac{3}{x+2} - \dfrac{6(x+2)}{1(x+2)}$$

$$= \dfrac{3-6(x+2)}{x+2}$$

$$= \dfrac{3-6x-12}{x+2}$$

$$= \dfrac{-6x-9}{x+2}$$

$$\text{or } -\dfrac{6x+9}{x+2} \text{ or } -\dfrac{3(2x+3)}{x+2}$$

13. Simplify: $\dfrac{4}{\left(\dfrac{2}{x}+8\right)}$

Solution

$$\dfrac{4}{\left(\dfrac{2}{x}+8\right)} = \dfrac{4(x)}{\left(\dfrac{2}{x}+8\right)(x)}$$

$$= \dfrac{4x}{2+8x}$$

$$= \dfrac{\not{2}(2x)}{\not{2}(1+4x)}$$

$$= \dfrac{2x}{1+4x}, \quad x \neq 0$$

14. Subtract and simplify: $\dfrac{2}{x+1} - \dfrac{2x}{x^2+2x+1}$

Solution

$$\dfrac{2}{x+1} - \dfrac{2x}{x^2+2x+1}$$

$$= \dfrac{2}{x+1} - \dfrac{2x}{(x+1)^2}$$

$$= \dfrac{2(x+1)}{(x+1)(x+1)} - \dfrac{2x}{(x+1)^2}$$

$$= \dfrac{2(x+1)-2x}{(x+1)^2}$$

$$= \dfrac{2x+2-2x}{(x+1)^2}$$

$$= \dfrac{2}{(x+1)^2}$$

15. Determine whether the given value of x is a solution of $\dfrac{x}{4} + \dfrac{2}{x} = \dfrac{3}{2}$, and explain your reasoning.

(a) $x = 1$ (b) $x = 2$ (c) $x = -\dfrac{1}{2}$ (d) $x = 4$

Solution

(a)
$$\frac{1}{4} + \frac{2}{1} \stackrel{?}{=} \frac{3}{2}$$
$$\frac{1}{4} + \frac{2(4)}{1(4)} \stackrel{?}{=} \frac{3(2)}{2(2)}$$
$$\frac{1}{4} + \frac{8}{4} \stackrel{?}{=} \frac{6}{4}$$
$$\frac{9}{4} \neq \frac{6}{4}$$

1 *is not* a solution.

(b)
$$\frac{2}{4} + \frac{2}{2} \stackrel{?}{=} \frac{3}{2}$$
$$\frac{1}{2} + \frac{2}{2} \stackrel{?}{=} \frac{3}{2}$$
$$\frac{3}{2} = \frac{3}{2}$$

2 *is* a solution.

(c)
$$\frac{-1/2}{4} + \frac{2}{-1/2} \stackrel{?}{=} \frac{3}{2}$$
$$\left(-\frac{1}{2} \div 4\right) + \left(2 \div -\frac{1}{2}\right) \stackrel{?}{=} \frac{3}{2}$$
$$\left(-\frac{1}{2} \cdot \frac{1}{4}\right) + \left(\frac{2}{1} \cdot -\frac{2}{1}\right) \stackrel{?}{=} \frac{3}{2}$$
$$-\frac{1}{8} + \frac{-4}{1} \stackrel{?}{=} \frac{3}{2}$$
$$-\frac{1}{8} + \frac{-4(8)}{1(8)} \stackrel{?}{=} \frac{3(4)}{2(4)}$$
$$-\frac{1}{8} - \frac{32}{8} \stackrel{?}{=} \frac{12}{8}$$
$$-\frac{33}{8} \neq \frac{12}{8}$$

$-\dfrac{1}{2}$ *is not* a solution.

(d)
$$\frac{4}{4} + \frac{2}{4} \stackrel{?}{=} \frac{3}{2}$$
$$1 + \frac{1}{2} \stackrel{?}{=} \frac{3}{2}$$
$$\frac{3}{2} = \frac{3}{2}$$

4 *is* a solution.

16. Solve: $5 + \dfrac{t}{3} = t + 2$

Solution

$$5 + \frac{t}{3} = t + 2$$
$$3\left(5 + \frac{t}{3}\right) = 3(t + 2)$$
$$3(5) + 3\left(\frac{t}{3}\right) = 3(t + 2)$$
$$15 + t = 3t + 6$$
$$15 = 2t + 6$$
$$9 = 2t$$
$$\frac{9}{2} = t$$

17. Solve: $\dfrac{5}{x+1} - \dfrac{1}{x} = \dfrac{3}{x}$

Solution

$$\frac{5}{x+1} - \frac{1}{x} = \frac{3}{x}$$

$$x(x+1)\left(\frac{5}{x+1} - \frac{1}{x}\right) = x(x+1)\left(\frac{3}{x}\right)$$

$$\frac{x(x+1)}{1} \cdot \frac{5}{x+1} - \frac{x(x+1)}{1} \cdot \frac{1}{x} = \frac{x(x+1)}{1} \cdot \frac{3}{x}$$

$$5x - (x+1) = 3(x+1), \quad x \neq 0, \; x \neq -1$$

$$5x - x - 1 = 3x + 3$$

$$4x - 1 = 3x + 3$$

$$x - 1 = 3$$

$$x = 4$$

18. Solve: $2\left(x + \dfrac{1}{x}\right) = 5$

Solution

$$2\left(x + \frac{1}{x}\right) = 5$$

$$2x + \frac{2}{x} = 5$$

$$x\left(2x + \frac{2}{x}\right) = x(5)$$

$$x(2x) + \frac{x}{1}\left(\frac{2}{x}\right) = 5x, \quad x \neq 0$$

$$2x^2 + 2 = 5x$$

$$2x^2 - 5x + 2 = 0$$

$$(2x - 1)(x - 2) = 0$$

$$2x - 1 = 0 \rightarrow 2x = 1 \rightarrow x = \frac{1}{2}$$

$$x - 2 = 0 \rightarrow x = 2$$

19. The capacity of a pump is 80 gallons per minute. Determine the time required to pump (a) 1 gallon, (b) x gallons, and (c) 16 gallons.

Solution

(a) $\dfrac{1}{80}$ min

(b) $x\left(\dfrac{1}{80}\right) = \dfrac{x}{80}$ min

(c) $16\left(\dfrac{1}{80}\right) = \dfrac{16}{80} = \dfrac{1}{5}$ min

20. Partners in a small business plan to buy new equipment for $18,000 by sharing the cost equally. To ease the financial burden, they look for three additional partners so that the cost per person will be reduced by $1000. Determine the number of partners currently in the group.

Solution

Verbal model: $\boxed{\begin{array}{c}\text{Current cost} \\ \text{per partner}\end{array}} - \boxed{\begin{array}{c}\text{Desired cost} \\ \text{per partner}\end{array}} = \boxed{1000}$

Labels: Total cost $= \$18,000$
Current number of partners $= n$
Desired number of partners $= n + 3$

Equation:

$$\frac{18,000}{n} - \frac{18,000}{n+3} = 1000$$

$$n(n+3)\left(\frac{18,000}{n} - \frac{18,000}{n+3}\right) = n(n+3)(1000)$$

$$(n+3)(18,000) - n(18,000) = 1000n(n+3), \quad n \neq 0, \ n \neq -3$$

$$18,000n + 54,000 - 18,000n = 1000n^2 + 3000n$$

$$54,000 = 1000n^2 + 3000n$$

$$0 = 1000n^2 + 3000n - 54,000$$

$$0 = 1000(n^2 + 3n - 54)$$

$$0 = 1000(n+9)(n-6)$$

$$0 \neq 1000$$

$$0 = n + 9 \rightarrow n = -9$$

$$0 = n - 6 \rightarrow n = 6$$

The number of partners must be a positive integer, so we discard the -9 answer. Thus, the current number of partners is 6.

Cumulative Test for Chapters 7–9

1. Given $f(x) = \frac{1}{2}x - 3$, find (a) $f(4)$, (b) $f(-2)$, and (c) $f\left(\frac{9}{2}\right)$.

Solution

(a) $f(4) = \frac{1}{2}(4) - 3$

$= 2 - 3$

$= -1$

(b) $f(-2) = \frac{1}{2}(-2) - 3$

$= -1 - 3$

$= -4$

(c) $f\left(\frac{9}{2}\right) = \frac{1}{2}\left(\frac{9}{2}\right) - 3$

$= \frac{9}{4} - 3$

$= \frac{9}{4} - \frac{12}{4}$

$= -\frac{3}{4}$

2. Plot $(-7, 0)$ and $(2, 6)$ and find the slope of the line through the points.

Solution

$$m = \frac{y_2 - y_1}{x_2 - x_1}$$

$$m = \frac{6 - 0}{2 - (-7)}$$

$$m = \frac{6}{9}$$

$$m = \frac{2}{3}$$

3. The slope of a line is $-\frac{1}{4}$ and a point on the line is $(2, 1)$. Find the coordinates of a second point on the line. Explain why there are many correct answers.

Solution

Slope: $-\frac{1}{4}$

Point: $(2, 1)$

Other points on the line include: $(6, 0)$, $(10, -1)$, $(-2, 2)$, $(-6, 3)$, etc.

4. Find an equation of the line through the point $\left(0, -\frac{3}{2}\right)$ with slope $m = \frac{5}{6}$.

Solution

$$y - y_1 = m(x - x_1)$$

$$y - \left(-\frac{3}{2}\right) = \frac{5}{6}(x - 0)$$

$$y + \frac{3}{2} = \frac{5}{6}x$$

$$6\left(y + \frac{3}{2}\right) = 6\left(\frac{5}{6}x\right)$$

$$6y + 9 = 5x$$

$$-5x + 6y + 9 = 0 \text{ or } 5x - 6y - 9 = 0$$

Note: The point $\left(0, -\frac{3}{2}\right)$ is the y-intercept, so the equation could be determined using the $y = mx + b$ slope-intercept form.

$$y = \frac{5}{6}x + \left(-\frac{3}{2}\right)$$

$$6(y) = 6\left(\frac{5}{6}x - \frac{3}{2}\right)$$

$$6y = 5x - 9$$

$$-5x + 6y + 9 = 0 \text{ or } 5x - 6y - 9 = 0$$

5. Sketch the lines $y = \frac{2}{3}x - 3$ and $y = -\frac{3}{2}x + 1$ and determine whether they are parallel, perpendicular, or neither.

Solution

$y = \frac{2}{3}x - 3, \qquad m = \frac{2}{3}$

$y = -\frac{3}{2}x + 1, \qquad m = -\frac{3}{2}$

The lines are perpendicular because the slopes are negative reciprocals of each other.

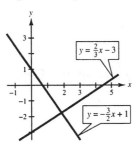

6. Sketch the lines $y = 2 - 0.4x$ and $y = -\frac{2}{5}x$ and determine whether they are parallel, perpendicular, or neither.

Solution

$y = 2 - 0.4x, \qquad m = -0.4$

$y = -\frac{2}{5}x, \qquad m = -\frac{2}{5} \text{ or } -0.4$

The lines are parallel because the slopes are the same.

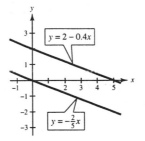

7. Sketch the graph of $3x - 2y + 8 = 0$.

Solution

$3x - 2y + 8 = 0$

$-2y = -3x - 8$

$y = \frac{-3}{-2}x - \frac{8}{-2}$

$y = \frac{3}{2}x + 4$

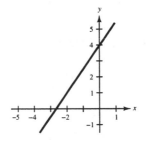

8. Sketch the graph of $y - 1 \geq 2(x + 3)$.

Solution

$y - 1 \geq 2(x + 3)$

$y - 1 \geq 2x + 6$

$y \geq 2x + 7$

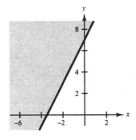

9. Solve the following system of equations graphically.

$x + 5y = 0$

$7x + 5y = 30$

Solution

$x + 5y = 0$

$7x + 5y = 30$

The solution is $(5, -1)$.

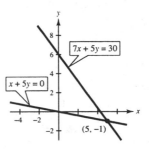

10. Solve the following system of equations by substitution.

$$x - 2y = -3$$
$$x + 5y = 4$$

Solution

$$x - 2y = -3 \rightarrow x = 2y - 3$$
$$x + 5y = 4$$

$$(2y - 3) + 5y = 4 \qquad \text{Replace } x \text{ by } 2y - 3 \text{ in second equation.}$$
$$2y - 3 + 5y = 4$$
$$7y - 3 = 4$$
$$7y = 7$$
$$y = 1 \qquad x = 2(1) - 3 = -1 \qquad \text{Replace } y \text{ by 1 in revised first equation.}$$

The solution is $(-1, 1)$.

11. Solve the following system of equations by elimination.

$$2x + y = 4$$
$$4x - 3y = 3$$

Solution

$$2x + y = 4 \rightarrow 6x + 3y = 12 \qquad \text{Multiply equation by 3.}$$
$$4x - 3y = 3 \rightarrow 4x - 3y = 3$$
$$\overline{10x = 15}$$
$$x = \frac{15}{10}$$
$$x = \frac{3}{2}$$

$$2\left(\frac{3}{2}\right) + y = 4 \qquad \text{Replace } x \text{ by } \tfrac{3}{2} \text{ in first equation.}$$
$$3 + y = 4$$
$$y = 1$$

The solution is $\left(\frac{3}{2}, 1\right)$.

12. Perform the indicated operation and simplify the result.

$$\frac{5x - 25}{x^2 - 25}$$

Solution

$$\frac{5x - 25}{x^2 - 25} = \frac{5(x - 5)}{(x + 5)(x - 5)} = \frac{5\cancel{(x - 5)}}{(x + 5)\cancel{(x - 5)}} = \frac{5}{x + 5}, \quad x \neq 5$$

13. Perform the indicated operation and simplify the result.

$$\frac{c}{c-1} \cdot \frac{c^2+9c-10}{c^3}$$

Solution

$$\frac{c}{c-1} \cdot \frac{c^2+9c-10}{c^3} = \frac{c(c+10)(c-1)}{(c-1)(c^3)}$$

$$= \frac{\cancel{c}(c+10)\cancel{(c-1)}}{\cancel{(c-1)}\cancel{(c)}(c^2)}$$

$$= \frac{c+10}{c^2}, \quad c \neq 1$$

14. Perform the indicated operation and simplify the result.

$$\frac{6}{(c-1)^2} \div \frac{8}{c^3-c^2}$$

Solution

$$\frac{6}{(c-1)^2} \div \frac{8}{c^3-c^2} = \frac{6}{(c-1)^2} \cdot \frac{c^3-c^2}{8}$$

$$= \frac{6(c^2)(c-1)}{(c-1)^2(8)}$$

$$= \frac{\cancel{2}(3)(c^2)\cancel{(c-1)}}{\cancel{(c-1)}(c-1)\cancel{(2)}(4)}$$

$$= \frac{3c^2}{4(c-1)}, \quad c \neq 0$$

15. Perform the indicated operation and simplify the result.

$$\frac{3}{x-2} + \frac{x}{4-x^2}$$

Solution

The denominators are $x-2$ and $4-x^2 = -1(x^2-4) = -1(x+2)(x-2)$. The least common denominator is $(x+2)(x-2)$.

$$\frac{3}{x-2} + \frac{4}{4-x^2} = \frac{3}{x-2} + \frac{x(-1)}{(4-x^2)(-1)}$$

$$= \frac{3}{x-2} + \frac{-x}{x^2-4}$$

$$= \frac{3(x+2)}{(x-2)(x+2)} + \frac{-x}{(x+2)(x-2)}$$

$$= \frac{3x+6-x}{(x+2)(x-2)}$$

$$= \frac{2x+6}{(x+2)(x-2)} \text{ or } \frac{2(x+3)}{(x+2)(x-2)}$$

16. Perform the indicated operation and simplify the result.

$$\frac{5}{x-1} - \frac{2}{x}$$

Solution

The denominators are $x-1$ and x. The least common denominator is $x(x-1)$.

$$\frac{5}{x-1} - \frac{2}{x} = \frac{5x}{x(x-1)} - \frac{2(x-1)}{x(x-1)}$$

$$= \frac{5x-2(x-1)}{x(x-1)} = \frac{5x-2x+2}{x(x-1)} = \frac{3x+2}{x(x-1)}$$

17. Perform the indicated operation and simplify the result.

$$\frac{\left(a - \dfrac{1}{a}\right)}{\left(\dfrac{1}{2} + \dfrac{1}{a}\right)}$$

Solution

$$\frac{\left(a - \dfrac{1}{a}\right)}{\left(\dfrac{1}{2} + \dfrac{1}{a}\right)} = \frac{\left(a - \dfrac{1}{a}\right) \cdot 2a}{\left(\dfrac{1}{2} + \dfrac{1}{a}\right) \cdot 2a} = \frac{(a)(2a) - \left(\dfrac{1}{a}\right)(2a)}{\left(\dfrac{1}{2}\right)(2a) + \left(\dfrac{1}{a}\right)(2a)} = \frac{2a^2 - 2}{a + 2}, \quad a \neq 0$$

This answer could also be written as $\dfrac{2(a^2 - 1)}{a + 2}$ or $\dfrac{2(a + 1)(a - 1)}{a + 2}$.

Alternate Method:

$$\frac{\left(a - \dfrac{1}{a}\right)}{\left(\dfrac{1}{2} + \dfrac{1}{a}\right)} = \frac{\dfrac{a \cdot a}{1 \cdot a} - \dfrac{1}{a}}{\dfrac{1 \cdot a}{2 \cdot a} + \dfrac{1 \cdot 2}{a \cdot 2}} = \frac{\dfrac{a^2 - 1}{a}}{\dfrac{a + 2}{2a}}$$

$$= \frac{a^2 - 1}{a} \div \frac{a + 2}{2a} = \frac{(a + 1)(a - 1)}{a} \cdot \frac{2a}{a + 2}$$

$$= \frac{(a + 1)(a - 1)(2)(d)}{d(a + 2)} = \frac{2(a + 1)(a - 1)}{a + 2}, \quad a \neq 0$$

18. The total cost of 10 gallons of regular gasoline and 12 gallons of premium gasoline is $31.88. Premium costs $0.20 more per gallon than regular. Find the price per gallon of each grade of gasoline.

Solution

Verbal model:

$$\boxed{10} \cdot \boxed{\begin{array}{c}\text{Cost per gallon of}\\ \text{regular gasoline}\end{array}} + \boxed{12} \cdot \boxed{\begin{array}{c}\text{Cost per gallon of}\\ \text{premium gasoline}\end{array}} = \boxed{\$31.88}$$

$$\boxed{\begin{array}{c}\text{Cost per gallon of}\\ \text{premium gasoline}\end{array}} = \boxed{\begin{array}{c}\text{Cost per gallon of}\\ \text{regular gasoline}\end{array}} + \boxed{\$0.20}$$

Labels: Cost per gallon of regular gasoline $= x$ (dollars)
Cost per gallon of premium gasoline $= y$ (dollars)

System of equations: $10x + 12y = 31.88$

$$y = x + 0.20$$

Solving by substitution: $10x + 12(x + 0.20) = 31.88$ Replace y by $x + 0.20$ in second equation

$$10x + 12x + 2.40 = 31.88$$

$$22x + 2.40 = 31.88$$

$$22x = 29.48$$

$$x = 1.34$$

$$y = x + 0.20 = 1.54$$

The regular gasoline costs $1.34 per gallon, and the premium gasoline costs $1.54 per gallon.

19. On the second half of a 200-mile trip, you average 10 more miles per hour than on the first half. What is your average speed on the second half of the trip if the total time for the trip is $4\frac{1}{2}$ hours?

Solution

Verbal model: $\boxed{\begin{array}{c}\text{Time for first} \\ \text{half of trip}\end{array}} + \boxed{\begin{array}{c}\text{Time for second} \\ \text{half of trip}\end{array}} = \dfrac{9}{2}$

Labels: Distance for first half of trip $= 100$ (miles)
Distance for second half of trip $= 100$ (miles)
Speed for first half of trip $= x$ (miles per hour)
Speed for second half of trip $= x + 10$

Time for first half of trip $= \dfrac{100}{x}$

Time for second half of trip $= \dfrac{100}{x + 10}$

Equation:

$$\frac{100}{x} + \frac{100}{x + 10} = \frac{9}{2}$$

$$2x(x + 10)\left(\frac{100}{x} + \frac{100}{x + 10}\right) = 2x(x + 10)\left(\frac{9}{2}\right)$$

$$2(x + 10)(100) + 2x(100) = x(x + 10)(9), \quad x \neq 0, x \neq -10$$

$$200(x + 10) + 200x = 9x(x + 10)$$

$$200x + 2000 + 200x = 9x^2 + 90x$$

$$400x + 2000 = 9x^2 + 90x$$

$$0 = 9x^2 - 310x - 2000$$

$$(9x + 50)(x - 40) = 0$$

$$9x + 50 = 0 \rightarrow 9x = -50 \rightarrow x = -\frac{50}{9} \qquad \text{(Discard.)}$$

$$x - 40 = 0 \rightarrow x = 40 \text{ and } x + 10 = 50$$

The average speed on the second half of the trip was 50 miles per hour.

20. A new employee takes twice as long as an experienced employee to complete a task. Together they can complete the task in 3 hours. Determine the time it takes them to do the task individually.

Solution

Verbal model:

$$\boxed{\begin{array}{c}\text{Rate of}\\ \text{new employee}\end{array}} + \boxed{\begin{array}{c}\text{Rate of}\\ \text{experienced employee}\end{array}} = \boxed{\begin{array}{c}\text{Rate when}\\ \text{working together}\end{array}}$$

Labels:

Experienced employee: time $= x$ (hours), rate $= 1/x$ (tasks per hour)

New employee: time $= 2x$ (hours), rate $= 1/(2x)$ (tasks per hour)

Together: time $= 3$ (hours), rate $= 1/3$ (tasks per hour)

Equation:

$$\frac{1}{2x} + \frac{1}{x} = \frac{1}{3}$$

$$6x\left(\frac{1}{2x} + \frac{1}{x}\right) = 6x\left(\frac{1}{3}\right)$$

$$6x\left(\frac{1}{2x}\right) + 6x\left(\frac{1}{x}\right) = 6x\left(\frac{1}{3}\right)$$

$$3 + 6 = 2x$$

$$9 = 2x$$

$$\frac{9}{2} = x$$

The experienced employee can complete the task in $\frac{9}{2}$ or $4\frac{1}{2}$ hours. The new employee would take 9 hours to complete the task.

CHAPTER TEN
Radical Expressions and Equations

10.1 | Square Roots and Radicals

7. Fill in the blank.

$9^2 = 81 \rightarrow$ A square root of 81 is ☐ .

Solution

9

9. Find the positive and negative square roots of the real number 36, if possible. (Do not use a calculator.)

Solution

Number: 36
Positive square root: 6
Negative square root: -6

11. Find the positive and negative square roots of the real number -16, if possible. (Do not use a calculator.)

Solution

A negative number has no square root in the real numbers.

13. Evaluate the expression $\sqrt{100}$, if possible. (Do not use a calculator.)

Solution

$\sqrt{100} = 10$

15. Evaluate the expression $-\sqrt{100}$, if possible. (Do not use a calculator.)

Solution

$-\sqrt{100} = -10$

17. Evaluate the expression $\sqrt{-100}$, if possible. (Do not use a calculator.)

Solution

A negative number has no square root in the real numbers.

19. Evaluate the expression $-\sqrt{\frac{1}{9}}$, if possible. (Do not use a calculator.)

Solution

$-\sqrt{\frac{1}{9}} = -\frac{1}{3}$

21. Evaluate the expression $\sqrt{-\frac{1}{25}}$, if possible. (Do not use a calculator.)

Solution

A negative number has no square root in the real numbers.

23. Evaluate the expression $\sqrt{0.04}$, if possible. (Do not use a calculator.)

Solution

$\sqrt{0.04} = 0.2$

25. Decide whether the number $\sqrt{15}$ is rational or irrational.

Solution

Irrational
(15 *is not* a perfect square.)

27. Decide whether the number $\sqrt{400}$ is rational or irrational.

Solution

Rational
(400 *is* a perfect square; $\sqrt{400} = 20$.)

29. Use a calculator to approximate the expression $\sqrt{43}$, if possible. Round the result to three decimal places.

Solution

$-\sqrt{43} \approx 6.557$

Keystrokes

43 $\boxed{\sqrt{}}$ Scientific

$\boxed{\sqrt{}}$ 43 $\boxed{\text{ENTER}}$ Graphing

31. Use a calculator to approximate the expression $\sqrt{\frac{95}{6}}$, if possible. Round the result to three decimal places.

Solution

$\sqrt{\frac{95}{6}} \approx 3.979$

Keystrokes

95 $\boxed{\div}$ 6 $\boxed{=}$ $\boxed{\sqrt{}}$ Scientific

$\boxed{\sqrt{}}$ $\boxed{(}$ 95 $\boxed{\div}$ 6 $\boxed{)}$ $\boxed{\text{ENTER}}$ Graphing

33. Use a calculator to approximate the expression $-\sqrt{10 \cdot 324}$, if possible. Round the result to three decimal places.

Solution

$\sqrt{10(324)} \approx -56.921$

Keystrokes

10 $\boxed{\times}$ 324 $\boxed{=}$ $\boxed{\sqrt{}}$ $\boxed{+/-}$ Scientific

$\boxed{(-)}$ $\boxed{\sqrt{}}$ $\boxed{(}$ 10 $\boxed{\times}$ 324 $\boxed{)}$ $\boxed{\text{ENTER}}$ Graphing

35. Use a calculator to approximate the expression $\dfrac{-4 - 3\sqrt{2}}{12}$, if possible. Round the result to three decimal places.

Solution

$\dfrac{-4 - 3\sqrt{2}}{12} \approx -0.687$

Keystrokes

4 $\boxed{+/-}$ $\boxed{-}$ 3 $\boxed{\times}$ 2 $\boxed{\sqrt{}}$ $\boxed{=}$ $\boxed{\div}$ 12 $\boxed{=}$ Scientific

$\boxed{(}$ $\boxed{(-)}$ 4 $\boxed{-}$ 3 $\boxed{\times}$ $\boxed{\sqrt{}}$ 2 $\boxed{)}$ $\boxed{\div}$ 12 $\boxed{\text{ENTER}}$ Graphing

37. Approximate $\sqrt{55}$ without using a calculator. Then check your estimate by using a calculator.

Solution

The number 55 lies between two integers that are perfect squares; $49 = 7^2$ and $64 = 8^2$. Therefore, $\sqrt{55}$ is between 7 and 8. Since 55 is closer to 49 than to 64, we could roughly approximate $\sqrt{55}$ to be 7.4. Using a calculator, we find $\sqrt{55} \approx 7.416$.

39. Approximate $\sqrt{150}$ without using a calculator. Then check your estimate by using a calculator.

Solution

The number 150 lies between two integers that are perfect squares; $144 = 12^2$ and $169 = 13^2$. Therefore, $\sqrt{150}$ is between 12 and 13. Since 150 is much closer to 144 than to 169, we could approximate $\sqrt{150}$ to be 12.2. Using a calculator, we find $\sqrt{150} = 12.247$.

41. *Geometry*   Use the formula $h = \frac{1}{2}\sqrt{3}s$ to find the height h of an equilateral triangle whose sides are of length $s = 5$ inches.

Solution

$$h = \frac{1}{2}\sqrt{3}s = \frac{1}{2}\sqrt{3(5)} = \frac{1}{2}\sqrt{15} \text{ or } \frac{\sqrt{15}}{2}$$

43. Evaluate the expression -5^2.

Solution

$-5^2 = -25$

Note: The base of the exponent is 5.

45. Evaluate the expression $-\left(\frac{1}{3}\right)^2\left(-\frac{1}{3}\right)^2$.

Solution

$$-\left(\tfrac{1}{3}\right)^2\left(-\tfrac{1}{3}\right)^2 = -\left(\tfrac{1}{9}\right)\left(\tfrac{1}{9}\right)$$
$$= -\tfrac{1}{81}$$

47. Solve the following system of equations.

$$3x - 4y = -15$$
$$2x + 5y = \;\;\; 13$$

Solution

$3x - 4y = -15$	Multiply equation by 2.	$6x - 8y = -30$	$3x - 4(3) = -15$
$2x + 5y = \;\; 13$	Multiply equation by -3.	$-6x - 15y = -39$	$3x - 12 = -15$
		$\overline{ - 23y = -69}$	$3x = -3$
		$y = 3$	$x = -1$
			$(-1, 3)$

49. *Geometry*   Find the area of the region of the figure shown in the textbook.

Solution

$$\text{Area} = (3x - 1)^2$$
$$= (3x)^2 - 2(3x)(1) + 1^2$$
$$= 9x^2 - 6x + 1$$

51. Find all square roots of 0. (Do not use a calculator.)

Solution

The number 0 has only one square root: 0.

53. Find all square roots of -9. (Do not use a calculator.)

Solution

The negative number -9 has no square root because there is no real number that can be multiplied by itself to obtain -9.

55. Find all square roots of 0.16. (Do not use a calculator.)

Solution

The positive number 0.16 has two squre roots: 0.4 and -0.4.

57. Find all square roots of $\frac{9}{16}$. (Do not use a calculator.)

Solution

The positive number $\frac{9}{16}$ has two square roots: $\frac{3}{4}$ and $-\frac{3}{4}$.

59. Evaluate the expression $\sqrt{49}$, if possible. (Do not use a calculator.)

Solution

$\sqrt{49} = 7$

61. Evaluate the expression $\sqrt{169}$, if possible. (Do not use a calculator.)

Solution

$\sqrt{169} = 13$

63. Evaluate the expression $-\sqrt{\frac{81}{121}}$, if possible. (Do not use a calculator.)

Solution

$-\sqrt{\frac{81}{121}} = -\frac{9}{11}$

65. Evaluate the expression $-\sqrt{0.0009}$, if possible. (Do not use a calculator.)

Solution

$-\sqrt{0.0009} = -0.03$

67. Evaluate the expression $\sqrt{32 - 7}$, if possible. (Do not use a calculator.)

Solution

$\sqrt{32 - 7} = \sqrt{25} = 5$

69. Evaluate the expression $\sqrt{2 \cdot 18}$, if possible. (Do not use a calculator.)

Solution

$\sqrt{2 \cdot 18} = \sqrt{36} = 6$

71. Use a calculator to approximate the expression $\sqrt{-632}$, if possible. Round your answer to three decimal places.

Solution

The negative number -632 has no square root in the real numbers.

73. Use a calculator to approximate $-\sqrt{517.8}$, if possible. Round your answer to three decimal places.

Solution

$-\sqrt{517.8} \approx -22.755$

Keystrokes

517.8 $\boxed{\sqrt{}}$ $\boxed{+/-}$ Scientific

$\boxed{(-)}$ $\boxed{\sqrt{}}$ 517.8 $\boxed{\text{ENTER}}$ Graphing

75. Use a calculator to approximate $16 - \sqrt{92.6}$, if possible. Round your answer to three decimal places.

Solution

$16 - \sqrt{92.6} \approx 6.377$

Keystrokes

16 $\boxed{-}$ 92.6 $\boxed{\sqrt{}}$ $\boxed{=}$ Scientific

16 $\boxed{-}$ $\boxed{\sqrt{}}$ 92.6 $\boxed{\text{ENTER}}$ Graphing

77. Use a calculator to approximate $(9 + \sqrt{45})/2$, if possible. Round your answer to three decimal places.

Solution

$\frac{9 + \sqrt{45}}{2} \approx 7.854$

Keystrokes

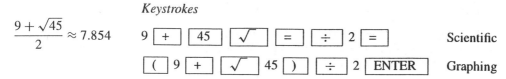

9 $\boxed{+}$ $\boxed{45}$ $\boxed{\sqrt{}}$ $\boxed{=}$ $\boxed{\div}$ 2 $\boxed{=}$ Scientific

$\boxed{(}$ 9 $\boxed{+}$ $\boxed{\sqrt{}}$ 45 $\boxed{)}$ $\boxed{\div}$ 2 $\boxed{\text{ENTER}}$ Graphing

79. Evaluate $\sqrt{b^2 - 4ac}$ for the values $a = 4$, $b = 5$, and $c = 1$. Give the exact value if possible. Otherwise, give an approximation to two decimal places.

Solution

$$\sqrt{b^2 - 4ac} = \sqrt{5^2 - 4(4)(1)}$$
$$= \sqrt{25 - 16} = \sqrt{9} = 3$$

81. Evaluate $\sqrt{b^2 - 4ac}$ for the values $a = 3$, $b = -7$, and $c = -6$. Give the exact value if possible. Otherwise, give an approximation to two decimal places.

Solution

$$\sqrt{b^2 - 4ac} = \sqrt{(-7)^2 - 4(3)(-6)}$$
$$= \sqrt{49 + 72} = \sqrt{121} = 11$$

83. *Geometry* Use the area $A = 27.04$ to find the length of the side of a square when $s = \sqrt{A}$.

Solution

$$s = \sqrt{A} = \sqrt{27.04} = 5.2$$

85. *Geometry* Use the area $A = 50.24$ to find the length of the radius of a circle when $r = \sqrt{A/\pi}$.

Solution

$$r = \sqrt{\frac{A}{\pi}} \approx \sqrt{\frac{50.24}{3.14}} = \sqrt{16} = 4$$

87. Approximate $\sqrt{70}$ without using a calculator. Then check your estimate by using a calculator.

Solution

The number 70 lies between two integers that are perfect squares: $64 = 8^2$ and $81 = 9^2$. Since 70 is closer to 64 than to 81, we could roughly approximate $\sqrt{70}$ to be 8.4. Using a calculator, we find $\sqrt{70} \approx 8.37$.

89. Approximate $\sqrt{130}$ without using a calculator. Then check your estimate by using a calculator.

Solution

The number 130 lies between two integers that are perfect squares: $121 = 11^2$ and $144 = 12^2$. Since 130 is closer to 121 than to 144, we could roughly approximate $\sqrt{130}$ to be 11.4. Using a calculator, we find $\sqrt{130} \approx 11.40$.

91. *Geometry* A square room has 529 square feet of floor space (see figure in textbook). What are the dimensions of the room?

Solution

$x = \sqrt{529} = 23$; The room is 23 feet long and 23 feet wide.

93. *Think About It* (a) Find all possible last digits of integers that are perfect squares. (For instance, the last digit of 81 is 1, and the last digit of 64 is 4.) (b) Using the results of part (a), is it possible that 5,788,942,862 is a perfect square?

Solution

(a) $0^2 = 0$
$1^2 = 1$
$2^2 = 4$
$3^2 = 9$
$4^2 = 16$
$5^2 = 25$
$6^2 = 36$
$7^2 = 49$
$8^2 = 64$
$9^2 = 81$

Writing the squares of the integers from 0 to 9, we see that all the last digits of these squares are from this list:

0, 1, 4, 5, 6, 9

Since all integers end in one of the 10 integers from 0 to 9, *every* perfect square integer must have a last digit from the list 0, 1, 4, 5, 6, 9.

(b) No, 5,788,942,862 could not be a perfect square because its last digit, 2, is *not* on the list of possible last digits. (The list is 0, 1, 4, 5, 6, 9.)

95. Use a graphing utility to graph the function $y = \sqrt{x}$.

Solution

97. Use a graphing utility to graph the funciton $y = \sqrt{x} + 2$.

Solution

99. *Graphical Interpretation* Describe the relationship of the graph $y = \sqrt{x}$ to the graphs in Exercises 96–98.

Solution

All the graphs have the same shape, but they are positioned differently. Comparing graphs (b), (c), and (d) with the graph of (a), we see that:

(b) the graph of $y = \sqrt{x - 2}$ is a shift of the first graph two units to the right;
(c) the graph of $y = \sqrt{x} + 2$ is a shift of the first graph two units upward;
(d) the graph of $y = -\sqrt{x}$ is a reflection of the first graph about the x-axis.

101. Use the result of Exercize 100 to determine $(\sqrt{a})^2$ where a is a nonnegative real number.

Solution

$(\sqrt{a})^2 = a$

10.2 Simplifying Radicals

7. Write $\sqrt{2} \cdot \sqrt{7}$ as a single radical.

Solution

$\sqrt{2} \cdot \sqrt{7} = \sqrt{14}$

9. Write $\sqrt{11} \cdot \sqrt{10}$ as a single radical.

Solution

$\sqrt{11} \cdot \sqrt{10} = \sqrt{110}$

11. Write $\sqrt{4 \cdot 15}$ as a product of two radicals and simplify.

Solution

$\sqrt{4 \cdot 15} = \sqrt{4} \cdot \sqrt{15} = 2\sqrt{15}$

13. Write $\sqrt{64 \cdot 11}$ as a product of two radicals and simplify.

Solution

$\sqrt{64 \cdot 11} = \sqrt{64} \cdot \sqrt{11} = 8\sqrt{11}$

15. Simplify $\sqrt{8}$. Explain how to use a calculator to check your result.

Solution

$\sqrt{8} = \sqrt{4 \cdot 2} = \sqrt{4} \cdot \sqrt{2} = 2\sqrt{2}$

To check your result, evaluate $\sqrt{8}$ and $2\sqrt{2}$ on your calculator. Both expressions are approximately equal to 2.828427.

17. Simplify $\sqrt{180}$. Explain how to use a calculator to check your result.

Solution

$$\sqrt{180} = \sqrt{36 \cdot 5} = \sqrt{36} \cdot \sqrt{5} = 6\sqrt{5}$$

Note: Here is another way to simplify this radical.

$$\sqrt{180} = \sqrt{9 \cdot 20} = \sqrt{9} \cdot \sqrt{20} = 3\sqrt{20} = 3\sqrt{4 \cdot 5} = 3\sqrt{4} \cdot \sqrt{5} = 3 \cdot 2\sqrt{5} = 6\sqrt{5}$$

This version gives the same result, but it takes more steps. In the first solution, we used the largest perfect square factor; this minimized the number of steps. To check your result, evaluate $\sqrt{180}$ and $6\sqrt{5}$ on your calculator. Both expressions are approximately equal to 13.4164.

19. Simplify $\sqrt{64x^3}$. Use absolute value signs if appropriate.

Solution

$$\sqrt{64x^3} = \sqrt{64 \cdot x^2 \cdot x}$$
$$= \sqrt{64} \cdot \sqrt{x^2} \cdot \sqrt{x}$$
$$= 8x\sqrt{x}$$

Note: If x were negative, x^3 would be negative and the original radical would be undefined in the real numbers. Thus, we can assume that the variable is nonnegative and absolute value signs are not necessary.

21. Simplify $\sqrt{200x^2y^4}$. Use absolute value signs if appropriate.

Solution

$$\sqrt{200x^2y^4} = \sqrt{100 \cdot 2 \cdot x^2 \cdot y^4}$$
$$= \sqrt{100} \cdot \sqrt{2} \cdot \sqrt{x^2} \cdot \sqrt{y^4}$$
$$= 10 \cdot \sqrt{2} \cdot |x| \cdot y^2$$
$$= 10|x|y^2\sqrt{2}$$

23. Write $\dfrac{\sqrt{39}}{\sqrt{15}}$ as a single radical. Then simplify the result.

Solution

$$\frac{\sqrt{39}}{\sqrt{15}} = \sqrt{\frac{39}{15}} = \sqrt{\frac{13}{5}}$$

25. Write $\dfrac{\sqrt{152}}{\sqrt{3}}$ as a single radical. Then simplify the result.

Solution

$$\frac{\sqrt{152}}{\sqrt{3}} = \sqrt{\frac{152}{3}}$$

27. Simplify $\dfrac{\sqrt{54}}{\sqrt{6}}$.

Solution

$$\frac{\sqrt{54}}{\sqrt{6}} = \sqrt{\frac{54}{6}} = \sqrt{9} = 3$$

29. Simplify $\sqrt{\frac{35}{4}}$.

Solution

$$\sqrt{\frac{35}{4}} = \frac{\sqrt{35}}{\sqrt{4}} = \frac{\sqrt{35}}{2} \text{ or } \frac{1}{2}\sqrt{35}$$

31. Simplify the expression $\sqrt{\frac{3}{25}}$.

Solution

$$\sqrt{\frac{3}{25}} = \frac{\sqrt{3}}{\sqrt{25}} = \frac{\sqrt{3}}{5}$$

33. Simplify the expression $\sqrt{\dfrac{18a^3}{2a}}$.

Solution

$$\sqrt{\frac{18a^3}{2a}} = \sqrt{9a^2} = \sqrt{9} \cdot \sqrt{a^2} = 3|a|$$

35. Rationalize the denominator in $\sqrt{\frac{1}{3}}$ and simplify.

Solution

$$\sqrt{\frac{1}{3}} = \frac{\sqrt{1}}{\sqrt{3}} = \frac{1}{\sqrt{3}} = \frac{1}{\sqrt{3}} \cdot \frac{\sqrt{3}}{\sqrt{3}} = \frac{\sqrt{3}}{\sqrt{9}} = \frac{\sqrt{3}}{3}$$

37. Rationalize the denominator in $\dfrac{1}{\sqrt{3}}$ and simplify.

Solution

$$\frac{1}{\sqrt{3}} = \frac{1}{\sqrt{3}} \cdot \frac{\sqrt{3}}{\sqrt{3}} = \frac{\sqrt{3}}{\sqrt{9}} = \frac{\sqrt{3}}{3}$$

39. Rationalize the denominator in $\dfrac{5}{\sqrt{10}}$ and simplify.

Solution

$$\frac{5}{\sqrt{10}} = \frac{5}{\sqrt{10}} \cdot \frac{\sqrt{10}}{\sqrt{10}} = \frac{5\sqrt{10}}{\sqrt{100}} = \frac{5\sqrt{10}}{10} = \frac{\sqrt{10}}{2}$$

41. Rationalize the denominator in $\dfrac{a^3}{\sqrt{a}}$ and simplify. (Assume that the variables are positive.)

Solution

$$\frac{a^3}{\sqrt{a}} = \frac{a^3}{\sqrt{a}} \cdot \frac{\sqrt{a}}{\sqrt{a}} = \frac{a^3\sqrt{a}}{\sqrt{a^2}} = \frac{a^3\sqrt{a}}{\sqrt{a}} = a^2\sqrt{a}$$

43. Rationalize the denominator in $\sqrt{\dfrac{3}{6x}}$ and simplify. (Assume that the variables are positive.)

Solution

$$\sqrt{\frac{3}{6x}} = \sqrt{\frac{1}{2x}}$$
$$= \frac{\sqrt{1}}{\sqrt{2x}}$$
$$= \frac{1}{\sqrt{2x}} \cdot \frac{\sqrt{2x}}{\sqrt{2x}}$$
$$= \frac{\sqrt{2x}}{\sqrt{4x^2}}$$
$$= \frac{\sqrt{2x}}{2x}$$

45. Rationalize the denominator in $\dfrac{\sqrt{2z}}{\sqrt{30z^3}}$ and simplify. (Assume that the variables are positive.)

Solution

$$\frac{\sqrt{2z}}{\sqrt{30z^3}} = \sqrt{\frac{2z}{30z^3}} = \sqrt{\frac{1}{15z^2}}$$
$$= \frac{\sqrt{1}}{\sqrt{15z^2}}$$
$$= \frac{1}{\sqrt{15z^2}} \cdot \frac{\sqrt{15}}{\sqrt{15}}$$
$$= \frac{\sqrt{15}}{\sqrt{225z^2}}$$
$$= \frac{\sqrt{15}}{15z}$$

47. *Geometry*   Find the area of the figure in the textbook. Round the result to two decimal places.

Solution

$$\text{Area} = (\text{Length})(\text{Width})$$
$$= (\sqrt{20})(\sqrt{10})$$
$$= \sqrt{200}$$
$$= \sqrt{100} \cdot \sqrt{2}$$
$$= 10\sqrt{2}$$
$$\approx 10(1.41421)$$
$$\approx 14.14$$

49. *Geometry*   Find the area of the figure in the textbook. Round the result to two decimal places.

Solution

$$\text{Area} = \tfrac{1}{2}(\text{Base})(\text{Height})$$
$$= \tfrac{1}{2}(6\sqrt{3})(4\sqrt{2})$$
$$= \tfrac{1}{2}(24\sqrt{6})$$
$$= 12\sqrt{6}$$
$$\approx 12(2.44949)$$
$$\approx 29.39$$

51. Solve the equation $0.60x = 24$ and check your solution.

Solution

$$0.60x = 24$$
$$x = \frac{24}{0.60}$$
$$x = 40$$

53. Solve the equation $t^2 - 5t = 0$ and check your solution.

Solution

$$t^2 - 5t = 0$$
$$t(t - 5) = 0$$
$$t = 0$$
$$t - 5 = 0 \rightarrow t = 5$$

The solutions are 0 and 5.

55. *Wholesale Cost* The selling price of an electric drill is $84.50. The markup rate is 30% of the wholesale cost. Find the wholesale cost.

Solution

Verbal model: [Wholesale cost] + 0.30 [Wholesale cost] = [Selling price]

Labels: Wholesale cost = x
Selling price = 84.50

Equation: $x + 0.30x = 84.50$

$1.30x = 84.50$

$x = 65$

The wholesale cost is $65.

57. Write $\sqrt{5} \cdot \sqrt{6}$ as a single radical.

Solution
$\sqrt{5} \cdot \sqrt{6} = \sqrt{30}$

59. Write $\sqrt{2} \cdot \sqrt{x}$ as a single radical.

Solution
$\sqrt{2} \cdot \sqrt{x} = \sqrt{2x}$

61. Write $\dfrac{\sqrt{6}}{\sqrt{3}}$ as a single radical.

Solution
$\dfrac{\sqrt{6}}{\sqrt{3}} = \sqrt{\dfrac{6}{3}} = \sqrt{2}$

63. Write $\dfrac{\sqrt{15x}}{\sqrt{5x}}$ as a single radical.

Solution
$\dfrac{\sqrt{15x}}{\sqrt{5x}} = \sqrt{\dfrac{15x}{5x}} = \sqrt{3}$

65. Simplify $\sqrt{27}$. Use absolute value signs if appropriate.

Solution
$\sqrt{27} = \sqrt{9 \cdot 3} = \sqrt{9} \cdot \sqrt{3} = 3\sqrt{3}$

67. Simplify $\sqrt{20}$. Use absolute value signs if appropriate.

Solution
$\sqrt{20} = \sqrt{4 \cdot 5} = \sqrt{4} \cdot \sqrt{5} = 2\sqrt{5}$

69. Simplify $\sqrt{300}$. Use absolute value signs if appropriate.

Solution
$\sqrt{300} = \sqrt{100 \cdot 3} = \sqrt{100} \cdot \sqrt{3} = 10\sqrt{3}$

71. Simplify $\sqrt{30{,}000}$. Use absolute value signs if appropriate.

Solution
$\sqrt{30{,}000} = \sqrt{10{,}000 \cdot 3} = \sqrt{10{,}000} \cdot \sqrt{3} = 100\sqrt{3}$

73. Simplify $\sqrt{4x^2}$. Use absolute value signs if appropriate.

Solution
$\sqrt{4x^2} = \sqrt{4 \cdot x^2}$
$= \sqrt{4} \cdot \sqrt{x^2}$
$= 2|x|$

75. Simplify $\sqrt{8a^4}$. Use absolute value signs if appropriate.

Solution
$\sqrt{8a^4} = \sqrt{4 \cdot 2 \cdot a^4}$
$= \sqrt{4} \cdot \sqrt{2} \cdot \sqrt{a^4}$
$= 2\sqrt{2}a^2 \text{ or } 2a^2\sqrt{2}$

77. Simplify $\sqrt{x^2 y^3}$. Use absolute value signs if appropriate.

Solution

$$\sqrt{x^2 y^3} = \sqrt{x^2 \cdot y^2 \cdot y}$$
$$= \sqrt{x^2} \cdot \sqrt{y^2} \cdot \sqrt{y}$$
$$= |x| y \sqrt{y}$$

79. Simplify $(7\sqrt{x})^2$. Use absolute value signs if appropriate.

Solution

$$(7\sqrt{x})^2 = (7\sqrt{x})(7\sqrt{x})$$
$$= 49\sqrt{x^2}$$
$$= 49x$$

81. Simplify $\dfrac{\sqrt{24}}{\sqrt{2}}$. Use absolute value signs if appropriate.

Solution

$$\frac{\sqrt{24}}{\sqrt{2}} = \sqrt{\frac{24}{2}} = \sqrt{12} = \sqrt{4} \cdot \sqrt{3} = 2\sqrt{3}$$

83. Simplify $\sqrt{\dfrac{3 \cdot 12}{27}}$. Use absolute value signs if appropriate.

Solution

$$\sqrt{\frac{3 \cdot 12}{27}} = \sqrt{\frac{12}{9}} = \frac{\sqrt{4}\sqrt{3}}{\sqrt{9}} = \frac{2\sqrt{3}}{3} \text{ or } \frac{2}{3}\sqrt{3}$$

Note: There are several other ways in which this radical could be correctly simplified, but the result will always be the same.

85. Simplify $\sqrt{\dfrac{12x^2}{25}}$. Use absolute value signs if appropriate.

Solution

$$\sqrt{\frac{12x^2}{25}} = \frac{\sqrt{12x^2}}{\sqrt{25}}$$
$$= \frac{\sqrt{4 \cdot 3 \cdot x^2}}{5}$$
$$= \frac{\sqrt{4} \cdot \sqrt{3} \cdot \sqrt{x^2}}{5}$$
$$= \frac{2|x|\sqrt{3}}{5}$$

87. Simplify $\dfrac{\sqrt{5u^2}}{\sqrt{4u^4}}$. Use absolute value signs if appropriate.

Solution

$$\frac{\sqrt{5u^2}}{\sqrt{4u^4}} = \sqrt{\frac{5u^2}{4u^4}}$$
$$= \sqrt{\frac{5}{4u^2}}$$
$$= \frac{\sqrt{5}}{\sqrt{4u^2}}$$
$$= \frac{\sqrt{5}}{2|u|}$$

89. Rationalize the denominator in $\dfrac{\sqrt{2}}{\sqrt{3}}$ and simplify.

Solution

$$\frac{\sqrt{2}}{\sqrt{3}} = \frac{\sqrt{2}}{\sqrt{3}} \cdot \frac{\sqrt{3}}{\sqrt{3}}$$
$$= \frac{\sqrt{6}}{\sqrt{9}}$$
$$= \frac{\sqrt{6}}{3}$$

91. Rationalize the denominator in $\sqrt{\dfrac{100}{11}}$ and simplify.

Solution

$$\sqrt{\frac{100}{11}} = \frac{\sqrt{100}}{\sqrt{11}}$$
$$= \frac{10}{\sqrt{11}} = \frac{10}{\sqrt{11}} \cdot \frac{\sqrt{11}}{\sqrt{11}}$$
$$= \frac{10\sqrt{11}}{\sqrt{121}} = \frac{10\sqrt{11}}{11}$$

93. Rationalize the denominator in $\dfrac{1}{\sqrt{y}}$ and simplify. (Assume that the variables are positive.)

Solution

$$\dfrac{1}{\sqrt{y}} = \dfrac{1}{\sqrt{y}} \cdot \dfrac{\sqrt{y}}{\sqrt{y}} = \dfrac{\sqrt{y}}{\sqrt{y^2}} = \dfrac{\sqrt{y}}{y}$$

95. Rationalize the denominator in $\sqrt{\dfrac{5}{x}}$ and simplify. (Assume that the variables are positive.)

Solution

$$\sqrt{\dfrac{5}{x}} = \dfrac{\sqrt{5}}{\sqrt{x}} = \dfrac{\sqrt{5}}{\sqrt{x}} \cdot \dfrac{\sqrt{x}}{\sqrt{x}} = \dfrac{\sqrt{5x}}{\sqrt{x^2}} = \dfrac{\sqrt{5x}}{x}$$

97. Rationalize the denominator in $\sqrt{\dfrac{t}{8}}$ and simplify. (Assume that the variables are positive.)

Solution

$$\sqrt{\dfrac{t}{8}} = \dfrac{\sqrt{t}}{\sqrt{8}} = \dfrac{\sqrt{t} \cdot \sqrt{2}}{\sqrt{8} \cdot \sqrt{2}} = \dfrac{\sqrt{2t}}{\sqrt{16}} = \dfrac{\sqrt{2t}}{4}$$

99. Rationalize the denominator in $\sqrt{\dfrac{2x}{3y}}$ and simplify. (Assume that the variables are positive.)

Solution

$$\sqrt{\dfrac{2x}{3y}} = \dfrac{\sqrt{2x}}{\sqrt{3y}} = \dfrac{\sqrt{2x}}{\sqrt{3y}} \cdot \dfrac{\sqrt{3y}}{\sqrt{3y}} = \dfrac{\sqrt{6xy}}{\sqrt{9y^2}} = \dfrac{\sqrt{6xy}}{3y}$$

101. *Graphical Reasoning* Use a graphing utility to graph the equations $y_1 = \sqrt{2} \cdot \sqrt{x}$ and $y_2 = \sqrt{2x}$ on the same screen. What inference can you make from the graphs?

Solution

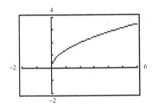

The two graphs are the same.

$$\sqrt{2}\sqrt{x} = \sqrt{2x}$$

103. *Graphical Reasoning* Use a graphing utility to graph the equations $y_1 = \dfrac{\sqrt{x}}{\sqrt{8}}$ and $y_2 = \dfrac{1}{4}\sqrt{2x}$ on the same screen. What inference can you make from the graphs?

Solution

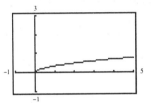

The two graphs are the same.

$$\dfrac{\sqrt{x}}{\sqrt{8}} = \dfrac{\sqrt{x}}{\sqrt{8}} \cdot \dfrac{\sqrt{2}}{\sqrt{2}} = \dfrac{\sqrt{2x}}{\sqrt{16}} = \dfrac{\sqrt{2x}}{4}$$

105. *Period of a Pendulum* Use the formula

$$t = 2\pi\sqrt{\dfrac{L}{32}}$$

which gives the period t (in seconds) of a simple pendulum whose length is L (in feet). How long will it take the performer in the figure in the textbook to swing through one period of the 12-foot trapeze?

Solution

$t = 2\pi\sqrt{\dfrac{L}{32}}$ and $L = 12$ ft

$$t = 2\pi\sqrt{\dfrac{12}{32}} = 2\pi\sqrt{\dfrac{6}{16}} = \dfrac{2\pi}{1} \cdot \dfrac{\sqrt{6}}{\sqrt{16}} = \dfrac{2\pi}{1} \cdot \dfrac{\sqrt{6}}{4} = \dfrac{\pi\sqrt{6}}{2}$$

$t \approx 3.85$

It will take approximately 3.85 seconds.

107. *Geometry*   Find the area of the rectangular desktop in the figure. (Round your result to two decimal places.)

Solution

Area = Length × Width

$$A = \left(29\sqrt{2}\right)\left(12\sqrt{5}\right)$$

$$= 29 \cdot 12 \cdot \sqrt{2} \cdot \sqrt{5}$$

$$= 348\sqrt{10}$$

$$A \approx 1100.47 \text{ square inches}$$

The area is approximately 1100.47 square inches or $\dfrac{1100.47}{144} \approx 7.64$ square feet.

109. Simplify $\sqrt{2.25 \times 10^6}$.

Solution

$$\sqrt{2.25 \times 10^6} = \sqrt{2.25} \times \sqrt{10^6}$$

$$= 1.5 \times 10^3 \text{ or } 1500$$

111. Simplify $\sqrt{1.69 \times 10^{-6}}$.

Solution

$$\sqrt{1.69 \times 10^{-6}} = \sqrt{1.69} \times \sqrt{10^{-6}}$$

$$= 1.3 \times 10^{-3} \text{ or } 0.0013$$

Mid-Chapter Quiz for Chapter 10

1. Determine if $\sqrt{\dfrac{3}{4}}$ is rational or irrational.

Solution

$$\sqrt{\dfrac{3}{4}} = \dfrac{\sqrt{3}}{\sqrt{4}} = \dfrac{\sqrt{3}}{2}$$

The number is irrational because 3 is not a perfect square.

2. Determine if $\sqrt{900}$ is rational or irrational.

Solution

$$\sqrt{900} = 30$$

The number is rational because 900 is a perfect square.

3. Evaluate the expression $\sqrt{121}$, if possible.

Solution

$$\sqrt{121} = 11$$

4. Evaluate the expression $-\sqrt{0.25}$, if possible.

Solution

$$-\sqrt{0.25} = -0.5$$

5. Evaluate the expression $\sqrt{3^2 + 4^2}$, if possible.

Solution

$$\sqrt{3^2 + 4^2} = \sqrt{9 + 16} = \sqrt{25} = 5$$

6. Evaluate the expression $\sqrt{-\dfrac{1}{16}}$, if possible.

Solution

$$\sqrt{-\dfrac{1}{16}}$$

This is not a real number. There is no real number that can be multiplied by itself to obtain $-\dfrac{1}{16}$.

7. Use a calculator to approximate $\sqrt{15.8}$. Round the result to three decimal places.

Solution

$\sqrt{15.8} \approx 3.975$

8. Use a calculator to approximate $\dfrac{-5 + \sqrt{25 + 20}}{10}$. Round the result to three decimal places.

Solution

$\dfrac{-5 + \sqrt{25 + 20}}{10} = \dfrac{-5 + \sqrt{45}}{10} = \dfrac{-5 + 3\sqrt{5}}{10} \approx 0.171$

9. Write $\sqrt{15} \cdot \sqrt{7}$ as a single radical.

Solution

$\sqrt{15} \cdot \sqrt{7} = \sqrt{15 \cdot 7} = \sqrt{105}$

10. Write $\dfrac{\sqrt{42}}{\sqrt{6}}$ as a single radical.

Solution

$\dfrac{\sqrt{42}}{\sqrt{6}} = \sqrt{\dfrac{42}{6}} = \sqrt{7}$

11. Simplify $\sqrt{45}$. Use absolute value signs if appropriate.

Solution

$\sqrt{45} = \sqrt{9}\sqrt{5} = 3\sqrt{5}$

12. Simplify $\sqrt{72x^2}$. Use absolute value signs if appropriate.

Solution

$\sqrt{72x^2} = \sqrt{36x^2}\sqrt{2} = 6|x|\sqrt{2}$

13. Simplify $\dfrac{\sqrt{600}}{\sqrt{12}}$. Use absolute value signs if appropriate.

Solution

$\dfrac{\sqrt{600}}{\sqrt{12}} = \dfrac{\sqrt{100}\sqrt{6}}{\sqrt{4}\sqrt{3}} = \dfrac{10}{2}\sqrt{\dfrac{6}{3}} = 5\sqrt{2}$

Alternate Method:

$\dfrac{\sqrt{600}}{\sqrt{12}} = \sqrt{\dfrac{600}{12}} = \sqrt{50} = \sqrt{25}\sqrt{2} = 5\sqrt{2}$

14. Simplify $\sqrt{\dfrac{90b^4}{2b^2}}$. Use absolute value signs if appropriate.

Solution

$\sqrt{\dfrac{90b^4}{2b^2}} = \sqrt{45b^2} = \sqrt{9b^2}\sqrt{5} = 3|b|\sqrt{5}$

15. Rationalize the denominator in $\sqrt{\dfrac{3}{2}}$ and simplify.

Solution

$\sqrt{\dfrac{3}{2}} = \dfrac{\sqrt{3}}{\sqrt{2}} \cdot \dfrac{\sqrt{2}}{\sqrt{2}} = \dfrac{\sqrt{6}}{\sqrt{4}} = \dfrac{\sqrt{6}}{2}$

16. Rationalize the denominator in $\dfrac{2}{\sqrt{12}}$ and simplify.

Solution

$\dfrac{2}{\sqrt{12}} = \dfrac{2}{\sqrt{4}\sqrt{3}} = \dfrac{2}{2\sqrt{3}} = \dfrac{1}{\sqrt{3}} \cdot \dfrac{\sqrt{3}}{\sqrt{3}} = \dfrac{\sqrt{3}}{\sqrt{9}} = \dfrac{\sqrt{3}}{3}$

17. Rationalize the denominator in $\dfrac{4a}{\sqrt{2a}}$ and simplify. (Assume that the variables are positive.)

Solution

$\dfrac{4a}{\sqrt{2a}} = \dfrac{4a}{\sqrt{2a}} \cdot \dfrac{\sqrt{2a}}{\sqrt{2a}} = \dfrac{4a\sqrt{2a}}{\sqrt{4a^2}} = \dfrac{4a\sqrt{2a}}{2a} = 2\sqrt{2a}$

18. Rationalize the denominator in $\sqrt{\dfrac{5a^3}{4a}}$ and simplify. (Assume that the variables are positive.)

Solution

$\sqrt{\dfrac{5a^3}{4a}} = \sqrt{\dfrac{5a^2}{4}} = \dfrac{\sqrt{a^2}\sqrt{5}}{\sqrt{4}} = \dfrac{a\sqrt{5}}{2}$

19. Use a graphing utility to graph the equations $y_1 = \sqrt{3} \cdot \sqrt{x}$ and $y_2 = \sqrt{3x}$. What inference can you make from the graphs?

Solution

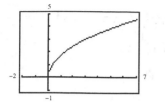

The graphs are the same.

$$\sqrt{3} \cdot \sqrt{x} = \sqrt{3x}$$

20. The base b of the triangle in the figure in the textbook is $b = \sqrt{(2s)^2 - s^2}$. Simplify the expression.

Solution

$$b = \sqrt{(2s)^2 - s^2}$$
$$= \sqrt{4s^2 - s^2}$$
$$= \sqrt{3s^2}$$
$$= \sqrt{s^2}\sqrt{3}$$
$$= s\sqrt{3}$$

10.3 | Operations with Radical Expressions

7. Simplify the expression $3\sqrt{5} - \sqrt{5}$.

Solution

$$3\sqrt{5} - \sqrt{5} = (3-1)\sqrt{5} = 2\sqrt{5}$$

9. Simplify the expression $2\sqrt{2} - 3\sqrt{5} + 8\sqrt{2}$.

Solution

$$2\sqrt{2} - 3\sqrt{5} + 8\sqrt{2} = (2\sqrt{2} + 8\sqrt{2}) - 3\sqrt{5}$$
$$= (2+8)\sqrt{2} - 3\sqrt{5}$$
$$= 10\sqrt{2} - 3\sqrt{5}$$

11. Simplify the expression $2\sqrt{50} + 12\sqrt{8}$.

Solution

$$2\sqrt{50} + 12\sqrt{8} = 2\sqrt{25 \cdot 2} + 12\sqrt{4 \cdot 2}$$
$$= 2 \cdot 5\sqrt{2} + 12 \cdot 2\sqrt{2}$$
$$= 10\sqrt{2} + 24\sqrt{2}$$
$$= (10+24)\sqrt{2}$$
$$= 34\sqrt{2}$$

13. Simplify the expression $\sqrt{9x} + \sqrt{36x}$.

Solution

$$\sqrt{9x} + \sqrt{36x} = \sqrt{9 \cdot x} + \sqrt{36 \cdot x}$$
$$= 3\sqrt{x} + 6\sqrt{x}$$
$$= (3+6)\sqrt{x}$$
$$= 9\sqrt{x}$$

15. Simplify the expression $\sqrt{a/4} - \sqrt{a/9}$.

Solution

$$\sqrt{\frac{a}{4}} - \sqrt{\frac{a}{9}} = \frac{\sqrt{a}}{\sqrt{4}} - \frac{\sqrt{a}}{\sqrt{9}}$$

$$= \frac{\sqrt{a}}{2} - \frac{\sqrt{a}}{3}$$

$$= \left(\frac{1}{2} - \frac{1}{3}\right)\sqrt{a} \quad \text{Note: } \frac{\sqrt{a}}{2} = \frac{1}{2}\sqrt{a}; \ \frac{\sqrt{a}}{3} = \frac{1}{3}\sqrt{a}$$

$$= \left(\frac{3}{6} - \frac{2}{6}\right)\sqrt{a}$$

$$= \frac{1}{6}\sqrt{a} \text{ or } \frac{\sqrt{a}}{6}$$

17. Simplify the expression $\sqrt{x^3 y} + 4\sqrt{xy}$.

Solution

$$\sqrt{x^3 y} + 4\sqrt{xy} = \sqrt{x^2 \cdot xy} + 4\sqrt{xy}$$

$$= |x|\sqrt{xy} + 4\sqrt{xy}$$

$$= (|x| + 4)\sqrt{xy}$$

19. Multiply and simplify $\sqrt{3} \cdot \sqrt{27}$.

Solution

$$\sqrt{3} \cdot \sqrt{27} = \sqrt{81} = 9$$

21. Multiply and simplify $\sqrt{6}(\sqrt{12} + 8)$.

Solution

$$\sqrt{6}(\sqrt{12} + 8) = \sqrt{6} \cdot \sqrt{12} + \sqrt{6} \cdot 8$$

$$= \sqrt{72} + 8\sqrt{6}$$

$$= \sqrt{36 \cdot 2} + 8\sqrt{6}$$

$$= 6\sqrt{2} + 8\sqrt{6}$$

23. Multiply and simplify $(\sqrt{2} - 1)(\sqrt{2} + 1)$.

Solution

$$(\sqrt{2} - 1)(\sqrt{2} + 1) = (\sqrt{2})^2 - (1)^2 \quad \text{Pattern: } (a+b)(a-b) = a^2 - b^2$$

$$= 2 - 1$$

$$= 1$$

25. Multiply and simplify $(\sqrt{13} + 2)^2$.

Solution

$$(\sqrt{13} + 2)^2 = (\sqrt{13} + 2)(\sqrt{13} + 2)$$

$$= \sqrt{13} \cdot \sqrt{13} + 2\sqrt{13} + 2\sqrt{13} + 4$$

$$= 13 + 4\sqrt{13} + 4$$

$$= 17 + 4\sqrt{13}$$

We could use the pattern for the square of a binomial.

$$(\sqrt{13} + 2)^2 = (\sqrt{13})^2 + 2(\sqrt{13})(2) + 2^2 \quad \text{Pattern: } (a+b)^2 = a^2 + 2ab + b^2$$

$$= 13 + 4\sqrt{13} + 4$$

$$= 17 + 4\sqrt{13}$$

27. Multiply and simplify $\left(3 + \sqrt{x}\right)^2$.

Solution

$$\left(3 + \sqrt{x}\right)^2 = \left(3 + \sqrt{x}\right)\left(3 + \sqrt{x}\right)$$
$$= 3 \cdot 3 + 3\sqrt{x} + 3\sqrt{x} + \sqrt{x} \cdot \sqrt{x}$$
$$= 9 + 6\sqrt{x} + x$$

We could use the pattern for the square of a binomial.

$$\left(3 + \sqrt{x}\right)^2 = 3^2 + 2(3)\left(\sqrt{x}\right) + \left(\sqrt{x}\right)^2 \qquad \text{Pattern: } (a + b)^2 = a^2 + 2ab + b^2$$
$$= 9 + 6\sqrt{x} + x$$

29. Multiply and simplify $\left(\sqrt{x} + 1\right)\left(\sqrt{x} - 3\right)$.

Solution

$$\overset{\text{F} \qquad \text{O} \qquad \text{I} \qquad \text{L}}{\left(\sqrt{x} + 1\right)\left(\sqrt{x} - 3\right) = \sqrt{x} \cdot \sqrt{x} - 3\sqrt{x} + 1\sqrt{x} - 3}$$
$$= x - 2\sqrt{x} - 3$$

31. Write the conjugate of $4 + \sqrt{3}$. Then find the product of the expression and its conjugate.

Solution

Expression	*Conjugate*	*Product*
$4 + \sqrt{3}$	$4 - \sqrt{3}$	$(4)^2 - \left(\sqrt{3}\right)^2 = 16 - 3 = 13$

33. Write the conjugate of $\sqrt{u} - \sqrt{2}$. Then find the product of the expression and its conjugate.

Solution

Expression	*Conjugate*	*Product*
$\sqrt{u} - \sqrt{2}$	$\sqrt{u} + \sqrt{2}$	$\left(\sqrt{u}\right)^2 - \left(\sqrt{2}\right)^2 = u - 2$

35. Rationalize the denominator in $\dfrac{4}{\sqrt{7} - 3}$.

Solution

$$\frac{4}{\sqrt{7} - 3} = \frac{4}{\sqrt{7} - 3} \cdot \frac{\left(\sqrt{7} + 3\right)}{\left(\sqrt{7} + 3\right)}$$
$$= \frac{4\left(\sqrt{7} + 3\right)}{\left(\sqrt{7}\right)^2 - (3)^2}$$
$$= \frac{4\left(\sqrt{7} + 3\right)}{7 - 9}$$
$$= \frac{4\left(\sqrt{7} + 3\right)}{-2}$$
$$= -2\left(\sqrt{7} + 3\right) \text{ or } -2\sqrt{7} - 6$$

37. Rationalize the denominator in $\dfrac{5x}{\sqrt{7} - \sqrt{2}}$.

Solution

$$\frac{5x}{\sqrt{7} - \sqrt{2}} = \frac{5x}{\sqrt{7} - \sqrt{2}} \cdot \frac{\sqrt{7} + \sqrt{2}}{\sqrt{7} + \sqrt{2}}$$
$$= \frac{5x\left(\sqrt{7} + \sqrt{2}\right)}{\left(\sqrt{7}\right)^2 - \left(\sqrt{2}\right)^2}$$
$$= \frac{5x\left(\sqrt{7} + \sqrt{2}\right)}{5}$$
$$= x\left(\sqrt{7} + \sqrt{2}\right) \text{ or } x\sqrt{7} + x\sqrt{2}$$

39. *The Golden Section* The ratio of the width of the Temple of Hephaestus to its height (see figure in textbook) is

$$\frac{w}{h} = \frac{2}{\sqrt{5}-1}.$$

This number is called the **golden section**. Early Greeks believed that the most aesthetically pleasing rectangles were those with sides of this ratio. Rationalize the denominator for this number. Approximate your answer, rounded to two decimal places.

Solution

$$\frac{2}{\sqrt{5}-1} = \frac{2}{\sqrt{5}-1} \cdot \frac{\sqrt{5}+1}{\sqrt{5}+1} = \frac{2(\sqrt{5}+1)}{(\sqrt{5})^2-(1)^2} = \frac{2(\sqrt{5}+1)}{5-1} = \frac{2(\sqrt{5}+1)}{4} = \frac{\sqrt{5}+1}{2} \approx 1.62$$

The Golden Section is approximately equal to 1.62.

41. Rewrite the following expression as a single fraction and simplify.

$$3 - \frac{1}{\sqrt{3}}$$

Solution

$$3 - \frac{1}{\sqrt{3}} = 3 - \frac{1}{\sqrt{3}} \cdot \frac{\sqrt{3}}{\sqrt{3}} = \frac{3}{1} - \frac{\sqrt{3}}{3} = \frac{9}{3} - \frac{\sqrt{3}}{3} = \frac{9-\sqrt{3}}{3}$$

Note: Here is another way to do this exercise.

$$3 - \frac{1}{\sqrt{3}} = \frac{3}{1} - \frac{1}{\sqrt{3}} = \frac{3\sqrt{3}}{\sqrt{3}} - \frac{1}{\sqrt{3}} = \frac{3\sqrt{3}-1}{\sqrt{3}} = \frac{3\sqrt{3}-1}{\sqrt{3}} \cdot \frac{\sqrt{3}}{\sqrt{3}}$$

$$= \frac{(3\sqrt{3}-1)\sqrt{3}}{3} = \frac{3\sqrt{9}-\sqrt{3}}{3} = \frac{3\cdot3-\sqrt{3}}{3} = \frac{9-\sqrt{3}}{3}$$

43. Rewrite the following expression as a single fraction and simplify.

$$\sqrt{50} - \frac{6}{\sqrt{2}}$$

Solution

$$\sqrt{50} - \frac{6}{\sqrt{2}} = \sqrt{25\cdot2} - \frac{6}{\sqrt{2}} \cdot \frac{\sqrt{2}}{\sqrt{2}} = 5\sqrt{2} - \frac{6\sqrt{2}}{2} = 5\sqrt{2} - 3\sqrt{2} = 2\sqrt{2}$$

45. *Geometry* Write expressions for the perimeter and area of the rectangle shown in the textbook. Then simplify the expressions.

Solution

Perimeter = 2(Length) + 2(Width)   Area = (Length)(Width)

$$= 2\sqrt{121x} + 2(5\sqrt{x}) \qquad\qquad = (\sqrt{121x})(5\sqrt{x})$$

$$= 2\sqrt{121}\sqrt{x} + 10\sqrt{x} \qquad\qquad = 5\sqrt{121}\sqrt{x^2}$$

$$= 2\cdot11\sqrt{x} + 10\sqrt{x} \qquad\qquad = 5\cdot11x$$

$$= 22\sqrt{x} + 10\sqrt{x} \qquad\qquad = 55x$$

$$= 32\sqrt{x}$$

47. Expand the expression $10(x - 1)$ by using the Distributive Property.

Solution

$10(x - 1) = 10x - 10$

49. Expand the expression $-\frac{1}{2}(4 - 6x)$ by using the Distributive Property.

Solution

$-\frac{1}{2}(4 - 6x) = -2 + 3x$

51. *Work Rate* One machine completes a job in r hours and another machine requires $\frac{5}{4}r$ hours. Find the individual times to complete the job if it takes 5 hours using both machines.

Solution

Verbal model: $\boxed{\begin{array}{c}\text{Rate of first} \\ \text{machine}\end{array}} + \boxed{\begin{array}{c}\text{Rate of second} \\ \text{machine}\end{array}} = \boxed{\begin{array}{c}\text{Combined} \\ \text{rate}\end{array}}$

Labels: Time for first machine $= r$ (hours)

Rate of first machine $= \dfrac{1}{r}$ (jobs per hours)

Time for second machine $= \dfrac{5}{4}r$ (hours)

Rate of second machine $= \dfrac{1}{\left(\frac{5}{4}r\right)} = 1 \div \dfrac{5r}{4} = 1\left(\dfrac{4}{5r}\right) = \dfrac{4}{5r}$

Combined time $= 5$ (hours)

Combined rate $= \dfrac{1}{5}$ (jobs per hour)

Equation: $\dfrac{1}{r} + \dfrac{1}{\frac{5}{4}r} = \dfrac{1}{5}$

$\dfrac{1}{r} + \dfrac{4}{5r} = \dfrac{1}{5}$

$5r\left(\dfrac{1}{r} + \dfrac{4}{5r}\right) = 5r\left(\dfrac{1}{5}\right)$

$5 + 4 = r$

$9 = r$ and $\dfrac{5}{4}r = \dfrac{45}{4}$ or 11.25

The first machine completes the job in 9 hours, and the second machine completes the job in 11.25 hours.

53. Simplify the expression $10\sqrt{11} + 8\sqrt{11}$.

Solution

$10\sqrt{11} + 8\sqrt{11} = 18\sqrt{11}$

55. Simplify the expression $\frac{2}{5}\sqrt{3} - \frac{6}{5}\sqrt{3}$.

Solution

$\dfrac{2}{5}\sqrt{3} - \dfrac{6}{5}\sqrt{3} = \left(\dfrac{2}{5} - \dfrac{6}{5}\right)\sqrt{3} = -\dfrac{4}{5}\sqrt{3}$

57. Simplify the expression $\sqrt{3} - 5\sqrt{7} - 12\sqrt{3}$.

Solution

$\sqrt{3} - 5\sqrt{7} - 12\sqrt{3} = \sqrt{3} - 12\sqrt{3} - 5\sqrt{7}$

$= (1 - 12)\sqrt{3} - 5\sqrt{7}$

$= -11\sqrt{3} - 5\sqrt{7}$

59. Simplify the expression $12\sqrt{8} - 3\sqrt{8}$.

Solution

$12\sqrt{8} - 3\sqrt{8} = (12 - 3)\sqrt{8} = 9\sqrt{8}$

$= 9\sqrt{4 \cdot 2} = 9 \cdot 2\sqrt{2}$

$= 18\sqrt{2}$

61. Simplify the expression $8\sqrt{75} + \sqrt{50} + \sqrt{2}$.

Solution

$$8\sqrt{75} + \sqrt{50} + \sqrt{2} = 8\sqrt{25}\sqrt{3} + \sqrt{25}\sqrt{2} + \sqrt{2}$$
$$= 8 \cdot 5\sqrt{3} + 5\sqrt{2} + \sqrt{2}$$
$$= 40\sqrt{3} + 6\sqrt{2}$$

63. Simplify the expression $5\sqrt{x} - 3\sqrt{x}$.

Solution

$$5\sqrt{x} - 3\sqrt{x} = (5 - 3)\sqrt{x} = 2\sqrt{x}$$

65. Simplify the expression $\sqrt{16b} + \sqrt{b}$.

Solution

$$\sqrt{16b} + \sqrt{b} = \sqrt{16}\sqrt{b} + \sqrt{b}$$
$$= 4\sqrt{b} + \sqrt{b}$$
$$= 5\sqrt{b}$$

67. Simplify the expression $\sqrt{45z} - \sqrt{125z}$.

Solution

$$\sqrt{45z} - \sqrt{125z} = \sqrt{9 \cdot 5z} - \sqrt{25 \cdot 5z}$$
$$= 3\sqrt{5z} - 5\sqrt{5z}$$
$$= (3 - 5)\sqrt{5z}$$
$$= -2\sqrt{5z}$$

69. Simplify the expression $3\sqrt{3y} - \sqrt{27y}$.

Solution

$$3\sqrt{3y} - \sqrt{27y} = 3\sqrt{3y} - \sqrt{9}\sqrt{3y}$$
$$= 3\sqrt{3y} - 3\sqrt{3y}$$
$$= 0$$

71. Multiply and simplify $\sqrt{2} \cdot \sqrt{8}$.

Solution

$$\sqrt{2} \cdot \sqrt{8} = \sqrt{16} = 4$$

73. Multiply and simplify $\sqrt{5} \cdot \sqrt{30}$.

Solution

$$\sqrt{5} \cdot \sqrt{30} = \sqrt{150} = \sqrt{25}\sqrt{6} = 5\sqrt{6}$$

75. Multiply and simplify $\sqrt{7}\left(1 - \sqrt{2}\right)$.

Solution

$$\sqrt{7}\left(1 - \sqrt{2}\right) = \sqrt{7} \cdot 1 - \sqrt{7} \cdot \sqrt{2}$$
$$= \sqrt{7} - \sqrt{14}$$

77. Multiply and simplify $(1 + \sqrt{11})(1 - \sqrt{11})$.

Solution

$$(1 + \sqrt{11})(1 - \sqrt{11}) = 1^2 - (\sqrt{11})^2 = 1 - 11 = -10$$

79. Multiply and simplify $\left(\sqrt{10} + \sqrt{5}\right)\left(\sqrt{10} - \sqrt{5}\right)$.

Solution

$$\left(\sqrt{10} + \sqrt{5}\right)\left(\sqrt{10} - \sqrt{5}\right) = \left(\sqrt{10}\right)^2 - \left(\sqrt{5}\right)^2 \qquad \text{Pattern: } (a + b)(a - b) = a^2 - b^2$$
$$= 10 - 5 = 5$$

81. Multiply and simplify $(3 + \sqrt{8})^2$.

Solution

$$
\begin{aligned}
(3 + \sqrt{8})^2 &= (3 + \sqrt{8})(3 + \sqrt{8}) \\
&= 9 + 3\sqrt{8} + 3\sqrt{8} + \sqrt{64} \\
&= 9 + 6\sqrt{8} + 8 \\
&= 17 + 6\sqrt{8} \\
&= 17 + 6\sqrt{4}\sqrt{2} \\
&= 17 + 6 \cdot 2\sqrt{2} \\
&= 17 + 12\sqrt{2}
\end{aligned}
$$

Alternate Method:

$$
\begin{aligned}
(3 + \sqrt{8})^2 &= (3 + 2\sqrt{2})^2 \\
&= 3^2 + 2(3)(2\sqrt{2}) + (2\sqrt{2})^2 \\
&= 9 + 12\sqrt{2} + 4(2) \\
&= 17 + 12\sqrt{2}
\end{aligned}
$$

83. Multiply and simplify $\left(\sqrt{2} + 1\right)\left(\sqrt{3} - 5\right)$.

Solution

$$
\left(\sqrt{2} + 1\right)\left(\sqrt{3} - 5\right) = \overset{\text{F}}{\sqrt{2} \cdot \sqrt{3}} \; \overset{\text{O}}{- 5\sqrt{2}} \; \overset{\text{I}}{+ 1\sqrt{3}} \; \overset{\text{L}}{- 1 \cdot 5} = \sqrt{6} - 5\sqrt{2} + \sqrt{3} - 5
$$

85. Multiply and simplify $\sqrt{x}\left(\sqrt{x} + 5\right)$.

Solution

$$
\sqrt{x}\left(\sqrt{x} + 5\right) = \sqrt{x} \cdot \sqrt{x} + \sqrt{x} \cdot 5 = x + 5\sqrt{x}
$$

87. Find the conjugate of $\sqrt{15} - \sqrt{7}$. Then find the product of the expression and its conjugate.

Solution

Expression	*Conjugate*	*Product*
$\sqrt{15} - \sqrt{7}$	$\sqrt{15} + \sqrt{7}$	$\left(\sqrt{15}\right)^2 - \left(\sqrt{7}\right)^2 = 15 - 7 = 8$

89. Find the conjugate of $\sqrt{x} - 4$. Then find the product of the expression and its conjugate.

Solution

Expression	*Conjugate*	*Product*
$\sqrt{x} - 4$	$\sqrt{x} + 4$	$\left(\sqrt{x}\right)^2 - (4)^2 = x - 16$

91. Rationalize the denominator in $\dfrac{5}{\sqrt{14} - 2}$.

Solution

$$\dfrac{5}{\sqrt{14} - 2} = \dfrac{5}{\sqrt{14} - 2} \cdot \dfrac{\sqrt{14} + 2}{\sqrt{14} + 2}$$

$$= \dfrac{5(\sqrt{14} + 2)}{(\sqrt{14})^2 - (2)^2}$$

$$= \dfrac{5(\sqrt{14} + 2)}{14 - 4}$$

$$= \dfrac{5(\sqrt{14} + 2)}{10}$$

$$= \dfrac{\sqrt{14} + 2}{2}$$

93. Rationalize the denominator in $\dfrac{\sqrt{5} + 1}{\sqrt{7} + 1}$.

Solution

$$\dfrac{\sqrt{5} + 1}{\sqrt{7} + 1} = \dfrac{\sqrt{5} + 1}{\sqrt{7} + 1} \cdot \dfrac{\sqrt{7} - 1}{\sqrt{7} - 1}$$

$$= \dfrac{(\sqrt{5} + 1)(\sqrt{7} - 1)}{(\sqrt{7})^2 - (1)^2}$$

$$= \dfrac{\sqrt{5} \cdot \sqrt{7} - \sqrt{5} + \sqrt{7} - 1}{7 - 1}$$

$$= \dfrac{\sqrt{35} - \sqrt{5} + \sqrt{7} - 1}{6}$$

95. Rationalize the denominator in $\dfrac{2x}{5 - \sqrt{3}}$.

Solution

$$\dfrac{2x}{5 - \sqrt{3}} = \dfrac{2x}{5 - \sqrt{3}} \cdot \dfrac{5 + \sqrt{3}}{5 + \sqrt{3}}$$

$$= \dfrac{2x(5 + \sqrt{3})}{(5)^2 - (\sqrt{3})^2} = \dfrac{2x(5 + \sqrt{3})}{25 - 3} = \dfrac{2x(5 + \sqrt{3})}{22} = \dfrac{x(5 + \sqrt{3})}{11} \text{ or } \dfrac{5x + x\sqrt{3}}{11}$$

97. Rationalize the denominator in $\dfrac{\sqrt{x} - 5}{2\sqrt{x} - 1}$.

Solution

$$\dfrac{\sqrt{x} - 5}{2\sqrt{x} - 1} = \dfrac{\sqrt{x} - 5}{2\sqrt{x} - 1} \cdot \dfrac{2\sqrt{x} + 1}{2\sqrt{x} + 1}$$

$$= \dfrac{(\sqrt{x} - 5)(2\sqrt{x} + 1)}{(2\sqrt{x})^2 - (1)^2} = \dfrac{2x + \sqrt{x} - 10\sqrt{x} - 5}{4x - 1} = \dfrac{2x - 9\sqrt{x} - 5}{4x - 1}$$

99. Use a graphing utility to graph $y_1 = \sqrt{8x} + \sqrt{2x}$ and $y_2 = 3\sqrt{2x}$ on the same screen. Use the graphs to verify the simplification.

Solution

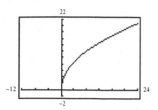

The two graphs are the same.

$$\sqrt{8x} + \sqrt{2x} = \sqrt{4}\sqrt{2x} + \sqrt{2x}$$
$$= 2\sqrt{2x} + \sqrt{2x}$$
$$= 3\sqrt{2x}$$

101. Use a graphing utility to graph the following equations on the same screen. Use the graphs to verify the simplification.

$$y_1 = \frac{2}{\sqrt{x+1}}, \quad y_2 = \frac{2\sqrt{x+1}}{x+1}$$

Solution

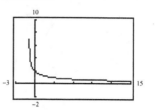

The two graphs are the same.

$$\frac{2}{\sqrt{x+1}} = \frac{2}{\sqrt{x+1}} \cdot \frac{\sqrt{x+1}}{\sqrt{x+1}} = \frac{2\sqrt{x+1}}{x+1}$$

103. Insert the correct symbol ($<$, $>$, or $=$) between $\sqrt{5} + \sqrt{3}$ and $\sqrt{5+3}$.

Solution

$$\sqrt{5} + \sqrt{3} > \sqrt{5+3}$$
$$\sqrt{5} + \sqrt{3} \approx 3.97$$
$$\sqrt{5+3} = \sqrt{8} \approx 2.83$$

105. Insert the correct symbol ($<$, $>$, or $=$) between 5 and $\sqrt{3^2 + 2^2}$.

Solution

$$5 > \sqrt{3^2 + 2^2}$$
$$\sqrt{3^2 + 2^2} = \sqrt{13} \approx 3.61$$

107. *Exploration*   Enter any positive real number in your calculator and repeatedly take the square root. What real number does the display appear to be approaching?

Solution

The display approaches 1.

Examples:	145	or	0.145
	12.042		0.381
	3.470		0.617
	1.863		0.786
	1.365		0.886
	1.168		0.941
	1.081		0.970
	1.040		0.985
	1.020		0.992
	1.010		0.996
	1.005		0.998
	1.002		0.999
	etc.		etc.

109. Write expressions for the perimeter and area of the rectangle shown in the textbook. Then simplify the expressions.

Solution

Perimeter $= 2(\text{Length}) + 2(\text{Width})$

$$= 2(3 + \sqrt{5}) + 2(7 - \sqrt{5})$$
$$= 6 + 2\sqrt{5} + 14 - 2\sqrt{5}$$
$$= 20$$

Area $= 2(\text{Length})(\text{Width})$

$$= (3 + \sqrt{5})(7 - \sqrt{5})$$
$$= 21 - 3\sqrt{5} + 7\sqrt{5} - \sqrt{25}$$
$$= 21 + 4\sqrt{5} - 5$$
$$= 16 + 4\sqrt{5}$$

10.4   Solving Equations and Applications

7. Is x a solution of the equation $\sqrt{x} - 6 = 0$ when (a) $x = -1$, (b) $x = -36$, (c) $x = 36$, and (d) $x = 6$?

Solution

(a) $x = -1$

$$\sqrt{-1} - 6 \overset{?}{=} 0$$
$$\sqrt{-1} - 6 \neq 0$$

$\sqrt{-1}$ is not a real number.

-1 is *not* a solution.

(b) $x = -36$

$$\sqrt{-36} - 6 \overset{?}{=} 0$$
$$\sqrt{-36} - 6 \neq 0$$

$\sqrt{-36}$ is not a real number.

-36 is *not* a solution.

(c) $x = 36$

$$\sqrt{36} - 6 \overset{?}{=} 0$$
$$6 - 6 = 0$$

36 *is* a solution.

(d) $x = 6$

$$\sqrt{6} - 6 \overset{?}{=} 0$$
$$\sqrt{6} - 6 \neq 0$$

6 is *not* a solution.

9. Is x a solution of the equation $x = \sqrt{2x + 3}$ when (a) $x = -1$, (b) $x = 2$, (c) $x = 8$, and (d) $x = 3$?

Solution

(a) $x = -1$

$$-1 \overset{?}{=} \sqrt{2(-1) + 3}$$
$$-1 \neq \sqrt{1}$$

-1 is *not* a solution.

(b) $x = 2$

$$2 \overset{?}{=} \sqrt{2(2) + 3}$$
$$2 \neq \sqrt{7}$$

2 is *not* a solution

(c) $x = 8$

$$8 \overset{?}{=} \sqrt{2(8) + 3}$$
$$8 \neq \sqrt{19}$$

8 is *not* a solution.

(d) $x = 3$

$$3 \overset{?}{=} \sqrt{2(3) + 3}$$
$$3 = \sqrt{9}$$

3 *is* a solution.

11. Solve the equation $\sqrt{x} = 10$.

Solution

$$\sqrt{x} = 10$$
$$\left(\sqrt{x}\right)^2 = (10)^2$$
$$x = 100$$

13. Solve $\sqrt{u} + 3 = 0$.

Solution

$$\sqrt{u} + 3 = 0$$
$$\sqrt{u} = -3$$
$$\left(\sqrt{u}\right)^2 = (-3)^2$$
$$u = 9 \qquad \text{Trial solution}$$

However, this answer does *not* check.

$$\sqrt{9} + 3 \overset{?}{=} 0$$
$$3 + 3 \neq 0$$

Thus, the equation has *no* solution.

15. Solve the equation $\sqrt{3x} - 4 = 6$.

Solution

$$\sqrt{3x} - 4 = 6$$
$$\sqrt{3x} = 10$$
$$(\sqrt{3x})^2 = 10^2$$
$$3x = 100$$
$$x = \frac{100}{3}$$

17. Solve the equation $\sqrt{3x - 2} = 4$.

Solution

$$\sqrt{3x - 2} = 4$$
$$\left(\sqrt{3x - 2}\right)^2 = 4^2$$
$$3x - 2 = 16$$
$$3x = 18$$
$$x = 6$$

19. Solve the equation $8 - \sqrt{t} = 3$.

Solution

$$8 - \sqrt{t} = 3$$
$$-\sqrt{t} = -5$$
$$\left(-\sqrt{t}\right)^2 = (-5)^2$$
$$t = 25$$

21. Solve the equation $\sqrt{x + 4} = \sqrt{2x + 1}$.

Solution

$$\sqrt{x + 4} = \sqrt{2x + 1}$$
$$\left(\sqrt{x + 4}\right)^2 = \left(\sqrt{2x + 1}\right)^2$$
$$x + 4 = 2x + 1$$
$$-x + 4 = 1$$
$$-x = -3$$
$$x = 3$$

23. Solve the equation $x = \sqrt{20 - x}$.

Solution

$$x = \sqrt{20 - x}$$
$$x^2 = (\sqrt{20 - x})^2$$
$$x^2 = 20 - x$$
$$x^2 + x - 20 = 0$$
$$(x + 5)(x - 4) = 0$$
$$x + 50 = 0 \rightarrow x = -5 \qquad \text{Extraneous}$$
$$x - 4 = 0 \rightarrow x = 4$$

25. Solve the equation $\sqrt{6x + 7} = x + 2$.

Solution

$$\sqrt{6x + 7} = x + 2$$
$$\left(\sqrt{6x + 7}\right)^2 = (x + 2)^2$$
$$6x + 7 = x^2 + 4x + 4$$
$$0 = x^2 - 2x - 3$$
$$0 = (x - 3)(x + 1)$$
$$x - 3 = 0 \rightarrow x = 3$$
$$x + 1 = 0 \rightarrow x = -1$$

27. Solve the equation $x(\sqrt{x} + 2) = 0$.

Solution

$$x(\sqrt{x} + 2) = 0$$

$$x = 0$$

$$\sqrt{x} + 2 = 0 \rightarrow \sqrt{x} = -2 \rightarrow (\sqrt{x})^2 = 4 \rightarrow x = 4 \qquad \text{Extraneous}$$

29. *Estimation* A construction worker drops a nail from a building and observes it strike a water puddle after approximately three seconds. Estimate the height of the worker. The time t in seconds for a free-falling object to fall d feet is given by

$$t = \sqrt{\frac{d}{16}}.$$

Solution

$$t = \sqrt{\frac{d}{16}}$$

$$3 = \sqrt{\frac{d}{16}}$$

$$3^2 = \left(\sqrt{\frac{d}{16}}\right)^2$$

$$9 = \frac{d}{16}$$

$$16 \cdot 9 = d$$

$$144 = d$$

The worker is approximately 144 feet high.

31. *Geometry* Solve for x (see figure in textbook). Round your answer to two decimal places.

Solution

$$c^2 = a^2 + b^2$$

$$x^2 = 12^2 + 5^2$$

$$x = \sqrt{12^2 + 5^2}$$

$$= \sqrt{144 + 25}$$

$$= \sqrt{169} = 13$$

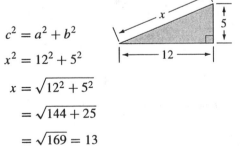

33. *Geometry* Solve for x (see figure in textbook). Round your answer to two decimal places.

Solution

$$c^2 = a^2 + b^2$$

$$x^2 = 12^2 + 8^2$$

$$= 144 + 64$$

$$= 208$$

$$x = \sqrt{208} = \sqrt{16 \cdot 13} = 4\sqrt{13}$$

$$x \approx 14.42$$

35. Find the distance between the points $(1, 2)$ and $(5, 5)$. Round your answer to two decimal places.

Solution

$d = \sqrt{(x_2 - x_1)^2 + (y_2 - y_1)^2}$

$d = \sqrt{(5 - 1)^2 + (5 - 2)^2}$

$ = \sqrt{4^2 + 3^2} = \sqrt{16 + 9} = \sqrt{25} = 5$

37. Find the distance between the points $(-4, 0)$ and $(2, 3)$. Round your answers to two decimal places.

Solution

$d = \sqrt{(x_2 - x_1)^2 + (y_2 - y_1)^2}$

$d = \sqrt{[2 - (-4)]^2 + (3 - 0)^2}$

$ = \sqrt{6^2 + 3^2} = \sqrt{36 + 9} = \sqrt{45}$

$ = \sqrt{9}\sqrt{5} = 3\sqrt{5} \approx 6.71$

39. *Height of a Ladder*   A ladder is 15 feet long, and the bottom of the ladder is 3 feet from the wall of a house (see figure in textbook). How far does the ladder reach up the side of the house?

Solution

$a^2 + b^2 = c^2$

$x^2 + 3^2 = 15^2$

$x^2 + 9 = 225$

$x^2 = 216$

$x = \sqrt{216} = \sqrt{36 \cdot 6} = 6\sqrt{6}$

$x \approx 14.70$

The ladder reaches approximately 14.70 feet up the side of the house.

41. Factor $x^2 - x$ completely.

Solution

$x^2 - x = x(x - 1)$

43. Factor $x^2 + x - 42$ completely.

Solution

$x^2 + x - 42 = (x + 7)(x - 6)$

45. *Retirement Plan*   You budget 5% of your annual income of $45,600 for an individual retirement plan. How much do you put in your plan each year?

Solution

Verbal model: $\boxed{\begin{array}{c}\text{Amount in budget}\\\text{for retirement}\end{array}} = 0.05 \boxed{\begin{array}{c}\text{Annual}\\\text{income}\end{array}}$

Labels: Amount in budget for retirement $= x$ (dollars)
                 Annual income $= 45,600$ (dollars)

Equation: $x = 0.05(45,600)$
                 $x = 2280$

The amount put in the retirement plan each year is $2280.

47. Solve the equation $\sqrt{y} - 5 = 0$.

Solution

$$\sqrt{y} - 5 = 0$$
$$\sqrt{y} = 5$$
$$\left(\sqrt{y}\right)^2 = 5^2$$
$$y = 25$$

51. Solve the equation $\sqrt{10x} = 100$.

Solution

$$\sqrt{10x} = 100$$
$$\left(\sqrt{10x}\right)^2 = (100)^2$$
$$10x = 10{,}000$$
$$x = 1000$$

55. Solve the equation $5\sqrt{x+1} = 6$.

Solution

$$5\sqrt{x+1} = 6$$
$$\left(5\sqrt{x+1}\right)^2 = 6^2$$
$$25(x+1) = 36$$
$$25x + 25 = 36$$
$$25x = 11$$
$$x = \tfrac{11}{25}$$

59. Solve the equation $\sqrt{x^2 + 5} = x + 1$.

Solution

$$\sqrt{x^2 + 5} = x + 1$$
$$\left(\sqrt{x^2 + 5}\right)^2 = (x + 1)^2$$
$$x^2 + 5 = x^2 + 2x + 1$$
$$5 = 2x + 1$$
$$4 = 2x$$
$$2 = x$$

49. Solve the equation $\sqrt{a + 3} = 20$.

Solution

$$\sqrt{a + 3} = 20$$
$$\left(\sqrt{a + 3}\right)^2 = (20)^2$$
$$a + 3 = 400$$
$$a = 397$$

53. Solve the equation $\sqrt{3y + 5} = 2$.

Solution

$$\sqrt{3y + 5} = 2$$
$$\left(\sqrt{3y + 5}\right)^2 = 2^2$$
$$3y + 5 = 4$$
$$3y = -1$$
$$y = -\tfrac{1}{3}$$

57. Solve the equation $\sqrt{1 - x} - 3 = 2$.

Solution

$$\sqrt{1 - x} - 3 = 2$$
$$\sqrt{1 - x} = 5$$
$$(\sqrt{1 - x})^2 = 5^2$$
$$1 - x = 25$$
$$-x = 24$$
$$x = -24$$

61. Solve the equation $\sqrt{5x - 1} = 3\sqrt{x}$.

Solution

$$\sqrt{5x - 1} = 3\sqrt{x}$$
$$\left(\sqrt{5x - 1}\right)^2 = \left(3\sqrt{x}\right)^2$$
$$5x - 1 = 9x$$
$$-1 = 4x$$
$$-\tfrac{1}{4} = x \qquad \text{Trial solution}$$

This answer does *not* check because when $x = -\tfrac{1}{4}$, the radicands are negative. Thus, the equation has *no* solution.

63. Solve the equation $\sqrt{3x - 4} = 2\sqrt{x}$.

Solution

$$\sqrt{3x - 4} = 2\sqrt{x}$$

$$(\sqrt{3x - 4})^2 = (2\sqrt{x})^2$$

$$3x - 4 = 4x$$

$$-4 = x \qquad \text{Trial solution}$$

This answer does *not* check because when $x = -4$, the radicands are negative. Thus, the equation has no solution.

65. Solve the equation $\sqrt{3t + 11} = 5\sqrt{t}$.

Solution

$$\sqrt{3t + 11} = 5\sqrt{t}$$

$$(\sqrt{3t + 11})^2 = (5\sqrt{t})^2$$

$$3t + 11 = 25t$$

$$-22t + 11 = 0$$

$$-22t = -11$$

$$t = \tfrac{1}{2}$$

67. Solve the equation $\sqrt{x} = \tfrac{1}{4}x + 1$.

Solution

$$\sqrt{x} = \tfrac{1}{4}x + 1$$

$$(\sqrt{x})^2 = \left(\tfrac{1}{4}x + 1\right)^2$$

$$x = \left(\tfrac{1}{4}x\right)^2 + 2\left(\tfrac{1}{4}x\right)(1) + 1^2 \qquad (a + b)^2 = a^2 + 2ab + b^2$$

$$x = \tfrac{1}{16}x^2 + \tfrac{1}{2}x + 1$$

$$16x = 16\left(\tfrac{1}{16}x^2 + \tfrac{1}{2}x + 1\right)$$

$$16x = x^2 + 8x + 16$$

$$0 = x^2 - 8x + 16$$

$$0 = (x - 4)^2$$

$$0 = x - 4 \rightarrow x = 4$$

69. Solve the equation $x = \sqrt{18 - 3x}$.

Solution

$$x = \sqrt{18 - 3x}$$

$$x^2 = (\sqrt{18 - 3x})^2$$

$$x^2 = 18 - 3x$$

$$x^2 + 3x - 18 = 0$$

$$(x + 6)(x - 3) = 0$$

$$x + 6 = 0 \rightarrow x = -6 \qquad \text{Extraneous}$$

$$x - 3 = 0 \rightarrow x = 3$$

71. *Graphical and Algebraic Approaches*   Use a graphing utility to graph the function $y = \sqrt{x-1} - 2$ and estimate its x-intercepts. Set $y = 0$ and solve the resulting radical equation. Compare the result with the x-intercepts of the graph.

Solution

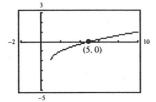

The y-intercept appears to be $(5, 0)$.

$$y = \sqrt{x-1} - 2$$
$$0 = \sqrt{x-1} - 2$$
$$2 = \sqrt{x-1}$$
$$2^2 = (\sqrt{x-1})^2$$
$$4 = x - 1$$
$$5 = x$$

This verifies that the x-intercept is $(5, 0)$.

73. *Graphical and Algebraic Approaches*   Use a graphing utility to graph the function $y = x - \sqrt{4x+5}$ and estimate its x-intercepts. Set $y = 0$ and solve the resulting radical equation. Compare the result with the x-intercepts of the graph.

Solution

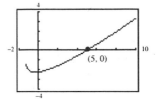

The y-intercept appears to be $(5, 0)$.

$$y = x - \sqrt{4x+5}$$
$$0 = x - \sqrt{4x+5}$$
$$-x = -\sqrt{4x+5}$$
$$(-x)^2 = (-\sqrt{4x+5})^2$$
$$x^2 = 4x + 5$$
$$x^2 - 4x - 5 = 0$$
$$(x-5)(x+1) = 0$$
$$x - 5 = 0 \rightarrow x = 5$$
$$x + 1 = 0 \rightarrow x = -1 \qquad \text{Extraneous}$$

This verifies that the x-intercept is $(5, 0)$.

75. *Geometry*   Using the figure in the textbook, solve for x. Round your answer to two decimal places.

Solution

$$a^2 + b^2 = c^2$$
$$x^2 + 4^2 = 7^2$$
$$x^2 + 16 = 49$$
$$x^2 = 33$$
$$x = \sqrt{33}$$
$$x \approx 5.74$$

77. *Geometry*   Using the figure in the textbook, solve for x. Round your answer to two decimal places.

Solution

$$c^2 = a^2 + b^2$$
$$x^2 = 10^2 + (x-2)^2$$
$$x^2 = 100 + x^2 - 4x + 4$$
$$x^2 = x^2 - 4x + 104$$
$$0 = -4x + 104$$
$$4x = 104$$
$$x = 26$$

79. Find the distance between the points $(4, 5)$ and $(7, 2)$. Round your result to two decimal places.

Solution

$$d = \sqrt{(x_2 - x_1)^2 + (y_2 - y_1)^2}$$

$$d = \sqrt{(7 - 4)^2 + (2 - 5)^2}$$

$$= \sqrt{3^2 + (-3)^2} = \sqrt{9 + 9} = \sqrt{18}$$

$$= \sqrt{9}\sqrt{2} = 3\sqrt{2} \approx 4.24$$

81. Find the distance between the points $(3, -2)$ and $(4, 6)$. Round your result to two decimal places.

Solution

$$d = \sqrt{(x_2 - x_1)^2 + (y_2 - y_1)^2}$$

$$d = \sqrt{(4 - 3)^2 + [6 - (-2)]^2}$$

$$= \sqrt{1^2 + 8^2} = \sqrt{1 + 64} = \sqrt{65} \approx 8.06$$

83. *Geometry* A volleyball court is 30 feet wide and 60 feet long. Find the length of the diagonal of the court.

Solution

$$c^2 = a^2 + b^2$$

$$c^2 = 30^2 + 60^2 = 900 + 3600 = 4500$$

$$c = \sqrt{4500} = \sqrt{900}\sqrt{5} = 30\sqrt{5} \approx 67.08$$

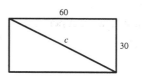

The length of the diagonal is $30\sqrt{5}$ feet, or approximately 67.08 feet.

85. *Geometry* A 12-foot plank is used to brace a basement wall during construction of a home. The plank is nailed to the wall 4 feet above the floor (see figure in textbook). Find the slope of the plank.

Solution

$$\text{Slope} = \frac{\text{Rise}}{\text{Run}}$$

$$\text{Rise} = 4 \text{ feet}$$

$$\text{Run} = x \text{ feet}$$

$$x^2 + 4^2 = 12^2$$

$$x^2 + 16 = 144$$

$$x^2 = 128$$

$$x = \sqrt{128} = \sqrt{64}\sqrt{2} = 8\sqrt{2}$$

$$\text{Slope} = \frac{\text{Rise}}{\text{Run}}$$

$$= \frac{4}{8\sqrt{2}} = \frac{1}{2\sqrt{2}} = \frac{1}{2\sqrt{2}} \cdot \frac{\sqrt{2}}{\sqrt{2}} = \frac{\sqrt{2}}{2(2)} = \frac{\sqrt{2}}{4}$$

The slope of the plank is $\dfrac{\sqrt{2}}{4}$.

87. *Velocity*   An object is dropped from a height of 30 feet. Find the velocity of the object when it strikes the ground. Use the equation for the velocity of a free-falling object, $v = \sqrt{2gh}$, where v is measured in feet per second, $g = 32$ feet per second squared, and h is the height in feet.

Solution

$$v = \sqrt{2gh}$$

$$v = \sqrt{2 \cdot 32 \cdot 30} = \sqrt{1920}$$

$$= \sqrt{64 \cdot 30} = 8\sqrt{30}$$

$$v \approx 43.82$$

The velocity is approximately 43.82 feet per second.

89. *Estimation*   An object strikes the ground with a velocity of 75 feet per second. Estimate the height from which it was dropped. Use the equation for the velocity of a free-falling object, $v = \sqrt{2gh}$, where v is measured in feet per second, $g = 32$ feet per second squared, and h is the height in feet.

Solution

$$v = \sqrt{2gh}$$

$$75 = \sqrt{2 \cdot 32 \cdot h}$$

$$75 = \sqrt{64h}$$

$$(75)^2 = \left(\sqrt{64h}\right)^2$$

$$5625 = 64h$$

$$\frac{5625}{64} = h$$

$$87.89 \approx h$$

The height was approximately 87.89 feet.

91. *Pendulum Length*   How long is the pendulum of a grandfather clock that has a period of two seconds? (See figure in textbook.) The time t in seconds for a pendulum of length L in feet to move through one complete cycle (its period) is given by

$$t = 2\pi\sqrt{\frac{L}{32}}.$$

Solution

$$t = 2\pi\sqrt{\frac{L}{32}}$$

$$2 = 2\pi\sqrt{\frac{L}{32}}$$

$$2^2 = \left(2\pi\sqrt{\frac{L}{32}}\right)^2$$

$$4 = 4\pi^2 \cdot \frac{L}{32} = \frac{\pi^2 L}{8}$$

$$8 \cdot 4 = \pi^2 L$$

$$32 = \pi^2 L$$

$$\frac{32}{\pi^2} = L$$

$$3.24 \approx L$$

The pendulum is approximately 3.24 feet long.

93. *Using a Model*   An airline offers daily flights between Chicago and Denver. The total monthly cost of these flights is modeled by $C = \sqrt{0.2x + 1}$ where C is measured in millions of dollars and x is measured in thousands of passengers. The total cost for a certain month was 2.5 million dollars. About how many passengers flew that month?

Solution

$$C = \sqrt{0.2x + 1}$$

$$2.5 = \sqrt{0.2x + 1}$$

$$(2.5)^2 = \left(\sqrt{0.2x + 1}\right)^2$$

$$6.25 = (0.2x + 1) = 0.2x + 1$$

$$5.25 = 0.2x$$

$$\frac{5.25}{0.2} = x$$

$$26.25 = x$$

Approximately 26.25 *thousand* passengers (or 26,250 passengers) flew that month.

95. *Geometry*   Use the formula $S = \pi r \sqrt{r^2 + h^2}$ to approximate the amount of material in an ice cream cone that has a height of 5 inches and a radius of 1.5 inches (see figure in textbook). Assume that the material in the cone is $\frac{1}{8}$ inch thick.

Solution

$S = \pi r \sqrt{r^2 + h^2}$

$S = \pi(1.5)\sqrt{(1.5)^2 + 5^2} = \pi(1.5)\sqrt{2.25 + 25} = \pi(1.5)\sqrt{27.25}$

$S \approx 24.60$

The surface area of the cone is approximately 24.60 square inches. Since the material in the cone is $\frac{1}{8}$-inch thick, the amount of material in the cone is approximately $\frac{1}{8}(24.60)$ or 3.07 cubic inches.

97. *Think About It*   Approximate the dimensions of the rectangle in Exercise 96 if the length of the diagonal is 12 inches. (See figure in textbook.) How do the dimensions change when the length of the diagonal increases? The length of the diagonal must be less than what real number?

Solution

The perimeter of the rectangle is 28 inches, so the sum of the length and width of the rectangle must be 14 inches. If the diagonal of the rectangle is 12 inches, the sum of the square of the length and the square of the width must be 144. Experimenting with pairs of numbers with a sum of 14, and looking for the sum of the squares to be near 144:

$8^2 + 6^2 = 64 + 36 = 100$

$9^2 + 5^2 = 81 + 25 = 106$

$10^2 + 4^2 = 100 + 16 = 116$

$11^2 + 3^2 = 121 + 9 = 130$

$12^2 + 2^2 = 144 + 4 = 148$

It appears we could try lengths between 11 and 12 inches:

$11.2^2 + 2.8^2 = 125.44 + 7.84 = 133.28$

$11.4^2 + 2.6^2 = 129.96 + 6.76 = 136.72$

$11.6^2 + 2.4^2 = 134.56 + 5.76 = 140.32$

$11.8^2 + 2.2^2 = 139.24 + 4.84 = 144.08$

The length and width of the rectangle would be approximately 11.8 inches and 2.2 inches. As the diagonal increases, the length of the rectangle increases and the width decreases. The diagonal must be less than 14 inches.

Note: The next chapter of the textbook describes methods of solving equations which would enable us to answer this question directly instead of using an experimental method.

Review Exercises for Chapter 10

1. Evaluate $\sqrt{121}$, if possible. (Do not use a calculator.)

Solution

$\sqrt{121} = 11$

3. Evaluate $\sqrt{1.44}$, if possible. (Do not use a calculator.)

Solution

$\sqrt{1.44} = 1.2$

5. Evaluate $\sqrt{-\frac{1}{16}}$, if possible. (Do not use a calculator.)

Solution

A negative number has no square root in the real numbers.

7. Evaluate $\sqrt{100 - 36}$, if possible. (Do not use a calculator.)

Solution

$\sqrt{100 - 36} = \sqrt{64} = 8$

9. Use a calculator to approximate $\sqrt{53}$. (Round your answer to two decimal places.)

Solution

Keystrokes

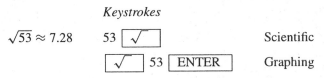

$\sqrt{53} \approx 7.28$

11. Use a calculator to approximate $\sqrt{\frac{7}{8}}$. (Round your answer to two decimal places.)

Solution

Keystrokes

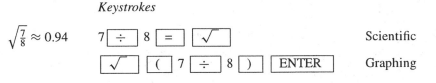

$\sqrt{\frac{7}{8}} \approx 0.94$

13. Use a calculator to approximate $\sqrt{9^2 - 4(2)(7)}$. (Round your answer to two decimal places.)

Solution

Keystrokes

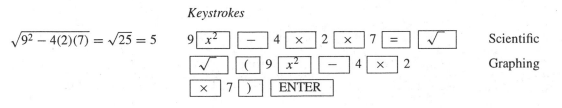

$\sqrt{9^2 - 4(2)(7)} = \sqrt{25} = 5$

15. Use a calculator to approximate $5\sqrt{23.5} - 13.2$. (Round your answer to two decimal places.)

Solution

Keystrokes

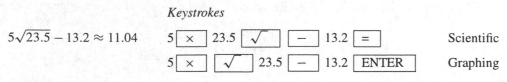

$5\sqrt{23.5} - 13.2 \approx 11.04$ 5 ☒ × 23.5 ☒ √ ☒ − 13.2 ☒ =   Scientific

5 ☒ × ☒ √ 23.5 ☒ − 13.2 ☒ ENTER   Graphing

17. Use the area $A = 92.16$ to find the length of the side of a square when $s = \sqrt{A}$.

Solution

$$s = \sqrt{A}$$
$$= \sqrt{92.16}$$
$$= 9.6$$

19. Use the area $A = 28.27$ to find the length of the radius of a circle when $r = \sqrt{A/\pi}$.

Solution

$$r = \sqrt{\frac{A}{\pi}}$$
$$\approx \sqrt{\frac{28.27}{3.14}}$$
$$\approx \sqrt{9}$$
$$\approx 3$$

21. Simplify $\sqrt{48}$.

Solution

$$\sqrt{48} = \sqrt{16 \cdot 3} = 4\sqrt{3}$$

23. Simplify $\sqrt{160}$.

Solution

$$\sqrt{160} = \sqrt{16 \cdot 10} = 4\sqrt{10}$$

25. Simplify $\sqrt{\frac{23}{9}}$.

Solution

$$\sqrt{\frac{23}{9}} = \frac{\sqrt{23}}{\sqrt{9}} = \frac{\sqrt{23}}{3}$$

27. Simplify $\sqrt{\frac{20}{9}}$.

Solution

$$\sqrt{\frac{20}{9}} = \frac{\sqrt{20}}{\sqrt{9}} = \frac{\sqrt{4 \cdot 5}}{3} = \frac{2\sqrt{5}}{3}$$

29. Simplify $\sqrt{36x^4}$.

Solution

$$\sqrt{36x^4} = \sqrt{36 \cdot x^4} = 6x^2$$

31. Simplify $\sqrt{4y^3}$.

Solution

$$\sqrt{4y^3} = \sqrt{4 \cdot y^2 \cdot y} = 2y\sqrt{y}$$

33. Simplify $\sqrt{0.04x^2y}$.

Solution

$$\sqrt{0.04x^2y} = \sqrt{0.04 \cdot x^2 \cdot y}$$
$$= 0.2|x|\sqrt{y}$$

35. Simplify $\sqrt{32a^3b}$.

Solution

$$\sqrt{32a^3b} = \sqrt{16 \cdot 2 \cdot a^2 \cdot ab} = 4|a|\sqrt{2ab}$$

37. Simplify $\sqrt{1.21 \times 10^8}$.

Solution

$$\sqrt{1.21 \times 10^8} = \sqrt{1.21} \times \sqrt{10^8}$$
$$= 1.1 \times 10^4$$

39. Simplify $\sqrt{2.25 \times 10^{-6}}$.

Solution

$$\sqrt{2.25 \times 10^{-6}} = \sqrt{2.25} \times \sqrt{10^{-6}}$$
$$= 1.5 \times 10^{-3}$$

41. Rationalize the denominator and simplify $\sqrt{\frac{3}{5}}$.

Solution

$$\sqrt{\frac{3}{5}} = \frac{\sqrt{3}}{\sqrt{5}} \cdot \frac{\sqrt{5}}{\sqrt{5}} = \frac{\sqrt{15}}{5}$$

43. Rationalize the denominator and simplify $\frac{4}{\sqrt{12}}$.

Solution

$$\frac{4}{\sqrt{12}} = \frac{4}{\sqrt{12}} \cdot \frac{\sqrt{3}}{\sqrt{3}} = \frac{4\sqrt{3}}{\sqrt{36}} = \frac{4\sqrt{3}}{6} = \frac{2\sqrt{3}}{3}$$

45. Rationalize the denominator and simplify $\frac{9}{\sqrt{7}-4}$.

Solution

$$\frac{9}{\sqrt{7}-4} = \frac{9}{\sqrt{7}-4} \cdot \frac{\sqrt{7}+4}{\sqrt{7}+4}$$

$$= \frac{9(\sqrt{7}+4)}{(\sqrt{7})^2 - 4^2}$$

$$= \frac{9(\sqrt{7}+4)}{7-16}$$

$$= \frac{9(\sqrt{7}+4)}{-9}$$

$$= -(\sqrt{7}+4) \text{ or } -\sqrt{7}-4$$

47. Rationalize the denominator simplify $\frac{3}{\sqrt{x}}$.

Solution

$$\frac{3}{\sqrt{x}} = \frac{3}{\sqrt{x}} \cdot \frac{\sqrt{x}}{\sqrt{x}} = \frac{3\sqrt{x}}{x}$$

49. Rationalize the denominator and simplify $\sqrt{\frac{11a}{b}}$.

Solution

$$\sqrt{\frac{11a}{b}} = \frac{\sqrt{11a}}{\sqrt{b}} \cdot \frac{\sqrt{b}}{\sqrt{b}} = \frac{\sqrt{11ab}}{|b|}$$

51. Rationalize the denominator and simplify $\frac{\sqrt{8x^2}}{\sqrt{2}}$.

Solution

$$\frac{\sqrt{8x^2}}{\sqrt{2}} = \sqrt{\frac{8x^2}{2}} = \sqrt{4x^2} = \sqrt{4 \cdot x^2} = 2|x|$$

53. Perform the indicated operations and simplify $7\sqrt{2} + 5\sqrt{2}$.

Solution

$$7\sqrt{2} + 5\sqrt{2} = (7+5)\sqrt{2} = 12\sqrt{2}$$

55. Perform the indicated operations and simplify $3\sqrt{20} - 10\sqrt{20}$.

Solution

$$3\sqrt{20} - 10\sqrt{20} = (3-10)\sqrt{20}$$

$$= -7\sqrt{4 \cdot 5} = -7 \cdot 2\sqrt{5} = -14\sqrt{5}$$

57. Perform the indicated operations and simplify $4\sqrt{48} + 2\sqrt{3} - 5\sqrt{12}$.

Solution

$$4\sqrt{48} + 2\sqrt{3} - 5\sqrt{12} = 4\sqrt{16 \cdot 3} + 2\sqrt{3} - 5\sqrt{4 \cdot 3}$$

$$= 4 \cdot 4\sqrt{3} + 2\sqrt{3} - 5 \cdot 2\sqrt{3}$$

$$= 16\sqrt{3} + 2\sqrt{3} - 10\sqrt{3}$$

$$= (16 + 2 - 10)\sqrt{3}$$

$$= 8\sqrt{3}$$

59. Perform the indicated operations and simplify $10\sqrt{y+3} - 3\sqrt{y+3}$.

Solution

$$10\sqrt{y+3} - 3\sqrt{y+3} = (10-3)\sqrt{y+3}$$

$$= 7\sqrt{y+3}$$

61. Perform the indicated operations and simplify $\sqrt{25x} + \sqrt{49x}$.

Solution

$$\sqrt{25x} + \sqrt{49x} = 5\sqrt{x} + 7\sqrt{x}$$
$$= (5+7)\sqrt{x}$$
$$= 12\sqrt{x}$$

63. Perform the indicated operations and simplify $\sqrt{3}(\sqrt{6} - 1)$.

Solution

$$\sqrt{3}(\sqrt{6} - 1) = \sqrt{18} - \sqrt{3}$$
$$= \sqrt{9}\sqrt{2} - \sqrt{3}$$
$$= 3\sqrt{2} - \sqrt{3}$$

65. Perform the indicated operations and simplify $(\sqrt{5} - 2)^2$.

Solution

$$(\sqrt{5} - 2)^2 = (\sqrt{5} - 2)(\sqrt{5} - 2)$$
$$= \sqrt{5} \cdot \sqrt{5} - 2\sqrt{5} - 2\sqrt{5} + 4$$
$$= 5 - 4\sqrt{5} + 4$$
$$= 9 - 4\sqrt{5}$$

We could use the pattern for the square of a binomial.

$$(\sqrt{5} - 2)^2 = (\sqrt{5})^2 - 2(\sqrt{5})(2) + 2^2 \qquad (a-b)^2 = a^2 - 2ab + b^2$$
$$= 5 - 4\sqrt{5} + 4$$
$$= 9 - 4\sqrt{5}$$

67. Perform the indicated operations and simplify $(\sqrt{8} + 2)(3\sqrt{2} - 1)$.

Solution

$$(\sqrt{8} + 2)(3\sqrt{2} - 1) = \sqrt{8} \cdot 3\sqrt{2} - \sqrt{8} + 2 \cdot 3\sqrt{2} - 2$$
$$= 3\sqrt{16} - \sqrt{4 \cdot 2} + 6\sqrt{2} - 2$$
$$= 3 \cdot 4 - 2\sqrt{2} + 6\sqrt{2} - 2$$
$$= 12 + 4\sqrt{2} - 2$$
$$= 10 + 4\sqrt{2}$$

69. Perform the indicated operations and simplify $\sqrt{x}(\sqrt{x} + 10)$.

Solution

$$\sqrt{x}(\sqrt{x} + 10) = \sqrt{x} \cdot \sqrt{x} + \sqrt{x} \cdot 10 = x + 10\sqrt{x}$$

71. Perform the indicated operations and simplify $\dfrac{3}{\sqrt{12}-3}$.

Solution

$$\dfrac{3}{\sqrt{12}-3} = \dfrac{3}{\sqrt{12}-3} \cdot \dfrac{\sqrt{12}+3}{\sqrt{12}+3}$$

$$= \dfrac{3(\sqrt{12}+3)}{(\sqrt{12})^2 - 3^2}$$

$$= \dfrac{3(\sqrt{12}+3)}{12-9}$$

$$= \dfrac{3(\sqrt{12}+3)}{3}$$

$$= \sqrt{12} + 3$$

$$= \sqrt{4 \cdot 3} + 3$$

$$= 2\sqrt{3} + 3$$

73. Perform the indicated operations and simplify $\dfrac{\sqrt{6}}{\sqrt{2}+\sqrt{3}}$.

Solution

$$\dfrac{\sqrt{6}}{\sqrt{2}+\sqrt{3}} = \dfrac{\sqrt{6}}{\sqrt{2}+\sqrt{3}} \cdot \dfrac{\sqrt{2}-\sqrt{3}}{\sqrt{2}-\sqrt{3}}$$

$$= \dfrac{\sqrt{12}-\sqrt{18}}{(\sqrt{2})^2 - (\sqrt{3})^2}$$

$$= \dfrac{\sqrt{4}\sqrt{3}-\sqrt{9}\sqrt{2}}{2-3}$$

$$= \dfrac{2\sqrt{3}-3\sqrt{2}}{-1}$$

$$= -2\sqrt{3} + 3\sqrt{2}$$

75. Perform the indicated operations and simplify $(\sqrt{x}-3) \div (\sqrt{x}+3)$.

Solution

$$(\sqrt{x}-3) \div (\sqrt{x}+3) = \dfrac{\sqrt{x}-3}{\sqrt{x}+3}$$

$$= \dfrac{\sqrt{x}-3}{\sqrt{x}+3} \cdot \dfrac{\sqrt{x}-3}{\sqrt{x}-3}$$

$$= \dfrac{(\sqrt{x}-3)(\sqrt{x}-3)}{(\sqrt{x})^2 - 3^2}$$

$$= \dfrac{\sqrt{x} \cdot \sqrt{x} - 3\sqrt{x} - 3\sqrt{x} + 9}{x-9}$$

$$= \dfrac{x - 6\sqrt{x} + 9}{x-9}$$

77. Solve $\sqrt{y} = 13$.

Solution

$$\sqrt{y} = 13$$

$$(\sqrt{y})^2 = 13^2$$

$$y = 169$$

79. Solve $\sqrt{x} + 2 = 0$.

Solution

$$\sqrt{x} + 2 = 0$$

$$\sqrt{x} = -2$$

$$(\sqrt{x})^2 = (-2)^2$$

$$x = 4 \qquad \text{Trial solution}$$

However, this answer does *not* check.

$$\sqrt{4} + 2 \overset{?}{=} 0$$

$$2 + 2 \neq 0$$

Thus, the equation has *no* solution.

81. Solve $\sqrt{t+3} = 4$.

Solution

$$\sqrt{t+3} = 4$$
$$\left(\sqrt{t+3}\right)^2 = 4^2$$
$$t+3 = 16$$
$$t = 13$$

83. Solve $\sqrt{2x-3} = \sqrt{x}$.

Solution

$$\sqrt{2x-3} = \sqrt{x}$$
$$\left(\sqrt{2x-3}\right)^2 = \left(\sqrt{x}\right)^2$$
$$2x-3 = x$$
$$x-3 = 0$$
$$x = 3$$

85. Solve $x = 2\sqrt{3-x}$.

Solution

$$x = 2\sqrt{3-x}$$
$$x^2 = (2\sqrt{3-x})^2$$
$$x^2 = 4(3-x)$$
$$x^2 = 12 - 4x$$
$$x^2 + 4x - 12 = 0$$
$$(x+6)(x-2) = 0$$
$$x+6 = 0 \Rightarrow x = -6 \quad \text{Extraneous}$$
$$x-2 = 0 \Rightarrow x = 2$$

87. Solve $\sqrt{2x} = \frac{1}{2}x + 1$.

Solution

$$\sqrt{2x} = \frac{1}{2}x + 1$$
$$\left(\sqrt{2x}\right)^2 = \left(\frac{1}{2}x + 1\right)^2$$
$$2x = \left(\frac{1}{2}x\right)^2 + 2\left(\frac{1}{2}x\right)(1) + 1^2 \quad (a+b)^2 = a^2 + 2ab + b^2$$
$$2x = \frac{1}{4}x^2 + x + 1$$
$$4(2x) = 4\left(\frac{1}{4}x^2 + x + 1\right)$$
$$8x = x^2 + 4x + 4$$
$$0 = x^2 - 4x + 4$$
$$0 = (x-2)^2$$
$$x-2 = 0 \quad \rightarrow \quad x = 2$$

89. Use a graphing utility to graph $y = \sqrt{x+3} - 4\sqrt{x}$. Use the graph to estimate the x-intercept of the graph. Set $y = 0$ and solve the resulting radical equation. Compare the result with the x-intercepts of the graph.

Solution

The x-intercept appears to be $(0.2, 0)$.

$$y = \sqrt{x+3} - 4\sqrt{x}$$

$$0 = \sqrt{x+3} - 4\sqrt{x}$$

$$4\sqrt{x} = \sqrt{x+3}$$

$$(4\sqrt{x})^2 = (\sqrt{x+3})^2$$

$$16x = x + 3$$

$$15x = 3$$

$$x = \tfrac{3}{15} = \tfrac{1}{5}$$

This verifies that the x-intercept is $\left(\tfrac{1}{5}, 0\right)$.

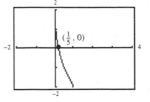

91. *Geometry*   Using the figure in the textbook, solve for x. Round your answer to two decimal places.

Solution

$$c^2 = a^2 + b^2$$

$$x^2 = (4\sqrt{3})^2 + 4^2 = 16(3) + 16 = 48 + 16 = 64$$

$$x = 8$$

93. *Geometry*   Using the figure in the textbook, solve for x. Round your answer to two decimal places.

Solution

$$a^2 + b^2 = c^2$$

$$x^2 + 3^2 = (3\sqrt{2})^2$$

$$x^2 + 9 = 9(2) = 18$$

$$x^2 = 9$$

$$x = 3$$

95. Find the distance between the two points on the graph in the textbook. Give the exact distance and the distance rounded to two decimal places.

Solution

$$d = \sqrt{(x_2 - x_1)^2 + (y_2 - y_1)^2}$$

$$d = \sqrt{[1 - (-3)]^2 + (5 - 1)^2}$$

$$= \sqrt{4^2 + 4^2} = \sqrt{16 + 16} = \sqrt{32} = \sqrt{16}\sqrt{2} = 4\sqrt{2} \approx 5.66$$

97. Find the distance between the two points on the graph in the textbook. Give the exact distance and the distance rounded to two decimal places.

Solution

$$d = \sqrt{(x_2 - x_1)^2 + (y_2 - y_1)^2}$$

$$d = \sqrt{[4 - (-3)]^2 + [1 - (-2)]^2}$$

$$= \sqrt{7^2 + 3^2} = \sqrt{49 + 9} = \sqrt{58} \approx 7.62$$

99. *Wire Length*   A guy wire on a radio tower is attached to the top of the tower and to an anchor 60 feet from the base of the tower (see figure in textbook). The tower is 100 feet high. How long is the wire?

Solution

$$c^2 = a^2 + b^2$$

$$x^2 = (60)^2 + (100)^2 = 3600 + 10,000 = 13,600$$

$$x = \sqrt{13,600} = \sqrt{400 \cdot 34} = 20\sqrt{34} \approx 116.62$$

The wire is approximately 116.62 feet long.

101. *Geometry*   Determine the length and width of a rectangle that has a perimeter of 70 inches and a diagonal length of 25 inches.

Solution

Verbal model:

$$\boxed{2} \cdot \boxed{\begin{array}{c}\text{Length of}\\ \text{rectangle}\end{array}} + \boxed{2} \cdot \boxed{\begin{array}{c}\text{Width of}\\ \text{rectangle}\end{array}} = \boxed{\begin{array}{c}\text{Perimeter of}\\ \text{rectangle}\end{array}}$$

$$\left(\boxed{\begin{array}{c}\text{Length of}\\ \text{rectangle}\end{array}}\right)^2 + \left(\boxed{\begin{array}{c}\text{Width of}\\ \text{rectangle}\end{array}}\right)^2 = \left(\boxed{\begin{array}{c}\text{Perimeter of}\\ \text{rectangle}\end{array}}\right)^2$$

Labels: Length of rectangle $= x$ (inches)
Width of rectangle $= y$ (inches)
Perimeter of rectangle $= 70$ (inches)
Diagonal of rectangle $= 25$ (inches)

System of equations: $2x + 2y = 70$

$$x^2 + y^2 = 25^2$$

Solving by substitution: $2x + 2y = 70 \quad \rightarrow \quad 2x = 70 - 2y \quad \rightarrow \quad x = 35 - y$

$$x^2 + y^2 = 25^2$$

$$(35 - y)^2 + y^2 = 25^2 \qquad \text{Replace } x \text{ by } 35 - y \text{ in second equation.}$$

$$1225 - 70y + y^2 + y^2 = 625$$

$$2y^2 - 70y + 600 = 0$$

$$y^2 - 35y + 300 = 0 \qquad \text{Divide both sides by 2.}$$

$$(y - 15)(y - 20) = 0$$

$$y - 15 = 0 \quad \rightarrow \quad y = 15 \text{ and } x = 35 - 15 = 20$$

$$y - 20 = 0 \quad \rightarrow \quad y = 20 \text{ and } x = 35 - 20 = 15$$

The dimensions of the rectangle are 20 inches by 15 inches.

103. *Pendulum Length*   The time t in seconds for a pendulum of length L in feet to go through one complete cycle (its period) is

$$t = 2\pi \sqrt{\frac{L}{32}}.$$

How long is the pendulum of a grandfather clock with a period of 1.75 seconds?

Solution

$$t = 2\pi \sqrt{\frac{L}{32}}$$

$$1.75 = 2\pi \sqrt{\frac{L}{32}}$$

$$(1.75)^2 = \left(2\pi \sqrt{\frac{L}{32}}\right)^2$$

$$(1.75)^2 = 4\pi^2 \cdot \frac{L}{32}$$

$$(1.75)^2 = \frac{\pi^2 L}{8}$$

$$8(1.75)^2 = \pi^2 L$$

$$\frac{8(1.75)^2}{\pi^2} = L$$

$$2.48 \approx L$$

The pendulum is approximately 2.48 feet long.

Test for Chapter 10

1. If possible, evaluate each of the following without a calculator. If it is not possible, explain why.

(a) $\sqrt{\dfrac{9}{16}}$
                                                           (b) $\sqrt{-36}$

Solution

(a) $\sqrt{\dfrac{9}{16}} = \dfrac{\sqrt{9}}{\sqrt{16}} = \dfrac{3}{4}$
                                 (b) A negative number has no square root in the real numbers.

2. Use a calculator to approximate $12 - \sqrt{322}$. Round the result to three decimal places.

Solution

Keystrokes

$12 - \sqrt{322} \approx -5.944$ 12 [−] 322 [√] [=]   Scientific

                                         12 [−] [√] 322 [ENTER]   Graphing

3. Simplify: $\sqrt{48}$

Solution

$\sqrt{48} = \sqrt{16 \cdot 3} = 4\sqrt{3}$

4. Simplify: $\sqrt{32x^2 y^3}$

Solution

$\sqrt{32x^2 y^3} = \sqrt{16 \cdot 2 \cdot x^2 \cdot y^2 \cdot y} = 4|x|y\sqrt{2y}$

5. Simplify: $\dfrac{\sqrt{128}}{\sqrt{2}}$

Solution

$\dfrac{\sqrt{128}}{\sqrt{2}} = \sqrt{\dfrac{128}{2}} = \sqrt{64} = 8$

6. Simplify: $\sqrt{\dfrac{3x^3}{y^4}}$

Solution

$\sqrt{\dfrac{3x^3}{y^4}} = \dfrac{\sqrt{3 \cdot x^2 \cdot x}}{\sqrt{y^4}} = \dfrac{x\sqrt{3x}}{y^2}$

7. Rationalize the denominator: $\dfrac{5}{\sqrt{15}}$

Solution

$\dfrac{5}{\sqrt{15}} = \dfrac{5}{\sqrt{15}} \cdot \dfrac{\sqrt{15}}{\sqrt{15}}$

$= \dfrac{5\sqrt{15}}{15}$

$= \dfrac{\sqrt{15}}{3}$

8. Rationalize the denominator: $\dfrac{10}{\sqrt{6}+1}$

Solution

$\dfrac{10}{\sqrt{6}+1} = \dfrac{10}{\sqrt{6}+1} \cdot \dfrac{\sqrt{6}-1}{\sqrt{6}-1}$

$= \dfrac{10(\sqrt{6}-1)}{(\sqrt{6})^2 - 1^2}$

$= \dfrac{10(\sqrt{6}-1)}{6-1}$

$= \dfrac{10(\sqrt{6}-1)}{5}$

$= 2(\sqrt{6}-1) \text{ or } 2\sqrt{6}-2$

9. Simplify: $10\sqrt{27} - 7\sqrt{12}$

Solution

$10\sqrt{27} - 7\sqrt{12} = 10\sqrt{9 \cdot 3} - 7\sqrt{4 \cdot 3}$

$= 10 \cdot 3\sqrt{3} - 7 \cdot 2\sqrt{3}$

$= 30\sqrt{3} - 14\sqrt{3}$

$= 16\sqrt{3}$

10. Simplify: $\sqrt{\dfrac{x}{4}} - 3\sqrt{\dfrac{x}{25}}$

Solution

$\sqrt{\dfrac{x}{4}} - 3\sqrt{\dfrac{x}{25}} = \dfrac{\sqrt{x}}{\sqrt{4}} - \dfrac{3\sqrt{x}}{\sqrt{25}}$

$= \dfrac{\sqrt{x}}{2} - \dfrac{3\sqrt{x}}{5}$

$= \dfrac{1}{2}\sqrt{x} - \dfrac{3}{5}\sqrt{x}$

$= \left(\dfrac{1}{2} - \dfrac{3}{5}\right)\sqrt{x}$

$= \left(\dfrac{5}{10} - \dfrac{6}{10}\right)\sqrt{x}$

$= -\dfrac{1}{10}\sqrt{x} \text{ or } -\dfrac{\sqrt{x}}{10}$

11. Multiply and simplify: $\sqrt{2}(\sqrt{8}-5)$

Solution

$$\sqrt{2}(\sqrt{8}-5) = \sqrt{2}\cdot\sqrt{8} - \sqrt{2}\cdot 5$$
$$= \sqrt{16} - 5\sqrt{2}$$
$$= 4 - 5\sqrt{2}$$

12. Multiply and simplify: $(\sqrt{6}-3)(\sqrt{6}+5)$

Solution

$$(\sqrt{6}-3)(\sqrt{6}+5) = \sqrt{6}\cdot\sqrt{6} + 5\sqrt{6} - 3\sqrt{6} - 15$$
$$= 6 + 2\sqrt{6} - 15$$
$$= -9 + 2\sqrt{6}$$

13. Multiply and simplify: $(\sqrt{5}+4)^2$

Solution

$$(\sqrt{5}+4)^2 = (\sqrt{5}+4)(\sqrt{5}+4)$$
$$= \sqrt{5}\cdot\sqrt{5} + 4\sqrt{5} + 4\sqrt{5} + 16$$
$$= 5 + 8\sqrt{5} + 16$$
$$= 21 + 8\sqrt{5}$$

We could use the pattern for the square of a binomial.

$$(\sqrt{5}+4)^2 = (\sqrt{5})^2 + 2(\sqrt{5})(4) + 4^2 \qquad (a+b)^2 = a^2 + 2ab + b^2$$
$$= 5 + 8\sqrt{5} + 16$$
$$= 21 + 8\sqrt{5}$$

14. Multiply and simplify: $(1-2\sqrt{x})(-2\sqrt{x})$

Solution

$$(1-2\sqrt{x})(-2\sqrt{x}) = -2\sqrt{x} + 4\sqrt{x^2}$$
$$= -2\sqrt{x} + 4x$$

15. In your own words, explain what *conjugate* means. Determine the conjugate of the real number $\sqrt{3}-5$. Then find the product of the number and its conjugate.

Solution

(Written answers will vary.)
Number: $\sqrt{3}-5$
Conjugate: $\sqrt{3}+5$
Product: $(\sqrt{3})^2 - 5^2 = 3 - 25 = -22$

16. Solve: $\sqrt{y} = 4$

Solution

$$\sqrt{y} = 4$$
$$(\sqrt{y})^2 = 4^2$$
$$y = 16$$

17. Solve: $2\sqrt{x+3} = 5$

Solution

$$2\sqrt{x+3} = 5$$
$$(2\sqrt{x+3})^2 = 5^2$$
$$4(x+3) = 25$$
$$4x + 12 = 25$$
$$4x = 13$$
$$x = \frac{13}{4}$$

18. Solve: $2\sqrt{2x} = x + 2$

Solution

$$2\sqrt{2x} = x + 2$$

$$\left(2\sqrt{2x}\right)^2 = (x + 2)^2$$

$$4 \cdot 2x = x^2 + 4x + 4$$

$$8x = x^2 + 4x + 4$$

$$0 = x^2 - 4x + 4$$

$$0 = (x - 2)^2$$

$$x - 2 = 0 \quad \rightarrow \quad x = 2$$

19. Solve for c in the figure.

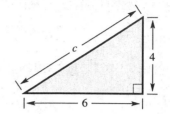

Solution

$$c^2 = a^2 + b^2$$

$$c^2 = 6^2 + 4^2$$

$$c^2 = 36 + 16$$

$$c^2 = 52$$

$$c = \sqrt{52}$$

$$c = \sqrt{4 \cdot 13}$$

$$c = 2\sqrt{13}$$

$$c \approx 7.21$$

Side c is approximately 7.21 units long.

20. The demand equation for a certain product is
$p = 100 - \sqrt{x - 25}$, where x is the number of units
demanded per day and p is the price per unit. Find the
demand when the price is $90.

Solution

$$p = 100 - \sqrt{x - 25}$$

$$90 = 100 - \sqrt{x - 25}$$

$$-10 = -\sqrt{x - 25}$$

$$10 = \sqrt{x - 25}$$

$$(10)^2 = \left(\sqrt{x - 25}\right)^2$$

$$100 = x - 25$$

$$125 = x$$

The demand is 125 units per day when the price is $90.

CHAPTER ELEVEN
Quadratic Functions and Equations

11.1 | Solution by Extracting Square Roots

7. Solve $4x^2 - 8x = 0$ by factoring.

Solution

$$4x^2 - 8x = 0$$
$$4x(x - 2) = 0$$
$$4x = 0 \rightarrow x = 0$$
$$x - 2 = 0 \rightarrow x = 2$$

The solutions are 0 and 2.

9. Solve $a^2 - 25 = 0$ by factoring.

Solution

$$a^2 - 25 = 0$$
$$(a + 5)(a - 5) = 0$$
$$a + 5 = 0 \rightarrow a = -5$$
$$a - 5 = 0 \rightarrow a = 5$$

The solutions are -5 and 5.

11. Solve $x^2 + 4x + 4 = 0$ by factoring.

Solution

$$x^2 + 4x + 4 = 0$$
$$(x + 2)(x + 2) = 0$$
$$x + 2 = 0 \rightarrow x = -2$$

The repeated solution is -2.

13. Solve $(x - 3)(x + 1) = 5$ by factoring.

Solution

$$(x - 3)(x + 1) = 5$$
$$x^2 - 2x - 3 = 5$$
$$x^2 - 2x - 8 = 0$$
$$(x - 4)(x + 2) = 0$$
$$x - 4 = 0 \rightarrow x = 4$$
$$x + 2 = 0 \rightarrow x = -2$$

The solutions are 4 and -2.

15. Solve $x^2 = 49$ by extracting square roots.

Solution

$$x^2 = 49$$
$$x = \pm\sqrt{49}$$
$$x = \pm 7$$

The solutions are 7 and -7.

17. Solve $9u^2 - 100 = 0$ by extracting square roots.

Solution

$$9u^2 - 100 = 0$$
$$9u^2 = 100$$
$$u^2 = \frac{100}{9}$$
$$u = \pm\sqrt{\frac{100}{9}}$$
$$u = \pm\frac{10}{3}$$

The solutions are $\frac{10}{3}$ and $-\frac{10}{3}$.

19. Solve $3x^2 = 363$ by extracting square roots.

Solution

$$3x^2 = 363$$

$$x^2 = 121$$

$$x = \pm\sqrt{121}$$

$$x = \pm 11$$

The solutions are 11 and -11.

21. Solve $\frac{1}{2}x^2 - 1 = 7$ by extracting square roots.

Solution

$$\frac{1}{2}x^2 - 1 = 7$$

$$\frac{1}{2}x^2 = 8$$

$$2\left(\frac{1}{2}x^2\right) = 2(8)$$

$$x^2 = 16$$

$$x = \pm\sqrt{16}$$

$$x = \pm 4$$

The solutions are 4 and -4.

23. Solve $(x+4)^2 = 144$ by extracting square roots.

Solution

$$(x+4)^2 = 144$$

$$x + 4 = \pm\sqrt{144}$$

$$x + 4 = \pm 12$$

$$x = -4 \pm 12$$

$$x = -4 + 12 = 8$$

$$x = -4 - 12 = -16$$

The solutions are 8 and -16.

25. Solve $t^2 = -4$ by extracting square roots.

Solution

This equation has no solution because there is no real number that can be multiplied by itself to obtain -4.

27. Solve $(x-3)^2 = 0.16$ by extracting square roots.

Solution

$$(x-3)^2 = 0.16$$

$$x - 3 = \pm\sqrt{0.16}$$

$$x - 3 = \pm 0.4$$

$$x = 3 \pm 0.4$$

$$x = 3 + 0.4 \rightarrow x = 3.4$$

$$x = 3 - 0.4 \rightarrow x = 2.6$$

The solutions are 3.4 and 2.6.

29. Solve $(y+2)^2 = 12$ by extracting square roots.

Solution

$$(y+2)^2 = 12$$

$$y + 2 = \pm\sqrt{12}$$

$$y + 2 = \pm 2\sqrt{3}$$

$$y = -2 \pm 2\sqrt{3}$$

The solutions are $-2 + 2\sqrt{3}$ and $-2 - 2\sqrt{3}$.

31. Solve $3x^2 - 75 = 0$ by factoring and extracting square roots. Which method do you prefer?

Solution

By factoring:

$$3x^2 = 75$$

$$3x^2 - 75 = 0$$

$$3(x^2 - 25) = 0$$

$$3(x + 5)(x - 5) = 0$$

$$3 \neq 0$$

$$x + 5 = 0 \rightarrow x = -5$$

$$x - 5 = 0 \rightarrow x = 5$$

By extracting square roots:

$$3x^2 = 75$$

$$x^2 = 25$$

$$x = \pm\sqrt{25}$$

$$x = \pm 5$$

The solutions are 5 and -5. Answers regarding the preferred method will vary.

33. Solve $2x^2 = 72$ by factoring and extracting square roots. Which method do you prefer?

Solution

By factoring:

$$2x^2 - 72 = 0$$

$$2(x^2 - 36) = 0$$

$$2(x + 6)(x - 6) = 0$$

$$2 \neq 0$$

$$x + 6 = 0 \rightarrow x = -6$$

$$x - 6 = 0 \rightarrow x = 6$$

By extracting square roots:

$$2x^2 - 72 = 0$$

$$2x^2 = 72$$

$$x^2 = 36$$

$$x = \pm\sqrt{36}$$

$$x = \pm 6$$

The solutions are 6 and -6. Answers regarding the preferred method will vary.

35. Solve $3x^2 + 2 = 56$ by extracting square roots. Use a calculator to approximate the solutions. Round the results to two decimal places.

Solution

$$3x^2 + 2 = 56$$

$$3x^2 = 54$$

$$x^2 = 18$$

$$x = \pm\sqrt{18}$$

$$x = \pm 3\sqrt{2}$$

The solutions are $3\sqrt{2}$ and $-3\sqrt{2}$; $3\sqrt{2} \approx 4.24$ and $-3\sqrt{2} \approx -4.24$.

37. Solve $(x + 5)^2 = 200$ by extracting square roots. Use a calculator to approximate the solutions. Round the results to two decimal places.

Solution

$$(x + 5)^2 = 200$$

$$x + 5 = \pm\sqrt{200}$$

$$x = -5 \pm \sqrt{200}$$

$$x = -5 \pm 10\sqrt{2}$$

The first solution, $-5 + 10\sqrt{2}$, is approximately equal to 9.14. The second solution, $-5 - 10\sqrt{2}$, is approximately equal to -19.14.

39. Solve $2(2x - 5)^2 = 16$ by extracting square roots. Use a calculator to approximate the solutions. Round the results to two decimal places.

Solution

$$2(2x - 5)^2 = 16$$

$$(2x - 5)^2 = 8$$

$$2x - 5 = \pm\sqrt{8}$$

$$2x - 5 = \pm 2\sqrt{2}$$

$$2x = 5 \pm 2\sqrt{2}$$

$$x = \frac{5 \pm 2\sqrt{2}}{2}$$

The first solution, $\dfrac{5 + 2\sqrt{2}}{2}$, is approximately equal to 3.91. The second solution, $\dfrac{5 - 2\sqrt{2}}{2}$, is approximately equal to 1.09.

41. Use a graphing utility to graph $y = x^2 - 9$. Use the graph to estimate the x-intercept of the graph. Then set $y = 0$ and solve the resulting equation. Compare the results with the x-intercepts of the graph.

Solution

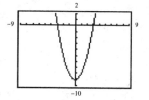

The x-intercepts appear to be $(-3, 0)$ and $(3, 0)$.

$$y = x^2 - 9$$

$$0 = x^2 - 9$$

$$0 = (x + 3)(x - 3)$$

$$x + 3 = 0 \rightarrow x = -3$$

$$x - 3 = 0 \rightarrow x = 3$$

This verifies that the x-intercepts are $(-3, 0)$ and $(3, 0)$.

43. Use a graphing utility to graph $y = 4 - (x - 1)^2$. Use the graph to estimate the x-intercept of the graph. Then set $y = 0$ and solve the resulting equation. Compare the results with the x-intercepts of the graph.

Solution

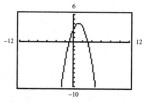

The x-intercepts appear to be $(-1, 0)$ and $(3, 0)$.

$$y = 4 - (x - 1)^2$$

$$0 = 4 - (x - 1)^2$$

$$(x - 1)^2 = 4$$

$$x - 1 = \pm\sqrt{4}$$

$$x - 1 = \pm 2$$

$$x = 1 \pm 2$$

$$x = 1 + 2 \rightarrow x = 3$$

$$x = 1 - 2 \rightarrow x = -1$$

This verifies that the x-intercepts are $(-1, 0)$ and $(3, 0)$.

45. *Geometry* An oil spill from an offshore drilling platform covers a circular region of approximately 10 square miles. Approximate the diameter of the region. (Use $\pi \approx 3.14$.)

    Solution

    $$A = \pi r^2$$

    $$10 = \pi r^2$$

    $$\frac{10}{\pi} = r^2$$

    $$\sqrt{\frac{10}{\pi}} = r$$

    $$\sqrt{\frac{10}{3.14}} \approx r$$

    $$2\sqrt{\frac{10}{3.14}} \approx d \qquad \text{Note: The diameter is twice the radius.}$$

    $$3.57 \approx d$$

    The diameter of the oil spill is approximately 3.57 miles.

47. Simplify $\frac{15}{25}$.

    Solution

    $$\frac{15}{25} = \frac{3\cancel{(5)}}{5\cancel{(5)}} = \frac{3}{5}$$

49. Solve the following system of equations.

    $$x + y = 5$$
    $$2x - 3y = 0$$

    Solution

    $$x + y = 5 \quad \rightarrow \quad -2x - 2y = -10 \qquad \text{Multiply equation by } -2.$$
    $$\underline{2x - 3y = 0 \quad \rightarrow \quad 2x - 3y = 0}$$
    $$-5y = -10$$
    $$y = 2$$

    $$x + 2 = 5 \qquad \text{Replace } y \text{ by 2 in first equation.}$$
    $$x = 3$$

    $(3, 2)$

51. *Geometry* Using the figure given in the textbook, solve for x.

Solution

$$2(\text{Length}) + 2(\text{Width}) = \text{Perimeter}$$

$$2(2x - 3) + 2(x) = 30$$

$$4x - 6 + 2x = 30$$

$$6x - 6 = 30$$

$$6x = 36$$

$$x = 6$$

The width of the rectangle is 6 meters.

53. Solve $y^2 - 3y = 0$ by factoring.

Solution

$$y^2 - 3y = 0$$

$$y(y - 3) = 0$$

$$y = 0$$

$$y - 3 = 0 \rightarrow y = 3$$

The solutions are 0 and 3.

55. Solve $16 - v^2 = 0$ by factoring.

Solution

$$16 - v^2 = 0$$

$$(4 + v)(4 - v) = 0$$

$$4 + v = 0 \rightarrow v = -4$$

$$4 - v = 0 \rightarrow v = 4$$

The solutions are -4 and 4.

57. Solve $x^2 - 8x + 16 = 0$ by factoring.

Solution

$$x^2 - 8x + 16 = 0$$

$$(x - 4)(x - 4) = 0$$

$$x - 4 = 0 \rightarrow x = 4$$

The repeated solution is 4.

59. Solve $x^2 - 5x + 6 = 0$ by factoring.

Solution

$$x^2 - 5x + 6 = 0$$

$$(x - 2)(x - 3) = 0$$

$$x - 2 = 0 \rightarrow x = 2$$

$$x - 3 = 0 \rightarrow x = 3$$

The solutions are 2 and 3.

61. Solve $(6 + x)(1 - x) = 10$ by factoring.

Solution

$$(6 + x)(1 - x) = 10$$

$$6 - 5x - x^2 = 10$$

$$-x^2 - 5x - 4 = 0$$

$$x^2 + 5x + 4 = 0 \qquad \text{Multiply both sides by } -1.$$

$$(x + 4)(x + 1) = 0$$

$$x + 4 = 0 \rightarrow x = -4$$

$$x + 1 = 0 \rightarrow x = -1$$

The solutions are -4 and -1.

63. Solve $5x^2 - 16x - 16 = 0$ by factoring.

Solution

$$5x^2 - 16x - 16 = 0$$

$$(5x + 4)(x - 4) = 0$$

$$5x + 4 = 0 \rightarrow 5x = -4 \rightarrow x = -\tfrac{4}{5}$$

$$x - 4 = 0 \rightarrow x = 4$$

The solutions are $-\tfrac{4}{5}$ and 4.

65. Solve $3z(z + 20) + 12(z + 20) = 0$ by factoring.

Solution

$$3z(z + 20) + 12(z + 20) = 0$$

$$(z + 20)(3z + 12) = 0$$

$$3(z + 20)(z + 4) = 0$$

$$3 \neq 0$$

$$z + 20 = 0 \rightarrow z = -20$$

$$z + 4 = 0 \rightarrow z = -4$$

The solutions are -20 and -4.

67. Solve $x^2 = 9$ by extracting square roots.

Solution

$$x^2 = 9$$

$$x = \pm\sqrt{9}$$

$$x = \pm 3$$

The solutions are 3 and -3.

69. Solve $9x^2 = 49$ by extracting square roots.

Solution

$$9x^2 = 49$$

$$x^2 = \frac{49}{9}$$

$$x = \pm\sqrt{\frac{49}{9}}$$

$$x = \pm\frac{7}{3}$$

The solutions are $\frac{7}{3}$ and $-\frac{7}{3}$.

71. Solve $u^2 - 100 = 0$ by extracting square roots.

Solution

$$u^2 - 100 = 0$$

$$u^2 = 100$$

$$u = \pm\sqrt{100}$$

$$u = \pm 10$$

The solutions are 10 and -10.

73. Solve $9x^2 + 1 = 0$ by extracting square roots.

Solution

$$9x^2 + 1 = 0$$

$$9x^2 = -1$$

$$x^2 = -\frac{1}{9}$$

The square of a real number cannot be negative. Thus, this equation has *no* real number solution.

75. Solve $2s^2 - 5 = 27$ by extracting square roots.

Solution

$$2s^2 - 5 = 27$$

$$2s^2 = 32$$

$$s^2 = 16$$

$$s = \pm\sqrt{16}$$

$$s = \pm 4$$

The solutions are 4 and -4.

77. Solve $\frac{1}{3}t^2 - 14 = 34$ by extracting square roots.

Solution

$$\frac{1}{3}t^2 - 14 = 34$$

$$\frac{1}{3}t^2 = 48$$

$$3\left(\frac{1}{3}t^2\right) = 3(48)$$

$$t^2 = 144$$

$$t = \pm\sqrt{144}$$

$$t = \pm 12$$

The solutions are 12 and -12.

79. Solve $(x - 8)^2 = 0.04$ by extracting square roots.

Solution

$$(x - 8)^2 = 0.04$$

$$x - 8 = \pm\sqrt{0.04}$$

$$x - 8 = \pm 0.2$$

$$x = 8 \pm 0.2$$

$$x = 8 + 0.2 = 8.2$$

$$x = 8 - 0.2 = 7.8$$

The solutions are 8.2 and 7.8.

81. Solve $4(x + 3)^2 = 25$ by extracting square roots.

Solution

$$4(x + 3)^2 = 25$$

$$(x + 3)^2 = \frac{25}{4}$$

$$x + 3 = \pm\sqrt{\frac{25}{4}}$$

$$x + 3 = \pm\frac{5}{2}$$

$$x = -3 \pm \frac{5}{2}$$

$$x = -3 + \frac{5}{2} = -\frac{1}{2}$$

$$x = -3 - \frac{5}{2} = -\frac{11}{2}$$

The solutions are $-\frac{1}{2}$ and $-\frac{11}{2}$.

83. Solve $(x - 1)^2 = 5$ by extracting square roots.

Solution

$$(x - 1)^2 = 5$$

$$x - 1 = \pm\sqrt{5}$$

$$x = 1 \pm \sqrt{5}$$

The solutions are $1 + \sqrt{5}$ and $1 - \sqrt{5}$.

85. Solve $(2x + 5)^2 = 8$ by extracting square roots.

Solution

$$(2x + 5)^2 = 8$$

$$2x + 5 = \pm\sqrt{8}$$

$$2x + 5 = \pm 2\sqrt{2}$$

$$2x = -5 \pm 2\sqrt{2}$$

$$x = \frac{-5 \pm 2\sqrt{2}}{2}$$

The solutions are $\dfrac{-5 + 2\sqrt{2}}{2}$ and $\dfrac{-5 - 2\sqrt{2}}{2}$.

87. Solve $2x^2 - 5 = 7$ by extracting square roots. Use a calculator to approximate the solutions. Round the results to two decimal places.

Solution

$$2x^2 - 5 = 7$$

$$2x^2 = 12$$

$$x^2 = 6$$

$$x = \pm\sqrt{6}$$

The first solution, $\sqrt{6}$, is approximately equal to 2.45. The second solution, $-\sqrt{6}$, is approximately equal to -2.45.

89. Solve $2(x - 3)^2 = 15$ by extracting square roots. Use a calculator to approximate the solutions. Round the results to two decimal places.

Solution

$$2(x - 3)^2 = 15$$

$$(x - 3)^2 = \frac{15}{2}$$

$$x - 3 = \pm\sqrt{\frac{15}{2}}$$

$$x - 3 = \pm\frac{\sqrt{15}}{\sqrt{2}} \cdot \frac{\sqrt{2}}{\sqrt{2}} = \pm\frac{\sqrt{30}}{2}$$

$$x = 3 \pm \frac{\sqrt{30}}{2} \text{ or } x = \frac{6 \pm \sqrt{30}}{2}$$

The first solution, $\dfrac{6 + \sqrt{30}}{2}$, is approximately equal to 5.74. The second solution, $\dfrac{6 - \sqrt{30}}{2}$, is approximately equal to 0.26.

91. Use a graphing utility to graph $y = x^2 - 4$. Use the graph to estimate the x-intercept of the graph. Then set $y = 0$ and solve the resulting equation. Compare the results with the x-intercepts of the graph.

Solution

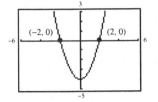

The x-intercepts appear to be $(-2, 0)$ and $(2, 0)$.

$$y = x^2 - 4$$
$$0 = x^2 - 4$$
$$0 = (x + 2)(x - 2)$$
$$x + 2 = 0 \rightarrow x = -2$$
$$x - 2 = 0 \rightarrow x = 2$$

This verifies that the x-intercepts are $(-2, 0)$ and $(2, 0)$.

93. Use a graphing utility to graph $y = 25 - (x - 3)^2$. Use the graph to estimate the x-intercept of the graph. Then set $y = 0$ and solve the resulting equation. Compare the results with the x-intercepts of the graph.

Solution

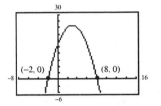

The x-intercepts appear to be $(-2, 0)$ and $(8, 0)$.

$$y = 25 - (x - 3)^2$$
$$0 = 25 - (x - 3)^2$$
$$(x - 3)^2 = 25$$
$$x - 3 = \pm\sqrt{25}$$
$$x - 3 = \pm 5$$
$$x = 3 \pm 5$$
$$x = 3 + 5 \rightarrow x = 8$$
$$x = 3 - 5 \rightarrow x = -2$$

This verifies that the x-intercepts are $(8, 0)$ and $(-2, 0)$.

95. *Geometry*   Using the figure shown in the textbook, solve for x.

Solution

$$(\text{Length})(\text{Width}) = \text{Area}$$
$$(x + 6)(x) = 16$$
$$x^2 + 6x = 16$$
$$x^2 + 6x - 16 = 0$$
$$(x + 8)(x - 2) = 0$$
$$x + 8 = 0 \rightarrow x = -8 \qquad \text{Discard the negative solution.}$$
$$x - 2 = 0 \rightarrow x = 2$$

The solution is 2.

97. *Geometry*   Using the figure shown in the textbook, solve for x.

Solution

$$\tfrac{1}{2}(\text{Base})(\text{Height}) = \text{Area}$$
$$\tfrac{1}{2}(8)(x^2 - 6) = 12$$
$$4(x^2 - 6) = 12$$
$$4x^2 - 24 = 12$$
$$4x^2 = 36$$
$$x^2 = 9$$
$$x = \pm\sqrt{9}$$
$$x = \pm 3$$

The solutions are 3 and -3. (*Note:* With either solution, the height of the triangle, $x^2 - 6$, is 3.)

99. *Falling Time* The height h (in feet) of an object dropped from a tower 64 feet high is modeled by $h = 64 - 16t^2$ where t measures the time in seconds (see figure in textbook). How long does it take the object to reach the ground?

Solution

$$h = 64 - 16t^2$$

$$0 = 64 - 16t^2 \qquad \text{The height is 0 when the object reaches the ground.}$$

$$16t^2 = 64$$

$$t^2 = \tfrac{64}{16} = 4$$

$$t = \pm\sqrt{4} = \pm 2$$

The answer $t = -2$ is extraneous in this application. Thus, it takes 2 seconds for the object to reach the ground. *Note:* The equation $0 = 64 - 16t^2$ could also be solved by factoring.

101. *Revenue* The revenue R (in dollars) when x units of a product are sold is modeled by

$$R = x\left(5 - \tfrac{1}{10}x\right), \ \ 0 < x < 25.$$

Determine the number of units that must be sold if the revenue is to be $60.

Solution

$$R = x\left(5 - \tfrac{1}{10}x\right), \ \ 0 < x < 25$$

$$60 = x\left(5 - \tfrac{1}{10}x\right)$$

$$60 = 5x - \tfrac{1}{10}x^2$$

$$10(60) = 10\left(5x - \tfrac{1}{10}x^2\right)$$

$$600 = 50x - x^2$$

$$x^2 - 50x + 600 = 0$$

$$(x - 30)(x - 20) = 0$$

$$x - 30 = 0 \rightarrow x = 30$$

$$x - 20 = 0 \rightarrow x = 20$$

The domain is $0 < x < 25$. Thus, the *only* solution is 20. To produce a revenue of $60, 20 units must be sold.

11.2 Solution by Completing the Square

7. Add a constant to the following expression to make it a perfect square trinomial.

$$x^2 + 10x + \boxed{}$$

Solution

$$x^2 + 10x + \boxed{25}$$

Half of 10 is 5, and $5^2 = 25$.

$$x^2 + 10x + 25 = (x + 5)^2$$

9. Add a constant to the following expression to make it a perfect square trinomial.

$$x^2 - \tfrac{4}{5}x + \boxed{}$$

Solution

$$x^2 - \tfrac{4}{5}x + \boxed{\tfrac{4}{25}}$$

Half of $-\tfrac{4}{5}$ is $-\tfrac{2}{5}$, and $\left(-\tfrac{2}{5}\right)^2 = \tfrac{4}{25}$.

$$x^2 - \tfrac{4}{5}x + \tfrac{4}{25} = \left(x - \tfrac{2}{5}\right)^2$$

11. Solve $x^2 - 4x = 0$ (a) by completing the square and (b) by factoring.

Solution

(a)
$$x^2 - 4x = 0$$
$$x^2 - 4x + 4 = 0 + 4$$
$$(x - 2)^2 = 4$$
$$x - 2 = \pm\sqrt{4}$$
$$x - 2 = \pm 2$$
$$x = 2 \pm 2$$
$$x = 2 + 2 = 4$$
$$x = 2 - 2 = 0$$

(b) $x^2 - 4x = 0$
$$x(x - 4) = 0$$
$$x = 0$$
$$x - 4 = 0 \quad \to \quad x = 4$$

The solutions are 0 and 4.

13. Solve $x^2 + 2x - 35 = 0$ (a) by completing the square and (b) by factoring.

Solution

(a) $x^2 + 2x - 35 = 0$
$$x^2 + 2x = 35$$
$$x^2 + 2x + 1 = 35 + 1$$
$$(x + 1)^2 = 36$$
$$x + 1 = \pm\sqrt{36}$$
$$x + 1 = \pm 6$$
$$x = -1 \pm 6$$
$$x = -1 + 6 = 5$$
$$x = -1 - 6 = -7$$

(b) $x^2 + 2x - 35 = 0$
$$(x + 7)(x - 5) = 0$$
$$x + 7 = 0 \quad \to \quad x = -7$$
$$x - 5 = 0 \quad \to \quad x = 5$$

The solutions are -7 and 5.

15. Solve $x^2 - 2x - 1 = 0$ by completing the square.

Solution

$x^2 - 2x - 1 = 0$

$x^2 - 2x = 1$

$x^2 - 2x + 1 = 1 + 1$ Half of -2 is -1, and $(-1)^2 = 1$.

$(x - 1)^2 = 2$

$x - 1 = \pm\sqrt{2}$

$x = 1 \pm \sqrt{2}$

The solutions are $1 + \sqrt{2}$ and $1 - \sqrt{2}$.

19. Solve $x^2 - 2x + 3 = 0$ by completing the square.

Solution

$x^2 - 2x + 3 = 0$

$x^2 - 2x = -3$

$x^2 - 2x + 1 = -3 + 1$ Half of -2 is -1, and $(-1)^2 = 1$.

$(x - 1)^2 = -2$

The square of a real number cannot be negative. Thus, this equation has *no* real number solution.

23. Solve $t^2 + 5t + 2 = 0$ by completing the square.

Solution

$t^2 + 5t + 2 = 0$

$t^2 + 5t = -2$

$t^2 + 5t + \dfrac{25}{4} = -2 + \dfrac{25}{4}$ Half of 5 is $\frac{5}{2}$, and $\left(\frac{5}{2}\right)^2 = \frac{25}{4}$.

$\left(t + \dfrac{5}{2}\right)^2 = -\dfrac{8}{4} + \dfrac{25}{4} = \dfrac{17}{4}$

$t + \dfrac{5}{2} = \pm\sqrt{\dfrac{17}{4}} = \pm\dfrac{\sqrt{17}}{2}$

$t = -\dfrac{5}{2} \pm \dfrac{\sqrt{17}}{2} = \dfrac{-5 \pm \sqrt{17}}{2}$

The solutions are $\dfrac{-5 + \sqrt{17}}{2}$ and $\dfrac{-5 - \sqrt{17}}{2}$.

17. Solve $u^2 - 4u - 1 = 0$ by completing the square.

Solution

$u^2 - 4u - 1 = 0$

$u^2 - 4u = 1$

$u^2 - 4u + 4 = 1 + 4$ Half of -4 is -2, and $(-2)^2 = 4$.

$(u - 2)^2 = 5$

$u - 2 = \pm\sqrt{5}$

$u = 2 \pm \sqrt{5}$

The solutions are $2 + \sqrt{5}$ and $2 - \sqrt{5}$.

21. Solve $x^2 - 8x - 2 = 0$ by completing the square.

Solution

$x^2 - 8x - 2 = 0$

$x^2 - 8x = 2$

$x^2 - 8x + 16 = 2 + 16$ Half of -8 is -4, and $(-4)^2 = 16$.

$(x - 4)^2 = 18$

$x - 4 = \pm\sqrt{18}$

$x - 4 = \pm 3\sqrt{2}$

$x = 4 \pm 3\sqrt{2}$

The solutions are $4 + 3\sqrt{2}$ and $4 - 3\sqrt{2}$.

25. Solve $3x^2 - 6x + 1 = 0$ by completing the square.

Solution

$3x^2 - 6x + 1 = 0$

$3x^2 - 6x = -1$

$x^2 - 2x = -\dfrac{1}{3}$

$x^2 - 2x + 1 = -\dfrac{1}{3} + 1$ Half of -2 is -1, and $(-1)^2 = 1$.

$(x - 1)^2 = -\dfrac{1}{3} + \dfrac{3}{3} = \dfrac{2}{3}$

$x - 1 = \pm\sqrt{\dfrac{2}{3}} = \pm\dfrac{\sqrt{2}}{\sqrt{3}} \cdot \dfrac{\sqrt{3}}{\sqrt{3}} = \pm\dfrac{\sqrt{6}}{3}$

$x = 1 \pm \dfrac{\sqrt{6}}{3} = \dfrac{3}{3} \pm \dfrac{\sqrt{6}}{3} = \dfrac{3 \pm \sqrt{6}}{3}$

The solutions are $\dfrac{3 + \sqrt{6}}{3}$ and $\dfrac{3 - \sqrt{6}}{3}$.

27. Solve $x^2 + 4x - 3 = 0$ by completing the square. Use a calculator to approximate the solution. Round the result to two decimal places.

Solution

$$x^2 + 4x - 3 = 0$$

$$x^2 + 4x = 3$$

$$x^2 + 4x + (2)^2 = 3 + 4$$

$$(x + 2)^2 = 7$$

$$x + 2 = \pm\sqrt{7}$$

$$x = -2 \pm \sqrt{7}$$

$$x = -2 + \sqrt{7} \rightarrow x \approx 0.65$$

$$x = -2 - \sqrt{7} \rightarrow x \approx -4.65$$

The solutions are approximately 0.65 and -4.65.

29. Solve $3y^2 - y - 1 = 0$ by completing the square. Use a calculator to approximate the solution. Round the result to two decimal places.

Solution

$$3y^2 - y - 1 = 0$$

$$3y^2 - y = 1$$

$$y^2 - \frac{1}{3}y = \frac{1}{3}$$

$$y^2 - \frac{1}{3}y + \left(\frac{1}{6}\right)^2 = \frac{1}{3} + \frac{1}{36}$$

$$\left(y - \frac{1}{6}\right)^2 = \frac{12}{36} + \frac{1}{36} = \frac{13}{36}$$

$$y - \frac{1}{6} = \pm\sqrt{\frac{13}{36}}$$

$$y - \frac{1}{6} = \pm\frac{\sqrt{13}}{6}$$

$$y = \frac{1}{6} \pm \frac{\sqrt{13}}{6}$$

$$y = \frac{1 \pm \sqrt{13}}{6}$$

$$y = \frac{1 + \sqrt{13}}{6} \rightarrow y \approx 0.77$$

$$y = \frac{1 - \sqrt{13}}{6} \rightarrow y \approx -0.43$$

The solutions are approximately 0.77 and -0.43.

31. Solve $\dfrac{x}{2} + \dfrac{1}{x} = 2$.

Solution

$$\frac{x}{2} + \frac{1}{x} = 2$$

$$2x\left(\frac{x}{2} + \frac{1}{x}\right) = 2x(2)$$

$$x^2 + 2 = 4x$$

$$x^2 - 4x + 2 = 0$$

$$x^2 - 4x = -2$$

$$x^2 - 4x + 4 = -2 + 4$$

$$(x - 2)^2 = 2$$

$$x - 2 = \pm\sqrt{2}$$

$$x = 2 \pm \sqrt{2}$$

The solutions are $2 + \sqrt{2}$ and $2 - \sqrt{2}$.

33. Solve $\sqrt{2x + 3} = x - 2$.

Solution

$$\sqrt{2x + 3} = x - 2$$

$$\left(\sqrt{2x + 3}\right)^2 = (x - 2)^2$$

$$2x + 3 = x^2 - 2(x)(2) + 2^2 = x^2 - 4x + 4$$

$$0 = x^2 - 6x + 1$$

$$-1 = x^2 - 6x$$

$$-1 + 9 = x^2 - 6x + 9$$

$$8 = (x - 3)^2$$

$$\pm\sqrt{8} = x - 3$$

$$\pm 2\sqrt{2} = x - 3$$

$$3 \pm 2\sqrt{2} = x$$

The answer of $3 + 2\sqrt{2}$ checks, but the second answer is extraneous. (If $x = 3 - 2\sqrt{2}$, $x \approx 0.172$. Then $\sqrt{2x + 3} \approx 1.828$, but $x - 2 \approx -1.828$.) Thus, the only solution is $3 + 2\sqrt{2}$.

35. Use a graphing utility to graph $y = x^2 - 4x + 2$. Graphically estimate the x-intercepts. Then set $y = 0$ and solve the resulting equation. Compare the results with the x-intercepts of the graph.

Solution

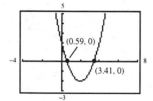

The x-intercepts appear to be approximately $(0.59, 0)$ and $(3.41, 0)$.

$$y = x^2 - 4x + 2$$

$$0 = x^2 - 4x + 2$$

$$x^2 - 4x = -2$$

$$x^2 - 4x + (-2)^2 = -2 + 4$$

$$(x - 2)^2 = 2$$

$$x - 2 = \pm\sqrt{2}$$

$$x = 2 \pm \sqrt{2}$$

$$x = 2 + \sqrt{2} \rightarrow x \approx 3.41$$

$$x = 2 - \sqrt{2} \rightarrow x \approx 0.59$$

These solutions verify the x-intercepts of approximately $(3.41, 0)$ and $(0.59, 0)$.

37. Use a graphing utility to graph $y = \sqrt{2x + 1} - x$. Graphically estimate the x-intercepts. Then set $y = 0$ and solve the resulting equation. Compare the results with the x-intercepts of the graph.

Solution

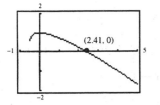

The x-intercept appears to be approximately $(2.41, 0)$.

$$y = \sqrt{2x + 1} - x$$
$$0 = \sqrt{2x + 1} - x$$
$$x = \sqrt{2x + 1}$$
$$x^2 = (\sqrt{2x + 1})^2 = 2x + 1$$
$$x^2 - 2x = 1$$
$$x^2 - 2x + (-1)^2 = 1 + 1$$
$$(x - 1)^2 = 2$$
$$x - 1 = \pm\sqrt{2}$$
$$x = 1 \pm \sqrt{2}$$
$$x = 1 + \sqrt{2} \rightarrow x \approx 2.41$$
$$x = 1 - \sqrt{2} \rightarrow x \approx -0.41 \quad \text{Extraneous}$$

This verifies that the x-intercept is approximately $(2.41, 0)$.

39. *Geometry*   Using the figure shown in the textbook, solve for x.

Solution

$$\tfrac{1}{2}(\text{Base})(\text{Height}) = \text{Area}$$
$$\tfrac{1}{2}(x)(x + 2) = 12$$
$$2\left[\tfrac{1}{2}(x)(x + 2)\right] = 2(12)$$
$$x(x + 2) = 24$$
$$x^2 + 2x - 24 = 0$$
$$(x + 6)(x - 4) = 0$$
$$x + 6 = 0 \rightarrow x = -6 \quad \text{Discard negative solution.}$$
$$x - 4 = 0 \rightarrow x = 4$$

The base of the triangle is 4 cm.

41. *Geometry* Using the figure shown in the textbook, solve for x.

Solution

$$(\text{Length})(\text{Width}) = \text{Area}$$

$$x(x - 4) = 16$$

$$x^2 - 4x = 16$$

$$x^2 - 4x + (-2)^2 = 16 + 4$$

$$(x - 2)^2 = 20$$

$$x - 2 = \pm\sqrt{20} = \pm\sqrt{4}\sqrt{5} = \pm 2\sqrt{5}$$

$$x = 2 \pm 2\sqrt{5}$$

$$x = 2 + 2\sqrt{5} \rightarrow x \approx 6.47$$

$$x = 2 - 2\sqrt{5} \rightarrow x \approx -2.47 \quad \text{Discard negative solution.}$$

The length of the rectangle is $2 + 2\sqrt{5}$ mm, or approximately 6.47 mm.

43. Determine an equation of the line passing through $(-3, 2)$ and $(5, 0)$.

Solution

$$m = \frac{y_2 - y_1}{x_2 - x_1}$$

$$m = \frac{0 - 2}{5 - (-3)}$$

$$m = \frac{-2}{8}$$

$$m = -\frac{1}{4}$$

$$y - y_1 = m(x - x_1)$$

$$y - 2 = -\frac{1}{4}(x + 3)$$

$$y - 2 = -\frac{1}{4}x - \frac{3}{4}$$

$$4y - 8 = -x - 3$$

$$x + 4y - 5 = 0$$

45. Determine an equation of the line passing through $(-4, -4)$ and $(-4, 6)$.

Solution

$$m = \frac{y_2 - y_1}{x_2 - x_1}$$

$$m = \frac{6 - (-4)}{-4 - (-4)}$$

$$m = \frac{10}{0}$$

The slope m is undefined, so the line is a vertical line. The equation of a vertical line has the form $x = a$. The x-coordinate of the given points is -4, so the equation is $x = -4$.

47. *Travel Reimbursement* A sales representative is reimbursed $125 per day for lodging and meals plus $0.32 per mile driven. Write a linear equation giving the daily cost C to the company in terms of x, the number of miles driven.

Solution

$$C = 0.32x + 125$$

49. Add a constant to the following expression to make it a perfect square trinomial.

$$y^2 - 24y + \boxed{}$$

Solution

$$y^2 - 24y + \boxed{144}$$

Half of -24 is -12, and $(-12)^2 = 144$.

$$y^2 - 24y + 144 = (y - 12)^2$$

51. Add a constant to the following expression to make it a perfect square trinomial.

$$t^2 + 3t + \boxed{}$$

Solution

$$t^2 + 3t + \boxed{\tfrac{9}{4}}$$

Half of 3 is $\tfrac{3}{2}$, and $\left(\tfrac{3}{2}\right)^2 = \tfrac{9}{4}$.

$$t^2 + 3t + \tfrac{9}{4} = \left(t + \tfrac{3}{2}\right)^2$$

53. Add a constant to the following expression to make it a perfect square trinomial.

$$x^2 + \tfrac{2}{7}x + \boxed{}$$

Solution

$$x^2 + \tfrac{2}{7}x + \boxed{\tfrac{1}{49}}$$

Half of $\tfrac{2}{7}$ is $\tfrac{1}{7}$, and $\left(\tfrac{1}{7}\right)^2 = \tfrac{1}{49}$.

$$x^2 + \tfrac{2}{7}x + \tfrac{1}{49} = \left(x + \tfrac{1}{7}\right)^2$$

55. Add a constant to the following expression to make it a perfect square trinomial.

$$y^2 - \tfrac{1}{5}y + \boxed{}$$

Solution

$$y^2 - \tfrac{1}{5}y + \boxed{\tfrac{1}{100}}$$

Half of $-\tfrac{1}{5}$ is $-\tfrac{1}{10}$, and $\left(-\tfrac{1}{10}\right)^2 = \tfrac{1}{100}$.

$$y^2 - \tfrac{1}{5}y + \tfrac{1}{100} = \left(y - \tfrac{1}{10}\right)^2$$

57. Solve $t^2 + 6t = 0$ (a) by completing the square and (b) by factoring.

Solution

(a)
$$t^2 + 6t = 0$$
$$t^2 + 6t + (3)^2 = 0 + 9$$
$$(t + 3)^2 = 9$$
$$t + 3 = \pm\sqrt{9} = \pm 3$$
$$t = -3 \pm 3$$
$$t = -3 + 3 \rightarrow t = 0$$
$$t = -3 - 3 \rightarrow t = -6$$

(b) $t^2 + 6t = 0$
$$t(t + 6) = 0$$
$$t = 0$$
$$t + 6 = 0 \rightarrow t = -6$$

The solutions are 0 and -6.

59. Solve $x^2 - 4x + 3 = 0$ (a) by completing the square and (b) by factoring.

Solution

(a) $x^2 - 4x + 3 = 0$
$$x^2 - 4x = -3$$
$$x^2 - 4x + 4 = -3 + 4$$
$$(x - 2)^2 = 1$$
$$x - 2 = \pm\sqrt{1} = \pm 1$$
$$x = 2 \pm 1$$
$$x = 2 + 1 = 3$$
$$x = 2 - 1 = 1$$

(b) $x^2 - 4x + 3 = 0$
$$(x - 3)(x - 1) = 0$$
$$x - 3 = 0 \rightarrow x = 3$$
$$x - 1 = 0 \rightarrow x = 1$$

The solutions are 3 and 1.

61. Solve $x^2 + 5x + 6 = 0$ (a) by completing the square and (b) by factoring.

Solution

(a) $\quad x^2 + 5x + 6 = 0$

$\quad\quad x^2 + 5x = -6$

$\quad x^2 + 5x + \left(\frac{5}{2}\right)^2 = -6 + \frac{25}{4}$

$\quad\quad \left(x + \frac{5}{2}\right)^2 = \frac{1}{4}$

$\quad\quad x + \frac{5}{2} = \pm\sqrt{\frac{1}{4}} = \pm\frac{1}{2}$

$\quad\quad x = -\frac{5}{2} \pm \frac{1}{2}$

$\quad\quad x = -\frac{5}{2} + \frac{1}{2} \rightarrow x = -2$

$\quad\quad x = -\frac{5}{2} - \frac{1}{2} \rightarrow x = -3$

The solutions are -2 and -3.

(b) $\quad x^2 + 5x + 6 = 0$

$\quad (x + 2)(x + 3) = 0$

$\quad\quad x + 2 = 0 \rightarrow x = -2$

$\quad\quad x + 3 = 0 \rightarrow x = -3$

63. Solve $2x^2 - 5x + 2 = 0$ (a) by completing the square and (b) by factoring.

Solution

(a) $\quad 2x^2 - 5x + 2 = 0$

$\quad\quad 2x^2 - 5x = -2$

$\quad\quad x^2 - \frac{5}{2}x = -1$

$\quad x^2 - \frac{5}{2}x + \left(-\frac{5}{4}\right)^2 = -1 + \frac{25}{16}$

$\quad\quad \left(x - \frac{5}{4}\right)^2 = \frac{9}{16}$

$\quad\quad x - \frac{5}{4} = \pm\sqrt{\frac{9}{16}} = \pm\frac{3}{4}$

$\quad\quad x = \frac{5}{4} \pm \frac{3}{4}$

$\quad\quad x = \frac{5}{4} + \frac{3}{4} \rightarrow x = 2$

$\quad\quad x = \frac{5}{4} - \frac{3}{4} \rightarrow x = \frac{1}{2}$

The solutions are 2 and $\frac{1}{2}$.

(b) $\quad 2x^2 - 5x + 2 = 0$

$\quad (2x - 1)(x - 2) = 0$

$\quad\quad 2x - 1 = 0 \rightarrow 2x = 1 \rightarrow x = \frac{1}{2}$

$\quad\quad x - 2 = 0 \rightarrow x = 2$

65. Solve $x^2 + 2x - 1 = 0$ by completing the square.

Solution

$x^2 + 2x - 1 = 0$

$\quad x^2 + 2x = 1$

$x^2 + 2x + 1 = 1 + 1$ $\quad$ Half of 2 is 1, and $1^2 = 1$.

$\quad (x + 1)^2 = 2$

$\quad\quad x + 1 = \pm\sqrt{2}$

$\quad\quad x = -1 \pm \sqrt{2}$

The solutions are $-1 + \sqrt{2}$ and $-1 - \sqrt{2}$.

67. Solve $y^2 + 14y + 17 = 0$ by completing the square.

Solution

$y^2 + 14y + 17 = 0$

$\quad y^2 + 14y = -17$

$y^2 + 14y + 49 = -17 + 49$ $\quad$ Half of 14 is 7, and $7^2 = 49$.

$\quad (y + 7)^2 = 32$

$\quad\quad y + 7 = \pm\sqrt{32} = \pm 4\sqrt{2}$

$\quad\quad y = -7 \pm 4\sqrt{2}$

The solutions are $-7 + 4\sqrt{2}$ and $-7 - 4\sqrt{2}$.

69. Solve $x^2 - x - 3 = 0$ by completing the square.

Solution

$$x^2 - x - 3 = 0$$

$$x^2 - x = 3$$

$$x^2 - x + \frac{1}{4} = 3 + \frac{1}{4} \qquad \text{Half of } -1 \text{ is } -\frac{1}{2}, \text{ and } \left(-\frac{1}{2}\right)^2 = \frac{1}{4}.$$

$$\left(x - \frac{1}{2}\right)^2 = \frac{12}{4} + \frac{1}{4} = \frac{13}{4}$$

$$x - \frac{1}{2} = \pm\sqrt{\frac{13}{4}} = \pm\frac{\sqrt{13}}{2}$$

$$x = \frac{1}{2} \pm \frac{\sqrt{13}}{2} = \frac{1 \pm \sqrt{13}}{2}$$

The solutions are $\dfrac{1 + \sqrt{13}}{2}$ and $\dfrac{1 - \sqrt{13}}{2}$.

71. Solve $v^2 + 3v - 10 = 0$ by completing the square.

Solution

$$v^2 + 3v - 10 = 0$$

$$v^2 + 3v = 10$$

$$v^2 + 3v + \frac{9}{4} = 10 + \frac{9}{4} \qquad \text{Half of } 3 \text{ is } \frac{3}{2}, \text{ and } \left(\frac{3}{2}\right)^2 = \frac{9}{4}.$$

$$\left(v + \frac{3}{2}\right)^2 = \frac{40}{4} + \frac{9}{4} = \frac{49}{4}$$

$$v + \frac{3}{2} = \pm\sqrt{\frac{49}{4}} = \pm\frac{7}{2}$$

$$v = -\frac{3}{2} \pm \frac{7}{2}$$

$$v = -\frac{3}{2} + \frac{7}{2} = \frac{4}{2} = 2$$

$$v = -\frac{3}{2} - \frac{7}{2} = \frac{-10}{2} = -5$$

The solutions are 2 and -5.

73. Solve $x^2 + 4x + 5 = 0$ by completing the square.

Solution

$$x^2 + 4x + 5 = 0$$

$$x^2 + 4x = -5$$

$$x^2 + 4x + 2^2 = -5 + 4 \qquad \text{Half of } 4 \text{ is } 2, \text{ and } 2^2 = 4.$$

$$(x + 2)^2 = -1$$

$$x + 2 = \pm\sqrt{-1} \qquad \text{Not a real number}$$

The equation has no real solution.

75. Solve $2x^2 + 6x - 5 = 0$ by completing the square.

Solution

$$2x^2 + 6x - 5 = 0$$

$$2x^2 + 6x = 5$$

$$x^2 + 3x = \frac{5}{2}$$

$$x^2 + 3x + \frac{9}{4} = \frac{5}{2} + \frac{9}{4} \qquad \text{Half of 3 is } \tfrac{3}{2},$$
$$\text{and } \left(\tfrac{3}{2}\right)^2 = \tfrac{9}{4}.$$

$$\left(x + \frac{3}{2}\right)^2 = \frac{10}{4} + \frac{9}{4} = \frac{19}{4}$$

$$x + \frac{3}{2} = \pm\sqrt{\frac{19}{4}} = \pm\frac{\sqrt{19}}{2}$$

$$x = -\frac{3}{2} \pm \frac{\sqrt{19}}{2} = \frac{-3 \pm \sqrt{19}}{2}$$

The solutions are $\dfrac{-3 + \sqrt{19}}{2}$ and $\dfrac{-3 - \sqrt{19}}{2}$.

77. Solve $5y^2 + 5y - 12 = 0$ by completing the square.

Solution

$$5y^2 + 5y - 12 = 0$$

$$5y^2 + 5y = 12$$

$$y^2 + y = \frac{12}{5}$$

$$y^2 + y + \frac{1}{4} = \frac{12}{5} + \frac{1}{4} \qquad \text{Half of 1 is } \tfrac{1}{2},$$
$$\text{and } \left(\tfrac{1}{2}\right)^2 = \tfrac{1}{4}.$$

$$\left(y + \frac{1}{2}\right)^2 = \frac{48}{20} + \frac{5}{20} = \frac{53}{20}$$

$$y + \frac{1}{2} = \pm\sqrt{\frac{53}{20}}$$

$$y + \frac{1}{2} = \pm\frac{\sqrt{53}}{\sqrt{20}} \cdot \frac{\sqrt{5}}{\sqrt{5}} = \pm\frac{\sqrt{265}}{\sqrt{100}} = \frac{\sqrt{265}}{10}$$

$$y = -\frac{1}{2} \pm \frac{\sqrt{265}}{10}$$

$$y = \frac{-5 \pm \sqrt{265}}{10}$$

The solutions are $\dfrac{-5 + \sqrt{265}}{10}$ and $\dfrac{-5 - \sqrt{265}}{10}$.

79. Solve $\frac{1}{2}x^2 + x - 1 = 0$ by completing the square.

Solution

$$\frac{1}{2}x^2 + x - 1 = 0$$

$$\frac{1}{2}x^2 + x = 1$$

$$x^2 + 2x = 2$$

$$x^2 + 2x + 1^2 = 2 + 1 \qquad \text{Half of 2 is 1, and } 1^2 = 1.$$

$$(x + 1)^2 = 3$$

$$x + 1 = \pm\sqrt{3}$$

$$x = -1 \pm \sqrt{3}$$

The solutions are $-1 + \sqrt{3}$ and $-1 - \sqrt{3}$.

81. Solve $x^2 - 6x + 7 = 0$ by completing the square. Use a calculator to approximate the solution. Round the result to two decimal places.

Solution

$$x^2 - 6x + 7 = 0$$

$$x^2 - 6x = -7$$

$$x^2 - 6x + (-3)^2 = -7 + 9$$

$$(x - 3)^2 = 2$$

$$x - 3 = \pm\sqrt{2}$$

$$x = 3 \pm \sqrt{2}$$

$$x = 3 + \sqrt{2} \rightarrow x \approx 4.41$$

$$x = 3 - \sqrt{2} \rightarrow x \approx 1.59$$

The solutions are approximately 4.41 and 1.59.

83. Solve $4z^2 - 4z - 3 = 0$ by completing the square. Use a calculator to approximate the solution. Round the result to two decimal places.

Solution

$$4z^2 - 4z - 3 = 0$$

$$4z^2 - 4z = 3$$

$$z^2 - z = \tfrac{3}{4}$$

$$z^2 - z + \left(-\tfrac{1}{2}\right)^2 = \tfrac{3}{4} + \tfrac{1}{4}$$

$$\left(z - \tfrac{1}{2}\right)^2 = 1$$

$$z - \tfrac{1}{2} = \pm\sqrt{1}$$

$$z - \tfrac{1}{2} = \pm 1$$

$$z = \tfrac{1}{2} \pm 1$$

$$z = \tfrac{1}{2} + 1 \rightarrow z = 1.50$$

$$z = \tfrac{1}{2} - 1 \rightarrow z = -0.50$$

The solutions are 1.50 and -0.50.

85. Solve $x + \dfrac{4}{x} = 6$.

Solution

$$x + \frac{4}{x} = 6$$

$$x\left(x + \frac{4}{x}\right) = 6x$$

$$x^2 + 4 = 6x$$

$$x^2 - 6x + 4 = 0$$

$$x^2 - 6x = -4$$

$$x^2 - 6x + (-3)^2 = -4 + 9$$

$$(x - 3)^2 = 5$$

$$x - 3 = \pm\sqrt{5}$$

$$x = 3 \pm \sqrt{5}$$

The solutions are $3 + \sqrt{5}$ and $3 - \sqrt{5}$.

87. Solve $\dfrac{3}{x - 2} = 2x$.

Solution

$$\frac{3}{x - 2} = 2x$$

$$(x - 2)\left(\frac{3}{x - 2}\right) = (x - 2)2x$$

$$3 = 2x^2 - 4x$$

$$\frac{3}{2} = x^2 - 2x$$

$$x^2 - 2x + (-1)^2 = \frac{3}{2} + 1$$

$$(x - 1)^2 = \frac{5}{2}$$

$$x - 1 = \pm\sqrt{\frac{5}{2}} = \pm\sqrt{\frac{10}{4}} = \pm\frac{\sqrt{10}}{2}$$

$$x = 1 \pm \frac{\sqrt{10}}{2} = \frac{2 \pm \sqrt{10}}{2}$$

The solutions are $\dfrac{2 + \sqrt{10}}{2}$ and $\dfrac{2 - \sqrt{10}}{2}$.

89. Solve $\sqrt{1 - 3x} = x$.

Solution

$$\sqrt{1 - 3x} = x$$

$$(\sqrt{1 - 3x})^2 = x^2$$

$$1 - 3x = x^2$$

$$0 = x^2 + 3x - 1$$

$$x^2 + 3x = 1$$

$$x^2 + 3x + \left(\frac{3}{2}\right)^2 = 1 + \left(\frac{3}{2}\right)^2$$

$$\left(x + \frac{3}{2}\right)^2 = 1 + \frac{9}{4} = \frac{13}{4}$$

$$x + \frac{3}{2} = \pm\sqrt{\frac{13}{4}} = \pm\frac{\sqrt{13}}{2}$$

$$x = -\frac{3}{2} \pm \frac{\sqrt{13}}{2}$$

$$x = \frac{-3 \pm \sqrt{13}}{2}$$

The solution of $\dfrac{-3 + \sqrt{13}}{2}$ checks, but the second solution of $\dfrac{-3 - \sqrt{13}}{2}$ is extraneous.

91. Solve $2\sqrt{x-1} = x - 4$.

Solution

$$2\sqrt{x-1} = x - 4$$

$$(2\sqrt{x-1})^2 = (x-4)^2$$

$$4(x-1) = x^2 - 8x + 16$$

$$4x - 4 = x^2 - 8x + 16$$

$$0 = x^2 - 12x + 20$$

$$0 = (x-10)(x-2)$$

$$x - 10 = 0 \rightarrow x = 10$$

$$x - 2 = 0 \rightarrow x = 2 \quad \text{Extraneous}$$

The solution is 10.

93. Solve $\sqrt{x^2 + 1} - \sqrt{2x} = 0$.

Solution

$$\sqrt{x^2 + 1} - \sqrt{2x} = 0$$

$$\sqrt{x^2 + 1} = \sqrt{2x}$$

$$(\sqrt{x^2 + 1})^2 = (\sqrt{2x})^2$$

$$x^2 + 1 = 2x$$

$$x^2 - 2x + 1 = 0$$

$$(x-1)(x-1) = 0$$

$$x - 1 = 0 \rightarrow x = 1$$

The solution is 1.

95. *Problem Solving*   Find two consecutive positive integers such that the sum of their squares is 85.

Solution

Verbal model:
$$\left(\boxed{\begin{array}{c}\text{First consecutive} \\ \text{positive integer}\end{array}} \right)^2 + \left(\boxed{\begin{array}{c}\text{Second consecutive} \\ \text{positive integer}\end{array}} \right)^2 = \boxed{85}$$

Labels: First consecutive positive integer $= n$
Second consecutive positive integer $= n + 1$

Equation:
$$n^2 + (n+1)^2 = 85$$

$$n^2 + n^2 + 2n + 1 = 85$$

$$2n^2 + 2n - 84 = 0$$

$$n^2 + n - 42 = 0 \quad \text{Divide both sides by 2.}$$

$$(n+7)(n-6) = 0$$

$$n + 7 = 0 \quad \rightarrow \quad n = -7 \text{ and } n + 1 = -6$$

$$n - 6 = 0 \quad \rightarrow \quad n = 6 \text{ and } n + 1 = 7$$

The problem specifies that the integers are *positive*, so we discard the answers of -7 and -6. Thus, the two consecutive positive integers are 6 and 7. *Note:* The equation *could* be solved by completing the square, but factoring is an easier method for this equation.

11.3   Solution by the Quadratic Formula

5. Write $x^2 = 3 - 2x$ in standard form.

Solution

$$x^2 = 3 - 2x$$

$$x^2 + 2x - 3 = 0$$

7. Write $x(4 - x) = 10$ in standard form.

Solution

$$x(4 - x) = 10$$

$$4x - x^2 = 10$$

$$-x^2 + 4x - 10 = 0$$

or

$$x^2 - 4x + 10 = 0$$

9. Use the discriminant to determine the number of real solutions of $2x^2 - 3x - 1 = 0$.

Solution

$$2x^2 - 3x - 1 = 0$$

$$a = 2, \ \ b = -3, \ \ c = -1$$

$$b^2 - 4ac = (-3)^2 - 4(2)(-1)$$

$$= 9 + 8$$

$$= 17$$

The *positive* discriminant indicates that the equation has *two* real number solutions.

11. Use the discriminant to determine the number of real solutions of $x^2 + 4x + 5 = 0$.

Solution

$$x^2 + 4x + 5 = 0$$

$$a = 1, \ \ b = 4, \ \ c = 5$$

$$b^2 - 4ac = 4^2 - 4(1)(5)$$

$$= 16 - 20$$

$$= -4$$

The *negative* discriminant indicates that the equation has *no* real number solutions.

13. Solve $x^2 - 11x + 30 = 0$ (a) by the Quadratic Formula and (b) by factoring. Which method is more efficient?

Solution

(a)

$$x^2 - 11x + 30 = 0$$

$$a = 1, \ \ b = -11, \ \ c = 30$$

$$x = \frac{-(-11) \pm \sqrt{(-11)^2 - 4(1)(30)}}{2(1)}$$

$$x = \frac{11 \pm \sqrt{121 - 120}}{2}$$

$$x = \frac{11 \pm \sqrt{1}}{2}$$

$$x = \frac{11 \pm 1}{2}$$

$$x = \frac{11 + 1}{2} = \frac{12}{2} = 6$$

$$x = \frac{11 - 1}{2} = \frac{10}{2} = 5$$

The solutions are 6 and 5.

(b) $x^2 - 11x + 30 = 0$

$$(x - 6)(x - 5) = 0$$

$$x - 6 = 0 \rightarrow x = 6$$

$$x - 5 = 0 \rightarrow x = 5$$

15. Solve $x^2 - 6x + 9 = 0$ (a) by the Quadratic Formula and (b) by factoring. Which method is more efficient?

Solution

(a)
$$x^2 - 6x + 9 = 0$$

$$a = 1, \quad b = -6, \quad c = 9$$

$$x = \frac{-(-6) \pm \sqrt{(-6)^2 - 4(1)(9)}}{2(1)}$$

$$x = \frac{6 \pm \sqrt{36 - 36}}{2}$$

$$x = \frac{6 \pm \sqrt{0}}{2}$$

$$x = \frac{6 \pm 0}{2}$$

$$x = \frac{6}{2} = 3$$

The repeated solution is 3.

(b)
$$x^2 - 6x + 9 = 0$$

$$(x - 3)(x - 3) = 0$$

$$x - 3 = 0 \rightarrow x = 3$$

17. Use the Quadratic Formula to solve $y^2 - 9y + 10 = 0$.

Solution

$$y^2 - 9y + 10 = 0$$

$$a = 1, \ b = -9, \ c = 10$$

$$y = \frac{-b \pm \sqrt{b^2 - 4ac}}{2a}$$

$$y = \frac{9 \pm \sqrt{(-9)^2 - 4(1)(10)}}{2(1)}$$

$$y = \frac{9 \pm \sqrt{81 - 40}}{2}$$

$$y = \frac{9 \pm \sqrt{41}}{2}$$

The solutions are $\dfrac{9 + \sqrt{41}}{2}$ and $\dfrac{9 - \sqrt{41}}{2}$.

19. Use the Quadratic Formula to solve $x^2 - 5x + 2 = 0$.

Solution

$$x^2 - 5x + 2 = 0$$

$$a = 1, \ b = -5, \ c = 2$$

$$x = \frac{-b \pm \sqrt{b^2 - 4ac}}{2a}$$

$$x = \frac{5 \pm \sqrt{(-5)^2 - 4(1)(2)}}{2(1)}$$

$$x = \frac{5 \pm \sqrt{25 - 8}}{2}$$

$$x = \frac{5 \pm \sqrt{17}}{2}$$

The solutions are $\dfrac{5 + \sqrt{17}}{2}$ and $\dfrac{5 - \sqrt{17}}{2}$.

21. Use the Quadratic Formula to solve $t^2 + t + 3 = 0$.

Solution

$t^2 + t + 3 = 0$

$$a = 1, \ b = 1, \ c = 3$$

$$t = \frac{-b \pm \sqrt{b^2 - 4ac}}{2a}$$

$$t = \frac{-1 \pm \sqrt{1^2 - 4(1)(3)}}{2(1)}$$

$$t = \frac{-1 \pm \sqrt{1 - 12}}{2}$$

$$t = \frac{-1 \pm \sqrt{-11}}{2} \quad \text{Not a real number.}$$

The equation has no real solution.

23. Use the Quadratic Formula to solve $x^2 = 3x + 1$.

Solution

$$x^2 = 3x + 1$$

$x^2 - 3x - 1 = 0$

$$a = 1, \ b = -3, \ c = -1$$

$$x = \frac{-b \pm \sqrt{b^2 - 4ac}}{2a}$$

$$x = \frac{3 \pm \sqrt{(-3)^2 - 4(1)(-1)}}{2(1)}$$

$$x = \frac{3 \pm \sqrt{9 + 4}}{2}$$

$$x = \frac{3 \pm \sqrt{13}}{2}$$

The solutions are $\dfrac{3 + \sqrt{13}}{2}$ and $\dfrac{3 - \sqrt{13}}{2}$.

25. Use the Quadratic Formula to solve $4x^2 - 4x - 1 = 0$.

Solution

$4x^2 - 4x - 1 = 0$

$$a = 4, \ b = -4, \ c = -1$$

$$x = \frac{-b \pm \sqrt{b^2 - 4ac}}{2a}$$

$$x = \frac{4 \pm \sqrt{(-4)^2 - 4(4)(-1)}}{2(4)}$$

$$x = \frac{4 \pm \sqrt{16 + 16}}{8}$$

$$x = \frac{4 \pm \sqrt{32}}{8}$$

$$x = \frac{4 \pm 4\sqrt{2}}{8}$$

$$x = \frac{\cancel{4}(1 \pm \sqrt{2})}{2\cancel{(4)}}$$

$$x = \frac{1 \pm \sqrt{2}}{2}$$

The solutions are $\dfrac{1 + \sqrt{2}}{2}$ and $\dfrac{1 - \sqrt{2}}{2}$.

27. Use the Quadratic Formula to solve $\frac{1}{2}x^2 + 2x - 3 = 0$.

Solution

$\dfrac{1}{2}x^2 + 2x - 3 = 0$

$x^2 + 4x - 6 = 0$

$$a = 1, \ b = 4, \ c = -6$$

$$x = \frac{-b \pm \sqrt{b^2 - 4ac}}{2a}$$

$$x = \frac{-4 \pm \sqrt{4^2 - 4(1)(-6)}}{2(1)}$$

$$x = \frac{-4 \pm \sqrt{16 + 24}}{2}$$

$$x = \frac{-4 \pm \sqrt{40}}{2}$$

$$x = \frac{-4 \pm 2\sqrt{10}}{2}$$

$$x = \frac{\cancel{2}(-2 \pm \sqrt{10})}{\cancel{2}}$$

$$x = -2 \pm \sqrt{10}$$

The solutions are $-2 + \sqrt{10}$ and $-2 - \sqrt{10}$.

29. Use the Quadratic Formula to solve $x^2 - 2.4x - 3.2 = 0$.

Solution

$$x^2 - 2.4x - 3.2 = 0$$

$$10x^2 - 24x - 32 = 0$$

$$5x^2 - 12x - 16 = 0$$

$$a = 5, \ b = -12, \ c = -16$$

$$x = \frac{-b \pm \sqrt{b^2 - 4ac}}{2a}$$

$$x = \frac{12 \pm \sqrt{(-12)^2 - 4(5)(-16)}}{2(5)}$$

$$x = \frac{12 \pm \sqrt{144 + 320}}{10}$$

$$x = \frac{12 \pm \sqrt{464}}{10}$$

$$x = \frac{12 \pm 4\sqrt{29}}{10}$$

$$x = \frac{\cancel{2}(6 \pm 2\sqrt{29})}{\cancel{2}(5)}$$

$$x = \frac{6 \pm 2\sqrt{29}}{5}$$

The solutions are $\dfrac{6 + 2\sqrt{29}}{5}$ and $\dfrac{6 - 2\sqrt{29}}{5}$.

31. Solve $(x - 3)^2 - 36 = 0$ by the most convenient method.

Solution

$$(x - 3)^2 - 36 = 0$$

$$(x - 3)^2 = 36$$

$$x - 3 = \pm\sqrt{36}$$

$$x - 3 = \pm 6$$

$$x = 3 \pm 6$$

$$x = 3 + 6 = 9$$

$$x = 3 - 6 = -3$$

The solutions are 9 and -3.

33. Solve $-2x^2 + 6x + 1 = 0$ by the most convenient method.

Solution

$$-2x^2 + 6x + 1 = 0$$

$$2x^2 - 6x - 1 = 0 \qquad \text{Multiply both sides by } -1.$$

$$a = 2, \ b = -6, \ c = -1$$

$$x = \frac{-(-6) \pm \sqrt{(-6)^2 - 4(2)(-1)}}{2(2)}$$

$$x = \frac{6 \pm \sqrt{36 + 8}}{4} = \frac{6 \pm \sqrt{44}}{4} = \frac{6 \pm 2\sqrt{11}}{4}$$

$$x = \frac{\cancel{2}(3 \pm \sqrt{11})}{\cancel{2}(2)}$$

$$x = \frac{3 \pm \sqrt{11}}{2}$$

The solutions are $\dfrac{3 + \sqrt{11}}{2}$ and $\dfrac{3 - \sqrt{11}}{2}$.

35. Solve $x^2 - 625 = 0$ by the most convenient method.

Solution

$$x^2 - 625 = 0 \qquad \text{(Factoring Method)}$$

$$(x + 25)(x - 25) = 0$$

$$x + 25 = 0 \to x = -25$$

$$x - 25 = 0 \to x = 25$$

$$x^2 - 625 = 0 \qquad \text{(Extracting square roots)}$$

$$x^2 = 625$$

$$x = \pm\sqrt{625}$$

$$x = \pm 25$$

The solutions are -25 and 25.

37. Solve $3x^2 - 14x + 4 = 0$ by the most convenient method. Use a calculator to approximate your solution to three decimal places.

Solution

$$3x^2 - 14x + 4 = 0$$

$$a = 3, \quad b = -14, \quad c = 4$$

$$x = \frac{-(-14) \pm \sqrt{(-14)^2 - 4(3)(4)}}{2(3)}$$

$$x = \frac{14 \pm \sqrt{196 - 48}}{6} = \frac{14 \pm \sqrt{148}}{6} = \frac{14 \pm 2\sqrt{37}}{6}$$

$$x = \frac{\cancel{2}(7 \pm \sqrt{37})}{\cancel{2}(3)}$$

$$x = \frac{7 \pm \sqrt{37}}{3}$$

$$x = \frac{7 + \sqrt{37}}{3} \approx 4.361$$

$$x = \frac{7 - \sqrt{37}}{3} \approx 0.306$$

The solutions are *approximately* 4.361 and 0.306.

39. Solve $-0.03x^2 + 2x = 0.5$ by the most convenient method. Use a calculator to approximate your solution to three decimal places.

Solution

$$-0.03x^2 + 2x - 0.5 = 0$$

$$0.03x^2 - 2x + 0.5 = 0$$

$$a = 0.03, \quad b = -2, \quad c = 0.5$$

$$x = \frac{-(-2) \pm \sqrt{(-2)^2 - 4(0.03)(0.5)}}{2(0.03)}$$

$$x = \frac{2 \pm \sqrt{4 - 0.06}}{0.06} = \frac{2 \pm \sqrt{3.94}}{0.06}$$

$$x = \frac{2 + \sqrt{3.94}}{0.06} \approx 66.416$$

$$x = \frac{2 - \sqrt{3.94}}{0.06} \approx 0.251$$

The solutions are *approximately* 66.416 and 0.251.

41. Solve $\dfrac{x^2}{3} - \dfrac{x}{2} = 1$.

Solution

$$\frac{x^2}{3} - \frac{x}{2} = 1$$

$$6\left(\frac{x^2}{3} - \frac{x}{2}\right) = 6(1)$$

$$2x^2 - 3x = 6$$

$$2x^2 - 3x - 6 = 0$$

$$a = 2, \quad b = -3, \quad c = -6$$

$$x = \frac{-(-3) \pm \sqrt{(-3)^2 - 4(2)(-6)}}{2(2)}$$

$$x = \frac{3 \pm \sqrt{9 + 48}}{4} = \frac{3 \pm \sqrt{57}}{4}$$

The solutions are $\dfrac{3 + \sqrt{57}}{4}$ and $\dfrac{3 - \sqrt{57}}{4}$.

43. Solve $\sqrt{4x + 3} = x - 1$.

Solution

$$\sqrt{4x + 3} = x - 1$$

$$\left(\sqrt{4x + 3}\right)^2 = (x - 1)^2$$

$$4x + 3 = x^2 - 2x + 1$$

$$-x^2 + 6x + 2 = 0$$

$$x^2 - 6x - 2 = 0$$

$$a = 1, \quad b = -6, \quad c = -2$$

$$x = \frac{-(-6) \pm \sqrt{(-6)^2 - 4(1)(-2)}}{2(1)}$$

$$x = \frac{6 \pm \sqrt{36 + 8}}{2} = \frac{6 \pm \sqrt{44}}{2} = \frac{6 \pm 2\sqrt{11}}{2}$$

$$x = \frac{\cancel{2}(3 \pm \sqrt{11})}{\cancel{2}} = 3 \pm \sqrt{11}$$

The answer of $3 + \sqrt{11}$ checks, but the second answer is extraneous. (If $x = 3 - \sqrt{11}$, $x \approx -0.317$. Then $\sqrt{4x + 3} \approx 1.317$ but $x - 1 \approx -1.317$.) Thus, the only solution is $3 + \sqrt{11}$.

45. *Falling Time* A ball is thrown upward with an initial velocity of 20 feet per second from a bridge that is 100 feet above the water. The height h (in feet) of the ball t seconds after it is thrown is modeled by $h = -16t^2 + 20t + 100$.

(a) Find the two times when the ball is 100 feet above the water level.

(b) Find the time when the ball strikes the water.

Solution

(a)
$$100 = -16t^2 + 20t + 100$$

$$16t^2 - 20t = 0$$

$$4t(4t - 5) = 0$$

$$4t = 0 \rightarrow \ t = 0$$

$$4t - 5 = 0 \rightarrow 4t = 5 \rightarrow t = \frac{5}{4}$$

The ball is 100 feet above the water level when $t = 0$ and when $t = \frac{5}{4}$. In other words, the ball is 100 feet above the water level *when* it is thrown and $\frac{5}{4}$ seconds *after* it is thrown.

(b)
$$0 = -16t^2 + 20t + 100$$

$$16t^2 - 20t - 100 = 0$$

$$4t^2 - 5t - 25 = 0 \qquad \text{Divide both sides by 4.}$$

$$a = 4, \ b = -5, \ c = -25$$

$$t = \frac{-(-5) \pm \sqrt{(-5)^2 - 4(4)(-25)}}{2(4)}$$

$$t = \frac{5 \pm \sqrt{25 + 400}}{8}$$

$$t = \frac{5 \pm \sqrt{425}}{8}$$

$$t = \frac{5 \pm 5\sqrt{17}}{8}$$

$$t = \frac{5 + 5\sqrt{17}}{8} \approx 3.20$$

$$t = \frac{5 - 5\sqrt{17}}{8} \approx -1.95$$

We choose the positive answer for this application and the solution is $t \approx 3.20$. Thus, the ball strikes the water approximately 3.20 seconds after it is thrown.

47. Solve $\dfrac{3}{x} + \dfrac{4}{3} = 1$.

Solution

$$\frac{3}{x} + \frac{4}{3} = 1$$

$$3x\left(\frac{3}{x} + \frac{4}{3}\right) = 3x(1)$$

$$3(3) + x(4) = 3x$$

$$9 + 4x = 3x$$

$$9 = -x$$

$$-9 = x$$

The solution is -9.

49. Solve $\dfrac{1-z}{3} - \dfrac{z}{2} = 2$.

Solution

$$\frac{1-z}{3} - \frac{z}{2} = 2$$

$$6\left(\frac{1-z}{3} - \frac{z}{2}\right) = 6(2)$$

$$2(1-z) - 3(z) = 12$$

$$2 - 2z - 3z = 12$$

$$2 - 5z = 12$$

$$-5z = 10$$

$$z = \frac{10}{-5}$$

$$z = -2$$

The solution is -2.

51. *Mixture Problem*   How many gallons of a 25% solution must be mixed with a 50% solution to obtain 10 gallons of a 40% solution?

Solution

Note: This mixture problem could be answered by writing and solving an equation in one variable or by writing and solving a system of two equations in two variables.

Using a system of equations:

$$
\begin{array}{llll}
x + \quad y = \quad 10 & \rightarrow & -25x - 25y = -250 & \text{Multiply equation by } -25. \\
0.25x + 0.50y = 0.40(10) & \rightarrow & \underline{25x + 50y = \quad 400} & \text{Multiply equation by } 100. \\
& & \quad\quad\quad\; 25y = \quad 150 & \\
& & \quad\quad\quad\quad\; y = \quad\; 6 &
\end{array}
$$

$x + 6 = 10$ Replace y by 6 in first equation.

$\quad x = 4$

$(4, 6)$

Using one equation:

$$0.25x + 0.50(10 - x) = 0.40(10)$$

$$0.25x + 5 - 0.50x = 4$$

$$5 - 0.25x = 4$$

$$-0.25x = -1$$

$$x = 4 \rightarrow 10 - x = 6$$

Four gallons of the 25% solution and six gallons of the 50% solution must be mixed to obtain ten gallons of the 40% solution.

53. Use the discriminant to decide how many real solutions $x^2 + 6x + 1 = 0$ has.

Solution

$x^2 + 6x + 1 = 0$

$a = 1, \ b = 6, \ c = 1$

$b^2 - 4ac = 6^2 - 4(1)(1)$

$\quad\quad\quad = 36 - 4$

$\quad\quad\quad = 32$

The *positive* discriminant indicates that the equation has *two* real number solutions.

55. Use the discriminant to decide how many real solutions $9x^2 - 12x + 4 = 0$ has.

Solution

$9x^2 - 12x + 4 = 0$

$a = 9, \ b = -12, \ c = 4$

$b^2 - 4ac = (-12)^2 - 4(9)(4)$

$\quad\quad\quad = 144 - 144$

$\quad\quad\quad = 0$

The discriminant of *zero* indicates that the equation has *one* (repeated) real number solution.

57. Solve $x^2 + 4x + 3 = 0$ (a) by the Quadratic Formula and (b) by factoring.

Solution

(a) $\quad x^2 + 4x + 3 = 0$

$a = 1, \ b = 4, \ c = 3$

$x = \dfrac{-4 \pm \sqrt{4^2 - 4(1)(3)}}{2(1)}$

$x = \dfrac{-4 \pm \sqrt{16 - 12}}{2}$

$x = \dfrac{-4 \pm \sqrt{4}}{2}$

$x = \dfrac{-4 \pm 2}{2}$

$x = \dfrac{-4 + 2}{2} = \dfrac{-2}{2} = -1$

$x = \dfrac{-4 - 2}{2} = \dfrac{-6}{2} = -3$

The solutions are -1 and -3.

(b) $\quad x^2 + 4x + 3 = 0$

$(x + 3)(x + 1) = 0$

$x + 3 = 0 \rightarrow x = -3$

$x + 1 = 0 \rightarrow x = -1$

59. Solve $4x^2 + 4x + 1 = 0$ (a) by the Quadratic Formula and (b) by factoring.

Solution

(a) $4x^2 + 4x + 1 = 0$

$a = 4, \ b = 4, \ c = 1$

$x = \dfrac{-4 \pm \sqrt{4^2 - 4(4)(1)}}{2(4)}$

$x = \dfrac{-4 \pm \sqrt{16 - 16}}{8}$

$x = \dfrac{-4 \pm \sqrt{0}}{8}$

$x = \dfrac{-4 \pm 0}{8}$

$x = \dfrac{-4}{8} = -\dfrac{1}{2}$

The repeated solution is $-\frac{1}{2}$.

(b) $4x^2 + 4x + 1 = 0$

$(2x + 1)(2x + 1) = 0$

$2x + 1 = 0 \rightarrow 2x = -1 \rightarrow x = -\dfrac{1}{2}$

61. Solve $x(2x + 1) = 3$ (a) by the Quadratic Formula and (b) by factoring.

Solution

(a) $x(2x + 1) = 3$

$2x^2 + x = 3$

$2x^2 + x - 3 = 0$

$a = 2, \ b = 1, \ c = -3$

$x = \dfrac{-1 \pm \sqrt{1^2 - 4(2)(-3)}}{2(2)}$

$x = \dfrac{-1 \pm \sqrt{1 + 24}}{4}$

$x = \dfrac{-1 \pm 5}{4}$

$x = \dfrac{-1 + 5}{4} \rightarrow x = 1$

$x = \dfrac{-1 - 5}{4} \rightarrow x = -\dfrac{3}{2}$

The solutions are 1 and $-\frac{3}{2}$.

(b) $x(2x + 1) = 3$

$2x^2 + x = 3$

$2x^2 + x - 3 = 0$

$(2x + 3)(x - 1) = 0$

$2x + 3 = 0 \rightarrow 2x = -3 \rightarrow x = -\dfrac{3}{2}$

$x - 1 = 0 \rightarrow x = 1$

63. Solve $4x^2 - 19x + 12 = 0$ (a) by the Quadratic Formula and (b) by factoring.

Solution

(a) $\qquad 4x^2 - 19x + 12 = 0$

$\qquad a = 4, \ b = -19, \ c = 12$

$$x = \frac{-(-19) \pm \sqrt{(-19)^2 - 4(4)(12)}}{2(4)}$$

$$x = \frac{19 \pm \sqrt{361 - 192}}{8}$$

$$x = \frac{19 \pm \sqrt{169}}{8}$$

$$x = \frac{19 \pm 13}{8}$$

$$x = \frac{19 + 13}{8} = \frac{32}{8} = 4$$

$$x = \frac{19 - 13}{8} = \frac{6}{8} = \frac{3}{4}$$

The solutions are 4 and $\frac{3}{4}$.

(b) $\qquad 4x^2 - 19x + 12 = 0$

$\qquad (4x - 3)(x - 4) = 0$

$$4x - 3 = 0 \rightarrow 4x = 3 \rightarrow x = \frac{3}{4}$$

$$x - 4 = 0 \rightarrow \ x = 4$$

65. Use the Quadratic Formula to solve $8x^2 - 10x + 3 = 0$.

Solution

$$8x^2 - 10x + 3 = 0$$

$a = 8, \ b = -10, \ c = 3$

$$x = \frac{-(-10) \pm \sqrt{(-10)^2 - 4(8)(3)}}{2(8)}$$

$$x = \frac{10 \pm \sqrt{100 - 96}}{16}$$

$$x = \frac{10 \pm \sqrt{4}}{16}$$

$$x = \frac{10 \pm 2}{16}$$

$$x = \frac{10 + 2}{16} = \frac{12}{16} = \frac{3}{4}$$

$$x = \frac{10 - 2}{16} = \frac{8}{16} = \frac{1}{2}$$

The solutions are $\frac{3}{4}$ and $\frac{1}{2}$.

67. Use the Quadratic Formula to solve $0.5x^2 - 0.8x + 0.3 = 0$.

Solution

$$0.5x^2 - 0.8x + 0.3 = 0$$

$a = 0.5, \ b = -0.8, \ c = 0.3$

$$x = \frac{-(-0.8) \pm \sqrt{(-0.8)^2 - 4(0.5)(0.3)}}{2(0.5)}$$

$$x = \frac{0.8 \pm \sqrt{0.64 - 0.6}}{1}$$

$$x = \frac{0.8 \pm \sqrt{0.04}}{1}$$

$$x = 0.8 \pm 0.2$$

$$x = 0.8 + 0.2 = 1$$

$$x = 0.8 - 0.2 = 0.6$$

The solutions are 1 and 0.6 (or 1 and $\frac{3}{5}$).

69. Use the Quadratic Formula to solve $2x^2 + 7x + 3 = 0$.

Solution

$$2x^2 + 7x + 3 = 0$$

$$a = 2, \quad b = 7, \quad c = 3$$

$$x = \frac{-7 \pm \sqrt{7^2 - 4(2)(3)}}{2(2)}$$

$$x = \frac{-7 \pm \sqrt{49 - 24}}{4}$$

$$x = \frac{-7 \pm \sqrt{25}}{4}$$

$$x = \frac{-7 \pm 5}{4}$$

$$x = \frac{-7 + 5}{4} = \frac{-2}{4} = -\frac{1}{2}$$

$$x = \frac{-7 - 5}{4} = \frac{-12}{4} = -3$$

The solutions are $-\frac{1}{2}$ and -3.

71. Use the Quadratic Formula to solve $t^2 + 5t + 6 = 0$.

Solution

$$t^2 + 5t + 6 = 0$$

$$a = 1, \quad b = 5, \quad c = 6$$

$$t = \frac{-5 \pm \sqrt{5^2 - 4(1)(6)}}{2(1)}$$

$$t = \frac{-5 \pm \sqrt{25 - 24}}{2}$$

$$t = \frac{-5 \pm \sqrt{1}}{2}$$

$$t = \frac{-5 \pm 1}{2}$$

$$t = \frac{-5 + 1}{2} = \frac{-4}{2} = -2$$

$$t = \frac{-5 - 1}{2} = \frac{-6}{2} = -3$$

The solutions are -2 and -3.

73. Use the Quadratic Formula to solve $x^2 - 6x + 7 = 0$.

Solution

$$x^2 - 6x + 7 = 0$$

$$a = 1, \quad b = -6, \quad c = 7$$

$$x = \frac{-(-6) \pm \sqrt{(-6)^2 - 4(1)(7)}}{2(1)}$$

$$x = \frac{6 \pm \sqrt{36 - 28}}{2}$$

$$x = \frac{6 \pm \sqrt{8}}{2}$$

$$x = \frac{6 \pm 2\sqrt{2}}{2}$$

$$x = \frac{\cancel{2}(3 \pm \sqrt{2})}{\cancel{2}}$$

$$x = 3 \pm \sqrt{2}$$

The solutions are $3 + \sqrt{2}$ and $3 - \sqrt{2}$.

75. Use the Quadratic Formula to solve $z^2 + 4z + 4 = 0$.

Solution

$$z^2 + 4z + 4 = 0$$

$$a = 1, \quad b = 4, \quad c = 4$$

$$z = \frac{-4 \pm \sqrt{4^2 - 4(1)(4)}}{2(1)}$$

$$z = \frac{-4 \pm \sqrt{16 - 16}}{2}$$

$$z = \frac{-4 \pm \sqrt{0}}{2}$$

$$z = \frac{-4}{2} = -2$$

The (repeated) solution is -2.

77. Use the Quadratic Formula to solve $5x^2 + 2x - 2 = 0$.

Solution

$$5x^2 + 2x - 2 = 0$$

$$a = 5, \quad b = 2, \quad c = -2$$

$$x = \frac{-2 \pm \sqrt{2^2 - 4(5)(-2)}}{2(5)}$$

$$x = \frac{-2 \pm \sqrt{4 + 40}}{10}$$

$$x = \frac{-2 \pm \sqrt{44}}{10}$$

$$x = \frac{-2 \pm 2\sqrt{11}}{10}$$

$$x = \frac{\cancel{2}(-1 \pm \sqrt{11})}{\cancel{2}(5)}$$

$$x = \frac{-1 \pm \sqrt{11}}{5}$$

The solutions are $\dfrac{-1 + \sqrt{11}}{5}$ and $\dfrac{-1 - \sqrt{11}}{5}$.

79. Use the Quadratic Formula to solve $2y^2 + y + 6 = 0$.

Solution

$$2y^2 + y + 6 = 0$$

$$a = 2, \quad b = 1, \quad c = 6$$

$$y = \frac{-1 \pm \sqrt{1^2 - 4(2)(6)}}{2(2)}$$

$$y = \frac{-1 \pm \sqrt{1 - 48}}{4}$$

$$y = \frac{-1 \pm \sqrt{-47}}{4}$$

A negative number has no square root in the real numbers. Thus, this equation has *no* real number solution.

81. Use the Quadratic Formula to solve $0.36s^2 - 0.12s + 0.01 = 0$.

Solution

$$0.36s^2 - 0.12s + 0.01 = 0$$

$$a = 0.36, \quad b = -0.12, \quad c = 0.01$$

$$s = \frac{-(-0.12) \pm \sqrt{(-0.12)^2 - 4(0.36)(0.01)}}{2(0.36)}$$

$$s = \frac{0.12 \pm \sqrt{0.0144 - 0.0144}}{0.72}$$

$$s = \frac{0.12 \pm \sqrt{0}}{0.72}$$

$$s = \frac{0.12}{0.72}$$

$$s = \frac{1}{6}$$

The (repeated) solution is $\frac{1}{6}$ (or $0.1\overline{6}$).

83. Solve $(x - 2)^2 = 9$ by the most convenient method.

Solution

$$(x - 2)^2 = 9$$

$$x - 2 = \pm\sqrt{9}$$

$$x - 2 = \pm 3$$

$$x = 2 \pm 3$$

$$x = 2 + 3 \rightarrow x = 5$$

$$x = 2 - 3 \rightarrow x = -1$$

The solutions are 5 and -1.

85. Solve $y^2 + 8y = 0$ by the most convenient method.

Solution

$$y^2 - 8y = 0$$

$$y(y + 8) = 0$$

$$y = 0$$

$$y + 8 = 0 \rightarrow y = -8$$

The solutions are 0 and -8.

87. Solve $10x^2 + x - 3 = 0$ by the most convenient method.

Solution

$$10x^2 + x - 3 = 0$$

$$(5x + 3)(2x - 1) = 0$$

$$5x + 3 = 0 \rightarrow 5x = -3 \rightarrow x = -\frac{3}{5}$$

$$2x - 1 = 0 \rightarrow 2x = 1 \rightarrow x = \frac{1}{2}$$

The solutions are $-\frac{3}{5}$ and $\frac{1}{2}$.

89. Solve $\dfrac{x+3}{2} - \dfrac{4}{x} = 2.$

Solution

$$\frac{x+3}{2} - \frac{4}{x} = 2$$

$$2x\left(\frac{x+3}{2} - \frac{4}{x}\right) = 2x(2)$$

$$x(x+3) - 2(4) = 4x$$

$$x^2 + 3x - 8 = 4x$$

$$x^2 - x - 8 = 0$$

$$a = 1, \ b = -1, \ c = -8$$

$$x = \frac{-(-1) \pm \sqrt{(-1)^2 - 4(1)(-8)}}{2(1)}$$

$$x = \frac{1 \pm \sqrt{1 + 32}}{2}$$

$$x = \frac{1 \pm \sqrt{33}}{2}$$

The solutions are $\dfrac{1 + \sqrt{33}}{2}$ and $\dfrac{1 - \sqrt{33}}{2}$.

91. Solve $\sqrt{x} = x - 5.$

Solution

$$\sqrt{x} = x - 5$$

$$(\sqrt{x})^2 = (x - 5)^2$$

$$x = x^2 - 10x + 25$$

$$0 = x^2 - 11x + 25$$

$$a = 1, \ b = -11, \ c = 25$$

$$x = \frac{-(-11) \pm \sqrt{(-11)^2 - 4(1)(25)}}{2(1)}$$

$$x = \frac{11 \pm \sqrt{121 - 100}}{2}$$

$$x = \frac{11 \pm \sqrt{21}}{2}$$

The solution $\dfrac{11 + \sqrt{21}}{2}$ checks, but the other solution $\dfrac{11 - \sqrt{21}}{2}$ is extraneous.

93. *Geometry* The area of the rectangle shown in the textbook is 58.14 in.^{2}. Use the Quadratic Formula to find its dimensions.

Solution

$$(\text{Length})(\text{Width}) = \text{Area}$$

$$(x + 6.3)(x) = 58.14$$

$$x^2 + 6.3x = 58.14$$

$$x^2 + 6.3x - 58.14 = 0$$

$$a = 1, \ b = 6.3, \ c = -58.14$$

$$x = \frac{-6.3 \pm \sqrt{6.3^2 - 4(1)(-58.14)}}{2(1)}$$

$$x = \frac{-6.3 \pm \sqrt{39.69 + 232.56}}{2}$$

$$x = \frac{-6.3 \pm \sqrt{272.25}}{2}$$

$$x = \frac{-6.3 \pm 16.5}{2}$$

$$x = \frac{-6.3 + 16.5}{2} \rightarrow x = 5.1 \rightarrow x + 6.3 = 11.4$$

$$x = \frac{-6.3 - 16.5}{2} \rightarrow x = -11.4 \qquad \text{Discard negative solution.}$$

The dimensions are 5.1 inches by 11.4 inches.

Mid-Chapter Quiz for Chapter 11

1. Solve $(x - 5)(4x + 15) = 0$.

Solution

$$(x - 5)(4x + 15) = 0$$

$$x - 5 = 0 \rightarrow x = 5$$

$$4x + 15 = 0 \rightarrow 4x = -15 \rightarrow x = -\tfrac{15}{4}$$

The solutions are 5 and $-\tfrac{15}{4}$.

2. Solve $x^2 = 400$.

Solution

$$x^2 = 400$$

$$x = \pm\sqrt{400}$$

$$x = \pm 20$$

The solutions are 20 and -20.

3. Solve $2x^2 - 50x = 0$ by factoring.

Solution

$$2x^2 - 50x = 0$$

$$2x(x - 25) = 0$$

$$2x = 0 \rightarrow x = 0$$

$$x - 25 = 0 \rightarrow x = 25$$

The solutions are 0 and 25.

4. Solve $9x^2 - 24x + 16 = 0$ by factoring.

Solution

$$9x^2 - 24x + 16 = 0$$

$$(3x - 4)(3x - 4) = 0$$

$$3x - 4 = 0 \rightarrow 3x = 4 \rightarrow x = \tfrac{4}{3}$$

The solution is $\tfrac{4}{3}$.

5. Solve $2x^2 + 9x - 35 = 0$ by factoring.

Solution

$$2x^2 + 9x - 35 = 0$$

$$(2x - 5)(x + 7) = 0$$

$$2x - 5 = 0 \rightarrow 2x = 5 \rightarrow x = \frac{5}{2}$$

$$x + 7 = 0 \rightarrow x = -7$$

The solutions are $\frac{5}{2}$ and -7.

6. Solve $x(x - 4) + 3(x - 4) = 0$ by factoring.

Solution

$$x(x - 4) + 3(x - 4) = 0$$

$$(x - 4)(x + 3) = 0$$

$$x - 4 = 0 \rightarrow x = 4$$

$$x + 3 = 0 \rightarrow x = -3$$

The solutions are 4 and -3.

7. Solve $x^2 - 2500 = 0$ by extracting square roots.

Solution

$$x^2 - 2500 = 0$$

$$x^2 = 2500$$

$$x = \pm\sqrt{2500}$$

$$x = \pm 50$$

The solutions are 50 and -50.

8. Solve $(z - 4)^2 - 81 = 0$ by extracting square roots.

Solution

$$(z - 4)^2 - 81 = 0$$

$$(z - 4)^2 = 81$$

$$z - 4 = \pm\sqrt{81}$$

$$z - 4 = \pm 9$$

$$z = 4 \pm 9$$

$$z = 4 + 9 \rightarrow z = 13$$

$$z = 4 - 9 \rightarrow z = -5$$

The solutions are 13 and -5.

9. Solve $y^2 + 6y - 11 = 0$ by completing the square.

Solution

$$y^2 + 6y - 11 = 0$$

$$y^2 + 6y = 11$$

$$y^2 + 6y + 3^2 = 11 + 9$$

$$(y + 3)^2 = 20$$

$$y + 3 = \pm\sqrt{20}$$

$$y + 3 = \pm 2\sqrt{5}$$

$$y = -3 \pm 2\sqrt{5}$$

The solutions are $-3 + 2\sqrt{5}$ and $-3 - 2\sqrt{5}$.

10. Solve $4u^2 + 12u - 1 = 0$ by completing the square.

Solution

$$4u^2 + 12u - 1 = 0$$

$$4u^2 + 12u = 1$$

$$u^2 + 3u = \frac{1}{4}$$

$$u^2 + 3u + \left(\frac{3}{2}\right)^2 = \frac{1}{4} + \frac{9}{4}$$

$$\left(u + \frac{3}{2}\right)^2 = \frac{10}{4}$$

$$u + \frac{3}{2} = \pm\sqrt{\frac{10}{4}}$$

$$u + \frac{3}{2} = \pm\frac{\sqrt{10}}{2}$$

$$u = -\frac{3}{2} \pm \frac{\sqrt{10}}{2}$$

$$u = \frac{-3 \pm \sqrt{10}}{2}$$

The solutions are $\dfrac{-3 + \sqrt{10}}{2}$ and $\dfrac{-3 - \sqrt{10}}{2}$.

11. Solve $x^2 + 3x + 1 = 0$ by the Quadratic Formula.

Solution

$x^2 + 3x + 1 = 0$

$$a = 1, \ b = 3, \ c = 1$$

$$x = \frac{-3 \pm \sqrt{3^2 - 4(1)(1)}}{2(1)}$$

$$x = \frac{-3 \pm \sqrt{9 - 4}}{2}$$

$$x = \frac{-3 \pm \sqrt{5}}{2}$$

The solutions are $\dfrac{-3 + \sqrt{5}}{2}$ and $\dfrac{-3 - \sqrt{5}}{2}$.

12. Solve $3x^2 - 4x - 10 = 0$ by the Quadratic Formula.

Solution

$3x^2 - 4x - 10 = 0$

$$a = 3, \ b = -4, \ c = -10$$

$$x = \frac{4 \pm \sqrt{(-4)^2 - 4(3)(-10)}}{2(3)}$$

$$x = \frac{4 \pm \sqrt{16 + 120}}{6}$$

$$x = \frac{4 \pm \sqrt{136}}{6}$$

$$x = \frac{4 \pm 2\sqrt{34}}{6}$$

$$x = \frac{\cancel{2}(2 \pm \sqrt{34})}{\cancel{2}(3)}$$

$$x = \frac{2 \pm \sqrt{34}}{3}$$

The solutions are $\dfrac{2 + \sqrt{34}}{3}$ and $\dfrac{2 - \sqrt{34}}{3}$.

13. Solve $x - \dfrac{24}{x} = 5$.

Solution

$$x - \frac{24}{x} = 5$$

$$x\left(x - \frac{24}{x}\right) = x(5)$$

$$x^2 - 24 = 5x$$

$$x^2 - 5x - 24 = 0$$

$$(x - 8)(x + 3) = 0$$

$$x - 8 = 0 \rightarrow x = 8$$

$$x + 3 = 0 \rightarrow x = -3$$

The solutions are 8 and -3.

14. Solve $\sqrt{x + 4} = x - 2$.

Solution

$$\sqrt{x + 4} = x - 2$$

$$(\sqrt{x + 4})^2 = (x - 2)^2$$

$$x + 4 = x^2 - 4x + 4$$

$$0 = x^2 - 5x$$

$$0 = x(x - 5)$$

$$x = 0 \qquad \text{Extraneous}$$

$$x - 5 = 0 \rightarrow x = 5$$

The solution is 5.

15. Use a graphing utility to graph $y = 3x^2 - x - 4$. Graphically estimate the x-intercepts. Then set $y = 0$ and solve the resulting equations. Compare the results with the x-intercepts of the graph.

Solution

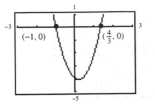

The x-intercepts appear to be $(-1, 0)$ and $\left(\frac{4}{3}, 0\right)$.

$$y = 3x^2 - x - 4$$

$$3x^2 - x - 4 = 0$$

$$(3x - 4)(x + 1) = 0$$

$$3x - 4 = 0 \rightarrow 3x = 4 \rightarrow x = \tfrac{4}{3}$$

$$x + 1 = 0 \rightarrow x = -1$$

This verifies that the x-intercepts are $(-1, 0)$ and $\left(\frac{4}{3}, 0\right)$.

16. Use a graphing utility to graph $y = \dfrac{10}{x^2 + 1} - 1$.

Graphically estimate the x-intercepts. Then set $y = 0$ and solve the resulting equations. Compare the results with the x-intercepts of the graph.

Solution

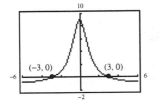

The x-intercepts appear to be $(3, 0)$ and $(-3, 0)$.

$$y = \frac{10}{x^2 + 1} - 1$$

$$0 = \frac{10}{x^2 + 1} - 1$$

$$(x^2 + 1)(0) = (x^2 + 1)\left(\frac{10}{x^2 + 1} - 1\right)$$

$$0 = 10 - (x^2 + 1)(1)$$

$$0 = 10 - x^2 - 1$$

$$x^2 = 9$$

$$x = \pm\sqrt{9}$$

$$x = \pm 3$$

This verifies that the x-intercepts are $(3, 0)$ and $(-3, 0)$.

17. Use a graphing utility to graph $y = \sqrt{x} - (2x - 3)$. Graphically estimate the x-intercepts. Then set $y = 0$ and solve the resulting equations. Compare the results with the x-intercepts of the graph.

Solution

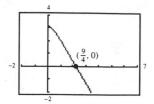

The x-intercept appears to be $\left(\frac{9}{4}, 0\right)$.

$$y = \sqrt{x} - (2x - 3)$$
$$0 = \sqrt{x} - (2x - 3)$$
$$0 = \sqrt{x} - 2x + 3$$
$$2x - 3 = \sqrt{x}$$
$$(2x - 3)^2 = (\sqrt{x})^2$$
$$4x^2 - 12x + 9 = x$$
$$4x^2 - 13x + 9 = 0$$
$$(4x - 9)(x - 1) = 0$$
$$4x - 9 = 0 \rightarrow 4x = 9 \rightarrow x = \tfrac{9}{4}$$
$$x - 1 = 0 \rightarrow x = 1 \qquad \text{Extraneous}$$

This verifies that the x-intercept is $\left(\frac{9}{4}, 0\right)$.

18. Use a graphing utility to graph $y = \sqrt{4 - x} - x$. Graphically estimate the x-intercepts. Then set $y = 0$ and solve the resulting equations. Compare the results with the x-intercepts of the graph.

Solution

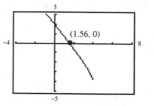

The x-intercept appears to be approximately $(1.56, 0)$.

$$y = \sqrt{4 - x} - x$$
$$0 = \sqrt{4 - x} - x$$
$$x = \sqrt{4 - x}$$
$$x^2 = (\sqrt{4 - x})^2$$
$$x^2 = 4 - x$$
$$x^2 + x - 4 = 0$$
$$a = 1, \, b = 1, \, c = -4$$
$$x = \frac{-1 \pm \sqrt{1^2 - 4(1)(-4)}}{2(1)}$$
$$x = \frac{-1 \pm \sqrt{1 + 16}}{2}$$
$$x = \frac{-1 + \sqrt{17}}{2}$$
$$x = \frac{-1 - \sqrt{17}}{2} \qquad \text{Extraneous}$$

This verifies that the x-intercept is $\left(\dfrac{-1 + \sqrt{17}}{2}, 0\right)$, or approximately $(1.56, 0)$.

19. On September 9, 1979, Kitty O'Neill dove 180 feet from a helicopter into a 30- by 60-foot air cushion for a TV film stunt. Her height h (in feet) after t seconds into the fall is modeled by $h = 180 - 16t^2$. How long was O'Neill in the air?

Solution

$$h = 180 - 16t^2$$

$$0 = 180 - 16t^2 \qquad \text{(Note: } h = 0 \text{ when she finishes her dive.)}$$

$$16t^2 = 180$$

$$t^2 = \frac{180}{16}$$

$$t^2 = \frac{45}{4}$$

$$t = \pm\sqrt{\frac{45}{4}}$$

$$t = \pm\frac{\sqrt{45}}{\sqrt{4}}$$

$$t = \pm\frac{3\sqrt{5}}{2}$$

$$t = \frac{3\sqrt{5}}{2} \qquad \text{Discard the negative solution.}$$

She was in the air for $\dfrac{3\sqrt{5}}{2}$ seconds, or approximately 3.35 seconds.

20. The surface area of the cube in the figure shown in the textbook is 150 square feet. Find the length of each edge.

Solution

The surface area of each face of the cube is x^2; the surface area of the entire cube is $6x^2$.

$$6x^2 = 150$$

$$x^2 = 25$$

$$x = \pm\sqrt{25}$$

$$x = \pm 5$$

$$x = 5 \qquad \text{Discard the negative solution.}$$

The length of each edge of the cube is 5 feet.

Graphing Quadratic Functions

7. Match $y = 5 - x$ with the correct graph. [The graphs are shown in the textbook and are labeled (a), (b), (c), (d), (e), and (f).]

Solution

Matches graph (d)

9. Match $y = x^2 + 1$ with the correct graph. [The graphs are shown in the textbook and are labeled (a), (b), (c), (d), (e), and (f).]

Solution

Matches graph (f)

11. Match $y = (x - 3)^2$ with the correct graph. [The graphs are shown in the textbook and are labeled (a), (b), (c), (d), (e), and (f).]

Solution

Matches graph (a)

13. Determine whether the parabola $y = 6x^2 + 2$ opens up or down.

Solution

$y = 6x^2 + 2$

$a = 6$

Since $a > 0$, the parabola opens *up*.

15. Determine whether the parabola $y = 6 - x(x + 1)$ opens up or down.

Solution

$y = 6 - x(x + 1)$

$y = 6 - x^2 - x$

$y = -x^2 - x + 6$

$a = -1$

Since $a < 0$, the parabola opens *down*.

17. Find the intercepts of the graph of $y = 16 - x^2$.

Solution

x-intercepts	*y-intercept*
Let $y = 0$.	Let $x = 0$.
$0 = 16 - x^2$	$y = 16 - 0^2$
$0 = (4 + x)(4 - x)$	$y = 16$
$4 + x = 0 \rightarrow x = -4$	$(0, 16)$
$4 - x = 0 \rightarrow x = 4$	
$(-4, 0)$ and $(4, 0)$	

The graph has two x-intercepts at $(-4, 0)$ and $(4, 0)$; the y-intercept is $(0, 16)$.

19. Find the intercepts of the graph of $y = x^2 - x - 6$.

Solution

x-intercepts	*y-intercept*
Let $y = 0$.	Let $x = 0$.
$0 = x^2 - x - 6$	$y = 0^2 - 0 - 6$
$0 = (x - 3)(x + 2)$	$y = -6$
$x - 3 = 0 \rightarrow x = 3$	$(0, -6)$
$x + 2 = 0 \rightarrow x = -2$	
$(3, 0)$ and $(-2, 0)$	

The graph has two x-intercepts at $(3, 0)$ and $(-2, 0)$; the y-intercept is $(0, -6)$.

21. Find the vertex of the parabola $y = -x^2 + 2$.

Solution

$y = -x^2 + 2$

$a = -1, \quad b = 0$

$x = -\dfrac{b}{2a} = -\dfrac{0}{2(-1)} = 0$

$y = -0^2 + 2 = 2$

The vertex is located at $(0, 2)$.

23. Find the vertex of the parabola $y = 2x^2 - 8x - 2$.

Solution

$y = 2x^2 - 8x - 2$

$a = 2, \ b = -8$

$x = -\dfrac{b}{2a}$

$x = -\dfrac{-8}{2(2)}$

$x = \dfrac{8}{4}$

$x = 2$

$y = 2(2)^2 - 8(2) - 2$

$y = -10$

The vertex is located at $(2, -10)$.

25. Sketch the graph of $y = x^2 - 1$. Identify the vertex and intercepts.

Solution

Leading coefficient test: Since $a > 0$, the parabola opens up.

Vertex: $(a = 1, \ b = 0)$ $x = -\dfrac{b}{2a} = -\dfrac{0}{2(1)} = 0$

$y = 0^2 - 1 = -1$

The vertex is located at $(0, -1)$.

y-intercept: (Let $x = 0$.) $y = 0^2 - 1 = -1$

The y-intercept is $(0, -1)$.

x-intercepts: (Let $y = 0$.) $0 = x^2 - 1$

$1 = x^2 \rightarrow x = \pm\sqrt{1} \rightarrow x = \pm 1$

The x-intercepts are $(1, 0)$ and $(-1, 0)$.

Additional solution points:

x	2	3	-2	-3
$y = x^2 - 1$	3	8	3	8
Points	$(2, 3)$	$(3, 8)$	$(-2, 3)$	$(-3, 8)$

27. Sketch the graph of $y = -x^2 + 4x$. Identify the vertex and intercepts.

Solution

Leading coefficient test: Since $a < 0$, the parabola opens down.

Vertex: $(a = -1,\ b = 4)$ $x = -\dfrac{b}{2a} = -\dfrac{4}{2(-1)} = -(-2) = 2$

$y = -2^2 + 4(2) = -4 + 8 = 4$

The vertex is located at $(2, 4)$.

y-intercept: (Let $x = 0$.) $y = -0^2 + 4 \cdot 0 = 0$; the y-intercept is $(0, 0)$.

x-intercepts: (Let $y = 0$.) $0 = -x^2 + 4x \rightarrow -x(x - 4) = 0$

$-x = 0 \rightarrow x = 0$

$x - 4 = 0 \rightarrow x = 4$

The x-intercepts are $(0, 0)$ and $(4, 0)$.

Additional solution points:

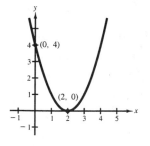

x	5	3	1	-1
$y = -x^2 + 4x$	-5	3	3	-5
Points	$(5, -5)$	$(3, 3)$	$(1, 3)$	$(-1, -5)$

29. Sketch the graph of $y = (x - 2)^2$. Identify the vertex and intercepts.

Solution

$y = (x - 2)^2$ or $y = x^2 - 4x + 4$

Leading coefficient test: Since $a > 0$, the parabola opens up.

Vertex: $(a = 1,\ b = -4)$ $x = -\dfrac{b}{2a} = -\dfrac{-4}{2(1)} = -(-2) = 2$

$y = (2 - 2)^2 = 0$

The vertex is located at $(2, 0)$.

y-intercept: (Let $x = 0$.) $y = (0 - 2)^2 = (-2)^2 = 4$; the y-intercept is $(0, 4)$.

x-intercepts: (Let $y = 0$.) $0 = (x - 2)^2 \rightarrow (x - 2) = 0$

$x - 2 = 0 \rightarrow x = 2$

The x-intercept is $(2, 0)$.

Additional solution points:

x	4	5	-1	1
$y = (x - 2)^2$	4	9	9	1
Points	$(4, 4)$	$(5, 9)$	$(-1, 9)$	$(1, 1)$

31. Sketch the graph of $y = -(x^2 + 4x + 2)$. Identify the vertex and intercepts.

Solution

$y = -(x^2 + 4x + 2)$ or $y = -x^2 - 4x - 2$

Leading coefficient test: Since $a < 0$, the parabola opens down.

Vertex: $(a = -1, \ b = -4)$ $x = -\dfrac{b}{2a} = -\dfrac{-4}{2(-1)} = -2$

$y = -(-2)^2 - 4(-2) - 2 = -4 + 8 - 2 = 2$
The vertex is located at $(-2, 2)$.

y-intercept: (Let $x = 0$.) $y = -0^2 - 4 \cdot 0 - 2 = -2$; the y-intercept is $(0, -2)$.

x-intercepts: (Let $y = 0$.) $0 = -x^2 - 4x - 2$ or $0 = x^2 + 4x + 2$

$$x = \frac{-4 \pm \sqrt{4^2 - 4(1)(2)}}{2(1)} = \frac{-4 \pm \sqrt{8}}{2}$$

$$x = \frac{-4 \pm 2\sqrt{2}}{2} = -2 \pm \sqrt{2}$$

The x-intercepts are $(-2 + \sqrt{2}, 0)$ and $(-2 - \sqrt{2}, 0)$.

Additional solution points:

x	-5	-4	-3	1
$y = -x^2 - 4x - 2$	-7	-2	1	-7
Points	$(-5, -7)$	$(-4, -2)$	$(-3, 1)$	$(1, -7)$

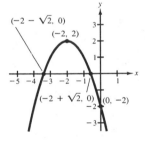

33. Use a graphing utility to graph $y = -x^2 + 6x$. Graphically approximate the vertex.

Solution

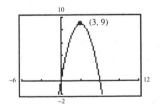

The vertex appears to be $(3, 9)$.

35. Use a graphing utility to graph $y = 6 - \frac{1}{2}(x - 4)^2$. Graphically approximate the vertex.

Solution

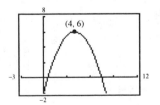

The vertex appears to be $(4, 6)$.

37. Sketch the parabola $y = -x^2 + 3$ and the horizontal line $y = 2$. Find any points of intersection.

Solution

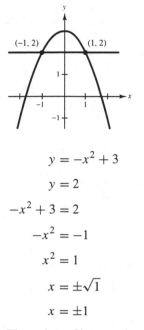

$$y = -x^2 + 3$$
$$y = 2$$
$$-x^2 + 3 = 2$$
$$-x^2 = -1$$
$$x^2 = 1$$
$$x = \pm\sqrt{1}$$
$$x = \pm 1$$

The points of intersection are $(1, 2)$ and $(-1, 2)$.

39. Sketch the parabola $y = \frac{1}{2}x^2 - 4x + 10$ and the horizontal line $y = 3$. Find any points of intersection.

Solution

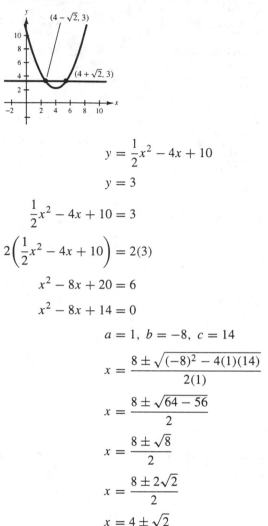

$$y = \frac{1}{2}x^2 - 4x + 10$$
$$y = 3$$
$$\frac{1}{2}x^2 - 4x + 10 = 3$$
$$2\left(\frac{1}{2}x^2 - 4x + 10\right) = 2(3)$$
$$x^2 - 8x + 20 = 6$$
$$x^2 - 8x + 14 = 0$$
$$a = 1, \ b = -8, \ c = 14$$
$$x = \frac{8 \pm \sqrt{(-8)^2 - 4(1)(14)}}{2(1)}$$
$$x = \frac{8 \pm \sqrt{64 - 56}}{2}$$
$$x = \frac{8 \pm \sqrt{8}}{2}$$
$$x = \frac{8 \pm 2\sqrt{2}}{2}$$
$$x = 4 \pm \sqrt{2}$$

The points of intersection are $(4 + \sqrt{2}, 3)$ and $(4 - \sqrt{2}, 3)$.

41. *Modeling a Punt*   A football player kicks a 41-yard punt. The path of the ball is modeled by $y = -0.035x^2 + 1.4x + 1$ where x and y are measured in yards. What was the maximum height of the ball? The ball was kicked toward midfield from the 18-yard line. Over which yard line was the ball at its maximum height?

Solution

$y = -0.035x^2 + 1.4x + 1$

$a = -0.035, \ b = 1.4$

$x = \dfrac{-b}{2a}$

$x = \dfrac{-1.4}{2(-0.035)}$

$x = \dfrac{-1.4}{-0.07}$

$x = 20$

$y = -0.035(20)^2 + 1.4(20) + 1$

$y = 15$

The vertex of the parabolic path was (20, 15). Therefore, the maximum height of the football was 15 yards. The ball reached this height 20 feet down the field from where it was kicked at the 18-yard line, so it reached this height over the 38-yard line.

43. Sketch the graph of $y = -\frac{1}{3}x + 3$.

Solution

45. Sketch the graph of $x + 5 = 0$.

Solution

47. *Height*   A 20-foot board leans against the side of a house. The bottom is 5 feet from the house. How far does the board reach up the side of the house?

Solution

$a^2 + b^2 = c^2$

$x^2 + 5^2 = 20^2$

$x^2 + 25 = 400$

$\qquad x^2 = 375$

$\qquad x = \pm\sqrt{375}$

$\qquad x = 5\sqrt{15}$ Discard the negative solution.

The board reaches $5\sqrt{15}$ feet, or approximately 19.36 feet up the side of the house.

49. Determine whether the parabola $y = 4 + 10x - x^2$ opens up or down.

Solution

$y = 4 + 10x - x^2$

$y = -x^2 + 10x + 4$

$a = -1$

Since $a < 0$, the parabola opens *down*.

51. Find the intercepts of $y = x^2 - 2x$.

Solution

(a) x-intercepts

    Let $y = 0$.

        $0 = x^2 - 2x$

        $0 = x(x - 2)$

        $x = 0$

    $x - 2 = 0 \rightarrow x = 2$

(b) y-intercepts

    Let $x = 0$.

        $y = 0^2 - 2(0)$

        $y = 0$

The x-intercepts are $(0, 0)$ and $(2, 0)$. The y-intercept is $(0, 0)$.

53. Find the intercepts of $y = x^2 - 2x + 3$.

Solution

(a) x-intercepts

    Let $y = 0$.

    $0 = x^2 - 2x + 3$

    $a = 1, b = -2, c = 3$

    $x = \dfrac{2 \pm \sqrt{(-2)^2 - 4(1)(3)}}{2(1)}$

    $x = \dfrac{2 \pm \sqrt{4 - 12}}{2}$

    $x = \dfrac{2 \pm \sqrt{-8}}{2}$ (These are not real numbers.)

(b) y-intercepts

    Let $x = 0$.

    $y = 0^2 - 2(0) + 3$

    $y = 3$

There are no x-intercepts. The y-intercept is $(0, 3)$.

55. Find the vertex of the parabola
$y = x^2 - 4x + 7$.

Solution

$y = x^2 - 4x + 7$

$a = 1, \quad b = -4$

$x = -\dfrac{b}{2a} = -\dfrac{-4}{2(1)} = -(-2) = 2$

$y = 2^2 - 4 \cdot 2 + 7 = 4 - 8 + 7 = 3$

The vertex is located at $(2, 3)$.

57. Find the vertex of the parabola $y = x(x - 4)$.

Solution

$y = x(x - 4)$

$y = x^2 - 4x$

$a = 1, \quad b = -4$

$x = -\dfrac{b}{2a} = -\dfrac{-4}{2(1)} = -(-2) = 2$

$y = 2^2 - 4 \cdot 2 = 4 - 8 = -4$

The vertex is located at $(2, -4)$.

59. Sketch the graph of $y = -x^2 + 9$. Identify the vertex and the intercepts.

Solution

$y = -x^2 + 9$

Leading coefficient test: Since $a < 0$, the parabola opens down.

Vertex: $(a = -1, \ b = 0)$ $x = -\dfrac{b}{2a} = -\dfrac{0}{2(-1)} = 0$

$y = -0^2 + 9 = 9$
The vertex is located at $(0, 9)$.

y-intercept: (Let $x = 0$.) $y = -0^2 + 9 = 9$; the y-intercept is $(0, 9)$.

x-intercepts: (Let $y = 0$.) $0 = -x^2 + 9$

$x^2 = 9 \rightarrow x = \pm\sqrt{9} \rightarrow x = \pm 3$

The x-intercepts are $(3, 0)$ and $(-3, 0)$.

Additional solution points:

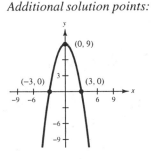

x	2	4	−2	−4
$y = -x^2 + 9$	5	−7	5	−7
Points	$(2, 5)$	$(4, -7)$	$(-2, 5)$	$(-4, -7)$

61. Sketch the graph of $y = x^2 - 6x$. Identify the vertex and the intercepts.

Solution

$y = x^2 - 6x$

Leading coefficient test: Since $a > 0$, the parabola opens up.

Vertex: $(a = 1, \ b = -6)$ $\quad x = -\dfrac{b}{2a} = -\dfrac{-6}{2(1)} = -(-3) = 3$

$y = 3^2 - 6 \cdot 3 = 9 - 18 = -9$

The vertex is located at $(3, -9)$.

y-intercept: (Let $x = 0$.) $\quad y = 0^2 - 6 \cdot 0 = 0$; the y-intercept is $(0, 0)$.

x-intercepts: (Let $y = 0$.) $\quad 0 = x^2 - 6x \rightarrow x(x - 6) = 0$

$x = 0$

$x - 6 = 0 \rightarrow x = 6$

The x-intercepts are $(0, 0)$ and $(6, 0)$.

Additional solution points:

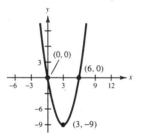

x	7	5	1	-1
$y = x^2 - 6x$	7	-5	-5	7
Points	$(7, 7)$	$(5, -5)$	$(1, -5)$	$(-1, 7)$

63. Sketch the graph of $y = -x^2 - 2x + 3$. Identify the vertex and the intercepts.

Solution

Leading coefficient test: Since $a < 0$, the parabola opens down.

Vertex: $(a = -1, \ b = -2)$ $\quad x = -\dfrac{b}{2a} = -\dfrac{-2}{2(-1)} = -1$

$y = -(-1)^2 - 2(-1) + 3 = -1 + 2 + 3 = 4$

The vertex is located at $(-1, 4)$.

y-intercept: (Let $x = 0$.) $\quad y = -0^2 - 2 \cdot 0 + 3 = 3$; the y-intercept is $(0, 3)$.

x-intercepts: (Let $y = 0$.) $\quad 0 = -x^2 - 2x + 3 \rightarrow x^2 + 2x - 3 = 0 \rightarrow (x + 3)(x - 1) = 0$

$x + 3 = 0 \rightarrow x = -3$

$x - 1 = 0 \rightarrow x = 1$

The x-intercepts are $(-3, 0)$ and $(1, 0)$.

Additional solution points:

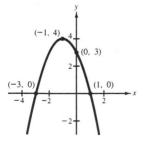

x	2	-2	-4
$y = -x^2 - 2x + 3$	-5	3	-5
Points	$(2, -5)$	$(-2, 3)$	$(-4, -5)$

65. Sketch the graph of $y = x^2 + 6x + 8$. Identify the vertex and the intercepts.

Solution

Leading coefficient test: Since $a > 0$, the parabola opens up.

Vertex: $(a = 1, \ b = 6)$

$$x = -\frac{b}{2a} = -\frac{6}{2(1)} = -3$$

$$y = (-3)^2 + 6(-3) + 8 = 9 - 18 + 8 = -1$$

The vertex is located at $(-3, -1)$.

y-intercept: (Let $x = 0$.) $\quad y = 0^2 + 6 \cdot 0 + 8 = 8$; the y-intercept is $(0, 8)$.

x-intercepts: (Let $y = 0$.) $\quad 0 = x^2 + 6x + 8 \to (x + 4)(x + 2) = 0$

$$x + 4 = 0 \to x = -4$$

$$x + 2 = 0 \to x = -2$$

The x-intercepts are $(-4, 0)$ and $(-2, 0)$.

Additional solution points:

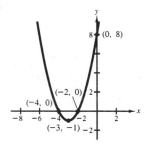

x	-1	-5	-6
$y = x^2 + 6x + 8$	3	3	8
Points	$(-1, 3)$	$(-5, 3)$	$(-6, 8)$

67. Sketch the graph of $y = x^2 - 4x + 1$. Identify the vertex and the intercepts.

Solution

Leading coefficient test: Since $a > 0$, the parabola opens up.

Vertex: $(a = 1, \ b = -4)$ $\quad x = -\dfrac{b}{2a} = -\dfrac{-4}{2(1)} = -(-2) = 2$

$$y = 2^2 - 4(2) + 1 = 4 - 8 + 1 = -3$$

The vertex is located at $(2, -3)$.

y-intercept: (Let $x = 0$.) $\quad y = 0^2 - 4 \cdot 0 + 1 = 1$; the y-intercept is $(0, 1)$.

x-intercepts: (Let $y = 0$.) $\quad 0 = x^2 - 4x + 1$

$$x = \frac{-(-4) \pm \sqrt{(-4)^2 - 4(1)(1)}}{2(1)} = \frac{4 \pm \sqrt{12}}{2}$$

$$x = \frac{4 \pm 2\sqrt{3}}{2} = 2 \pm \sqrt{3}$$

The x-intercepts are $(2 + \sqrt{3}, \ 0)$ and $(2 - \sqrt{3}, \ 0)$.

Additional solution points:

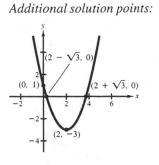

x	4	5	-1
$y = x^2 - 4x + 1$	1	6	6
Points	$(4, 1)$	$(5, 6)$	$(-1, 6)$

69. Sketch the graph of $y = \frac{1}{2}x^2 - 4x + 6$. Identify the vertex and the intercepts.

Solution

Leading coefficient test: Since $a > 0$, the parabola opens up.

Vertex: $\left(a = \frac{1}{2},\ b = -4\right)$ $x = -\dfrac{b}{2a} = -\dfrac{-4}{2\left(\frac{1}{2}\right)} = -\dfrac{-4}{1} = 4$

$$y = \frac{1}{2}(4)^2 - 4(4) + 6 = 8 - 16 + 6 = -2$$

The vertex is located at $(4, -2)$.

y-intercept: (Let $x = 0$.) $y = \dfrac{1}{2} \cdot 0^2 - 4 \cdot 0 + 6 = 6$; the y-intercept is $(0, 6)$.

x-intercepts: (Let $y = 0$.) $0 = \dfrac{1}{2}x^2 - 4x + 6$ or $0 = x^2 - 8x + 12$

$$(x - 6)(x - 2) = 0$$

$$x - 6 = 0 \rightarrow x = 6 \text{ and } x - 2 = 0 \rightarrow x = 2$$

The x-intercepts are $(6, 0)$ and $(2, 0)$.

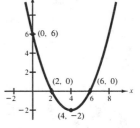

Additional solution points:

x	8	7	5	1
$y = \frac{1}{2}x^2 - 4x + 6$	6	$2\frac{1}{2}$	$-1\frac{1}{2}$	$2\frac{1}{2}$
Points	$(8, 6)$	$\left(7, 2\frac{1}{2}\right)$	$\left(5, -1\frac{1}{2}\right)$	$\left(1, 2\frac{1}{2}\right)$

71. Sketch the graph of $y = 2x^2 + 8x + 9$. Identify the vertex and the intercepts.

Solution

Leading coefficient test: Since $a > 0$, the parabola opens up.

Vertex: $(a = 2,\ b = 8)$ $x = -\dfrac{b}{2a} = -\dfrac{8}{2(2)} = -2$

$$y = 2(-2)^2 + 8(-2) + 9 = 8 - 16 + 9 = 1$$

The vertex is located at $(-2, 1)$.

y-intercept: (Let $x = 0$.) $y = 2 \cdot 0^2 + 8 \cdot 0 + 9 = 9$; the y-intercept is $(0, 9)$.

x-intercepts: (Let $y = 0$.) $0 = 2x^2 + 8x + 9$

$$x = \frac{-8 \pm \sqrt{8^2 - 4(2)(9)}}{2(2)} = \frac{-8 \pm \sqrt{-8}}{4}$$

There are *no* x-intercepts.

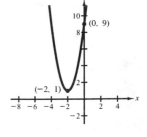

Additional solution points:

x	-1	-3	-4
$y = 2x^2 + 8x + 9$	3	3	9
Points	$(-1, 3)$	$(-3, 3)$	$(-4, 9)$

73. Sketch the graph of $y = -4x^2 + 8x$. Identify the vertex and the intercepts.

Solution

Leading coefficient test: Since $a < 0$, the parabola opens down.

Vertex: ($a = -4$, $b = 8$) $x = -\dfrac{b}{2a} = -\dfrac{8}{2(-4)} = -(-1) = 1$

$y = -4(1)^2 + 8(1) = -4 + 8 = 4$
The vertex is located at $(1, 4)$.

y-intercept: (Let $x = 0$.) $y = -4 \cdot 0^2 + 8 \cdot 0 = 0$; the y-intercept is $(0, 0)$.

x-intercepts: (Let $y = 0$.) $0 = -4x^2 + 8x \rightarrow -4x(x - 2) = 0$

$-4x = 0 \rightarrow x = 0$ and $x - 2 = 0 \rightarrow x = 2$

The x-intercepts are $(0, 0)$ and $(2, 0)$.

Additional solution points:

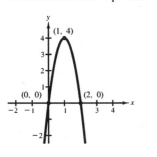

x	3	-1
$y = -4x^2 + 8x$	-12	-12
Points	$(3, -12)$	$(1, -12)$

75. Use a graphing utility to graph $y = \frac{1}{4}x^2 - x$. Graphically approximate the vertex.

Solution

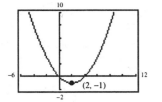

The vertex appears to be $(2, -1)$.

77. Use a graphing utility to graph $y = x^2 + 4x + 3$. Graphically approximate the vertex.

Solution

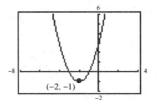

The vertex appears to be $(-2, -1)$.

79. Use a graphing utility to graph $y = 1 + 2x - \frac{1}{2}x^2$. Graphically approximate the vertex.

Solution

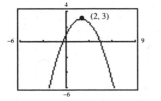

The vertex appears to be $(2, 3)$.

81. Use a graphing utility to graph $y = 0.1(x^2 + 4x + 14)$. Graphically approximate the vertex.

Solution

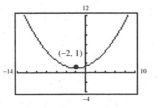

The vertex appears to be $(-2, 1)$.

83. *Tossing a Ball*   The height y (in feet) of a ball thrown on a parabolic path is modeled by $y = -\frac{1}{10}x^2 + 2x + 4$ where x is the horizontal distance (in feet) from where the ball was thrown (see figure in textbook).

(a) From what height was the ball thrown?

(b) What was the maximum height?

(c) How far away was the ball when it struck the ground?

Solution

(a) When $x = 0$, $y = -\frac{1}{10} \cdot 0^2 + 2(0) + 4 = 4$. The ball was thrown from a height of 4 feet.

(b) The maximum height is at the vertex of the parabola.

$$\text{Vertex: } x = -\frac{b}{2a} = -\frac{2}{2\left(-\frac{1}{10}\right)} = -\frac{2}{-\frac{1}{5}} = 10$$

$$y = -\frac{1}{10}(10)^2 + 2(10) + 4$$

$$y = -10 + 20 + 4 = 14$$

The vertex is (10, 14), so the maximum height of the ball was 14 feet.

(c) When the ball struck the ground, the height $y = 0$.

$$0 = -\frac{1}{10}x^2 + 2x + 4$$

$$10(0) = 10\left(-\frac{1}{10}x^2 + 2x + 4\right)$$

$$0 = -x^2 + 20x + 40$$

$$x^2 - 20x - 40 = 0$$

$$x = \frac{-(-20) \pm \sqrt{(-20)^2 - 4(1)(-40)}}{2(1)}$$

$$x = \frac{20 \pm \sqrt{560}}{2} = \frac{20 \pm 4\sqrt{35}}{2}$$

$$x = 10 \pm 2\sqrt{35} \qquad \text{Choose the positive answer.}$$

$$x = 10 + 2\sqrt{35} \approx 21.8$$

The ball struck the ground approximately 21.8 feet from where it was thrown.

85. Use a graphing utility to graph the functions on the same screen. Describe the relationship between the graphs.

(a) $y = x^2$ (b) $y = x^2 + 3$ (c) $y = x^2 - 3$ (d) $y = x^2 - 8$

Solution

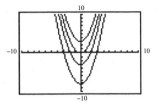

The graphs have the same shape but they are shifted up or down.

(a) The vertex is at the origin.

(b) The graph is shifted three units up from (a); the vertex is $(0, 3)$.

(c) The graph is shifted three units down from (a); the vertex is $(0, -3)$.

(d) The graph is shifted eight units down from (a); the vertex is $(0, -8)$.

87. *Exploration* Graph $y = (x - 2)^2 - 2$. Then describe how the vertex can be determined from the completed square form of the equation.

Solution

$$y = (x - 2)^2 - 2$$

$$y = x^2 - 4x + 4 - 2$$

$$y = x^2 - 4x + 2$$

Vertex: $x = -\dfrac{b}{2a} = -\dfrac{-4}{2(1)} = -(-2) = 2$

$$y = (2)^2 - 4(2) + 2 = 4 - 8 + 2 = -2$$

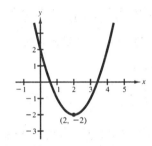

The vertex is located at $(2, -2)$. When the equation is written in the form $y = (x - h)^2 + k$, the vertex is located at (h, k).

89. Complete the square for the right side of $y = x^2 - 10x + 26$ and write it in the form of Exercise 87. What is the vertex of the parabola?

Solution

$y = x^2 - 10x + 26$ We want to rewrite this equation in the form $y = (x - h)^2 + k$.

$y = (x^2 - 10x) + 26$ Since the *left* side of the equation should remain unchanged, we will *not* subtract the 26 from both sides of the equation. Instead, we group the terms involving x's and leave the constant 26 outside the parentheses.

$y = (x^2 - 10x + 25) + 26 - 25$ Now we complete the square. Half of -10 is -5, and $(-5)^2 = 25$. Instead of adding 25 to *both* sides of the equation, we *add and subtract* 25 on the right side of the equation. Note that $(x^2 - 10x + 25) + 26 - 25 = x^2 - 10x + 26$.

$y = (x - 5)^2 + 1$ We rewrite $x^2 - 10x + 25$ as $(x - 5)^2$ and combine the constants outside the parentheses.

Vertex: $(5, 1)$ When the equation is written in the form $y = (x - h)^2 + k$, the vertex is (h, k).

11.5 | Applications of Quadratic Equations

7. The product of two consecutive positive integers is 132. Find the integers.

Solution

Verbal model: $\boxed{\text{First consecutive positive integer}} \cdot \boxed{\text{Second consecutive positive integer}} = \boxed{132}$

Labels: First consecutive positive integer $= n$
Second consecutive positive integer $= n + 1$

Equation:
$$n(n + 1) = 132$$
$$n^2 + n = 132$$
$$n^2 + n - 132 = 0$$
$$(n + 12)(n - 11) = 0$$
$$n + 12 = 0 \rightarrow n = -12 \text{ and } n + 1 = -11$$
$$n - 11 = 0 \rightarrow n = 11 \text{ and } n + 1 = 12$$

The problem specifies that these are *positive* integers, so we discard the negative answers. The two consecutive positive integers are 11 and 12.

9. *Falling Time* The height h (in feet) of a falling object at any time t (in seconds) is modeled by $h = h_0 - 16t^2$ where $h_0 = 1600$ is the intial height. Find the time necessary for the object to fall to ground level.

Solution

$$h = h_0 - 16t^2$$
$$h_0 = 1600 \text{ and } h = 0$$
$$0 = 1600 - 16t^2$$
$$0 = 16(100 - t^2)$$
$$0 = 16(10 + t)(10 - t)$$
$$16 \neq 0$$
$$10 + t = 0 \rightarrow t = -10$$
$$10 - t = 0 \rightarrow t = 10$$

We choose the positive answer for this application. Thus, the time required for the object to fall to ground level is 10 seconds.

11. *Falling Time*   The height h (in feet) of a falling object at any time t (in seconds) is modeled by $h = h_0 - 16t^2$ where $h_0 = 550$ (height of the Washington Monument) is the initial height. Find the time necessary for the object to fall to ground level.

Solution

$$h = h_0 - 16t^2$$

$$h_0 = 550 \text{ and } h = 0$$

$$0 = 550 - 16t^2$$

$$16t^2 = 550$$

$$t^2 = \frac{550}{16}$$

$$t = \pm\sqrt{\frac{550}{16}}$$

$$t = \pm\frac{\sqrt{25 \cdot 22}}{\sqrt{16}}$$

$$t = \pm\frac{5\sqrt{22}}{4} \approx \pm 5.86$$

We choose the positive answer for this application. Thus, the time required for the object to fall to ground level is approximately 5.86 seconds.

13. *Geometry*   Find the area of a rectangle with width $0.6l$, length l, and a perimeter of 64 in.

Solution

Formula: 2(Length) + 2(Width) = Perimeter

$$2l + 2(0.6l) = 64$$

$$2l + 1.2l = 64$$

$$3.2l = 64$$

$$l = 20$$

$$w = 0.6l = (0.6)(20) = 12$$

Formula: (Length)(Width) = Area

$$(20)(12) = A$$

$$240 = A$$

Width	Length	Perimeter	Area
12 in.	20 in.	64 in.	240 in.^{2}

15. *Geometry* Find the perimeter of a rectangle with width $\frac{1}{4}l$, length l, and area 100 in^2.

Solution

Formula: (Length)(Width) = Area

$$(l)\left(\tfrac{1}{4}l\right) = 100$$

$$\tfrac{1}{4}l^2 = 100$$

$$4\left(\tfrac{1}{4}l^2\right) = 4(100)$$

$$l^2 = 400$$

$$l = \pm\sqrt{400}$$

$$l = 20 \qquad \text{Discard the negative answer.}$$

$$w = \tfrac{1}{4}l = \tfrac{1}{4}(20) = 5$$

Formula: 2(Length) + 2(Width) = Perimeter

$$2(20) + 2(5) = P$$

$$40 + 10 = P$$

$$50 = P$$

Width	Length	Perimeter	Area
5 in.	20 in.	50 in.	100 in.^{2}

17. *Geometry* The height of a triangle is three times its base and the area of the triangle is 864 square inches. Find the dimensions of the triangle.

Solution

Verbal model: $\boxed{\tfrac{1}{2}}$ $\cdot$ $\boxed{\text{Height of triangle}}$ $\cdot$ $\boxed{\text{Base of triangle}}$ = $\boxed{\text{Area of triangle}}$

Labels: Base of triangle = x (inches)
Height of triangle = $3x$ (inches)
Area of triangle = 864 (square inches)

Equation:

$$\tfrac{1}{2}(3x)(x) = 864$$

$$\tfrac{3}{2}x^2 = 864$$

$$2\left(\tfrac{3}{2}x^2\right) = 2(864)$$

$$3x^2 = 1728$$

$$x^2 = 576$$

$$x = \pm\sqrt{576}$$

$$x = 24 \qquad \text{Choose positive answer.}$$

$$3x = 72$$

The height of the triangle is 72 inches and the base is 24 inches.

19. *Geometry* To add more space to your yard, you purchase an additional 4 feet along the side of your property (see figure shown in the textbook). The area of the lot is now 9600 square feet. What are the dimensions of the new lot?

Solution

Verbal model: $\boxed{\text{Length of property}} + \boxed{\text{Width of property}} = \boxed{\text{Area of property}}$

Labels: Length of property $= x + 4$ (feet)
Width of property $= x$ (feet)
Area of property $= 9600$ (square feet)

Equation:
$$(x + 4)(x) = 9600$$
$$x^2 + 4x = 9600$$
$$x^2 + 4x - 9600 = 0$$
$$(x + 100)(x - 96) = 0$$
$$x + 100 = 0 \rightarrow x = -100 \qquad \text{Exclude negative solution.}$$
$$x - 96 = 0 \rightarrow x = 96 \text{ and } x + 4 = 100$$

The lot is 96 feet wide and 100 feet long.

21. *Delivery Distance* You are delivering pizza to Offices B and C (see figure in the textbook) and are required to keep a log of all mileages between stops. You forget to look at the odometer at stop B, but after getting to stop C you record the total distance traveled from the pizza shop as 14 miles. The return distance on the beltway from C to A is 10 miles. The route forms a right triangle. Find the distance from A to B.

Solution

Verbal model: $\left(\boxed{\begin{array}{c}\text{Distance from}\\ \text{A to B}\end{array}} \right)^2 + \left(\boxed{\begin{array}{c}\text{Distance from}\\ \text{B to C}\end{array}} \right)^2 = \left(\boxed{\begin{array}{c}\text{Distance from}\\ \text{C to A}\end{array}} \right)^2$

Labels: Distance from A to B $= x$ (miles)
Distance from B to C $= 14 - x$ (miles)
Distance from C to A $= 10$ (miles)

Equation:
$$x^2 + (14 - x)^2 = 10^2$$
$$x^2 + 196 - 28x + x^2 = 100$$
$$2x^2 - 28x + 196 = 100$$
$$2x^2 - 28x + 96 = 0$$
$$x^2 - 14x + 48 = 0$$
$$(x - 6)(x - 8) = 0$$
$$x - 6 = 0 \rightarrow x = 6 \text{ and } 14 - x = 8$$
$$x - 8 = 0 \rightarrow x = 8 \text{ and } 14 - x = 6$$

The problem specifies that the distance from A to B is greater than the distance from B to C, so we choose the second answer. Thus, the distance from A to B is 8 miles.

23. *Bus Fares*   A science club charters a bus for $360 to attend a science fair. In order to lower the bus fare per member, the club invites nonmembers to go along. After ten nonmembers join the trip, the fare per person is decreased by $3. How many people are going on the excursion?

Solution

Verbal model:
Cost per person in larger group	$=$	Cost per person in smaller group	$-$	$3

Labels:
Total cost $= \$360$

Number of persons in smaller group $= x$

Cost per person in smaller group $= \dfrac{360}{x}$ (dollars)

Number of persons in larger group $= x + 10$

Cost per person in larger group $= \dfrac{360}{x + 10}$ (dollars)

Equation:

$$\frac{360}{x + 10} = \frac{360}{x} - 3$$

$$x(x + 10)\left(\frac{360}{x + 10}\right) = x(x + 10)\left(\frac{360}{x} - 3\right)$$

$$360x = 360(x + 10) - 3x(x + 10)$$

$$360x = 360x + 3600 - 3x^2 - 30x$$

$$3x^2 + 30x - 3600 = 0$$

$$x^2 + 10x - 1200 = 0$$

$$(x + 40)(x - 30) = 0$$

$$x + 40 = 0 \rightarrow x = -40 \text{ and } x + 10 = -30$$

$$x - 30 = 0 \rightarrow x = 30 \text{ and } x + 10 = 40$$

We choose the positive answers for this application. Thus, there are 40 people going on the trip.

25. *Air Speed*   An airline runs a commuter flight between two cities that are 480 miles apart. If the average speed of the planes could be increased by 20 miles per hour, the travel time would be decreased by 20 minutes. What air speed is required to obtain this decrease in travel time?

Solution

Verbal model: $\boxed{\begin{array}{c}\text{Travel time}\\\text{at faster speed}\end{array}} = \boxed{\begin{array}{c}\text{Travel time}\\\text{at slower speed}\end{array}} - \boxed{\dfrac{1}{3}}$

Labels: Distance of travel $= 480$ (miles)

Slower speed $= x$ (miles per hour)

Travel time at slower speed $= \dfrac{480}{x}$ (hours)

Faster speed $= x + 20$ (miles per hour)

Travel time at faster speed $= \dfrac{480}{x + 20}$ (hours)   *Note:* 20 minutes $= \dfrac{1}{3}$ hour

Equation: $\dfrac{480}{x + 20} = \dfrac{480}{x} - \dfrac{1}{3}$

$$3x(x + 20)\left(\dfrac{480}{x + 20}\right) = 3x(x + 20)\left(\dfrac{480}{x} - \dfrac{1}{3}\right)$$

$$3x(480) = 3(x + 20)(480) - x(x + 20)$$

$$1440x = 1440(x + 20) - x^2 - 20x$$

$$1440x = 1440x + 28{,}800 - x^2 - 20x$$

$$x^2 + 20x - 28{,}800 = 0$$

$$(x + 180)(x - 160) = 0$$

$$x + 180 = 0 \rightarrow x = -180 \text{ and } x + 20 = -160$$

$$x - 160 = 0 \rightarrow x = 160 \text{ and } x + 20 = 180$$

We choose the positive answers for this application. Thus, the faster airspeed required is 180 miles per hour.

27. *Work Rate*   Working together, two people can complete a task in 4 hours. Working alone, one person takes 6 hours longer than the other. How long would it take each person to do the task alone?

Solution

Verbal model: $\boxed{\begin{array}{c}\text{Rate for}\\\text{faster person}\end{array}} + \boxed{\begin{array}{c}\text{Rate for}\\\text{slower person}\end{array}} = \boxed{\begin{array}{c}\text{Rate for both}\\\text{working together}\end{array}}$

Labels: Both people: time = 4 (hours); rate = 1/4 (tasks per hour)
Faster person: time = x (hours); rate = $1/x$ (tasks per hour)
Slower person: time = $x + 6$ (hours); rate = $1/(x + 6)$ (tasks per hour)

Equation:
$$\frac{1}{x} + \frac{1}{x + 6} = \frac{1}{4}$$

$$4x(x + 6)\left(\frac{1}{x}\right) + 4x(x + 6)\left(\frac{1}{x + 6}\right) = 4x(x + 6)\left(\frac{1}{4}\right)$$

$$4(x + 6) + 4x = x(x + 6)$$

$$4x + 24 + 4x = x^2 + 6x$$

$$8x + 24 = x^2 + 6x$$

$$0 = x^2 - 2x - 24$$

$$0 = (x - 6)(x + 4)$$

$$x - 6 = 0 \rightarrow x = \ \ \ 6 \text{ and } x + 6 = 12$$

$$x + 4 = 0 \rightarrow x = -4 \text{ and } x + 6 = 2$$

We choose the positive answer for x in this application. Thus, the faster person could complete the job in 6 hours, and the slower person could complete the job in 12 hours.

29. *Compound Interest*   Find the interest rate r. The amount $A = \$1123.60$ in an account earning r percent compounded annually for 2 years is given by $A = P(1 + r)^2$, where $P = \$1000$ is the original investment.

Solution

$$A = P(1 + r)^2$$

$$1123.60 = 1000(1 + r)^2$$

$$\frac{1123.60}{1000} = \frac{1000(1 + r)^2}{1000}$$

$$1.1236 = (1 + r)^2$$

$$\pm\sqrt{1.1236} = 1 + r$$

$$\pm 1.06 = 1 + r$$

$$-1 \pm 1.06 = r$$

$$-1 + 1.06 = r \qquad \text{Exclude negative solution.}$$

$$0.06 = r$$

The interest rate is 6%.

31. *Compound Interest*   Find the interest rate r. The amount $A = \$235.44$ in an account earning r percent compounded annually for 2 years is given by $A = P(1 + r)^2$, where $P = \$200$ is the original investment.

Solution

$$A = P(1 + r)^2$$

$$235.44 = 200(1 + r)^2$$

$$\frac{235.44}{200} = \frac{200(1 + r)^2}{200}$$

$$1.1722 = (1 + r)^2$$

$$\pm\sqrt{1.1722} = 1 + r$$

$$\pm 1.085 \approx 1 + r$$

$$-1 \pm 1.085 \approx r$$

$$-1 + 1.085 \approx r \qquad \text{Exclude negative solution.}$$

$$0.085 \approx r$$

The interest rate is 8.5%.

33. Find $-2x^2(5x^3)$.

Solution

$-2x^2(5x^3) = -10x^5$

35. Find $(x + 7)^2$.

Solution

$(x + 7)^2 = x^2 + 14x + 49$

37. *List Price*   A computer was discounted 15% off the list price to sell for \$1450. What was the list price?

Solution

Verbal model: | List price | $-$ | Discount | $=$ | Sale price |

Labels: List price $= x$ (dollars)
Discount $= 0.15x$ (dollars)
Sale price $= 1450$ (dollars)

Equation: $x - 0.15x = 1450$

$$0.85x = 1450$$

$$x = \frac{1450}{0.85}$$

$$x \approx 1705.88$$

The list price was approximately \$1705.88.

39. *Problem Solving*   Find two positive integers satisfying the requirement, "the product of two consecutive integers is 240."

Solution

Verbal model: | First consecutive positive integer | $\cdot$ | Second consecutive positive integer | $=$ | 420 |

Labels: First consecutive positive integer $= n$
Second consecutive positive integer $= n + 1$

Equation: $$n(n + 1) = 420$$

$$n^2 + n = 420$$

$$n^2 + n - 420 = 0$$

$$(n + 21)(n - 20) = 0$$

$$n + 21 = 0 \rightarrow n = -21 \text{ and } n + 1 = -20$$

$$n - 20 = 0 \rightarrow n = 20 \text{ and } n + 1 = 21$$

The problem specifies that these are *positive* integers, so we discard the negative answers. The two consecutive positive integers are 20 and 21.

41. *Problem Solving* Find two positive integers satisfying the requirement, "the product of two consecutive even integers is 168."

Solution

Verbal model: $\boxed{\text{First consecutive positive even integer}} \cdot \boxed{\text{Second consecutive positive even integer}} = \boxed{168}$

Labels: First consecutive positive even integer $= 2n$
Second consecutive positive even integer $= 2n + 2$

Equation:
$$2n(2n + 2) = 168$$
$$4n^2 + 4n = 168$$
$$4n^2 + 4n - 168 = 0$$
$$n^2 + n - 42 = 0 \qquad \text{Divide both sides by 4.}$$
$$(n + 7)(n - 6) = 0$$
$$n + 7 = 0 \rightarrow n = -7 \text{ and } 2n = -14, \ 2n + 2 = -12$$
$$n - 6 = 0 \rightarrow n = \ \ 6 \text{ and } 2n = 12, \ 2n + 2 = 14$$

The problem specifies that the integers are *positive*, so we discard the negative answers. The two consecutive positive even integers are 12 and 14.

43. *Problem Solving* Find two positive integers satisfying the requirement, "the sum of the squares of two consecutive integers is 113."

Solution

Verbal model: $\left(\boxed{\text{First consecutive positive integer}}\right)^2 + \left(\boxed{\text{Second consecutive positive integer}}\right)^2 = \boxed{113}$

Labels: First consecutive positive integer $= n$
Second consecutive positive integer $= n + 1$

Equation:
$$n^2 + (n + 1)^2 = 113$$
$$n^2 + n^2 + 2n + 1 = 113$$
$$2n^2 + 2n + 1 = 113$$
$$2n^2 + 2n - 112 = 0$$
$$n^2 + n - 56 = 0 \qquad \text{Divide both sides by 2.}$$
$$(n + 8)(n - 7) = 0$$
$$n + 8 = 0 \rightarrow n = -8 \text{ and } n + 1 = -7$$
$$n - 7 = 0 \rightarrow n = \ \ 7 \text{ and } n + 1 = 8$$

The problem specifies that the integers are *positive*, so we discard the negative answers. The two consecutive positive integers are 7 and 8.

45. *Geometry* Find the area of a rectangle with width w, length $w + 4$, and a perimeter of 56 km.

Solution

Formula: 2(Length) + 2(Width) = Perimeter

$$2(w + 4) + 2(w) = 56$$

$$2w + 8 + 2w = 56$$

$$4w + 8 = 56$$

$$4w = 48$$

$$w = 12$$

$$l = w + 4 = 12 + 4 = 16$$

Formula: (Length)(Width) = Area

$$(16)(12) = A$$

$$192 = A$$

Width	Length	Perimeter	Area
12 km	16 km	56 km	192 km^2

47. *Geometry* Find the perimeter of a rectangle with width w, length $2w$, and area of 50 ft^2.

Solution

Formula: (Length)(Width) = Area

$$(2w)(w) = 50$$

$$2w^2 = 50$$

$$w^2 = 25$$

$$w = \pm\sqrt{25}$$

$$w = \pm 5$$

$$w = 5 \qquad \text{Choose the positive answer.}$$

$$l = 2w = 10$$

Formula: 2(Length) + 2(Width) = Perimeter

$$2 \cdot 10 + 2 \cdot 5 = P$$

$$20 + 10 = P$$

$$30 = P$$

Width	Length	Perimeter	Area
5 ft	10 ft	30 ft	50 ft^2

49. *Geometry*   Find the perimeter of a rectangle with width $l - 10$, length l, and area 75 m^2.

Solution

Formula: (Length)(Width) = Area

$$l(l - 10) = 75$$

$$l^2 - 10l = 75$$

$$l^2 - 10l - 75 = 0$$

$$(l - 15)(l + 5) = 0$$

$$l - 15 = 0 \rightarrow l = 15 \text{ and } \quad l - 10 = 5$$

$$l + 5 = 0 \rightarrow l = -5 \qquad \text{Choose positive answer.}$$

Formula: 2(Length) + 2(Width) = Perimeter

$$2(15) + 2(5) = P$$

$$30 + 10 = P$$

$$40 = P$$

Width	Length	Perimeter	Area
5 m	15 m	40 m	75 m^2

51. *Geometry*   You have 200 feet of fencing to enclose two adjacent rectangular corrals (see figure in textbook). The total area of the enclosed region is 1400 square feet. What are the dimensions of each corral? (The corrals are the same size.)

Solution

Verbal model: 2 $\boxed{\text{Length of corral}}$ $\cdot$ $\boxed{\text{Width of corral}}$ = $\boxed{\text{Area of enclosed region}}$

Labels: Length of corral = x (feet)

Width of corral = $\dfrac{200 - 4x}{3}$ (feet)

Area of enclosed region = 1440 (square feet)

Equation:

$$2(x)\left(\frac{200 - 4x}{3}\right) = 1400$$

$$3(2)(x)\left(\frac{200 - 4x}{3}\right) = 3(1400)$$

$$2x(200 - 4x) = 4200$$

$$400x - 8x^2 = 4200$$

$$-8x^2 + 400x - 4200 = 0$$

$$8x^2 - 400x + 4200 = 0$$

$$x^2 - 50x + 525 = 0$$

$$(x - 35)(x - 15) = 0$$

$$x - 35 = 0 \rightarrow x = 35 \text{ and } \frac{200 - 4x}{3} = 20$$

$$x - 15 = 0 \rightarrow x = 15 \text{ and } \frac{200 - 4x}{3} = 46\frac{2}{3}$$

The dimensions of each corral could be 35 feet by 20 feet or 15 feet by $46\frac{2}{3}$ feet (or 15 feet by 46 feet 8 inches).

53. *Geometry*   The height of a triangle is one-third its base and the area of the triangle is 24 square inches. Find the dimensions of the triangle.

Solution

Verbal model: $\boxed{\frac{1}{2}} \cdot \boxed{\text{Height of triangle}} \cdot \boxed{\text{Base of triangle}} = \boxed{\text{Area of triangle}}$

Labels: Base of triangle $= x$ (inches)
Height of triangle $= \frac{1}{3}x$ (inches)
Area of triangle $= 24$ (square inches)

Equation: $\frac{1}{2}\left(\frac{1}{3}x\right)(x) = 24$

$\frac{1}{6}x^2 = 24$

$6\left(\frac{1}{6}x^2\right) = 6(24)$

$x^2 = 144$

$x = \pm\sqrt{144}$

$x = 12$ Choose positive answer.

$\frac{1}{3}x = \frac{1}{3}(12) = 4$

The base of the triangle is 12 inches and the height is 4 inches.

55. *Length of a Ladder*   A 20-foot ladder is leaning against a building. The ladder must reach a point 19 feet above the ground. Determine the distance from the base of the ladder to the building.

Solution

$a^2 + b^2 = c^2$

$a^2 + 19^2 = 20^2$

$a^2 + 361 = 400$

$a^2 = 39$

$a = \pm\sqrt{39}$

$a \approx 6.24$ Discard negative solution.

The base of the ladder is approximately 6.24 feet from the building.

57. *Work Rate* A farmer has two combines. Combine B is known to take 2 hours longer than Combine A to harvest a field. Using both machines, it takes 4 hours to harvest the field. How long would it take to harvest the field using each machine individually?

Solution

Verbal model: $\boxed{\text{Rate for Combine A}} + \boxed{\text{Rate for Combine B}} = \boxed{\text{Rate of the two together}}$

Labels: Time for Combine A to complete job $= x$ (hours)

Rate for Combine A $= \dfrac{1}{x}$ (jobs per hour)

Time for Combine B to complete job $= x + 2$ (hours)

Rate for Combine B $= \dfrac{1}{x+2}$ (jobs per hour)

Time for the two combines to complete job $= 4$ (hours)

Rate for the two together $= \dfrac{1}{4}$ (jobs per hour)

Equation:

$$\frac{1}{x} + \frac{1}{x+2} = \frac{1}{4}$$

$$4x(x+2)\left(\frac{1}{x} + \frac{1}{x+2}\right) = 4x(x+2)\left(\frac{1}{4}\right)$$

$$4(x+2)(1) + 4x(1) = x(x+2)$$

$$4x + 8 + 4x = x^2 + 2x$$

$$8x + 8 = x^2 + 2x$$

$$0 = x^2 - 6x - 8$$

$$a = 1,\ b = -6,\ c = -8$$

$$x = \frac{-b \pm \sqrt{b^2 - 4ac}}{2a}$$

$$x = \frac{6 \pm \sqrt{(-6)^2 - 4(1)(-8)}}{2(1)}$$

$$x = \frac{6 \pm \sqrt{68}}{2}$$

$$x = \frac{6 + \sqrt{68}}{2} \approx 7.12 \rightarrow x + 2 \approx 9.12$$

$$x = \frac{6 - \sqrt{68}}{2} \approx -1.12 \qquad \text{Discard negative solution.}$$

Combine A would take approximately 7.12 hours and Combine B would take approximately 9.12 hours to harvest the field.

59. *Venture Capital*   Sixty thousand dollars is needed to begin a small business. The cost will be divided equally among investors. Some have made a commitment to invest. By finding five more investors, the amount required from each would decrease by $1000. How many have made a commitment to invest in the business?

Solution

Verbal model: | Current price per investor | $- 1000 =$ | Price per investor with five more investors |

Labels: Current number of investors $= x$

Current price per investor $= \dfrac{60,000}{x}$ (dollars)

Price per investor with five more investors $= \dfrac{60,000}{x+5}$ (dollars)

Equation:

$$\frac{60,000}{x} - 1000 = \frac{60,000}{x+5}$$

$$x(x+5)\left(\frac{60,000}{x} - 1000\right) = x(x+5)\left(\frac{60,000}{x+5}\right)$$

$$(x+5)(60,000) - 1000x(x+5) = x(60,000)$$

$$60,000x + 300,000 - 1000x^2 - 5000x = 60,000x$$

$$-1000x^2 - 5000x + 300,000 = 0$$

$$x^2 + 5x - 300 = 0$$

$$(x+20)(x-15) = 0$$

$$x + 20 = 0 \rightarrow x = -20 \qquad \text{Discard negative solution.}$$

$$x - 15 = 0 \rightarrow x = 15$$

Fifteen people have made a commitment to invest in the business.

Review Exercises for Chapter 11

1. Solve $x^2 + 10x = 0$ by factoring.

Solution

$$x^2 + 10x = 0$$

$$x(x + 10) = 0$$

$$x = 0$$

$$x + 10 = 0 \rightarrow x = -10$$

The solutions are 0 and -10.

3. Solve $5z(z + 1) - 8(z + 1) = 0$ by factoring.

Solution

$$5z(z + 1) - 8(z + 1) = 0$$

$$(z + 1)(5z - 8) = 0$$

$$z + 1 = 0 \rightarrow z = -1$$

$$5z - 8 = 0 \rightarrow 5z = 8 \rightarrow z = \tfrac{8}{5}$$

The solutions are -1 and $\tfrac{8}{5}$.

5. Solve $4y^2 - 25 = 0$ by factoring.

Solution
$$4y^2 - 25 = 0$$
$$(2y + 5)(2y - 5) = 0$$
$$2y + 5 = 0 \rightarrow 2y = -5 \rightarrow y = -\tfrac{5}{2}$$
$$2y - 5 = 0 \rightarrow 2y = 5 \rightarrow y = \tfrac{5}{2}$$

The solutions are $\tfrac{5}{2}$ and $-\tfrac{5}{2}$.

7. Solve $x^2 + x + \tfrac{1}{4} = 0$ by factoring.

Solution
$$x^2 + x + \tfrac{1}{4} = 0$$
$$\left(x + \tfrac{1}{2}\right)\left(x + \tfrac{1}{2}\right) = 0$$
$$x + \tfrac{1}{2} = 0 \rightarrow x = -\tfrac{1}{2}$$

The repeated solution is $x = -\tfrac{1}{2}$.

Note: We could multiply both sides of the equation by 4 and then factor.
$$4\left(x^2 + x + \tfrac{1}{4}\right) = 4(0)$$
$$4x^2 + 4x + 1 = 0$$
$$(2x + 1)(2x + 1) = 0$$
$$2x + 1 = 0 \rightarrow 2x = -1 \rightarrow x = -\tfrac{1}{2}$$

9. Solve $x(x - 3) = 4$ by factoring.

Solution
$$x(x - 3) = 4$$
$$x^2 - 3x = 4$$
$$x^2 - 3x - 4 = 0$$
$$(x - 4)(x + 1) = 0$$
$$x - 4 = 0 \rightarrow x = 4$$
$$x + 1 = 0 \rightarrow x = -1$$

The solutions are 4 and -1.

11. Solve $x^2 - x - 56 = 0$ by factoring.

Solution
$$x^2 - x - 56 = 0$$
$$(x - 8)(x + 7) = 0$$
$$x - 8 = 0 \rightarrow x = 8$$
$$x + 7 = 0 \rightarrow x = -7$$

The solutions are 8 and -7.

13. Solve $x^2 = 625$ by extracting square roots.

Solution
$$x^2 = 625$$
$$x = \pm\sqrt{625}$$
$$x = \pm 25$$

The solutions are 25 and -25.

15. Solve $y^2 + 8 = 0$ by extracting square roots.

Solution
$$y^2 + 8 = 0$$
$$y^2 = -8$$

The square of a real number cannot be negative. Thus, this equation has *no* real number solution.

17. Solve $(x - 15)^2 = 400$ by extracting square roots.

Solution

$$(x - 15)^2 = 400$$
$$x - 15 = \pm\sqrt{400}$$
$$x - 15 = \pm 20$$
$$x = 15 \pm 20$$
$$x = 15 + 20 = 35$$
$$x = 15 - 20 = -5$$

The solutions are 35 and -5.

21. Solve $x^2 - x - 1 = 0$ by completing the square.

Solution

$$x^2 - x - 1 = 0$$
$$x^2 - x = 1$$
$$x^2 - x + \frac{1}{4} = 1 + \frac{1}{4} \quad \text{Half of } -1 \text{ is } -\frac{1}{2}$$
$$\text{and } \left(-\frac{1}{2}\right)^2 = \frac{1}{4}.$$
$$\left(x - \frac{1}{2}\right)^2 = \frac{5}{4}$$
$$x - \frac{1}{2} = \pm\sqrt{\frac{5}{4}}$$
$$x - \frac{1}{2} = \pm\frac{\sqrt{5}}{\sqrt{4}}$$
$$x - \frac{1}{2} = \pm\frac{\sqrt{5}}{2}$$
$$x = \frac{1}{2} \pm \frac{\sqrt{5}}{2}$$
$$x = \frac{1 \pm \sqrt{5}}{2}$$

The solutions are $\dfrac{1 + \sqrt{5}}{2}$ and $\dfrac{1 - \sqrt{5}}{2}$.

19. Solve $x^2 - 6x - 1 = 0$ by completing the square.

Solution

$$x^2 - 6x - 1 = 0$$
$$x^2 - 6x = 1$$
$$x^2 - 6x + 9 = 1 + 9 \quad \text{Half of } -6 \text{ is } -3$$
$$\text{and } (-3)^2 = 9.$$
$$(x - 3)^2 = 10$$
$$x - 3 = \pm\sqrt{10}$$
$$x = 3 \pm \sqrt{10}$$

The solutions are $3 + \sqrt{10}$ and $3 - \sqrt{10}$.

23. Solve $t^2 + 3t + 1 = 0$ by completing the square.

Solution

$$t^2 + 3t + 1 = 0$$
$$t^2 + 3t = -1$$
$$t^2 + 3t + \frac{9}{4} = -1 + \frac{9}{4} \quad \text{Half of 3 is } \frac{3}{2}$$
$$\text{and } \left(\frac{3}{2}\right)^2 = \frac{9}{4}.$$
$$\left(t + \frac{3}{2}\right)^2 = -\frac{4}{4} + \frac{9}{4}$$
$$\left(t + \frac{3}{2}\right)^2 = \frac{5}{4}$$
$$t + \frac{3}{2} = \pm\sqrt{\frac{5}{4}}$$
$$t + \frac{3}{2} = \pm\frac{\sqrt{5}}{\sqrt{4}}$$
$$t + \frac{3}{2} = \pm\frac{\sqrt{5}}{2}$$
$$t = -\frac{3}{2} \pm \frac{\sqrt{5}}{2}$$
$$t = \frac{-3 \pm \sqrt{5}}{2}$$

The solutions are $\dfrac{-3 + \sqrt{5}}{2}$ and $\dfrac{-3 - \sqrt{5}}{2}$.

25. Solve $y^2 + y - 42 = 0$ by using the Quadratic Formula.

Solution

$$y^2 + y - 42 = 0$$

$$a = 1, \quad b = 1, \quad c = -42$$

$$y = \frac{-1 \pm \sqrt{1^2 - 4(1)(-42)}}{2(1)}$$

$$y = \frac{-1 \pm \sqrt{169}}{2}$$

$$y = \frac{-1 \pm 13}{2}$$

$$y = \frac{-1 + 13}{2} = \frac{12}{2} = 6$$

$$y = \frac{-1 - 13}{2} = \frac{-14}{2} = -7$$

The solutions are 6 and -7.

27. Solve $2y^2 + y - 42 = 0$ by using the Quadratic Formula.

Solution

$$2y^2 + y - 42 = 0$$

$$a = 2, \quad b = 1, \quad c = -42$$

$$y = \frac{-1 \pm \sqrt{1^2 - 4(2)(-42)}}{2(2)}$$

$$y = \frac{-1 \pm \sqrt{337}}{4}$$

The solutions are

$$\frac{-1 + \sqrt{337}}{4} \text{ and } \frac{-1 - \sqrt{337}}{4}.$$

29. Solve $0.3t^2 - 2t + 1 = 0$ by using the Quadratic Formula.

Solution

$$0.3t^2 - 2t + 1 = 0$$

$$a = 0.3, \quad b = -2, \quad c = 1$$

$$t = \frac{-(-2) \pm \sqrt{(-2)^2 - 4(0.3)(1)}}{2(0.3)}$$

$$t = \frac{2 \pm \sqrt{2.8}}{0.6}$$

$$t = \frac{2 + \sqrt{2.8}}{0.6} \approx 6.12$$

$$t = \frac{2 - \sqrt{2.8}}{0.6} \approx 0.54$$

The solutions are *approximately* 6.12 and 0.54.
Note: If you multiply both sides of the original equation by 10 to obtain $3t^2 - 20t + 10 = 0$, the solutions will be in the *equivalent* form

$$t = \frac{10 \pm \sqrt{70}}{3}.$$

31. Solve $v^2 = 250$ by using the Quadratic Formula.

Solution

$$v^2 = 250$$

$$v = \pm\sqrt{250}$$

$$v = \pm\sqrt{25 \cdot 10}$$

$$v = \pm 5\sqrt{10}$$

The solutions are $5\sqrt{10}$ and $-5\sqrt{10}$.

33. Solve $4x^2 + 4x + 1 = 0$ by using the Quadratic Formula.

Solution

$$4x^2 + 4x + 1 = 0$$

$$(2x + 1)(2x + 1) = 0$$

$$2x + 1 = 0 \rightarrow 2x = -1 \rightarrow x = -\tfrac{1}{2}$$

The repeated solution is $-\tfrac{1}{2}$.

35. Solve $50 - (x - 6)^2 = 0$ by using the Quadratic Formula.

Solution

$$50 - (x - 6)^2 = 0$$

$$-(x - 6)^2 = -50$$

$$(x - 6)^2 = 50 \qquad \text{Multiply both sides by } -1.$$

$$x - 6 = \pm\sqrt{50}$$

$$x - 6 = \pm\sqrt{25 \cdot 2}$$

$$x - 6 = \pm 5\sqrt{2}$$

$$x = 6 \pm 5\sqrt{2}$$

The solutions are $6 + 5\sqrt{2}$ and $6 - 5\sqrt{2}$.

37. Solve $c^2 - 6c + 6 = 0$ by using the Quadratic Formula.

Solution

$$c^2 - 6c + 6 = 0$$

$$c = \frac{-(-6) \pm \sqrt{(-6)^2 - 4(1)(6)}}{2(1)}$$

$$c = \frac{6 \pm \sqrt{12}}{2}$$

$$c = \frac{6 \pm 2\sqrt{3}}{2}$$

$$c = \frac{2(3 \pm \sqrt{3})}{2}$$

$$c = 3 \pm \sqrt{3}$$

The solutions are $3 + \sqrt{3}$ and $3 - \sqrt{3}$.

39. Solve $-x^2 + 3x + 70 = 0$ by using the Quadratic Formula.

Solution

$$-x^2 + 3x + 70 = 0$$

$$x^2 - 3x - 70 = 0 \qquad \text{Multiply both sides by } -1.$$

$$(x - 10)(x + 7) = 0$$

$$x - 10 = 0 \rightarrow x = 10$$

$$x + 7 = 0 \rightarrow x = -7$$

The solutions are 10 and -7.

41. Solve $\dfrac{1}{x} + \dfrac{1}{x+1} = \dfrac{1}{2}$ by using the Quadratic Formula.

Solution

$$\frac{1}{x} + \frac{1}{x+1} = \frac{1}{2}$$

$$2x(x+1)\left(\frac{1}{x}\right) + 2x(x+1)\left(\frac{1}{x+1}\right) = 2x(x+1)\left(\frac{1}{2}\right)$$

$$2(x+1) + 2x = x(x+1)$$

$$2x + 2 + 2x = x^2 + x$$

$$4x + 2 = x^2 + x$$

$$0 = x^2 - 3x - 2$$

$$x = \frac{-(-3) \pm \sqrt{(-3)^2 - 4(1)(-2)}}{2(1)}$$

$$x = \frac{3 \pm \sqrt{17}}{2}$$

The solutions are $\dfrac{3 + \sqrt{17}}{2}$ and $\dfrac{3 - \sqrt{17}}{2}$.

43. Sketch the graph of $y = (x-4)^2$. Identify the vertex and any intercepts.

Solution

$y = (x-4)^2$ or $y = x^2 - 8x + 16$

Leading coefficient test: Since $a > 0$, the parabola opens up.

Vertex: $(a = 1,\ b = -8)$ $x = -\dfrac{b}{2a} = -\dfrac{-8}{2(1)} = -(-4) = 4$

$y = 4^2 - 8 \cdot 4 + 16 = 0$
The vertex is located at $(4, 0)$.

y-intercept: (Let $x = 0$.) $y = 0^2 - 8 \cdot 0 + 16 = 16$; the y-intercept is $(0, 16)$.

x-intercept: (Let $y = 0$.) $0 = (x-4)^2$

$x - 4 = 0 \rightarrow x = 4$

The x-intercept is $(4, 0)$.

Additional solution points:

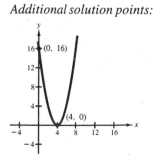

x	8	6	2	1
$y = (x-4)^2$	16	4	4	9
Points	$(8, 16)$	$(6, 4)$	$(2, 4)$	$(1, 9)$

45. Sketch the graph of $y = 3 - (x - 4)^2$. Identify the vertex and any intercepts.

Solution

$$y = 3 - (x - 4)^2 \text{ or } y = 3 - (x^2 - 8x + 16)$$

$$y = 3 - x^2 + 8x - 16$$

$$y = -x^2 + 8x - 13$$

Leading coefficient test: Since $a < 0$, the parabola opens down.

Vertex: $(a = -1,\ b = 8)$ $x = -\dfrac{b}{2a} = -\dfrac{8}{2(-1)} = -(-4) = 4$

$$y = -4^2 + 8 \cdot 4 - 13 = -16 + 32 - 13 = 3$$

The vertex is located at $(4, 3)$.

y-intercept: (Let $x = 0$.) $y = -0^2 + 8 \cdot 0 - 13 = -13$; the y-intercept is $(0, -13)$.

x-intercepts: (Let $y = 0$.) $0 = -x^2 + 8x - 13$

$$x^2 - 8x + 13 = 0$$

$$x = \frac{-(-8) \pm \sqrt{(-8)^2 - 4(1)(13)}}{2(1)}$$

$$x = \frac{8 \pm \sqrt{12}}{2} = \frac{8 \pm 2\sqrt{3}}{2} = 4 \pm \sqrt{3}$$

The x-intercepts are $\left(4 + \sqrt{3}, 0\right)$ and $\left(4 - \sqrt{3}, 0\right)$.

Additional solution points:

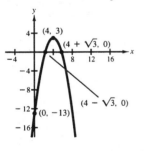

x	8	7	6	2
$y = 3 - (x - 4)^2$	-13	-6	-1	-1
Points	$(8, -13)$	$(7, -6)$	$(6, -1)$	$(2, -1)$

47. Sketch the graph of $y = -x^2 + 3x$. Identify the vertex and any intercepts.

Solution

Leading coefficient test: Since $a < 0$, the parabola opens down.

Vertex: ($a = -1$, $b = 3$)
$$x = -\frac{b}{2a} = -\frac{3}{2(-1)} = -\left(-\frac{3}{2}\right) = \frac{3}{2}$$

$$y = -\left(\frac{3}{2}\right)^2 + 3\left(\frac{3}{2}\right) = -\frac{9}{4} + \frac{9}{2} = \frac{9}{4}$$

The vertex is located at $\left(\frac{3}{2}, \frac{9}{4}\right)$.

y-intercept: (Let $x = 0$.) $y = -0^2 + 3 \cdot 0 = 0$; the y-intercept is $(0, 0)$.

x-intercepts: (Let $y = 0$.) $0 = -x^2 + 3x \rightarrow 0 = -x(x - 3)$

$-x = 0 \rightarrow x = 0$

$x - 3 = 0 \rightarrow x = 3$

The x-intercepts are $(0, 0)$ and $(3, 0)$.

Additional solution points:

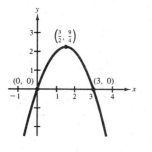

x	4	5	-2	-1
$y = -x^2 + 3x$	-4	-10	-10	-4
Points	$(4, -4)$	$(5, -10)$	$(-2, -10)$	$(-1, -4)$

49. Sketch the graph of $y = \frac{1}{5}x^2 - 2x + 4$. Identify the vertex and any intercepts.

Solution

Leading coefficient test:

Vertex: $\left(a = \dfrac{1}{5},\ b = -2\right)$

Since $a > 0$, the parabola opens up.

$$x = -\frac{b}{2a} = -\frac{-2}{2(1/5)} = 5$$

$$y = \frac{1}{5}(5^2) - 2(5) + 4 = 5 - 10 + 4 = -1$$

The vertex is located at $(5, -1)$.

y-intercept: (Let $x = 0$.)

$$y = \frac{1}{5} \cdot 0^2 - 2 \cdot 0 + 4 = 4;\ \text{the } y\text{-intercept is } (0, 4).$$

x-intercepts: (Let $y = 0$.)

$$0 = \frac{1}{5}x^2 - 2x + 4 = x^2 - 10x + 20$$

$$x = \frac{-(-10) \pm \sqrt{(-10)^2 - 4(1)(20)}}{2(1)} = \frac{10 \pm \sqrt{20}}{2} = \frac{10 \pm 2\sqrt{5}}{2} = 5 \pm \sqrt{5}$$

The x-intercepts are $(5 + \sqrt{5}, 0)$ and $(5 - \sqrt{5}, 0)$.

Additional solution points:

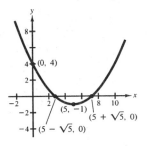

x	10	8	1	-2
$y = \frac{1}{5}x^2 - 2x + 4$	4	0.8	2.2	8.8
Points	$(10, 4)$	$(8, 0.8)$	$(1, 2.2)$	$(-2, 8.8)$

51. Sketch the graph of $y = 2x^2 + 4x + 5$. Identify the vertex and any intercepts.

Solution

$y = 2x^2 + 4x + 5$

Leading coefficient test:

Since $a > 0$, the parabola opens up.

Vertex: $(a = 2, \ b = 4)$

$x = -\dfrac{b}{2a} = -\dfrac{4}{2(2)} = -1$

$y = 2(-1)^2 + 4(-1) + 5 = 2 - 4 + 5 = 3$

The vertex is located at $(-1, 3)$.

y-intercept: (Let $x = 0$.) $y = 2 \cdot 0^2 + 4 \cdot 0 + 5 = 5$; the y-intercept is $(0, 5)$.

x-intercepts: (Let $y = 0$.) $0 = 2x^2 + 4x + 5$

$x = \dfrac{-4 \pm \sqrt{4^2 - 4(2)(5)}}{2(2)} = \dfrac{-4 \pm \sqrt{-24}}{4}$

There are *no* x-intercepts.

Additional solution points:

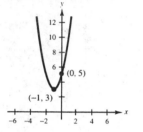

x	1	-2	-3
$y = 2x^2 + 4x + 5$	11	5	11
Points	$(1, 11)$	$(-2, 5)$	$(-3, 11)$

53. Sketch the graph of $y = -x^2 + 4x - 3$. Identify the vertex and any intercepts.

Solution

Leading coefficient test:

Since $a < 0$, the parabola opens down.

Vertex: $\left(a = -1, \ b = 4 \right)$

$x = -\dfrac{b}{2a} = -\dfrac{4}{2(-1)} = -(-2) = 2$

$y = -(2)^2 + 4(2) - 3 = 1$

The vertex is located at $(2, 1)$.

y-intercept: (Let $x = 0$.) $y = -0^2 + 4(0) - 3 = -3$; the y-intercept is $(0, -3)$.

x-intercepts: (Let $y = 0$.) $0 = -x^2 + 4x - 3 = x^2 - 4x + 3 = (x - 3)(x - 1)$

$x - 3 = 0 \rightarrow x = 3$

$x - 1 = 0 \rightarrow x = 1$

The x-intercepts are $(3, 0)$ and $(1, 0)$.

Additional solution points:

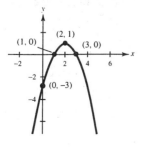

x	-1	4	5
$y = -x^2 + 4x - 3$	-8	-3	-8
Points	$(-1, -8)$	$(4, -3)$	$(5, -8)$

55. Use a graphing utility to graph $y = 3 - x^2$. Graphically approximate the vertex.

Solution

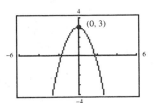

The vertex appears to be $(0, 3)$. The y-intercept is also $(0, 3)$. The x-intercepts are approximately $(1.73, 0)$ and $(-1.73, 0)$.

57. Use a graphing utility to graph $y = (x - 1)^2 - 2$. Graphically approximate the vertex.

Solution

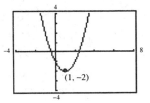

The vertex appears to be $(1, -2)$. The y-intercept is $(0, -1)$. The x-intercepts are approximately $(2.41, 0)$ and $(-0.41, 0)$.

59. Use a graphing utility to graph $y = x^2 - 6x + 5$. Graphically approximate the vertex.

Solution

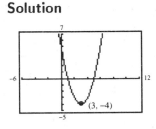

The vertex appears to be $(3, -4)$. The y-intercept is $(0, 5)$. The x-intercepts are $(5, 0)$ and $(1, 0)$.

61. *Problem Solving* Find two consecutive positive integers whose product is 240.

Solution

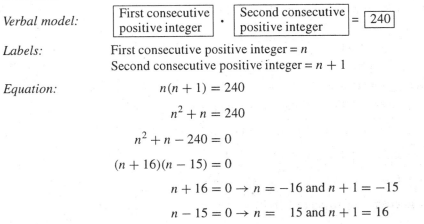

Verbal model: $\boxed{\begin{array}{c}\text{First consecutive}\\\text{positive integer}\end{array}} \cdot \boxed{\begin{array}{c}\text{Second consecutive}\\\text{positive integer}\end{array}} = \boxed{240}$

Labels: First consecutive positive integer $= n$
Second consecutive positive integer $= n + 1$

Equation:
$$n(n + 1) = 240$$
$$n^2 + n = 240$$
$$n^2 + n - 240 = 0$$
$$(n + 16)(n - 15) = 0$$
$$n + 16 = 0 \rightarrow n = -16 \text{ and } n + 1 = -15$$
$$n - 15 = 0 \rightarrow n = 15 \text{ and } n + 1 = 16$$

The problem specifies that these are *positive* integers, so we discard the negative answers. The two consecutive positive integers are 15 and 16.

63. *Falling Time* The height h (in feet) of an object that is dropped 48 feet above the ground is given by $h = 48 - 16t^2$, where t is the time in seconds. How long does it take for the object to hit the ground?

Solution

$$h = 48 - 16t^2 \text{ and } h = 0$$

$$0 = 48 - 16t^2$$

$$16t^2 = 48$$

$$t^2 = 3$$

$$t = \pm\sqrt{3}$$

$$t \approx \pm 1.73$$

We choose the positive answer for this application. Thus, the object strikes the ground in approximately 1.73 seconds.

65. *Geometry* The perimeter of a rectangle of length l and width w is 40 feet (see figure in textbook).
(a) Show that $w = 20 - l$.
(b) Show that the area A is given by $A = lw = l(20 - l)$.
(c) Complete the table shown in the textbook by using the equation from part (b).
(d) Sketch the graph of the function $A = l(20 - l)$.
(e) Of all rectangles with perimeters of 40 feet, which has the maximum area? How can you tell?

Solution

(a) *Formula:*

$$2(\text{Length}) + 2(\text{Width}) = \text{Perimeter}$$

$$2l + 2w = 40$$

$$2w = 40 - 2l$$

$$w = \frac{40 - 2l}{2}$$

$$w = 20 - l$$

(b) *Formula:*

$$\text{Area} = (\text{Length})(\text{Width})$$

$$A = lw$$

$$A = l(20 - l) \qquad \text{Replace } w \text{ by } 20 - l.$$

(c)

l	2	4	6	8	10	12	14	16	18
A	36	64	84	96	100	96	84	64	36

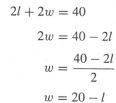

$2(20 - 2) = 2(18) = 36$ $12(20 - 12) = 12(8) = 96$

$4(20 - 4) = 4(16) = 64$ $14(20 - 14) = 14(6) = 84$

$6(20 - 6) = 6(14) = 84$ $16(20 - 16) = 16(4) = 64$

$8(20 - 8) = 8(12) = 96$ $18(20 - 18) = 18(2) = 36$

$10(20 - 10) = 10(10) = 100$

65. —CONTINUED—

(d)

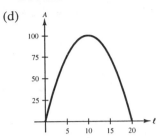

(e) The area is maximum when $l = 10$ and
$A = 100$. (These values are the coordinates
of the vertex of the parabola.) When
$l = 10$, $w = 20 - 10 = 10$. Thus, the area
is maximum when the rectangle is 10 feet by
10 feet.

67. *Ticket Sales*   A Little League baseball team paid $72 for a block of tickets to a ball game. The block contained three more tickets than the team needed. By inviting three more people to attend (and share in the cost), the team lowered the price per ticket by $1.20. How many people are going to the game?

Solution

Verbal model: $\boxed{\begin{array}{c}\text{Cost per person} \\ \text{in larger group}\end{array}} = \boxed{\begin{array}{c}\text{Cost per person} \\ \text{in smaller group}\end{array}} - \boxed{\$1.20}$

Labels: Total cost = $72
Number of people in smaller group = x
Cost per person in smaller group = $72/x$ (dollars)
Number of people in larger group = $x + 3$
Cost per person in larger group = $72/(x + 3)$ (dollars)

Equation:
$$\frac{72}{x+3} = \frac{72}{x} - 1.20$$

$$x(x+3)\left(\frac{72}{x+3}\right) = x(x+3)\left(\frac{72}{x}\right) - x(x+3)(1.20)$$

$$72x = 72(x+3) - 1.20x(x+3)$$

$$72x = 72x + 216 - 1.20x^2 - 3.60x$$

$$1.20x^2 + 3.60x - 216 = 0$$

$$12x^2 + 36x - 2160 = 0 \qquad \text{Multiply both sides by 10.}$$

$$x^2 + 3x - 180 = 0 \qquad \text{Divide both sides by 12.}$$

$$(x + 15)(x - 12) = 0$$

$$x + 15 = 0 \quad \rightarrow \quad x = -15 \text{ and } x + 3 = -12$$

$$x - 12 = 0 \quad \rightarrow \quad x = 12 \text{ and } x + 3 = 15$$

We choose the positive answer for this application. Thus, the number of people going to the game is 15.

69. *Work Rate* Working together, two people can complete a task in 10 hours. Working alone, one person takes 4 hours longer than the other to complete the task. Working alone, how long would it take each person to complete the task?

Solution

Verbal model:

$$\boxed{\begin{array}{c}\text{Rate of}\\\text{faster person}\end{array}} + \boxed{\begin{array}{c}\text{Rate of}\\\text{slower person}\end{array}} = \boxed{\begin{array}{c}\text{Rate of two}\\\text{working together}\end{array}}$$

Labels:

Both people: time = 10 (hours); rate = $\dfrac{1}{10}$ (task per hour)

Faster person: time = x (hours); rate = $\dfrac{1}{x}$ (task per hour)

Slower person: time = $x + 4$ (hours); rate = $\dfrac{1}{x+4}$ (task per hour)

Equation:

$$\frac{1}{x} + \frac{1}{x+4} = \frac{1}{10}$$

$$10x(x+4)\left(\frac{1}{x}\right) + 10x(x+4)\left(\frac{1}{x+4}\right) = 10x(x+4)\left(\frac{1}{10}\right)$$

$$10(x+4) + 10x = x(x+4)$$

$$10x + 40 + 10x = x^2 + 4x$$

$$20x + 40 = x^2 + 4x$$

$$0 = x^2 - 16x - 40$$

$$x = \frac{-(-16) \pm \sqrt{(-16)^2 - 4(1)(-40)}}{2(1)}$$

$$x = \frac{16 \pm \sqrt{416}}{2}$$

$$x = \frac{16 \pm 4\sqrt{26}}{2}$$

$$x = 8 \pm 2\sqrt{26}$$

$$x = 8 + 2\sqrt{26} \qquad \text{Choose the positive answer.}$$

$$x \approx 18.2 \text{ and } x + 4 \approx 22.2$$

The faster person could complete the task in approximately 18.2 hours and the slower person could complete the task in approximately 22.2 hours.

71. *World Gold Production*   From 1970 to 1990, the annual world gold production G, in thousands of ounces, can be modeled by $G = 47,974 - 2446t + 167t^2$ where t is the year, with $t = 0$ corresponding to 1970. The graph of this model is shown in the figure in the textbook. During which years between 1970 and 1990 was the world production of gold decreasing? During which years was the production increasing? How are these questions related to the vertex of the graph?

Solution

The production of gold was decreasing from 1970 until sometime in 1977. The production of gold was increasing from sometime in 1977 until 1990. The vertex indicates when the production of gold was at a minimum. The first coordinate of the vertex is located at

$$t = \frac{-b}{2a}$$

$$t = \frac{2446}{2(167)}$$

$$t \approx 7.3$$

Thus, the minumum occurred approximately 7.3 years after 1970.

Test for Chapter 11

1. Solve $x^2 - 144 = 0$ by extracting square roots.

Solution
$$x^2 - 144 = 0$$
$$x^2 = 144$$
$$x = \pm\sqrt{144}$$
$$x = \pm 12$$

The solutions are 12 and -12.

2. Solve $(x + 2)^2 - 16 = 0$ by extracting square roots.

Solution
$$(x + 2)^2 - 16 = 0$$
$$(x + 2)^2 = 16$$
$$x + 2 = \pm\sqrt{16}$$
$$x + 2 = \pm 4$$
$$x = -2 \pm 4$$
$$x = -2 + 4 \rightarrow x = 2$$
$$x = -2 - 4 \rightarrow x = -6$$

The solutions are 2 and -6.

3. Solve $x(x + 3) - 10(x + 3) = 0$ by factoring.

Solution
$$x(x + 3) - 10(x + 3) = 0$$
$$(x + 3)(x - 10) = 0$$
$$x + 3 = 0 \rightarrow x = -3$$
$$x - 10 = 0 \rightarrow x = 10$$

The solutions are -3 and 10.

4. Solve $2x^2 + x - 15 = 0$ by factoring.

Solution
$$2x^2 + x - 15 = 0$$
$$(2x - 5)(x + 3) = 0$$
$$2x - 5 = 0 \rightarrow 2x = 5 \rightarrow x = \tfrac{5}{2}$$
$$x + 3 = 0 \rightarrow x = -3$$

The solutions are $\tfrac{5}{2}$ and -3.

5. Solve $t^2 - 6t + 7 = 0$ by completing the square.

Solution

$$t^2 - 6t + 7 = 0$$

$$t^2 - 6t = -7$$

$$t^2 - 6t + 9 = -7 + 9 \qquad \text{Half of } -6 \text{ is } -3, \text{ and } (-3)^2 = 9.$$

$$(t - 3)^2 = 2$$

$$t - 3 = \pm\sqrt{2}$$

$$t = 3 \pm \sqrt{2}$$

The solutions are $3 + \sqrt{2}$ and $3 - \sqrt{2}$.

6. Solve $3z^2 + 9z + 5 = 0$ by completing the square.

Solution

$$3z^2 + 9z + 5 = 0$$

$$3z^2 + 9z = -5$$

$$z^2 + 3z = -\frac{5}{3}$$

$$z^2 + 3z + \frac{9}{4} = -\frac{5}{3} + \frac{9}{4} \qquad \text{Half of 3 is } \frac{3}{2}, \left(\frac{3}{2}\right)^2 = \frac{9}{4}.$$

$$\left(z + \frac{3}{2}\right)^2 = -\frac{20}{12} + \frac{27}{12}$$

$$\left(z + \frac{3}{2}\right)^2 = \frac{7}{12}$$

$$z + \frac{3}{2} = \pm\sqrt{\frac{7}{12}}$$

$$z + \frac{3}{2} = \pm\frac{\sqrt{7}}{\sqrt{12}} \cdot \frac{\sqrt{3}}{\sqrt{3}}$$

$$z + \frac{3}{2} = \pm\frac{\sqrt{21}}{\sqrt{36}}$$

$$z + \frac{3}{2} = \pm\frac{\sqrt{21}}{6}$$

$$z = -\frac{3}{2} \pm \frac{\sqrt{21}}{6}$$

$$z = -\frac{9}{6} \pm \frac{\sqrt{21}}{6}$$

$$z = \frac{-9 \pm \sqrt{21}}{6}$$

The solutions are $\dfrac{-9 + \sqrt{21}}{6}$ and $\dfrac{-9 - \sqrt{21}}{6}$.

7. Solve $x^2 - x - 3 = 0$ by using the Quadratic Formula.

Solution

$$x^2 - x - 3 = 0$$

$$a = 1, \quad b = -1, \quad c = -3$$

$$x = \frac{-(-1) \pm \sqrt{(-1)^2 - 4(1)(-3)}}{2(1)}$$

$$x = \frac{1 \pm \sqrt{13}}{2}$$

The solutions are $\dfrac{1 + \sqrt{13}}{2}$ and $\dfrac{1 - \sqrt{13}}{2}$.

8. Solve $2u^2 + 4u + 1 = 0$ by using the Quadratic Formula.

Solution

$$2u^2 + 4u + 1 = 0$$

$$a = 2, \quad b = 4, \quad c = 1$$

$$u = \frac{-4 \pm \sqrt{4^2 - 4(2)(1)}}{2(2)}$$

$$u = \frac{-4 \pm \sqrt{8}}{4}$$

$$u = \frac{-4 \pm 2\sqrt{2}}{4}$$

$$u = \frac{\cancel{2}(-2 \pm \sqrt{2})}{\cancel{2}(2)}$$

$$u = \frac{-2 \pm \sqrt{2}}{2}$$

The solutions are $\dfrac{-2 + \sqrt{2}}{2}$ and $\dfrac{-2 - \sqrt{2}}{2}$.

9. Solve $\dfrac{1}{x + 1} - \dfrac{1}{x - 2} = 1$.

Solution

$$\frac{1}{x + 1} - \frac{1}{x - 2} = 1$$

$$(x + 1)(x - 2)\left(\frac{1}{x + 1}\right) - (x + 1)(x - 2)\left(\frac{1}{x - 2}\right) = (x + 1)(x - 2)(1)$$

$$(x - 2) - (x + 1) = (x + 1)(x - 2)$$

$$x - 2 - x - 1 = x^2 - x - 2$$

$$-3 = x^2 - x - 2$$

$$0 = x^2 - x + 1$$

$$x = \frac{-(-1) \pm \sqrt{(-1)^2 - 4(1)(1)}}{2(1)}$$

$$x = \frac{1 \pm \sqrt{-3}}{2}$$

A negative number has no square root in the real numbers. Thus, the equation has *no* real number solution.

10. Solve $\sqrt{2x} = x - 1$.

Solution

$$\sqrt{2x} = x - 1$$

$$\left(\sqrt{2x}\right)^2 = (x-1)^2$$

$$2x = x^2 - 2x + 1$$

$$0 = x^2 - 4x + 1$$

$$x = \frac{-(-4) \pm \sqrt{(-4)^2 - 4(1)(1)}}{2(1)}$$

$$x = \frac{4 \pm \sqrt{12}}{2}$$

$$x = \frac{4 \pm 2\sqrt{3}}{2}$$

$$x = \frac{2(2 \pm \sqrt{3})}{2}$$

$$x = 2 \pm \sqrt{3}$$

The answer of $2 + \sqrt{3}$ checks, but the second answer is extraneous. (If $x = 2 - \sqrt{3}$, $x \approx 0.27$. Then $\sqrt{2x} \approx 0.73$, but $x - 1 \approx -0.73$.) Thus, the only solution is $2 + \sqrt{3}$.

11. Determine whether the parabola $y = -2x^2 + 4$ opens up or down. Then find the coordinates of the vertex.

Solution

$$y = -2x^2 + 4$$

$$a = -2, \ b = 0$$

$$x = -\frac{b}{2a}$$

$$x = -\frac{0}{1(-2)}$$

$$x = 0$$

$$y = -2(0)^2 + 4$$

$$y = 4$$

Because a is negative, the parabola opens *down*. The vertex is $(0, 4)$.

12. Determine whether the parabola $y = 5 - 2x - x^2$ opens up or down. Then find the coordinates of the vertex.

Solution

$$y = 5 - 2x - x^2$$

$$y = -x^2 - 2x + 5$$

$$a = -1, \ b = -2$$

$$x = -\frac{b}{2a}$$

$$x = -\frac{-2}{2(-1)}$$

$$x = -\frac{-2}{-2}$$

$$x = -1$$

$$y = 5 - 2(-1) - (-1)^2$$

$$y = 5 + 2 - 1$$

$$y = 6$$

Because a is negative, the parabola opens *down*. The vertex is $(-1, 6)$.

13. Determine whether the parabola $y = (x - 2)^2 + 3$ opens up or down. Then find the coordinates of the vertex.

Solution

$$y = (x - 2)^2 + 3$$

$$y = x^2 - 4x + 4 + 3$$

$$y = x^2 - 4x + 7$$

$$a = 1, \ b = -4$$

$$x = -\frac{b}{2a}$$

$$x = -\frac{-4}{2(1)}$$

$$x = \frac{4}{2}$$

$$x = 2$$

$$y = (2 - 2)^2 + 3$$

$$y = 3$$

Because a is positive, the parabola opens *up*. The vertex is $(2, 3)$.

14. Explain how to find the x-intercepts of a quadratic equation and demonstrate with the graph of $y = x^2 - 8x + 12$.

Solution

To find the x-intercepts of a quadratic equation, set $y = 0$ and solve for x.

$$y = x^2 - 8x + 12$$

Let $y = 0$.

$$0 = x^2 - 8x + 12$$

$$0 = (x - 6)(x - 2)$$

$$x - 6 = 0 \rightarrow x = 6$$

$$x - 2 = 0 \rightarrow x = 2$$

The x-intercepts are $(6, 0)$ and $(2, 0)$.

15. Sketch the graph of $y = x^2 - 4x$.

Solution

Leading coefficient test: Since $a > 0$, the parabola opens up.

Vertex: $(a = 1, \ b = -4)$
$$x = -\frac{b}{2a} = -\frac{-4}{2(1)} = -(-2) = 2$$

$$y = 2^2 - 4 \cdot 2 = 4 - 8 = -4$$

The vertex is located at $(2, -4)$.

y-intercept: (Let $x = 0$.) $y = 0^2 - 4 \cdot 0 = 0$
The y-intercept is $(0, 0)$.

x-intercepts: (Let $y = 0$.)
$$0 = x^2 - 4x = x(x - 4)$$

$$x = 0$$

$$x - 4 = 0 \rightarrow x = 4$$

The x-intercepts are $(0, 0)$ and $(4, 0)$.

Additional solution points:

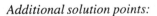

x	5	3	1	-1
$y = x^2 - 4x$	5	-3	-3	5
Points	$(5, 5)$	$(3, -3)$	$(1, -3)$	$(-1, 5)$

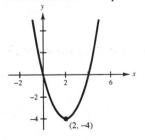

$(2, -4)$

16. Find two consecutive positive integers whose product is 420.

Solution

Verbal model: $\boxed{\begin{array}{c}\text{First consecutive}\\\text{positive integer}\end{array}} \cdot \boxed{\begin{array}{c}\text{Second consecutive}\\\text{positive integer}\end{array}} = \boxed{420}$

Labels: First consecutive positive integer $= n$
Second consecutive positive integer $= n + 1$

Equation:
$$n(n + 1) = 420$$
$$n^2 + n = 420$$
$$n^2 + n - 420 = 0$$
$$(n + 21)(n - 20) = 0$$
$$n + 21 = 0 \rightarrow n = -21 \text{ and } n + 1 = -20$$
$$n - 20 = 0 \rightarrow n = 20 \text{ and } n + 1 = 21$$

The problem specifies that the integers are *positive*, so we discard the negative answers. The two consecutive positive integers are 20 and 21.

17. The height of a triangle is three times the length of its base. The area of the triangle is 54 square inches. Find the dimensions of the triangle.

Solution

Formula: $\boxed{\dfrac{1}{2}} \cdot \boxed{\begin{array}{c}\text{Base of}\\\text{triangle}\end{array}} \cdot \boxed{\begin{array}{c}\text{Height of}\\\text{triangle}\end{array}} = \boxed{\begin{array}{c}\text{Area of}\\\text{triangle}\end{array}}$

Labels: Base of triangle $= x$ (inches)
Height of triangle $= 3x$ (inches)
Area of triangle $= 54$ (square inches)

Equation: $\dfrac{1}{2}(x)(3x) = 54$

$$\frac{3}{2}x^2 = 54$$
$$2\left(\frac{3}{2}\right)x^2 = 2(54)$$
$$3x^2 = 108$$
$$x^2 = 36$$
$$x = \pm\sqrt{36}$$
$$x = \pm 6 \qquad \text{Choose the positive answer.}$$
$$x = 6 \text{ and } 3x = 18$$

The base of the triangle is 6 inches and the height is 18 inches.

18. The length of a rectangle is 10 inches greater than the width of the rectangle. The area of the rectangle is 96 square inches. Use a quadratic equation to find its dimensions. Show your work.

Solution

Formula: $\boxed{\begin{array}{c}\text{Length of}\\\text{rectangle}\end{array}} \cdot \boxed{\begin{array}{c}\text{Width of}\\\text{rectangle}\end{array}} = \boxed{\begin{array}{c}\text{Area of}\\\text{rectangle}\end{array}}$

Labels: Width of rectangle $= x$ (inches)
Length of rectangle $= x + 10$ (inches)
Area of rectangle $= 96$ (square inches)

Equation: $(x + 10)(x) = 96$

$$x^2 + 10x = 96$$

$$x^2 + 10x - 96 = 0$$

$$(x + 16)(x - 6) = 0$$

$$x + 16 = 0 \rightarrow x = -16 \text{ and } x + 10 = -6$$

$$x - 6 = 0 \rightarrow x = \quad 6 \text{ and } x + 10 = 16$$

We choose the positive solutions for this application. Thus, the dimensions of the rectangle are 6 inches by 16 inches.

19. Together, two people can mow a lawn in 6 hours. When working alone, it takes one person 5 hours longer than the other. Find the time required for each person to mow the lawn alone.

Solution

Verbal model: $\boxed{\begin{array}{c}\text{Rate of}\\\text{faster person}\end{array}} + \boxed{\begin{array}{c}\text{Rate of}\\\text{slower person}\end{array}} = \boxed{\begin{array}{c}\text{Rate of two}\\\text{persons together}\end{array}}$

Labels: Two persons: time $= 6$ (hours), rate $= 1/6$ (lawn per hour)
Faster person: time x (hours), rate $= 1/x$ (lawn per hour)
Slower person: time $= x + 5$ (hours), rate $= 1/(x + 5)$ (lawn per hour)

Equation: $$\frac{1}{x} + \frac{1}{x + 5} = \frac{1}{6}$$

$$6x(x + 5)\left(\frac{1}{x}\right) + 6x(x + 5)\left(\frac{1}{x + 5}\right) = 6x(x + 5)\left(\frac{1}{6}\right)$$

$$6(x + 5) + 6x = x(x + 5)$$

$$6x + 30 + 6x = x^2 + 5x$$

$$12x + 30 = x^2 + 5x$$

$$0 = x^2 - 7x - 30$$

$$0 = (x - 10)(x + 3)$$

$$x - 10 = 0 \rightarrow x = 10 \quad \text{and } x + 5 = 15$$

$$x + 3 = 0 \rightarrow x = -3 \quad \text{Discard the negative answer.}$$

The faster person needed 10 hours to mow the lawn, and the slower person needed 15 hours.

20. Amtrak runs a train between two cities that are 360 miles apart. If the average speed could be increased by $7\frac{1}{2}$ miles per hour, the travel time would be decreased by 40 minutes. What average speed is required to obtain this decrease in travel time?

Solution

Verbal model: $\boxed{\begin{array}{c}\text{Travel time at}\\\text{slower speed}\end{array}} - \boxed{\begin{array}{c}\text{Travel time at}\\\text{faster speed}\end{array}} = \boxed{\dfrac{2}{3}}$

Labels: Slower speed: distance $= 360$ (miles), rate $= x$ (miles per hour)

$$\text{time} = \frac{360}{x} \text{ (hours)}$$

Faster speed: distance $= 360$ (miles), rate $= x + 7.5$ (miles per hour)

$$\text{time} = \frac{360}{x + 7.5} \text{ (hours)} \qquad \textit{Note: } 40 \text{ minutes} = \frac{2}{3} \text{ hour}$$

Equation:

$$\frac{360}{x} - \frac{360}{x + 7.5} = \frac{2}{3}$$

$$3x(x + 7.5)\left(\frac{360}{x}\right) - 3x(x + 7.5)\left(\frac{360}{x + 7.5}\right) = 3x(x + 7.5)\left(\frac{2}{3}\right)$$

$$3(x + 7.5)(360) - 3x(360) = x(x + 7.5)(2)$$

$$1080(x + 7.5) - 1080x = 2x(x + 7.5)$$

$$1080x + 8100 - 1080x = 2x^2 + 15x$$

$$8100 = 2x^2 + 15x$$

$$0 = 2x^2 + 15x - 8100$$

$$x = \frac{-15 \pm \sqrt{15^2 - 4(2)(-8100)}}{2(2)}$$

$$x = \frac{-15 \pm \sqrt{65{,}025}}{4}$$

$$x = \frac{-15 \pm 255}{4}$$

$$x = \frac{-15 + 255}{4} = \frac{240}{4} = 60 \text{ and } x + 7.5 = 67.5$$

$$x = \frac{-15 - 255}{4} = \frac{-270}{4} = -67.5 \text{ and } x + 7.5 = -60$$

We choose the positive solutions for this application. Thus, an average speed of 67.5 miles per hour is required. *Note:* The problem could also be solved by factoring.

$$2x^2 + 15x - 8100 = 0$$

$$(2x + 135)(x - 60) = 0$$

$$2x + 135 = 0 \rightarrow 2x = -135 \rightarrow x = -67.5 \text{ and } x + 7.5 = -60$$

$$x - 60 = 0 \rightarrow x = 60 \text{ and } x + 7.5 = 67.5$$